Atmospheric Chemistry and Physics

Atmospheric Chemistry and Physics

Edited by **Bruce Mullan**

SYRAWOOD
PUBLISHING HOUSE
New York

Published by Syrawood Publishing House,
750 Third Avenue, 9th Floor,
New York, NY 10017, USA
www.syrawoodpublishinghouse.com

Atmospheric Chemistry and Physics
Edited by Bruce Mullan

International Standard Book Number: 978-1-68286-247-6 (Hardback)

Printed in China.

Contents

	Preface	**VII**
Chapter 1	**A critical evaluation of proxy methods used to estimate the acidity of atmospheric particles** C. J. Hennigan, J. Izumi, A. P. Sullivan, R. J. Weber and A. Nenes	**1**
Chapter 2	**Temperature profiling of the atmospheric boundary layer with rotational Raman lidar during the HD(CP)^2 Observational Prototype Experiment** E. Hammann, A. Behrendt, F. Le Mounier and V. Wulfmeyer	**17**
Chapter 3	**Trends of ozone total columns and vertical distribution from FTIR observations at eight NDACC stations around the globe** C. Vigouroux, T. Blumenstock, M. Coffey, Q. Errera, O. García, N. B. Jones, J. W. Hannigan, F. Hase, B. Liley, E. Mahieu, J. Mellqvist, J. Notholt, M. Palm, G. Persson, M. Schneider, C. Servais, D. Smale, L. Thölix and M. De Mazière	**32**
Chapter 4	**Biomass burning related ozone damage on vegetation over the Amazon forest: a model sensitivity study** F. Pacifico, G. A. Folberth, S. Sitch, J. M. Haywood, L. V. Rizzo, F. F. Malavelle and P. Artaxo	**51**
Chapter 5	**Impact of pollution controls in Beijing on atmospheric oxygenated volatile organic compounds (OVOCs) during the 2008 Olympic Games: observation and modeling implications** Y. Liu, B. Yuan, X. Li, M. Shao, S. Lu, Y. Li, C.-C. Chang, Z. Wang, W. Hu, X. Huang, L. He, L. Zeng, M. Hu and T. Zhu	**65**
Chapter 6	**Two years of near real-time chemical composition of submicron aerosols in the region of Paris using an Aerosol Chemical Speciation Monitor (ACSM) and a multi-wavelength Aethalometer** J. -E. Petit, O. Favez, J. Sciare, V. Crenn, R. Sarda-Estève, N. Bonnaire, G. Močnik, J. -C. Dupont, M. Haeffelin and E. Leoz-Garziandia	**83**
Chapter 7	**Gaseous products and secondary organic aerosol formation during long term oxidation of isoprene and methacrolein** L. Brégonzio-Rozier, F. Siekmann, C. Giorio, E. Pangui, S. B. Morales, B. Temime-Roussel, A. Gratien, V. Michoud, S. Ravier, M Cazaunau, A. Tapparo, A. Monod and J. -F. Doussin	**104**

Chapter 8 **Long-term real-time chemical characterization of submicron aerosols at Montsec (southern Pyrenees, 1570m a.s.l.)** **120**
A. Ripoll, M. C. Minguillón, J. Pey, J. L. Jimenez, D. A. Day, Y. Sosedova, F. Canonaco, A. S. H. Prévôt, X. Querol and A. Alastuey

Chapter 9 **Properties and evolution of biomass burning organic aerosol from Canadian boreal forest fires** **137**
M. D. Jolleys, H. Coe, G. McFiggans, J. W. Taylor, S. J. O'Shea, M. Le Breton, S. J.-B. Bauguitte, S. Moller, P. Di Carlo, E. Aruffo, P. I. Palmer, J. D. Lee, C. J. Percival and M. W. Gallagher

Chapter 10 **Intercomparison of vertically resolved merged satellite ozone data sets: interannual variability and long-term trends** **156**
F. Tummon, B. Hassler, N. R. P. Harris, J. Staehelin, W. Steinbrecht, J. Anderson, G. E. Bodeker, A. Bourassa, S. M. Davis, D. Degenstein, S. M. Frith, L. Froidevaux, E. Kyrölä, M. Laine, C. Long, A. A. Penckwitt, C. E. Sioris, K. H. Rosenlof, C. Roth, H.-J. Wang and J. Wild

Chapter 11 **Stratospheric and mesospheric HO_2 observations from the Aura Microwave Limb Sounder** **179**
L. Millán, S. Wang, N. Livesey, D. Kinnison, H. Sagawa and Y. Kasai

Chapter 12 **Mesoscale modeling study of the interactions between aerosols and PBL meteorology during a haze episode in Jing–Jin–Ji (China) and its nearby surrounding region – Part 1: Aerosol distributions and meteorological features** **193**
H. Wang, M. Xue, X. Y. Zhang, H. L. Liu, C. H. Zhou, S. C. Tan, H. Z. Che, B. Chen and T. Li

Chapter 13 **Mesoscale modelling study of the interactions between aerosols and PBL meteorology during a haze episode in China Jing–Jin–Ji and its near surrounding region – Part 2: Aerosols' radiative feedback effects** **212**
H. Wang, G. Y. Shi, X. Y. Zhang, S. L. Gong, S. C. Tan, B. Chen, H. Z. Che and T. Li

Chapter 14 **Dicarboxylic acids, ketocarboxylic acids, α-dicarbonyls, fatty acids and benzoic acid in $PM_{2.5}$ aerosol collected during CAREBeijing-2007: an effect of traffic restriction on air quality** **223**
K. F. Ho, R.-J. Huang, K. Kawamura, E. Tachibana, S. C. Lee, S. S. H. Ho, T. Zhu and L. Tian

Chapter 15 **Analysis of the ozone profile specifications in the WRF-ARW model and their impact on the simulation of direct solar radiation** **236**
A. Montornès, B. Codina and J. W. Zack

Permissions

List of Contributors

Preface

Atmospheric chemistry and physics are two extremely significant branches of the interdisciplinary field of meteorology. This book is focused on the applications of atmospheric chemistry and physics for atmospheric modelling and evaluation. Some of the crucial concepts covered in this extensive book revolve around ozone damage, temperature profiling, aerosols, mesoscale modelling, etc. The innovative case studies presented in this book will provide in-depth knowledge of these fields. It will also further the scope of research in these areas. This book will serve as a reference text for students as well as professionals.

This book is a comprehensive compilation of works of different researchers from varied parts of the world. It includes valuable experiences of the researchers with the sole objective of providing the readers (learners) with a proper knowledge of the concerned field. This book will be beneficial in evoking inspiration and enhancing the knowledge of the interested readers.

In the end, I would like to extend my heartiest thanks to the authors who worked with great determination on their chapters. I also appreciate the publisher's support in the course of the book. I would also like to deeply acknowledge my family who stood by me as a source of inspiration during the project.

Editor

1

A critical evaluation of proxy methods used to estimate the acidity of atmospheric particles

C. J. Hennigan[1], J. Izumi[1], A. P. Sullivan[2], R. J. Weber[3], and A. Nenes[3,4,5]

[1]Department of Chemical, Biochemical and Environmental Engineering, University of Maryland, Baltimore County, Baltimore, MD, USA
[2]Department of Atmospheric Science, Colorado State University, Fort Collins, CO, USA
[3]School of Earth and Atmospheric Sciences, Georgia Institute of Technology, Atlanta, GA, USA
[4]School of Chemical and Biomolecular Engineering, Georgia Institute of Technology, Atlanta, GA, USA
[5]Foundation for Research and Technology – Hellas, Patras, Greece

Correspondence to: C. J. Hennigan (hennigan@umbc.edu)

Abstract. Given significant challenges with available measurements of aerosol acidity, proxy methods are frequently used to estimate the acidity of atmospheric particles. In this study, four of the most common aerosol acidity proxies are evaluated and compared: (1) the ion balance method, (2) the molar ratio method, (3) thermodynamic equilibrium models, and (4) the phase partitioning of ammonia. All methods are evaluated against predictions of thermodynamic models and against direct observations of aerosol–gas equilibrium partitioning acquired in Mexico City during the Megacity Initiative: Local and Global Research Objectives (MILAGRO) study. The ion balance and molar ratio methods assume that any deficit in inorganic cations relative to anions is due to the presence of H^+ and that a higher H^+ loading and lower cation / anion ratio both correspond to increasingly acidic particles (i.e., lower pH). Based on the MILAGRO measurements, no correlation is observed between H^+ levels inferred with the ion balance and aerosol pH predicted by the thermodynamic models and NH_3–NH_4^+ partitioning. Similarly, no relationship is observed between the cation / anion molar ratio and predicted aerosol pH. Using only measured aerosol chemical composition as inputs without any constraint for the gas phase, the E-AIM (Extended Aerosol Inorganics Model) and ISORROPIA-II thermodynamic equilibrium models tend to predict aerosol pH levels that are inconsistent with the observed NH_3–NH_4^+ partitioning. The modeled pH values from both E-AIM and ISORROPIA-II run with gas + aerosol inputs agreed well with the aerosol pH predicted by the phase partitioning of ammonia. It appears that (1) thermodynamic models constrained by gas + aerosol measurements and (2) the phase partitioning of ammonia provide the best available predictions of aerosol pH. Furthermore, neither the ion balance nor the molar ratio can be used as surrogates for aerosol pH, and previously published studies with conclusions based on such acidity proxies may need to be reevaluated. Given the significance of acidity for chemical processes in the atmosphere, the implications of this study are important and far reaching.

1 Introduction

The acidity of atmospheric particles is a critical parameter that affects air quality and the health of aquatic and terrestrial ecosystems. Acute and chronic exposures to acidic particles have been linked to deleterious effects in people, although the underlying physiological mechanisms are unclear (Gwynn et al., 2000; Dockery et al., 1996). The deposition of acidic gases and particles has been known for decades to damage freshwater and terrestrial ecosystems (Schindler, 1988; Johnson et al., 2008). While the trends in emissions are promising in the US and western Europe, ecosystem recovery from the effects of acid deposition is a slow process that can take decades (Likens et al., 1996; Stoddard et al., 1999). This may be a source of emerging environmental crisis in places such as China, where acid deposition is increasing due to rapid industrialization (Pan et al., 2013; Cao et al., 2013). Particle acidity also affects global biogeochemical cycles by controlling the solubility – and thus bioavailability – of limiting nutrients that are delivered through atmospheric deposition in many marine environments (Meskhidze et al., 2003, 2005; Nenes et al., 2011). This has important implications for marine primary productivity, the carbon cycle, and even climate (Mahowald, 2011).

Particle acidity is also a critical factor that influences many chemical processes in the atmosphere. The oxidation of S(IV) to S(VI) in liquid water, the primary pathway of sulfate formation, is highly sensitive to pH (Chameides, 1984). Halogen chemistry is strongly influenced by particle acidity, which has direct implications for the oxidation of volatile organic compounds (VOCs) and ozone formation in marine and coastal regions (e.g., Keene et al., 1998; Sander and Crutzen, 1996; Pszenny et al., 2004). Recent evidence has demonstrated that aerosol pH is also a critical parameter influencing halogen chemistry in continental locations (Brown et al., 2013; Thornton et al., 2010; Young et al., 2013). Further, aerosol acidity directly affects the deposition and lifetime of many compounds in the atmosphere through its influence on the gas–particle partitioning of semi-volatile species, including ammonia (NH_3), nitric acid (HNO_3), and organic acids (Ahrens et al., 2012; Keene et al., 2004). Aerosol acidity may affect secondary organic aerosol (SOA) formation, as well (e.g., Gaston et al., 2014; Surratt et al., 2007), although the atmospheric importance of this phenomenon remains highly uncertain (Peltier et al., 2007; Takahama et al., 2006; Tanner et al., 2009; Zhang et al., 2007; Xu et al., 2015).

Despite its significance, aerosol acidity remains very poorly constrained in the atmosphere (Keene et al., 1998). All direct measurements employ filter sampling (e.g., Jang et al., 2008; Keene et al., 2002; Koutrakis et al., 1988), which is both labor intensive and limited by poor time resolution. Measurements are also challenged by the non-conservative nature of H^+: due to buffering effects and the partial dissociation of weak acids, H^+ concentrations do not scale in proportion to the level of dilution (e.g., in aqueous filter extracts). Certain methods are also susceptible to sampling artifacts, which can greatly increase the uncertainty of an inherently challenging measurement (Pathak et al., 2004). Due to these limitations, indirect methods are frequently employed to estimate the acidity of atmospheric particles. These methods include (1) the ion balance method, (2) the molar ratio, (3) thermodynamic equilibrium models, and (4) the phase partitioning of semi-volatile species (HCl, NH_3, HNO_3). The purpose of this study is to evaluate and compare the proxy methods most commonly used to estimate aerosol acidity.

2 Methods to infer pH

Before proceeding with this analysis, it is necessary to define the different physical quantities commonly described by the term "aerosol acidity". First, it is used to represent the pH of an aerosol particle or distribution. The pH represents the hydrogen ion activity in an aqueous solution (Stumm and Morgan, 1996):

$$\mathrm{pH} = -\log(\gamma \cdot x_{\mathrm{H^+}}), \tag{1}$$

where γ is the hydrogen ion activity coefficient and $x_{\mathrm{H^+}}$ is the aqueous mole fraction of dissociated H^+. The presence of aerosol water is implicit in this definition since free H^+ cannot exist in solid particles. Second, "aerosol acidity" is commonly used to describe the loading of protons in atmospheric particles, in units of nmol m^{-3} or neq m^{-3}. This definition can take several forms, including aerosol strong acidity (H^+ contributed by strong acids that dissociate completely at most pH levels), free acidity (dissociated H^+), or total acidity (includes free H^+ and the undissociated H^+ bound to weak acids), typically defined by the measurement approach (Keene et al., 2004; Lawrence and Koutrakis, 1996). The major difference between aerosol pH and the proton loading is that pH is the H^+ concentration per liquid water volume (i.e., aerosol water) while the aerosol proton loading is the H^+ concentration per unit volume of air. Aerosol pH is the parameter of interest for the atmospheric phenomena described above, but the proton loading is often treated as a surrogate for pH. This is a critical distinction that will be discussed in detail below, especially in relation to the appropriate use of each parameter for the analysis of chemical processes in the atmosphere.

2.1 Ion balance method

The ion balance method is commonly employed to estimate the proton loading in atmospheric particles. This method is based upon the principle of electroneutrality and assumes that any deficit in measured cationic charge compared to measured anionic charge is due to the presence of protons, according to

$$[\mathrm{H^+}] = \sum n_i\,[\mathrm{anion}_i] - \sum n_i\,[\mathrm{cation}_i], \tag{2}$$

where n_i is the charge of species, i, and $[\mathrm{anion}_i]$ and $[\mathrm{cation}_i]$ are the molar concentrations of anion and cation species, respectively. If the sum of measured cations exceeds that of the measured anions, then the difference is attributed to OH^-. H^+ levels under an anion deficit are then calculated from the inferred $[OH^-]$ using the water dissociation constant, K_W. Most applications of the ion balance use inorganic ions only, even though organic acids can be important to the interpretation of aerosol acidity in diverse locations (Lawrence and Koutrakis, 1996; Metzger et al., 2006; Trebs et al., 2005), especially at relatively low acidities where organic acids dissociate and contribute to the ion balance. Neglecting this effect will lead to biases in the inferred H^+. Organic compounds can also form salt complexes with inorganic species (e.g., ammonium oxalate) (Reid et al., 1998; Paciga et al., 2014), further indicating the importance of organic acids in the ion balance.

2.2 Molar ratio

While the ion balance method is used to estimate the absolute proton loading in atmospheric particles, the molar ratio is independent of absolute concentrations. The molar ratio is a ratio of the total molar concentration of measured inorganic cations to the measured inorganic anions:

$$\text{molar ratio} = \sum(\text{cations})/\sum(\text{anions}). \quad (3)$$

It is also frequently employed as an equivalence (charge) ratio. The concept was first introduced in thermodynamic models to define major ions and composition domains (e.g., Pilinis and Seinfeld, 1987; Nenes et al., 1998; Kim et al., 1993; Fountoukis and Nenes, 2007), but never to infer the levels of acidity and pH. Thermodynamic models using the major species/composition domain approach (e.g., Fountoukis and Nenes, 2007) consider the possibility that aerosol species may volatilize enough to affect the ratio at equilibrium. Furthermore, the degree of dissociation of species such as H_2SO_4 / HSO_4^- / SO_4^{2-}, HNO_3 / NO_3^-, HCl / Cl^-, and NH_3 / NH_4^+ can affect the value of the ratio. In subsequent studies, however, the molar ratio has been treated as a proxy for acidity, with lower ratios corresponding to particles with the highest levels of acidity (lowest pH) (e.g., Kerminen et al., 2001). Molar ratios that yield a charge balance (i.e., equivalence ratios of unity or greater) are assumed for fully neutralized aerosol. Two common simplifications of the molar ratio approach are often applied when the concentrations of crustal elements are relatively low (e.g., Zhang et al., 2007):

$$\text{molar ratio} = NH_4^+/(Cl^- + NO_3^- + 2^*SO_4^{2-}) \quad (4)$$

and (e.g., Peltier et al., 2007; Tanner et al., 2009; Froyd et al., 2010):

$$\text{molar ratio} = NH_4^+/(2^*SO_4^{2-}). \quad (5)$$

2.3 Thermodynamic equilibrium models

Multiple thermodynamic equilibrium models have been developed to predict the behavior – most commonly the phase partitioning, liquid water content, and chemical speciation – of inorganic aerosol precursors. Previous studies have performed detailed comparisons and have explored the causes of disagreement among different thermodynamic equilibrium models (Ansari and Pandis, 1999; Zhang et al., 2000). We do not attempt to repeat this exercise. Instead, thermodynamic models are considered as one method to estimate the acidity of atmospheric particles. Two models are used in the present analysis: ISORROPIA-II (Fountoukis and Nenes, 2007; Nenes et al., 1999) and the Extended Aerosol Inorganics Model (E-AIM) (Clegg et al., 1998, 2003; Wexler and Clegg, 2002). ISORROPIA-II was designed with high computational efficiency to facilitate its incorporation in large-scale models and has seen wide usage. E-AIM is computationally expensive but is considered the benchmark since it employs few assumptions in its calculation of aerosol inorganic behavior (Zaveri et al., 2008).

Two applications of each model are considered (Fountoukis et al., 2009): (a) "forward" (or "closed") mode calculations, in which inputs to the model include T, relative humidity (RH), and the total (gas + aerosol) concentrations of aerosol precursors in the air parcel, and, (b) "reverse" (or "open") calculations, in which inputs to the model include T, RH, and the concentration of aerosol species. The output of both calculations is the concentration of species in the gas and aerosol (solid/liquid) phases, pH, and aerosol water. Highly time-resolved measurements of aerosol composition (e.g., via PILS-IC or AMS) are frequently conducted without the corresponding gas-phase aerosol precursor measurements (HCl, HNO_3, NH_3). Under this condition, it is conceptually straightforward to run the thermodynamic models in reverse mode, and this approach is frequently applied to analyze ambient and experimental data. In this analysis, we consider aerosol acidity predictions using both methods.

2.4 Phase partitioning

Aerosol pH can also be estimated from the phase partitioning of certain semi-volatile compounds, such as HNO_3 / NO_3^-, NH_3 / NH_4^+, and HCl / Cl^- (e.g., Keene et al., 2004; Meskhidze et al., 2003). The H^+ concentration in aqueous particles can be calculated assuming that the system is in equilibrium (using NH_3 / NH_4^+ as an example):

$$NH_3(g) \leftrightarrow NH_3(aq) \qquad K_H, \quad (R1)$$

$$NH_3(aq) + H_2O \leftrightarrow NH_4^+ + OH^- \qquad K_b. \quad (R2)$$

Under this approach, both the gas-phase and aerosol ionic components are measured, and the liquid H^+ concentration can be calculated after combining the equilibrium expressions from Reactions (R1) and (R2):

$$\{H^+\} = \frac{K_w\{NH_4^+\}}{K_H K_b p_{NH_3}}, \quad (6)$$

where $\{H^+\}$ is the activity of H^+ in atmospheric particles, K_H is the temperature-dependent Henry's law constant, K_b is the temperature-dependent base dissociation constant, K_w is the temperature-dependent water dissociation constant, p_{NH_3} is the gas-phase partial pressure of ammonia, and $\{NH_4^+\}$ is the activity of aqueous aerosol ammonium. The aerosol liquid water content is required to derive $\{NH_4^+\}$ (e.g., to convert from $\mu g\,m^{-3}$ to mol L^{-1}). The H^+ and NH_4^+ activity coefficients can also be calculated from thermodynamic equilibrium models in order to convert concentrations to activities, although the simplifying assumption of $\gamma = 1$ is sometimes employed, with satisfactory results (Fountoukis and Nenes, 2007).

Table 1. Overview of MILAGRO measurements used as modeling inputs.

Measurement overview					
Species	Measurement dates (2006)	Study n	Mean concentration $\pm$ 1σ ($\mu g\,m^{-3}$)	5th percentile ($\mu g\,m^{-3}$)	95th percentile ($\mu g\,m^{-3}$)
Na^+	10 Mar–1 Apr	998	0.31 ± 0.14	0.18	0.45
NH_4^+	10 Mar–1 Apr	1159	1.69 ± 1.36	0.25	4.43
Ca^{2+}	10 Mar–1 Apr	1654	0.53 ± 0.57	0.02[b]	1.64
Mg^{2+}	10 Mar–1 Apr	1051	0.15 ± 0.07	0.08	0.28
Cl^-	10 Mar–1 Apr	1603	0.27 ± 0.69	0.02[b]	1.23
NO_3^-	10 Mar–1 Apr	1634	2.93 ± 2.66	0.37	8.20
SO_4^{2-}	10 Mar–1 Apr	1652	3.74 ± 2.30	1.63	8.05
NH_3[a]	21 Mar–31 Mar	9594	25.2 ± 15.3	7.9	56.1
HNO_3[a]	17 Mar–30 Mar	430	2.43 ± 1.48	0.71	5.11

[a] Concentrations in ppb.
[b] This value represents the measurement limit of detection (LOD).

Table 2. Overview of modeling outputs and criteria.

Modeling overview					
Model	n	Average $[H^+]$ ($nmol\,m^{-3}$)	Median $[H^+]$ ($nmol\,m^{-3}$)	Average pH	Median pH
ISORROPIA-II, reverse[1]	438	19.39	10.34	1.98	−0.02
E-AIM, reverse[1]	460	15.77	6.56	2.36	0.14
ISORROPIA-II, forward[2]	89	0.008	0.005	3.31	3.38
E-AIM, forward[2]	39	0.015	0.012	3.24	3.32

[1] Output limited to samples meeting these criteria: (1) NH_4^+, NO_3^-, and $SO_4^{2-} > 0$; (2) RH $\geq$ 40 %; and (3) modeled aerosol liquid water content > 0.
[2] Output limited to samples meeting these criteria: (1) NH_4^+, NO_3^-, SO_4^{2-}, NH_3, and $HNO_3 > 0$; (2) RH $\geq$ 40 %; and (3) modeled aerosol liquid water content > 0.

2.5 Evaluation data set

We evaluate the above methods using ground-based data collected during the Megacity Initiative: Local and Global Research Objectives (MILAGRO) campaign at the T1 site in Mexico City (Molina et al., 2010). The measurements spanned 10 March–1 April 2006 (Table 1). Inorganic $PM_{2.5}$ composition (Na^+, NH_4^+, K^+, Mg^{2+}, Ca^{2+}, Cl^-, NO_3^-, and SO_4^{2-}) was measured with a Particle-into-Liquid Sampler coupled to a dual ion chromatograph (Hennigan et al., 2008). Six-minute integrated measurements were conducted every 15 min. Given that $PM_{2.5}$ composition only was measured, no size–acidity dependence could be elucidated (Keene et al., 2002). Ammonia was measured with a quantum cascade laser spectrometer with 1 min sample time resolution (Aerodyne Research, Inc.). Nitric acid was measured with 5 min time resolution via thermal-dissociation laser-induced fluorescence of nitrogen oxides (Day et al., 2002; Farmer et al., 2011). All chloride was assumed to reside in the particle phase (i.e., gas-phase HCl was effectively 0), which was determined to be a valid assumption given the high concentrations of gas-phase NH_3 (Fountoukis et al., 2009).

The chemical measurements, along with ambient temperature and RH, were used as inputs into ISORROPIA-II and E-AIM. Fountoukis et al. (2009) evaluated the equilibrium partitioning of semi-volatile compounds at T1 using ISORROPIA-II. We do not duplicate that effort here: instead, we focus solely on particle acidity and the comparison across the different proxy methods described above. For the reverse model runs, the 1 min met data were averaged to the aerosol sampling times and were used in conjunction with the aerosol concentrations as model inputs. Given the differences in sample timing, the forward model runs used as inputs the total gas + aerosol concentrations and met data averaged to common 10 min sample times. Output from the reverse model runs was used for the present analysis only if the following criteria were met for a given sample: (1) NH_4^+, NO_3^-, and SO_4^{2-} aerosol measurements were operational; (2) RH > 40 %; and (3) modeled aerosol liquid water > 0. Output from the forward model runs was used

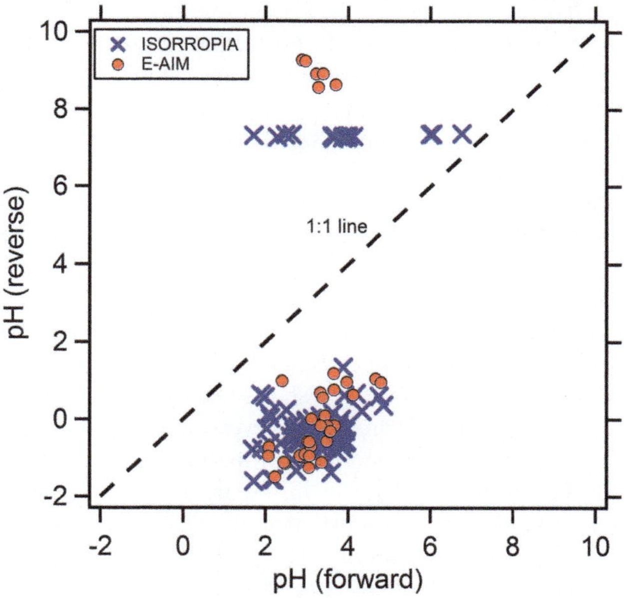

Figure 1. Comparison of pH predicted using the reverse and forward modes of ISORROPIA and E-AIM.

for the present analysis only if the above criteria were met, and both the NH_3 and HNO_3 measurements were operational for a given sampling interval. These constraints explain the differences in sample numbers between the forward and reverse simulations and between E-AIM and ISORROPIA (Table 2). ISORROPIA-II model runs were performed in "metastable" mode where the aerosol is only in the aqueous phase and can be supersaturated (http://isorropia.eas.gatech.edu/). ISORROPIA treats the Na^+–NH_4^+–K^+–Ca^{2+}–Mg^{2+}–Cl^-–NO_3^-–SO_4^{2-} system. For conditions of excess cations, ISORROPIA-II assumes that bicarbonate (HCO_3^-) and carbonate (CO_3^{2-}) account for the deficit, and a pH limit of 7 is imposed. Model version AIM-IV was used for RH conditions greater than 60 % ($n = 348$, Table 2), and model version AIM-II was used for RH conditions between 40 and 60 % ($n = 112$, Table 2) (http://www.aim.env.uea.ac.uk/aim/aim.php). AIM-IV treats the Na^+–NH_4^+–Cl^-–NO_3^-–SO_4^{2-} system, while AIM-II treats NH_4^+–NO_3^-–SO_4^{2-}. For AIM, measured cations not treated by the model were accounted for as equivalent sodium (Na*) in AIM-IV and equivalent ammonium (NH_4^*) in AIM-II. Cl^- was accounted for as equivalent sulfate (SO_4*) in AIM-II. Ion balance is required for all simulations in E-AIM: any cation deficit was balanced using H^+, while any anion deficit was balanced using OH^- since these are also required model inputs. E-AIM assumes that OH^- balances any excess cations, rather than carbonate, and thus has regions of higher predicted aerosol pH compared with ISORROPIA.

E-AIM model runs were performed in solid + liquid mode where salts precipitate once the aqueous solution becomes saturated. Overall, strong similarities between E-AIM and ISORROPIA suggest that the choice of metastable vs. stable (solid + liquid) mode did not appreciably affect the predicted aerosol pH levels. Given the importance of aerosol liquid water, data below 40 % RH were excluded from the analysis. During the MILAGRO study, this was approximately 50 % of the total measurement period. Further, points above 40 % RH with modeled aerosol liquid water content of 0 were excluded from the analysis since a pH cannot be derived for these samples. This was the cause of differences in sample numbers between the E-AIM and ISORROPIA simulations (Table 2).

3 Results and discussion

3.1 Thermodynamic equilibrium models

Thermodynamic equilibrium models are frequently used to estimate aerosol acidity. Prior studies have observed and discussed large differences in aerosol acidity predicted by different models (Ansari and Pandis, 1999; Yao et al., 2006); we do not revisit this analysis but instead seek to understand some of the limitations and uncertainties of using thermodynamic equilibrium models to predict aerosol pH. Figure 1 shows that a large source of uncertainty is tied to the availability of gas-phase data and whether the model is run in the forward (gas + aerosol inputs) or reverse (aerosol inputs only) mode. Note that the number of forward-mode predictions was less than the reverse-mode predictions due to availability of coincident aerosol- and gas-phase measurements. For ISORROPIA and E-AIM, the median differences between the models run in forward and reverse modes were 3.5 and 3.1 pH units, respectively. This finding is consistent with a modeling study of aerosol pH in Hong Kong as well (Yao et al., 2006). Other parameters, such as aerosol liquid water, have much closer agreement between the forward and reverse modes (not shown), given that it is largely driven by total aerosol mass and, thus, is much less sensitive to gas-phase species.

The large differences in the forward- and reverse-mode predictions of aerosol pH seen in Fig. 1 come about for several reasons. Upon specification of the aerosol species and surrounding RH and T, thermodynamic models first determine the aerosol pH and liquid water content (assuming that a liquid phase can exist), followed by computing the concentration of gas-phase semi-volatile compounds in equilibrium with the aerosol (e.g., NH_3, HNO_3, and HCl). The aerosol pH is largely driven by electroneutrality in the aqueous phase, as any imbalance between charges from cations and anions needs to be balanced by H^+ and OH^-:

$$[H^+] + \sum_{n+} [X^+] = [OH^-] + \sum_{n-} [Y^-],$$

$$\text{or} [H^+] + I_b - [OH^-] = 0, \tag{7}$$

where $I_b = \sum_{n+} [X^+] - \sum_{n-} [Y^-]$ is the ion balance parameter; n^+ and n^- are the number of positively and nega-

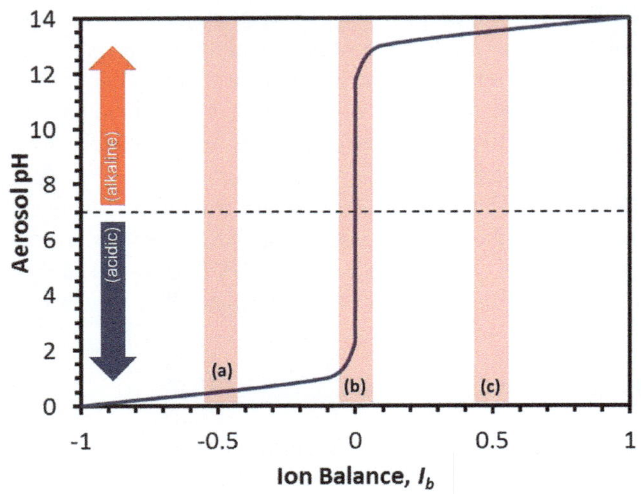

Figure 2. Aerosol pH as a function of the parameter I_b (see Eq. 7 for explanation). The data used here are hypothetical: values of I_b ranging from −1 to 1 at a constant aerosol loading were input under conditions of constant T and RH to demonstrate the extreme sensitivity of aerosol pH at I_b values near 0.0 (region b).

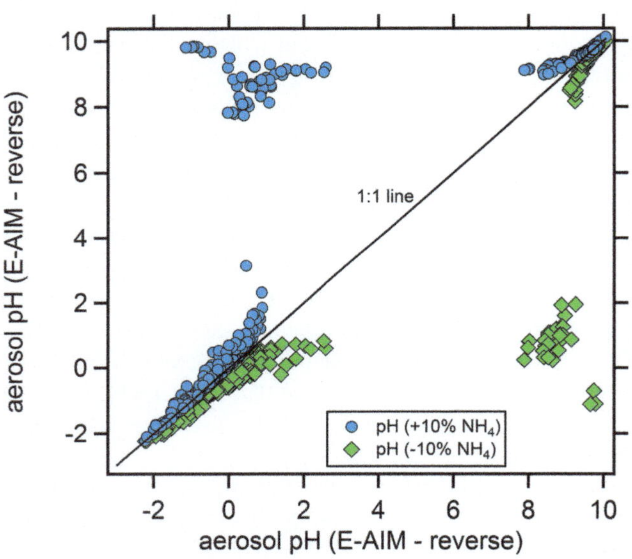

Figure 3. Sensitivity of aerosol pH predicted with E-AIM (reverse mode) to small changes in the input aerosol NH_4^+ concentration.

tively charged ionic species, respectively; and $[X^+]$ is the concentration of a species in the aqueous phase, in gram equivalents (geq). In the case where the aerosol is acidic (i.e., pH <7 or $I_b < -\sqrt[2]{K_w}$, with $K_w = [OH^-][H^+]$), $[OH^-]$ contributes negligibly to I_b, and $[H^+] \cong -I_b$. Similarly, when the aerosol is alkaline (i.e., pH > 7 or $I_b > \sqrt[2]{K_w}$), $[H^+]$ contributes negligibly to I_b, $[OH^-] \cong I_b$, and $[H^+] = K_w / [OH^-]$.

Given the above, one can construct a diagram that relates aerosol pH to I_b; this is shown in Fig. 2. It is noteworthy that pH changes considerably over a narrow range of I_b, when the value of I_b is close to 0. For acidic and very alkaline aerosol, uncertainty in I_b – shown as the (a) and (c) regions of Fig. 2 – may introduce a 0.5–1.0 pH unit bias in predicted pH. In region (b), however, a small uncertainty in I_b leads to shifts in pH that span effectively 10 pH units (or more). This uncertainty may come from either uncertainty in the measurements, themselves, or from approximations such as the exclusion of minor species (e.g., crustal elements) from the analysis. The data used in Fig. 2 are hypothetical: values of I_b ranging from −1 to 1 at a constant aerosol loading were input under conditions of constant T and RH to generate the predicted pH. However, Fig. 3 shows that for the MILAGRO data set, the predicted aerosol pH (reverse mode) is extremely sensitive to minor uncertainties in the measurement inputs and, thus, to uncertainties in I_b. Figure 3 shows the sensitivity in predicted aerosol pH under the reverse-mode calculation to ±10 % changes in the aerosol NH_4^+ concentration. The aerosol pH differed by more than 1.0 pH unit for 18 % of the data when NH_4^+ increased by 10 %. Likewise, aerosol pH differed by more than 1.0 pH unit for 12 % of the data when NH_4^+ decreased by 10 %. Similar sensitivities were also observed for ±10 % changes in aerosol NO_3^- and SO_4^{2-} inputs. Using gas + aerosol inputs strongly constrains the effects of measurement errors and therefore is thought to give a more accurate representation of aerosol partitioning and composition (Fountoukis et al., 2009; Guo et al., 2014). Our results support this thought and indicate that reverse-mode calculations of aerosol H^+ and pH should likely be avoided for the interpretation of experimental data. The exception to this recommendation would be a system that contains very low concentrations of semi-volatiles. Without accompanying gas-phase data to constrain the thermodynamic models, an alternate approach that may yield a more accurate representation of aerosol pH is the use of aerosol concentrations as input in forward-mode calculations. In the southeastern US, Guo et al. (2014) report that this approach led to biases in aerosol pH of ∼ 1 pH unit, which is considerably lower than the bias observed in Fig. 1 if we assume that the forward-mode predictions are accurate (see more discussion on this point below).

3.2 Phase partitioning

Figure 4 shows aerosol pH predicted by ammonia phase partitioning vs. aerosol pH predicted by E-AIM and ISORROPIA run in the forward and reverse modes. To predict aerosol pH from the NH_3 / NH_4^+ phase equilibrium (Eq. 6), aerosol liquid water was taken from the forward model output of ISORROPIA. The temperature-dependent K_H and K_b are from Chameides (1984), while the temperature-dependent K_w is from Stumm and Morgan (1996). We have used liquid concentrations, not activities in the application of Eq. (6) (i.e., activity coefficients for H^+ and NH_4^+ are assumed to be unity), following the approach of Keene

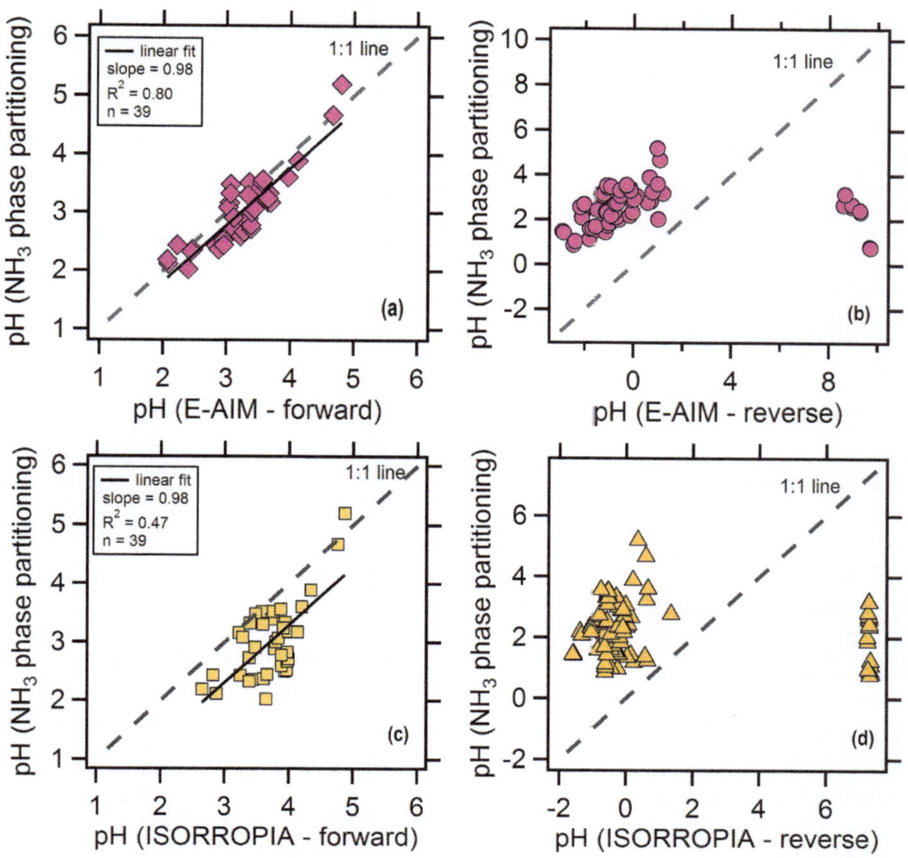

Figure 4. Aerosol pH predicted by the phase partitioning of ammonia compared to pH predicted by **(a)** E-AIM in the forward mode, **(b)** E-AIM in the reverse mode, **(c)** ISORROPIA in the forward mode, and **(d)** ISORROPIA in the reverse mode.

et al. (2004). The implications of these assumptions are discussed below. The best agreement between the phase-partitioning approach and the models was found for both forward model applications (Fig. 4a and c). A slope of 0.98 and high R^2 value (0.80) indicate excellent agreement between E-AIM and the phase-partitioning approach (Fig. 4a). The median difference between these methods was only 0.4 pH unit. Likewise, good agreement was observed between ISORROPIA-II in the forward mode and the phase-partitioning pH (Fig. 4c, slope = 0.98 and $R^2 = 0.47$).

Differences between the NH_3 / NH_4^+ phase-partitioning predictions and either thermodynamic model are possibly due to differences in equilibrium constants and/or differences in the activity coefficients, although the former is more likely. That is evident from the strong agreement seen in Fig. 4a; since E-AIM employs activity coefficient calculations for all species, our calculation using concentrations instead of activities does not appear to systematically bias the phase-partitioning pH predictions. A detailed characterization of uncertainties and sensitivities of the various methods to differences in equilibrium constants and activity coefficients is beyond the scope of this study but should be explored in the future.

The above results were contrasted by very poor agreement between the reverse models and phase-partitioning predictions of aerosol pH (Fig. 4b and d). The median difference between aerosol pH predicted by NH_3 phase partitioning and E-AIM run in the reverse mode was 3.5 pH units ($n = 72$). Similarly, the median difference between aerosol pH predicted by NH_3 phase partitioning and ISORROPIA-II run in reverse mode was 3.1 pH units ($n = 72$). These large differences are consistent with the large differences observed between the forward and reverse predictions of pH (Fig. 1). These results are also consistent with large discrepancies (on the order of 1–4 pH units) in aerosol pH between E-AIM run in the reverse mode and the phase-partitioning approach for a study in Hong Kong (Yao et al., 2006). As discussed above, gas + aerosol inputs constrain the effects of measurement uncertainty on thermodynamic models, and thus the large differences observed in Fig. 4b and d provide further support that the reverse model applications are challenged to accurately predict aerosol pH. For the MILAGRO data set, approximately half of all reverse model runs predicted pH values less than 0 (44 % for E-AIM, 51 % for ISORROPIA). This is inconsistent with the observed aerosol ammonium fraction (NH_4^+ / (NH_3+ NH_4^+)), which was only

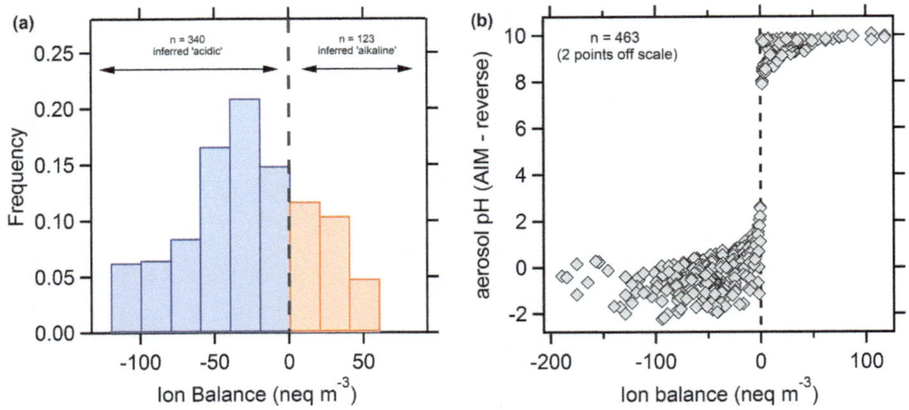

Figure 5. **(a)** Frequency distribution of the aerosol ion balance during MILAGRO, and **(b)** aerosol pH predicted by E-AIM in the reverse mode vs. the ion balance. The dashed lines at 0 serve as a visual reference for neutral particles.

0.2, on average, over the study period. Under conditions typical of the MILAGRO study (293 K, 10 μg m^{-3} aerosol liquid water, total NH_3 of 10–30 μg m^{-3}), if the aerosol pH is less than 0, thermodynamic calculations predict that essentially all of the NH_3 will partition to the particle phase at equilibrium. The fact that gas-phase NH_3 was abundant (Table 1) suggests either that (1) the system is not at equilibrium or that (2) aerosol pH >0. Fountoukis et al. (2009) found that the equilibration timescales for NH_3 / NH_4^+ (approximately 10 min) were on par with the measurement integration timescales during MILAGRO, strongly suggesting that the assumption of equilibrium is valid. This further supports the conclusion above that the reverse-mode models vastly overstate the acidity (underestimate pH) of the Mexico City aerosol.

These results suggest that the two best proxy methods for estimating aerosol pH are (1) thermodynamic equilibrium models run using gas + aerosol inputs, and (2) the phase partitioning of ammonia. Under certain conditions, the phase partitioning of HNO_3 / NO_3^- gives similar results to that of ammonia (Young et al., 2013), so this is another feasible approach. It should be noted that Young et al. (2013) caution against the use of HCl / Cl^- partitioning to estimate pH on the basis of large (order of magnitude) uncertainty in the Henry's law constant for HCl. It should also be noted that Keene et al. (2004) observed relatively poor agreement between pH predicted by ammonia and nitric acid phase partitioning. This may be due to the explicit treatment of activity coefficients (as was done by Young et al., 2013), or to differences in other factors, such as relative ammonia levels, since this can introduce major differences in thermodynamic model predictions (Ansari and Pandis, 1999). Other limitations of the phase-partitioning approach should be considered, depending on the environment. This includes long equilibration timescales for coarse particles and the presence of insoluble or partially soluble salts (Jacobson, 1999; Fridlind and Jacobson, 2000; Meng and Seinfeld, 1996).

3.3 Ion balance vs. aerosol pH

Figure 5a shows a probability distribution for the ion balance during MILAGRO. All of the measured inorganic species (Sect. 2.5) were included in the calculation according to Eq. (2), but dissociated organic acids were not included since they were not measured. Approximately 75 % of the paired cation–anion measurements had a cation deficit and would thus be inferred as acidic according to the ion balance. Figure 5b shows the aerosol pH predicted by E-AIM (reverse mode) vs. the inorganic ion balance. The data follow the traditional titration curve shape, with those points having a negative ion balance (anions > cations) all having predicted pH below 3.0 and points having a positive ion balance (cations > anions) all having predicted pH above 7.8.

The ion balance proxy method assumes that conditions of high aerosol H^+ loading (in nmol m^{-3}) correspond to an aerosol distribution with a low (acidic) pH (e.g., Feng et al., 2012; Zhang et al., 2012a). Figure 6 shows the relationship between predicted aerosol pH and the H^+ concentration inferred from the ion balance. An anion deficit in the ion balance is assumed to be due to OH^- (Eq. 2) – implying extremely low H^+ levels – so only data with a negative ion balance were included in this analysis ($n = 340$). Figure 6 shows no correlation at all between the H^+ loading and the predicted aerosol pH using either the forward (Fig. 6a) or reverse models (Fig. 6b). Note the differences in pH predicted by the forward and reverse models, as observed in Fig. 1. Numerous points in the lowest H^+ quartile have pH levels below the median pH of the upper H^+ quartile (Fig. 6b). Similarly, for a given H^+ loading, a wide pH range of ~2–3 pH units is typically predicted. This wide pH range is observed across all H^+ loadings, even at the high and low ends, because aerosol liquid water content and the H^+ activity coefficient can differ dramatically for a given H^+ level. The ion balance method is insensitive to either of these factors that are critical in determining aerosol pH (Eq. 1). Pathak et al. (2009) found a simi-

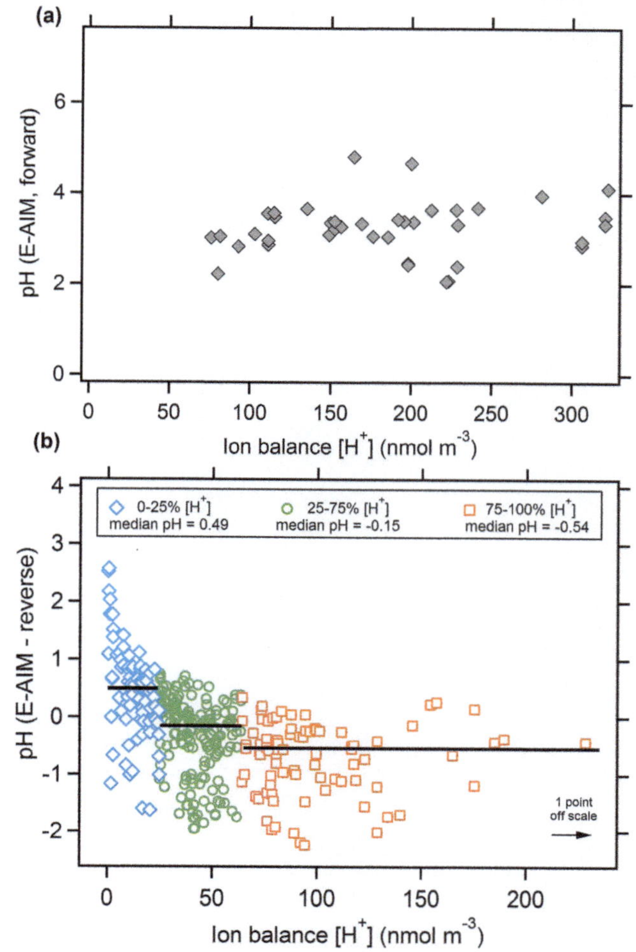

Figure 6. **(a)** Aerosol pH predicted by E-AIM in the forward mode vs. the H^+ concentration derived from the ion balance, and **(b)** aerosol pH predicted by E-AIM in the reverse mode vs. the H^+ concentration derived from the ion balance. Only points with a cation charge deficit (Σ(anions) > Σ(cations)) are included in the analysis in **(b)**. The horizontal black bars in **(b)** represent the medians corresponding to the values in the legend.

lar disconnect between H^+ inferred from the ion balance and predicted aerosol pH for measurements in four cities across China.

A further problem with the ion balance is that it is unable to distinguish between free and undissociated H^+ (e.g., protons associated with bisulfate (HSO_4^-)); Keene et al., 2004). This limitation becomes very important when the solution pH approaches the pK_a of any major ion that is associated with acidity, including HNO_3, HCl, and NH_3. The pK_a of HSO_4^- is 1.99 (Stumm and Morgan, 1996), and at pH levels below 3 the bisulfate–sulfate equilibrium begins to shift appreciably towards the protonated form, implying that the ion balance overestimates H^+. This is illustrated in Fig. 7a and b, which shows the H^+ levels inferred from the ion balance compared to the H^+ loadings predicted by E-AIM and ISORROPIA (both run in reverse mode), respectively. Although the thermodynamic models use electroneutrality to derive H^+, they account for partial dissociation, which explains why the ion balance gives H^+ levels that are ~45–65 % higher than the models. The discrepancy in pH predicted by the forward- and reverse-mode thermodynamic calculations (Figs. 1 and 6) also implies large differences in the predicted aerosol H^+ loadings. Figure 7c and d shows H^+ levels inferred from the ion balance compared to H^+ levels predicted by E-AIM and ISORROPIA, both run in forward mode. The forward-mode thermodynamic calculations predict H^+ levels that are orders of magnitude lower than either the ion balance or reverse-mode calculations, consistent with the large differences in predicted pH shown in Fig. 1. This is also consistent with large discrepancies in H^+ levels in Pittsburgh, PA, predicted by forward (Takahama et al., 2006) and reverse (Zhang et al., 2007) thermodynamic equilibrium model simulations. Although the measurement periods for these studies in Pittsburgh did not overlap, the similar NH_4^+ and SO_4^{2-} concentrations suggest that the large differences in aerosol H^+ were likely due to differences in the forward and reverse model simulations, not in the actual pH levels of the aerosol.

Finally, another major limitation in estimating the aerosol H^+ loading from the ion balance is high uncertainty in H^+ that comes about from the propagation of measurement error. Figure 8a shows the relative ion balance uncertainty (%) vs. the ion balance loading (nmol m^{-3}). Note that the points with a negative ion balance (orange trace in Fig. 8a) are inferred as the H^+ loading, and so this trace also represents the uncertainty in H^+ from the ion balance. The ion balance uncertainty was calculated using a standard propagation-of-error technique (Harris, 1999):

$$u_{IB} = (u^2_{Na^+} + u^2_{NH_4^+} + u^2_{Mg^{2+}} + u^2_{Ca^{2+}} + u^2_{K^+} + u^2_{Cl^-} + u^2_{NO_3^-} + u^2_{SO_4^{2-}})^{\frac{1}{2}}, \quad (8)$$

where u_{IB} is the absolute uncertainty in the ion balance (in neq m^{-3}), and the terms on the right-hand side represent the absolute uncertainties in each inorganic species in neq m^{-3} (Na^+, NH_4^+, Mg^{2+}, Ca^{2+}, K^+, Cl^-, NO_3^-, and SO_4^{2-}), calculated assuming 10 % relative uncertainty for each species. Figure 8a shows that the relative ion balance (and [H^+]) uncertainty (u_{IB}/[ion balance]) grows with decreasing ion balance or [H^+], as one would expect. A frequency distribution of the ion balance uncertainty from the MILAGRO study shows that approximately 40 % of the ion balance and [H^+] calculations had an associated uncertainty higher than 50 %, and more than 20 % of the ion balance and [H^+] calculations had an associated uncertainty higher than 100 % (Fig. 8b).

These results demonstrate numerous problems with the ion balance and strongly suggest that it is inadequate to represent the pH of atmospheric particles. The ion balance may provide a qualitative indication of acidic conditions when anions far exceed cations, but the variable effects of liquid water, the buffering action of HSO_4^- / SO_4^{2-}, and the effect of species

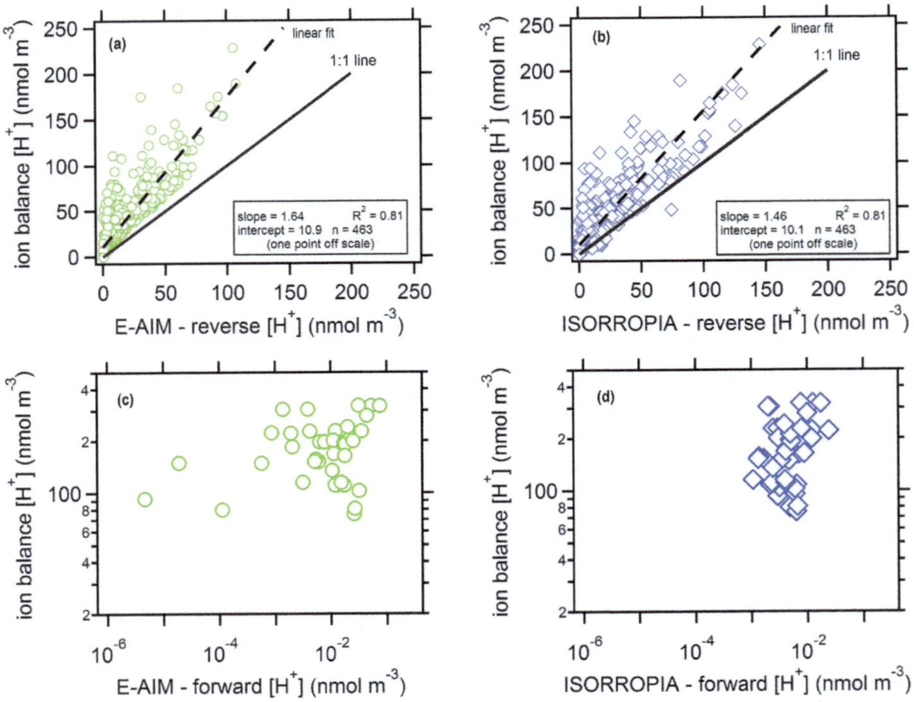

Figure 7. H^+ levels inferred from the ion balance compared to H^+ levels predicted by **(a)** E-AIM in the reverse mode, **(b)** ISORROPIA in the reverse mode, **(c)** E-AIM in the forward mode, and **(d)** ISORROPIA in the forward mode. Note the different log scales for the x and y axes in **(c)** and **(d)**.

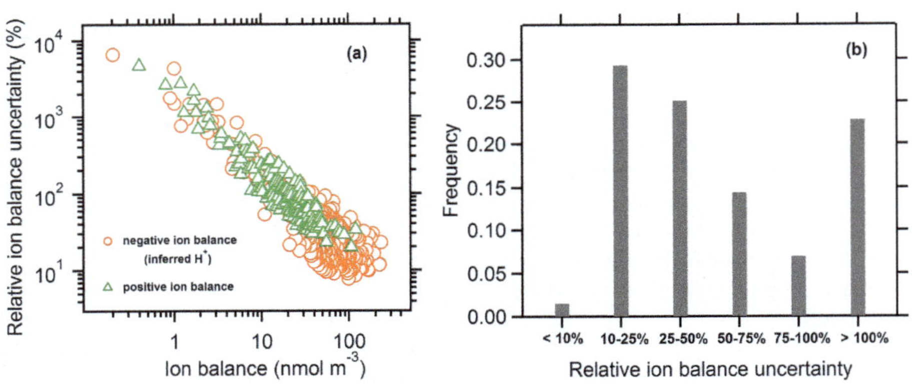

Figure 8. (a) Relative ion balance uncertainty (%) as a function of the ion balance level. Note that the orange trace represents $[H^+]$, and so this trace also represents the uncertainty in $[H^+]$ from the ion balance. **(b)** Frequency distribution in the relative ion balance uncertainty.

activity coefficients in concentrated particles preclude its use for any quantitative means. Even when all ionic components – including the dissociated ions of organic acids – are accounted for and measured with good precision, the ion balance is unlikely to qualitatively distinguish alkaline particles from mildly acidic particles due to the propagation of uncertainties in the aerosol composition measurements. This recommendation against the use of the ion balance as a proxy for aerosol pH is supported by prior work (Keene and Savoie, 1998; Winkler, 1986; Guo et al., 2014). The H^+ concentration derived from the ion balance is often just a surrogate for sulfate, especially in environments where sulfate is the dominant inorganic species (Pathak et al., 2009; Zhang et al., 2007). It is possible – likely even – to simultaneously have particles with a low concentration of H^+ (in nmol m^{-3}) that are highly acidic (low pH). Studies that infer the effects of particle acidity on atmospheric chemical processes using the inorganic ion balance are likely flawed. For example, correlations between SOA and aerosol H^+ concentrations from the ion balance have been interpreted as evidence of acid-catalyzed SOA formation (Feng et al., 2012; Budisulistiorini et al., 2013; Pathak et al., 2011; Nguyen et al., 2014; Zhang et

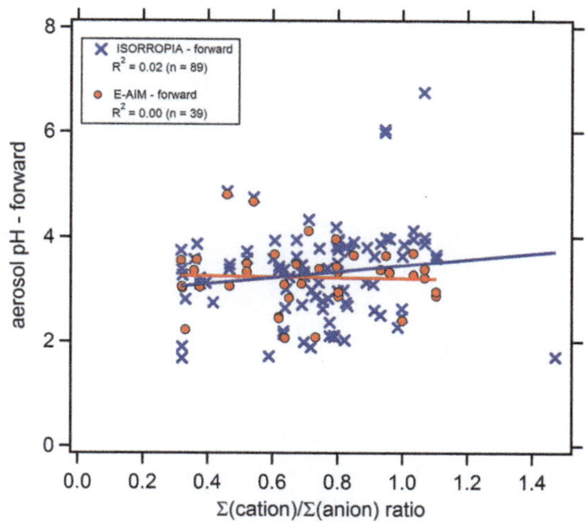

Figure 9. Aerosol pH predicted by the forward-mode equilibrium models compared to the Σ(cations) / Σ(anions) charge ratio.

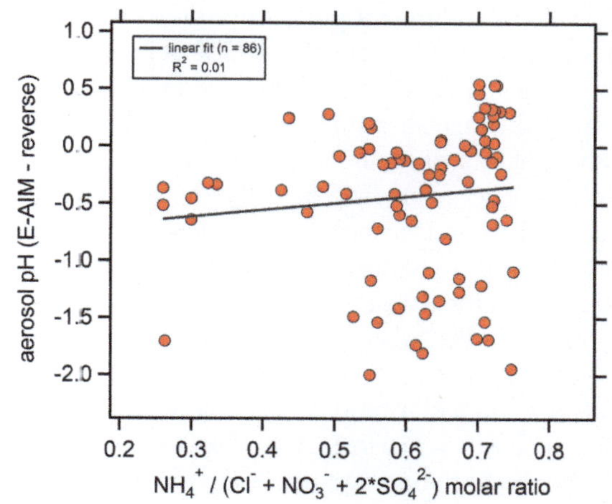

Figure 10. Predicted aerosol pH vs. the $NH_4^+ / (Cl^- + NO_3^- + 2*SO_4^{2-})$ molar ratio, using only data where the ratio is less than 0.75 (region inferred as highly acidic) and where $\Sigma(NH_4^+ + Cl^- + NO_3^- + SO_4^{2-})$ accounts for greater than 95 % of measured inorganic aerosol mass (i.e., with low crustal concentrations).

al., 2012a). These studies may have misinterpreted their data, since sulfate may actually be a limiting factor in isoprene and monoterpene SOA formation (Xu et al., 2015). H^+ derived from the ion balance is a surrogate for sulfate, and many factors lead to correlations between sulfate and SOA in regional pollution (Sun et al., 2011; Zhang et al., 2012b).

3.4 Molar ratio vs. pH

Similar to applications of the ion balance, the molar ratio is frequently used as a proxy for aerosol pH. Molar ratios that yield a charge balance (i.e., equivalence ratios of unity or greater) are assumed for fully neutralized aerosol, while decreasing cation / anion ratios are assumed to represent decreasing aerosol pH (He et al., 2012; Kerminen et al., 2001; Huang et al., 2013). The data and assumptions that underlie the molar ratio method are the same as those used to calculate the ion balance. Thus, many of the same limitations apply to the molar ratio method. The exclusion of minor ionic species, including crustal elements and dissociated organic acids, can significantly bias the results (Cao et al., 2013; Jacobson, 1999; Trebs et al., 2005; Ziemba et al., 2007; Moya et al., 2002). The propagation of analytical uncertainty can also create problems with the signal-to-noise ratio for many samples that will challenge the interpretations of the molar ratio. Finally, the molar ratio does not account for the effects of aerosol water or species activities on particle acidity. Given the similar assumptions and limitations, it should be unsurprising that $[H^+]$ from the ion balance was highly correlated with the cation / anion molar ratio ($R^2 = 0.66$, not shown) for the MILAGRO data set. Similar to the ion balance results, no correlation was observed between the molar ratio and forward-mode predictions of aerosol pH (Fig. 9). Zhang et al. (2007) and Behera et al. (2014) demonstrated similar problems translating the molar ratio into aerosol pH for studies in Pittsburgh and Singapore, respectively. Even under the limiting case when (1) crustal elements were low (where NH_4^+, Cl^-, NO_3^-, and SO_4^{2-} together accounted for greater than 95 % of inorganic aerosol ion mass) and (2) the aerosol regime was inferred to be highly acidic (samples with $NH_4^+ / (Cl^- + NO_3^- + 2*SO_4^{2-})$ molar ratios less than 0.75), there was no correlation between the molar ratio and predicted pH (Fig. 10).

These results indicate that the molar ratio is not a suitable proxy to infer the acidity of atmospheric particles. This applies to all variations of the molar ratio (Eqs. 3, 4, or 5). When the majority of inorganic species (> 95 %) are accounted for, the molar ratio is able to distinguish alkaline particles from acidic particles with good reliability; however, it is unable to provide any measure – even qualitative – of the degree of aerosol acidity. The lack of relationship between the molar ratio and predicted aerosol pH strongly suggests problems with studies that have used the molar ratio as a proxy for pH. For example, studies that have attempted to characterize the occurrence of acid-catalyzed SOA formation in the atmosphere may have incorrectly interpreted the aerosol acidity using a molar ratio approach (e.g., Tanner et al., 2009; Peltier et al., 2007; Zhang et al., 2007; Froyd et al., 2010).

4 Conclusions

We have carried out an analysis of proxy methods used to estimate the pH of atmospheric particles. The ion balance method, which is used to infer the aerosol H^+ loading, showed no correlation with aerosol pH predicted by multiple independent metrics. This indicates that conditions of increasing H^+ loading do not necessarily correspond to decreasing (i.e., more acidic) aerosol pH. Likewise, the cation / anion molar ratio (and related metrics) showed no relationship with different predictions of aerosol pH. When species accounting for greater than 95 % of inorganic aerosol mass were included in the analysis, the molar ratio appears to reliably distinguish acidic from alkaline particles; however, the molar ratio should not be treated as a surrogate for aerosol pH. The molar ratio actually showed a strong relationship with the ion balance. A major reason for the inability of these methods to represent aerosol pH is that both neglect the effects of aerosol water and partial dissolution of ions and acids on pH, in accordance with previously published studies (Keene and Savoie, 1998; Winkler, 1986). Further, the ion balance and molar ratio can also be severely limited by signal-to-noise ratio due to analytical uncertainties. These results strongly discourage the use of the ion balance or molar ratio for pH or H^+ inference.

These results also suggest that thermodynamic equilibrium models require both gas + aerosol inputs for accurate predictions of H^+ and pH. Two independent models – E-AIM and ISORROPIA – performed similarly and predict much lower pH (more acidic particles) in the reverse modes using aerosol inputs only. The aerosol pH levels predicted by both reverse models do not agree with either the pH predictions using gas + aerosol inputs or the phase partitioning of ammonia. The models in reverse mode predict highly acidic particles for extended periods of time (i.e., pH <0), despite high concentrations of gas-phase ammonia and relatively short equilibration timescales. Further, the models in reverse mode are highly sensitive to uncertainty in the measurement inputs: small deviations in major aerosol species on the order of common aerosol measurement uncertainties can induce changes in predicted pH that exceed 10 pH units. This recommendation is consistent with other studies that show much better model performance when aerosol + gas inputs are used (Fountoukis et al., 2009; Guo et al., 2014).

Thermodynamic equilibrium models in forward mode had very good agreement with pH predicted by the phase partitioning of ammonia. These methods are largely, but not completely, independent, as the aerosol liquid water required for the phase-partitioning calculation is obtained from the same thermodynamic models. In addition, the use of activity coefficients from thermodynamic models as inputs in the phase-partitioning calculations would further couple the two methods. This has been avoided in the present analysis by using concentrations instead of activities. Thermodynamic equilibrium models are generally quite skilled in predicting aerosol liquid water (Khlystov et al., 2005), so this likely introduces minimal uncertainty to the present analysis. In regions where organics are high, their contribution to aerosol liquid water (and hence pH) should be investigated (Guo et al., 2014). The fact that these largely independent methods agreed so closely provides the basis for the recommendation of their use to estimate the pH of atmospheric particles. As with the application of thermodynamic equilibrium models, the phase-partitioning approach makes the implicit assumption that the gas/particle system is at equilibrium. This appears to be a good assumption for the Mexico City data set (Fountoukis et al., 2009), but it can be a poor assumption when coarse particles are abundant, such as in the marine environment (Fridlind and Jacobson, 2000). The phase-partitioning approach is also likely limited in ammonia-poor environments where all or most of the ammonia resides in the aerosol phase. In such environments, the phase partitioning of nitric acid may be a good alternative if both nitric acid and aerosol nitrate concentrations are high enough (Young et al., 2013; Meskhidze et al., 2003). Uncertainty in the Henry's law constant of other compounds, especially HCl, may limit the application of the phase-partitioning approach beyond NH_3 / NH_4^+ and HNO_3 / NO_3^- (Young et al., 2013).

Our recommendation for the use of the phase-partitioning approach and forward equilibrium model calculations to best predict aerosol pH contradicts the recommendations in a similar study (Yao et al., 2006). Both our study (from Mexico City) and the Yao et al. (2006) study (Hong Kong) agree that E-AIM run in the reverse mode yields predicted aerosol pH levels significantly lower than the phase partitioning or forward model predictions. We differ in our interpretation of the results: Yao et al. (2006) conclude that the phase-partitioning approach and forward model calculations erroneously assume that the gas–aerosol system has reached equilibrium, even though their analysis is based upon 12 and 24 h $PM_{2.5}$ measurements. This sampling time should far exceed the equilibration time for submicron particles, which is on the order of seconds to minutes (Meng and Seinfeld, 1996). In addition, our comparisons in Fig. 4a and c only include the subset of data where RH exceeds 60 %, a region in which the thermodynamic predictions – and assumption of equilibrium – become more accurate (Moya et al., 2002). Fountoukis et al. (2009) examined the equilibrium assumption for $PM_{2.5}$ during MILAGRO and found it to be valid. Thus, we believe that the forward thermodynamic models and phase-partitioning approach are far more accurate in their predictions of pH than the reverse models run in closed mode without the additional constraint of gas-phase data.

These recommendations should be evaluated in other environments with different chemical characteristics and meteorology than was observed during MILAGRO. In particular, the recommendations should be evaluated in ammonia-poor environments, since thermodynamic equilibrium models often diverge in their predictions under such conditions (Ansari and Pandis, 1999). Despite their uncertainties, thermodynamic models and phase partitioning provide the best methods to estimate the pH of fine atmospheric particles. Other widely used metrics – ion balance and ion ratios – are misleading and should be avoided beyond the establishment of general alkalinity or acidity. Hence, conclusions that are sensitive to aerosol pH but that are based upon the ion balance or molar ratio may need revision.

Acknowledgements. The authors gratefully acknowledge the use of NH_3 data from M. L. Fischer (Environmental Energy Technologies Division, Lawrence Berkeley National Laboratory) and HNO_3 data from R. C. Cohen (Department of Chemistry, University of California, Berkeley) and D. K. Farmer (Department of Chemistry, Colorado State University). We also gratefully acknowledge useful conversations with J. G. Murphy (Department of Chemistry, University of Toronto). C. J. Hennigan acknowledges support from the NSF through award AGS-1464458. R. J. Weber acknowledges support from the NSF through award 1242258. A. Nenes acknowledges support from the EPA STAR program through award R835410, and from a Georgia Power Faculty Scholar Chair and Cullen-Peck Faculty Fellowship awarded by the Georgia Institute of Technology.

Edited by: A. Kiendler-Scharr

References

Ahrens, L., Harner, T., Shoeib, M., Lane, D. A., and Murphy, J. G.: Improved Characterization of Gas-Particle Partitioning for Per- and Polyfluoroalkyl Substances in the Atmosphere Using Annular Diffusion Denuder Samplers, Environ. Sci. Technol., 46, 7199–7206, doi:10.1021/Es300898s, 2012.

Ansari, A. S. and Pandis, S. N.: An analysis of four models predicting the partitioning of semivolatile inorganic aerosol components, Aerosol. Sci. Tech., 31, 129–153, doi:10.1080/027868299304200, 1999.

Behera, S., Cheng, J., and Balasubramanian, R.: In situ acidity and pH of size-fractionated aerosols during a recent smoke-haze episode in Southeast Asia, Environ. Geochem. Hlth., 1–17, doi:10.1007/s10653-014-9660-1, 2014.

Brown, S. S., Thornton, J. A., Keene, W. C., Pszenny, A. A. P., Sive, B. C., Dube, W. P., Wagner, N. L., Young, C. J., Riedel, T. P., Roberts, J. M., VandenBoer, T. C., Bahreini, R., Ozturk, F., Middlebrook, A. M., Kim, S., Hubler, G., and Wolfe, D. E.: Nitrogen, Aerosol Composition, and Halogens on a Tall Tower (NACHTT): Overview of a wintertime air chemistry field study in the front range urban corridor of Colorado, J. Geophys. Res.-Atmos., 118, 8067–8085, doi:10.1002/Jgrd.50537, 2013.

Budisulistiorini, S. H., Canagaratna, M. R., Croteau, P. L., Marth, W. J., Baumann, K., Edgerton, E. S., Shaw, S. L., Knipping, E. M., Worsnop, D. R., Jayne, J. T., Gold, A., and Surratt, J. D.: Real-Time Continuous Characterization of Secondary Organic Aerosol Derived from Isoprene Epoxydiols in Downtown Atlanta, Georgia, Using the Aerodyne Aerosol Chemical Speciation Monitor, Environ. Sci. Technol., 47, 5686–5694, doi:10.1021/Es400023n, 2013.

Cao, J. J., Tie, X. X., Dabberdt, W. F., Jie, T., Zhao, Z. Z., An, Z. S., Shen, Z. X., and Feng, Y. C.: On the potential high acid deposition in northeastern China, J. Geophys. Res.-Atmos., 118, 4834–4846, doi:10.1002/Jgrd.50381, 2013.

Chameides, W. L.: The photochemistry of a remote marine stratiform cloud, J. Geophys. Res.-Atmos., 89, 4739–4755, 10.1029/JD089iD03p04739, 1984.

Clegg, S. L., Brimblecombe, P., and Wexler, A. S.: Thermodynamic model of the system H^+-NH_4^+-SO_4^{2-}-NO_3^--H_2O at tropospheric temperatures, J. Phys. Chem. A, 102, 2137–2154, doi:10.1021/Jp973042r, 1998.

Clegg, S. L., Seinfeld, J. H., and Edney, E. O.: Thermodynamic modelling of aqueous aerosols containing electrolytes and dissolved organic compounds. II. An extended Zdanovskii-Stokes-Robinson approach, J. Aerosol Sci., 34, 667–690, doi:10.1016/s0021-8502(03)00019-3, 2003.

Day, D. A., Wooldridge, P. J., Dillon, M. B., Thornton, J. A., and Cohen, R. C.: A thermal dissociation laser-induced fluorescence instrument for in situ detection of NO_2, peroxy nitrates, alkyl nitrates, and HNO_3, J. Geophys. Res.-Atmos., 107, 4046, doi:10.1029/2001jd000779, 2002.

Dockery, D. W., Cunningham, J., Damokosh, A. I., Neas, L. M., Spengler, J. D., Koutrakis, P., Ware, J. H., Raizenne, M., and Speizer, F. E.: Health effects of acid aerosols on North American children: Respiratory symptoms, Environ. Health Persp., 104, 500–505, doi:10.2307/3432990, 1996.

Farmer, D. K., Perring, A. E., Wooldridge, P. J., Blake, D. R., Baker, A., Meinardi, S., Huey, L. G., Tanner, D., Vargas, O., and Cohen, R. C.: Impact of organic nitrates on urban ozone production, Atmos. Chem. Phys., 11, 4085–4094, doi:10.5194/acp-11-4085-2011, 2011.

Feng, J. L., Guo, Z. G., Zhang, T. R., Yao, X. H., Chan, C. K., and Fang, M.: Source and formation of secondary particulate matter in $PM_{2.5}$ in Asian continental outflow, J. Geophys. Res.-Atmos., 117, D03302, doi:10.1029/2011jd016400, 2012.

Fountoukis, C. and Nenes, A.: ISORROPIA II: a computationally efficient thermodynamic equilibrium model for K^+–Ca^{2+}–Mg^{2+}–NH_4^+–Na^+–SO_4^{2-}–NO_3^-–Cl^-–H_2O aerosols, Atmos. Chem. Phys., 7, 4639–4659, doi:10.5194/acp-7-4639-2007, 2007.

Fountoukis, C., Nenes, A., Sullivan, A., Weber, R., Van Reken, T., Fischer, M., Matías, E., Moya, M., Farmer, D., and Cohen, R. C.: Thermodynamic characterization of Mexico City aerosol during MILAGRO 2006, Atmos. Chem. Phys., 9, 2141–2156, doi:10.5194/acp-9-2141-2009, 2009.

Fridlind, A. M. and Jacobson, M. Z.: A study of gas-aerosol equilibrium and aerosol pH in the remote marine boundary layer during the First Aerosol Characterization Experiment (ACE 1), J. Geophys. Res.-Atmos., 105, 17325–17340, doi:10.1029/2000jd900209, 2000.

Froyd, K. D., Murphy, S. M., Murphy, D. M., de Gouw, J. A., Eddingsaas, N. C., and Wennberg, P. O.: Contribution of isoprene-derived organosulfates to free tropospheric aerosol mass, P. Natl. Acad. Sci. USA, 107, 21360–21365, doi:10.1073/pnas.1012561107, 2010.

Gaston, C. J., Riedel, T. P., Zhang, Z., Gold, A., Surratt, J. D., and Thornton, J. A.: Reactive Uptake of an Isoprene-derived Epoxydiol to Submicron Aerosol Particles, Environ. Sci. Technol., 48, 11178–11186, doi:10.1021/es5034266, 2014.

Guo, H., Xu, L., Bougiatioti, A., Cerully, K. M., Capps, S. L., Hite, J. R., Carlton, A. G., Lee, S.-H., Bergin, M. H., Ng, N. L., Nenes, A., and Weber, R. J.: Particle water and pH in the southeastern United States, Atmos. Chem. Phys. Discuss., 14, 27143–27193, doi:10.5194/acpd-14-27143-2014, 2014.

Gwynn, R. C., Burnett, R. T., and Thurston, G. D.: A time-series analysis of acidic particulate matter and daily mortality and morbidity in the Buffalo, New York, region, Environ. Health Persp., 108, 125–133, doi:10.2307/3454510, 2000.

Harris, D. C.: Quantitative Chemical Analysis, 5th Ed., edited by: Freeman, W. H. and Company, New York, 1999.

He, K., Zhao, Q., Ma, Y., Duan, F., Yang, F., Shi, Z., and Chen, G.: Spatial and seasonal variability of $PM_{2.5}$ acidity at two Chinese megacities: insights into the formation of secondary inorganic aerosols, Atmos. Chem. Phys., 12, 1377–1395, doi:10.5194/acp-12-1377-2012, 2012.

Hennigan, C. J., Sullivan, A. P., Fountoukis, C. I., Nenes, A., Hecobian, A., Vargas, O., Peltier, R. E., Case Hanks, A. T., Huey, L. G., Lefer, B. L., Russell, A. G., and Weber, R. J.: On the volatility and production mechanisms of newly formed nitrate and water soluble organic aerosol in Mexico City, Atmos. Chem. Phys., 8, 3761–3768, doi:10.5194/acp-8-3761-2008, 2008.

Huang, Y., Li, L., Li, J., Wang, X., Chen, H., Chen, J., Yang, X., Gross, D. S., Wang, H., Qiao, L., and Chen, C.: A case study of the highly time-resolved evolution of aerosol chemical and optical properties in urban Shanghai, China, Atmos. Chem. Phys., 13, 3931–3944, doi:10.5194/acp-13-3931-2013, 2013.

Jacobson, M. Z.: Studying the effects of calcium and magnesium on size-distributed nitrate and ammonium with EQUISOLV II, Atmos. Environ., 33, 3635–3649, doi:10.1016/S1352-2310(99)00105-3, 1999.

Jang, M., Cao, G., and Paul, J.: Colorimetric particle acidity analysis of secondary organic aerosol coating on submicron acidic aerosols, Aerosol. Sci. Tech., 42, 409–420, doi:10.1080/02786820802154861, 2008.

Johnson, A. H., Moyer, A., Bedison, J. E., Richter, S. L., and Willig, S. A.: Seven Decades of Calcium Depletion in Organic Horizons of Adirondack Forest Soils, Soil Sci. Soc. Am. J., 72, 1824–1830, doi:10.2136/sssaj2006.0407, 2008.

Keene, W. C. and Savoie, D. L.: The pH of deliquesced sea-salt aerosol in polluted marine air, Geophys. Res. Lett., 25, 2181–2184, doi:10.1029/98gl01591, 1998.

Keene, W. C., Sander, R., Pszenny, A. A. P., Vogt, R., Crutzen, P. J., and Galloway, J. N.: Aerosol pH in the marine boundary layer: A review and model evaluation, J. Aerosol Sci., 29, 339–356, doi:10.1016/S0021-8502(97)10011-8, 1998.

Keene, W. C., Pszenny, A. A. P., Maben, J. R., and Sander, R.: Variation of marine aerosol acidity with particle size, Geophys. Res. Lett., 29, 1101, doi:10.1029/2001gl013881, 2002.

Keene, W. C., Pszenny, A. A. P., Maben, J. R., Stevenson, E., and Wall, A.: Closure evaluation of size-resolved aerosol pH in the New England coastal atmosphere during summer, J. Geophys. Res.-Atmos., 109, D23307, doi:10.1029/2004jd004801, 2004.

Kerminen, V. M., Hillamo, R., Teinila, K., Pakkanen, T., Allegrini, I., and Sparapani, R.: Ion balances of size-resolved tropospheric aerosol samples: implications for the acidity and atmospheric processing of aerosols, Atmos. Environ., 35, 5255–5265, doi:10.1016/S1352-2310(01)00345-4, 2001.

Khlystov, A., Stanier, C. O., Takahama, S., and Pandis, S. N.: Water content of ambient aerosol during the Pittsburgh air quality study, J. Geophys. Res.-Atmos., 110, D07s10, doi:10.1029/2004jd004651, 2005.

Kim, Y. P., Seinfeld, J. H., and Saxena, P.: Atmospheric Gas Aerosol Equilibrium 1. Thermodynamic Model, Aerosol. Sci. Tech., 19, 157–181, doi:10.1080/02786829308959628, 1993.

Koutrakis, P., Wolfson, J. M., and Spengler, J. D.: An Improved Method for Measuring Aerosol Strong Acidity: Results from a Nine-month Study in St. Louis, Missouri and Kingston, Tennessee, Atmos. Environ., 22, 157–162, 1988.

Lawrence, J. and Koutrakis, P.: Measurement and speciation of gas and particulate phase organic acidity in an urban environment: 2. Speciation, J. Geophys. Res.-Atmos., 101, 9171–9184, doi:10.1029/95jd03357, 1996.

Likens, G. E., Driscoll, C. T., and Buso, D. C.: Long-term effects of acid rain: Response and recovery of a forest ecosystem, Science, 272, 244–246, doi:10.1126/science.272.5259.244, 1996.

Mahowald, N.: Aerosol Indirect Effect on Biogeochemical Cycles and Climate, Science, 334, 794–796, doi:10.1126/science.1207374, 2011.

Meng, Z. Y. and Seinfeld, J. H.: Time scales to achieve atmospheric gas-aerosol equilibrium for volatile species, Atmos. Environ., 30, 2889–2900, doi:10.1016/1352-2310(95)00493-9, 1996.

Meskhidze, N., Chameides, W. L., Nenes, A., and Chen, G.: Iron mobilization in mineral dust: Can anthropogenic SO_2 emissions affect ocean productivity?, Geophys. Res. Lett., 30, 2085, doi:10.1029/2003gl018035, 2003.

Meskhidze, N., Chameides, W. L., and Nenes, A.: Dust and pollution: A recipe for enhanced ocean fertilization?, J. Geophys. Res.-Atmos., 110, D03301, doi:10.1029/2004jd005082, 2005.

Metzger, S., Mihalopoulos, N., and Lelieveld, J.: Importance of mineral cations and organics in gas-aerosol partitioning of reactive nitrogen compounds: case study based on MINOS results, Atmos. Chem. Phys., 6, 2549–2567, doi:10.5194/acp-6-2549-2006, 2006.

Molina, L. T., Madronich, S., Gaffney, J. S., Apel, E., de Foy, B., Fast, J., Ferrare, R., Herndon, S., Jimenez, J. L., Lamb, B., Osornio-Vargas, A. R., Russell, P., Schauer, J. J., Stevens, P. S., Volkamer, R., and Zavala, M.: An overview of the MILAGRO 2006 Campaign: Mexico City emissions and their transport and transformation, Atmos. Chem. Phys., 10, 8697–8760, doi:10.5194/acp-10-8697-2010, 2010.

Moya, M., Pandis, S. N., and Jacobson, M. Z.: Is the size distribution of urban aerosols determined by thermodynamic equilibrium? An application to Southern California, Atmos. Environ., 36, 2349–2365, doi:10.1016/S1352-2310(01)00549-0, 2002.

Nenes, A., Pandis, S. N., and Pilinis, C.: ISORROPIA: A new thermodynamic equilibrium model for multiphase multicomponent inorganic aerosols, Aquat. Geochem., 4, 123–152, doi:10.1023/a:1009604003981, 1998.

Nenes, A., Pandis, S. N., and Pilinis, C.: Continued development and testing of a new thermodynamic aerosol module for urban and regional air quality models, Atmos. Environ., 33, 1553–1560, doi:10.1016/S1352-2310(98)00352-5, 1999.

Nenes, A., Krom, M. D., Mihalopoulos, N., Van Cappellen, P., Shi, Z., Bougiatioti, A., Zarmpas, P., and Herut, B.: Atmospheric acidification of mineral aerosols: a source of bioavailable phosphorus for the oceans, Atmos. Chem. Phys., 11, 6265–6272, doi:10.5194/acp-11-6265-2011, 2011.

Nguyen, Q. T., Christensen, M. K., Cozzi, F., Zare, A., Hansen, A. M. K., Kristensen, K., Tulinius, T. E., Madsen, H. H., Christensen, J. H., Brandt, J., Massling, A., Nøjgaard, J. K., and Glasius, M.: Understanding the anthropogenic influence on formation of biogenic secondary organic aerosols in Denmark via analysis of organosulfates and related oxidation products, Atmos. Chem. Phys., 14, 8961–8981, doi:10.5194/acp-14-8961-2014, 2014.

Paciga, A. L., Riipinen, I., and Pandis, S. N.: Effect of Ammonia on the Volatility of Organic Diacids, Environ. Sci. Technol., 48, 13769–13775, doi:10.1021/es5037805, 2014.

Pan, Y. P., Wang, Y. S., Tang, G. Q., and Wu, D.: Spatial distribution and temporal variations of atmospheric sulfur deposition in Northern China: insights into the potential acidification risks, Atmos. Chem. Phys., 13, 1675–1688, doi:10.5194/acp-13-1675-2013, 2013.

Pathak, R. K., Yao, X. H., and Chan, C. K.: Sampling artifacts of acidity and ionic species in $PM_{2.5}$, Environ. Sci. Technol., 38, 254–259, doi:10.1021/Es0342244, 2004.

Pathak, R. K., Wu, W. S., and Wang, T.: Summertime $PM_{2.5}$ ionic species in four major cities of China: nitrate formation in an ammonia-deficient atmosphere, Atmos. Chem. Phys., 9, 1711–1722, doi:10.5194/acp-9-1711-2009, 2009.

Pathak, R. K., Wang, T., Ho, K. F., and Lee, S. C.: Characteristics of summertime $PM_{2.5}$ organic and elemental carbon in four major Chinese cities: Implications of high acidity for water-soluble organic carbon (WSOC), Atmos. Environ., 45, 318–325, doi:10.1016/j.atmosenv.2010.10.021, 2011.

Peltier, R. E., Sullivan, A. P., Weber, R. J., Wollny, A. G., Holloway, J. S., Brock, C. A., de Gouw, J. A., and Atlas, E. L.: No evidence for acid-catalyzed secondary organic aerosol formation in power plant plumes over metropolitan Atlanta, Georgia, Geophys. Res. Lett., 34, L06801, doi:10.1029/2006gl028780, 2007.

Pilinis, C. and Seinfeld, J. H.: Continued development of a general equilibrium model for inorganic multicomponent atmospheric aerosols, Atmos. Environ., 21, 2453–2466, doi:10.1016/0004-6981(87)90380-5, 1987.

Pszenny, A. A. P., Moldanová, J., Keene, W. C., Sander, R., Maben, J. R., Martinez, M., Crutzen, P. J., Perner, D., and Prinn, R. G.: Halogen cycling and aerosol pH in the Hawaiian marine boundary layer, Atmos. Chem. Phys., 4, 147–168, doi:10.5194/acp-4-147-2004, 2004.

Reid, J. S., Hobbs, P. V., Ferek, R. J., Blake, D. R., Martins, J. V., Dunlap, M. R., and Liousse, C.: Physical, chemical, and optical properties of regional hazes dominated by smoke in Brazil, J. Geophys. Res.-Atmos., 103, 32059–32080, doi:10.1029/98jd00458, 1998.

Sander, R. and Crutzen, P. J.: Model study indicating halogen activation and ozone destruction in polluted air masses transported to the sea, J. Geophys. Res.-Atmos., 101, 9121–9138, doi:10.1029/95jd03793, 1996.

Schindler, D. W.: Effects of Acid-Rain on Fresh-Water Ecosystems, Science, 239, 149–157, doi:10.1126/science.239.4836.149, 1988.

Stoddard, J. L., Jeffries, D. S., Lukewille, A., Clair, T. A., Dillon, P. J., Driscoll, C. T., Forsius, M., Johannessen, M., Kahl, J. S., Kellogg, J. H., Kemp, A., Mannio, J., Monteith, D. T., Murdoch, P. S., Patrick, S., Rebsdorf, A., Skjelkvale, B. L., Stainton, M. P., Traaen, T., van Dam, H., Webster, K. E., Wieting, J., and Wilander, A.: Regional trends in aquatic recovery from acidification in North America and Europe, Nature, 401, 575–578, doi:10.1038/44114, 1999.

Stumm, W. and Morgan, J. J.: Aquatic chemistry : chemical equilibria and rates in natural waters, Wiley, New York, 1996.

Sun, Y.-L., Zhang, Q., Schwab, J. J., Demerjian, K. L., Chen, W.-N., Bae, M.-S., Hung, H.-M., Hogrefe, O., Frank, B., Rattigan, O. V., and Lin, Y.-C.: Characterization of the sources and processes of organic and inorganic aerosols in New York city with a high-resolution time-of-flight aerosol mass apectrometer, Atmos. Chem. Phys., 11, 1581–1602, doi:10.5194/acp-11-1581-2011, 2011.

Surratt, J. D., Lewandowski, M., Offenberg, J. H., Jaoui, M., Kleindienst, T. E., Edney, E. O., and Seinfeld, J. H.: Effect of acidity on secondary organic aerosol formation from isoprene, Environ. Sci. Technol., 41, 5363–5369, doi:10.1021/Es0704176, 2007.

Takahama, S., Davidson, C. I., and Pandis, S. N.: Semicontinuous measurements of organic carbon and acidity during the Pittsburgh air quality study: Implications for acid-catalyzed organic aerosol formation, Environ. Sci. Technol., 40, 2191–2199, doi:10.1021/Es050856+, 2006.

Tanner, R. L., Olszyna, K. J., Edgerton, E. S., Knipping, E., and Shaw, S. L.: Searching for evidence of acid-catalyzed enhancement of secondary organic aerosol formation using ambient aerosol data, Atmos. Environ., 43, 3440–3444, doi:10.1016/j.atmosenv.2009.03.045, 2009.

Thornton, J. A., Kercher, J. P., Riedel, T. P., Wagner, N. L., Cozic, J., Holloway, J. S., Dube, W. P., Wolfe, G. M., Quinn, P. K., Middlebrook, A. M., Alexander, B., and Brown, S. S.: A large atomic chlorine source inferred from mid-continental reactive nitrogen chemistry, Nature, 464, 271–274, doi:10.1038/Nature08905, 2010.

Trebs, I., Metzger, S., Meixner, F. X., Helas, G. N., Hoffer, A., Rudich, Y., Falkovich, A. H., Moura, M. A. L., da Silva, R. S., Artaxo, P., Slanina, J., and Andreae, M. O.: The NH_4^+-NO_3^--Cl^--SO_4^{2-}-H_2O aerosol system and its gas phase precursors at a pasture site in the Amazon Basin: How relevant are mineral cations and soluble organic acids?, J. Geophys. Res.-Atmos., 110, D07303, doi:10.1029/2004jd005478, 2005.

Wexler, A. S. and Clegg, S. L.: Atmospheric aerosol models for systems including the ions H^+, NH_4^+, Na^+, SO_4^{2-}, NO_3^-,Cl^-, Br^-, and H_2O, J. Geophys. Res.-Atmos., 107, 4207, doi:10.1029/2001jd000451, 2002.

Winkler, P.: Relations Between Aerosol Acidity and Ion Balance, in: Chemistry of Multiphase Atmospheric Systems, edited by: Jaeschke, W., 6, Springer-Verlag, New York, 269–298, 1986.

Xu, L., Guo, H., Boyd, C. M., Klein, M., Bougiatioti, A., Cerully, K. M., Hite, J. R., Isaacman-VanWertz, G., Kreisberg, N. M., Knote, C., Olson, K., Koss, A., Goldstein, A. H., Hering, S. V., de Gouw, J., Baumann, K., Lee, S.-H., Nenes, A., Weber, R. J., and Ng, N. L.: Effects of anthropogenic emissions on aerosol formation from isoprene and monoterpenes in the southeastern United States, P. Natl. Acad. Sci. USA, 112, 37–42, doi:10.1073/pnas.1417609112, 2015.

Yao, X. H., Ling, T. Y., Fang, M., and Chan, C. K.: Comparison of thermodynamic predictions for in situ pH in $PM_{2.5}$, Atmos. Environ., 40, 2835–2844, doi:10.1016/j.atmosenv.2006.01.006, 2006.

Young, A. H., Keene, W. C., Pszenny, A. A. P., Sander, R., Thornton, J. A., Riedel, T. P., and Maben, J. R.: Phase partitioning of soluble trace gases with size-resolved aerosols in near-surface continental air over northern Colorado, USA, during winter, J. Geophys. Res.-Atmos., 118, 9414–9427, doi:10.1002/Jgrd.50655, 2013.

Zaveri, R. A., Easter, R. C., Fast, J. D., and Peters, L. K.: Model for Simulating Aerosol Interactions and Chemistry (MOSAIC), J. Geophys. Res.-Atmos., 113, D13204, doi:10.1029/2007jd008782, 2008.

Zhang, H. F., Worton, D. R., Lewandowski, M., Ortega, J., Rubitschun, C. L., Park, J. H., Kristensen, K., Campuzano-Jost, P., Day, D. A., Jimenez, J. L., Jaoui, M., Offenberg, J. H., Kleindienst, T. E., Gilman, J., Kuster, W. C., de Gouw, J., Park, C., Schade, G. W., Frossard, A. A., Russell, L., Kaser, L., Jud, W., Hansel, A., Cappellin, L., Karl, T., Glasius, M., Guenther, A., Goldstein, A. H., Seinfeld, J. H., Gold, A., Kamens, R. M., and Surratt, J. D.: Organosulfates as Tracers for Secondary Organic Aerosol (SOA) Formation from 2-Methyl-3-Buten-2-ol (MBO) in the Atmosphere, Environ. Sci. Technol., 46, 9437–9446, doi:10.1021/Es301648z, 2012a.

Zhang, Q., Jimenez, J. L., Worsnop, D. R., and Canagaratna, M.: A case study of urban particle acidity and its influence on secondary organic aerosol, Environ. Sci. Technol., 41, 3213–3219, doi:10.1021/Es061812j, 2007.

Zhang, X., Liu, Z., Hecobian, A., Zheng, M., Frank, N. H., Edgerton, E. S., and Weber, R. J.: Spatial and seasonal variations of fine particle water-soluble organic carbon (WSOC) over the southeastern United States: implications for secondary organic aerosol formation, Atmos. Chem. Phys., 12, 6593–6607, doi:10.5194/acp-12-6593-2012, 2012b.

Zhang, Y., Seigneur, C., Seinfeld, J. H., Jacobson, M., Clegg, S. L., and Binkowski, F. S.: A comparative review of inorganic aerosol thermodynamic equilibrium modules: similarities, differences, and their likely causes, Atmos. Environ., 34, 117–137, doi:10.1016/S1352-2310(99)00236-8, 2000.

Ziemba, L. D., Fischer, E., Griffin, R. J., and Talbot, R. W.: Aerosol acidity in rural New England: Temporal trends and source region analysis, J. Geophys. Res.-Atmos., 112, D10s22, doi:10.1029/2006jd007605, 2007.

2

Temperature profiling of the atmospheric boundary layer with rotational Raman lidar during the HD(CP)^2 Observational Prototype Experiment

E. Hammann[1], A. Behrendt[1], F. Le Mounier[2], and V. Wulfmeyer[1]

[1]University of Hohenheim, Institute of Physics and Meteorology, Garbenstrasse 30, 70599 Stuttgart, Germany
[2]Laboratoire de Météorologie Dynamique, C.N.R.S, Paris, France

Correspondence to: E. Hammann (eva.hammann@uni-hohenheim.de)

Abstract. The temperature measurements of the rotational Raman lidar of the University of Hohenheim (UHOH RRL) during the High Definition of Clouds and Precipitation for advancing Climate Prediction (HD(CP)^2) Observation Prototype Experiment (HOPE) in April and May 2013 are discussed. The lidar consists of a frequency-tripled Nd:YAG laser at 355 nm with 10 W average power at 50 Hz, a two-mirror scanner, a 40 cm receiving telescope, and a highly efficient polychromator with cascading interference filters for separating four signals: the elastic backscatter signal, two rotational Raman signals with different temperature dependence, and the vibrational Raman signal of water vapor. The main measurement variable of the UHOH RRL is temperature. For the HOPE campaign, the lidar receiver was optimized for high and low background levels, with a novel switch for the passband of the second rotational Raman channel. The instrument delivers atmospheric profiles of water vapor mixing ratio as well as particle backscatter coefficient and particle extinction coefficient as further products. As examples for the measurement performance, measurements of the temperature gradient and water vapor mixing ratio revealing the development of the atmospheric boundary layer within 25 h are presented. As expected from simulations, a reduction of the measurement uncertainty of 70 % during nighttime was achieved with the new low-background setting. A two-mirror scanner allows for measurements in different directions. When pointing the scanner to low elevation, measurements close to the ground become possible which are otherwise impossible due to the non-total overlap of laser beam and receiving telescope field of view in the near range. An example of a low-level temperature measurement is presented which resolves the temperature gradient at the top of the stable nighttime boundary layer 100 m above the ground.

1 Introduction

In recent years, different techniques for measuring the atmospheric temperature profile with lidar have been developed, namely the rotational Raman technique, the integration technique (using elastic and Raman signals), and the resonance fluorescence technique, as well as the high-spectral-resolution lidar (HSRL) technique and differential absorption lidar (DIAL) (see Behrendt, 2005, for an overview). For daytime measurements in the troposphere, rotational Raman lidar (RRL) is presently the most reliable technique. Its capabilities in providing temperature profiles with high temporal and spatial resolution and low systematic and noise errors during night- and daytime even within aerosol layers and thin clouds are superior so far to all other techniques, particularly where measurements from the surface to the lower troposphere are concerned (Behrendt and Reichardt, 2000; Behrendt et al., 2002, 2004; Di Girolamo et al., 2004; Arshinov et al., 2005; Radlach et al., 2008).

Most rotational Raman systems operate at wavelengths of 532 or 355 nm, the second and third harmonic wavelengths of Nd:YAG lasers. UV systems are able to perform daytime measurements with lower uncertainties due to the higher backscatter cross section and less solar background (Zeyn et al., 1996; Behrendt, 2005). The rotational Raman lidars of

NASA's Goddard Space Flight Center (Di Girolamo et al., 2004), of the University of Hohenheim (UHOH; Radlach et al., 2008), of the University of Basilicata (Di Girolamo et al., 2009), of Xi'an University (Mao et al., 2009), and of Hampton University (Su et al., 2013) all operate in the UV with interference-filter polychromators. Rotational Raman lidars at 532 nm show lower performance during daytime but reach a larger range at night than an UV system due to the higher laser power available at 532 nm compared to 355 nm, higher efficiency in signal separation, and lower atmospheric extinction. Some of the systems at 532 nm are also based on interference filters (Behrendt and Reichardt, 2000; Behrendt et al., 2002, 2004; Achtert et al., 2013), and some employ double-grating polychromators (Balin et al., 2004; Arshinov et al., 2005).

Daytime temperature measurements are the main focus of the RRL of UHOH. But besides temperature, also the particle backscatter coefficient and the particle extinction coefficient can be measured independently. Furthermore, the system was extended recently with a water vapor Raman channel. For water vapor measurements, two different lidar techniques are available: the DIAL technique and the Raman lidar technique. While water vapor Raman lidar uses the vibrational Raman backscatter signals of water vapor (e.g., Melfi et al., 1969; Whiteman et al., 1992; Turner and Goldsmith, 1999; Leblanc and McDermid, 2008), the DIAL technique (Schotland, 1974; Wulfmeyer and Bösenberg, 1998; Behrendt et al., 2009; Wagner et al., 2013) relies on the different absorption of water vapor at two nearby wavelengths. In contrast to water vapor (WV) DIAL, whose self-calibrating property leads to a high accuracy of the measurements (Bhawar et al., 2011), a water vapor Raman lidar has to be calibrated and shows also lower performance during daytime. However, the laser transmitter is less complex, and, if a certain lidar system already contains rotational Raman channels, only one more detection channel is needed to derive in addition the water vapor mixing ratio (Behrendt et al., 2002). Thus it was decided to extend the UHOH RRL with a water vapor channel. Furthermore, this capability permits relative humidity measurements, which are, e.g., useful for aerosol (Wulfmeyer and Feingold, 2000) and convection initiation studies (Behrendt et al., 2011; Corsmeier et al., 2011). In parallel to the RRL, the UHOH has developed also a water vapor DIAL (Wagner et al., 2013), from which recent measurements are presented in Muppa et al. (2014).

Within the High Definition of Clouds and Precipitation for advancing Climate Prediction (HD(CP)^2) project, a new model for high-resolution weather forecasts will be developed (Stevens and Bony, 2013) and other model systems will be tested (Schwitalla and Wulfmeyer, 2014). To verify the model, high-resolution data sets are required. The HD(CP)^2 Observation Prototype Experiment (HOPE), which provides such a data set, took place during April and May 2013 in the area around the Research Center Jülich in northwest Germany. The area already had an existing infrastructure of dense standard meteorological instrumentation.

Both systems of UHOH were operated together in the HOPE field campaign and thus provided a synergetic data set of thermodynamic properties of the atmosphere. The lidars were located at one supersite (50°53′50.55″ N, 6°27′50.27″ E, 110 m above sea level) close to the village of Hambach together with the KITcube, an instrument suite of the Karlsruhe Institute of Technology (KIT) (Kalthoff et al., 2013). The site was on a hillside and elevated above the surroundings by 10 to 15 m. The RRL and the WV DIAL of UHOH were collocated with a Doppler lidar from KIT to acquire a complete data set of temperature, water vapor content, and vertical wind for the determination of fluxes of sensible and latent heat (e.g., Behrendt et al., 2011). It was also the launch site for radiosoundings.

The purpose of this paper is to introduce the measurement performance of the UHOH RRL during HOPE and present highlights of the measurements. One highlight was the application of a novel switch for low- and high-background RRL. During the 18 intensive observation periods more than 200 h of measurements were collected. Measurements took place in the time between sunrise and sunset and were stopped in the case of rain or continuing dense cloud cover. In addition there were two night-long measurements. On one day, RHI (range–height indicator) scans were performed. During one day and one night, the lidar was pointed at a low elevation above the ground. Water vapor measurement started at the beginning of May; therefore only 100 h of water vapor data is available. Case studies using the combination of data from the instrument set – and also from the other two supersites – are in preparation and will be presented later based on the results shown here.

The main research interest of our institute is land–surface–atmosphere feedback, which requires measurements of land–surface exchange, the surface layer, the atmospheric boundary layer, and the lower free troposphere. For investigating not only mean profiles or mean three-dimensional fields but also the turbulent features of the convective boundary layer, instruments providing data with high temporal and spatial resolution in conditions of high-background light are needed. The rotational Raman lidar of UHOH was optimized for such conditions. A new technical feature, which was implemented during the HOPE campaign, was a switch to optimize the performance of the temperature measurements for low- and high-background conditions. Detailed simulations were performed which showed that it is favorable to use another pair of filter center wavelengths in low-background conditions compared to cases with high solar background. The change between the settings can be made in a few minutes. Thus, it was possible to switch easily between the settings and acquire continuous time periods of data which included such changes.

This paper is structured as follows: in Sect. 2 the new setup of the UHOH RRL is explained. Also the simulations for two

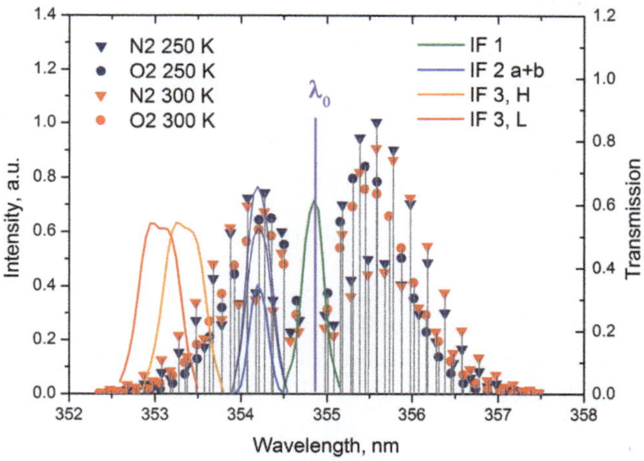

Figure 1. Rotational Raman spectrum of atmospheric nitrogen and oxygen for an excitation wavelength λ_0 of 354.83 nm and for two temperatures (250 and 300 K). Transmission curves of the interference filters are shown for comparison. Although the Stokes lines ($\lambda > \lambda_0$) are more intense, anti-Stokes lines ($\lambda < \lambda_0$) are used by us to avoid possible measurement errors due to fluorescence (Immler et al. 2005). For IF3, the filter positions for both the low- and high-background settings are shown (L and H, respectively).

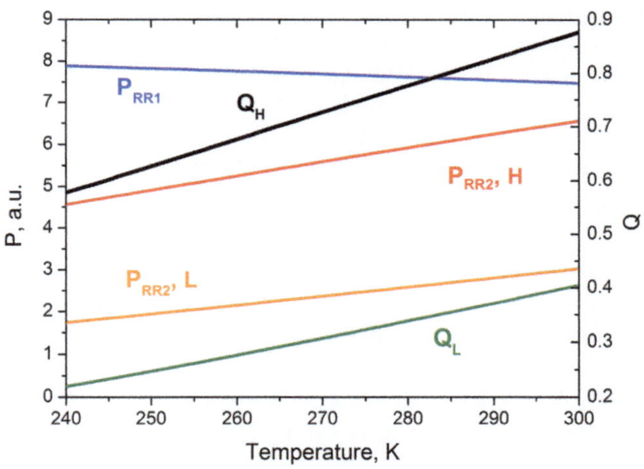

Figure 2. Simulated signal intensities for the filter configurations shown in Fig. 1: rotational Raman signal intensities for the H and L setting and corresponding ratios $Q = P_{\mathrm{RR2}}/P_{\mathrm{RR1}}$.

settings for the second rotational Raman channel are presented. Section 3 discusses the experimental results of the new system. Section 4 gives a short summary.

2 Methods and performance simulations

2.1 Methods

The rotational Raman lidar of UHOH makes use of the rotational Raman technique for deriving atmospheric temperature profiles (Cooney, 1972). Two parts of the rotational Raman spectrum of air with different temperature dependency are acquired (see Fig. 1). By taking the ratio of the two signals (Fig. 2) and calibrating it, the temperature is obtained without further assumptions about the state of the atmosphere. Figure 2 shows how the rotational Raman signals P_{RR1} and P_{RR2} depend on the temperature. For P_{RR2} there are the two settings shown in Fig. 1 with different central wavelengths CWL2: one for low-background (L) and one for high-background (H) conditions. The setting L results in a lower signal intensity in the second Raman channel. Also the ratio Q between P_{RR2} and P_{RR1} is lower and has a different slope than the one for the H setting. There are several formulas in use for the calibration (Behrendt and Reichardt, 2000; Behrendt, 2005). For temperature measurements up to a few kilometers altitude, the following equation can be used:

$$Q(T) = \exp\left(a - \frac{b}{T}\right), \tag{1}$$

where Q is the ratio between the background-corrected signals in the Raman channels RR2 and RR1, and a and b are calibration constants. This equation is exact for the ratio of two single rotational Raman lines. If several rotational Raman lines are extracted by the two channels, more complicated equations with more constants are needed when temperature measurements are made over a larger range of temperatures (Behrendt and Reichardt, 2000; Behrendt, 2005). This is not the purpose here, so this equation can still be used and results in a high accuracy of the inversion. Therefore the atmospheric temperature can be derived from Q by rearranging Eq. (1) to

$$T = \frac{b}{a - \ln(Q)}. \tag{2}$$

The statistical error of the temperature measurements can be determined from the signal intensities of the photon-counting data and applying Poisson statistics. For a signal with count number s, the 1σ statistical error is

$$\Delta s = \sqrt{s}. \tag{3}$$

This results in a noise error for the temperature ΔT (Behrendt et al., 2002) of

$$\Delta T = \frac{\partial T}{\partial Q} Q \sqrt{\frac{P_{\mathrm{RR1}} + P_{\mathrm{B1}}}{P_{\mathrm{RR1}}^2} + \frac{P_{\mathrm{RR2}} + P_{\mathrm{B2}}}{P_{\mathrm{RR2}}^2}}. \tag{4}$$

P_{RR1} is the background-corrected signal in the first Raman channel, P_{B1} the background in this channel, and P_{RR2} and P_{B2} are the same for the second Raman channel. This equation is valid if ΔP_{B} is 0, which is a valid approximation if the background is calculated from a high number of bins. One can see from Eq. (4) that the error ΔT scales with

$$\Delta T \sim \frac{1}{\sqrt{P \tau A \eta}}, \tag{5}$$

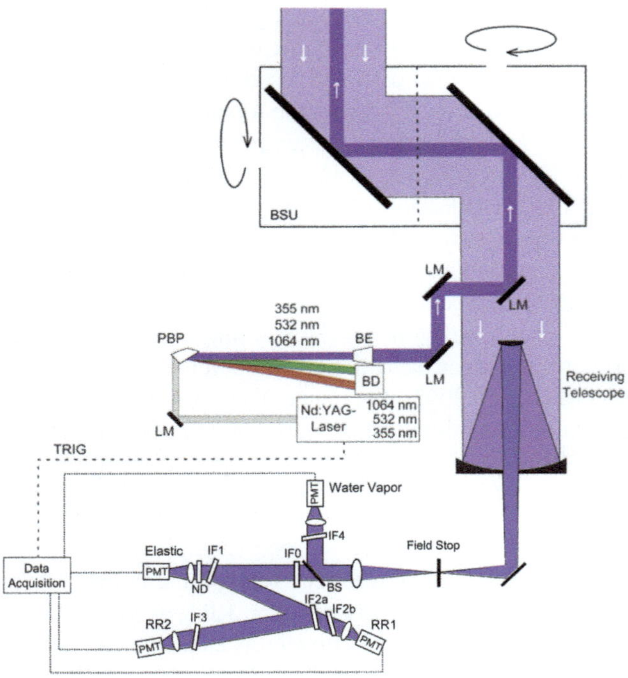

Figure 3. Setup of the rotational Raman lidar of the University of Hohenheim. LM stands for laser mirror, and the wavelengths are separated by a Pellin–Broca prism. Whereas 532 and 1064 nm are sent to a beam dump (BD), 355 nm is expanded by a beam expander (BE) and sent to the atmosphere through the beam steering unit (BSU). Angles in the polychromator are enlarged for clarity. IF0 is a daylight oppressing filter (see Table 1 for details). IF1 to IF4 are the interference filters, BS is a beam splitter to separate the vibrational Raman lines from the elastic backscattered and rotational Raman signal. In front of the elastic channel a neutral density filter (ND) is mounted. Photomultipliers are marked with PMT.

where P is the laser power, t the integration time, A the telescope area, and η the overall detection efficiency.

The statistical error ΔT_{Grad} of a temperature gradient measurements is then

$$\Delta T_{\mathrm{Grad}} = \frac{\sqrt{2}\Delta T}{\Delta r}, \tag{6}$$

with Δr being the range interval over which the gradient was calculated.

The two rotational Raman signals can also be used to form a temperature-independent reference signal P_{RR} for the determination of the particle backscatter coefficient, the particle extinction coefficient, or the water vapor mixing ratio (Behrendt et al., 2002, 2004).

The water vapor mixing ratio (MR) at a distance r is calculated with

$$\mathrm{MR}(r) = C \cdot \frac{P_{\mathrm{WV}}(r)}{P_{\mathrm{RR}}(r)} \Gamma(r), \tag{7}$$

where C is a calibration constant, and P_{WV} and P_{RR} are the background-corrected signals of the vibrational Raman line of water vapor and the rotational Raman lines of nitrogen and oxygen, respectively (Whiteman et al., 1992; Sherlock et al., 1999). $\Gamma(r)$ is a correction for the different atmospheric extinction Γ at the two wavelengths λ_{RR} and $\lambda_{\mathrm{H_2O}}$:

$$\Gamma(r) = \frac{\exp\left[-\int_{r0}^{r} \alpha\left(\lambda_{\mathrm{RR}}, r'\right) \mathrm{d}r'\right]}{\exp\left[-\int_{r0}^{r} \alpha\left(\lambda_{\mathrm{H_2O}}, r'\right) \mathrm{d}r'\right]}. \tag{8}$$

To determine the calibration constant C, a collocated radiosounding can be used. It was found that a correction for the different extinction coefficients of the two wavelengths is negligible for ranges of up to a few kilometers because the difference is less than 0.1 % up to 3 km.

The statistical error of the water vapor measurements can be calculated with Poisson statistics, and one gets

$$\Delta\mathrm{MR} = C \frac{P_{\mathrm{WV}}}{P_{\mathrm{RR}}} \sqrt{\frac{P_{\mathrm{WV}} + P_{\mathrm{BWV}}}{P_{\mathrm{WV}}^2} + \frac{P_{\mathrm{RR}} + P_{\mathrm{BRR}}}{P_{\mathrm{RR}}^2}}. \tag{9}$$

While the error analysis based on Poisson statistics determines the so-called shot noise errors, the total statistical error can be estimated with an analysis of the turbulent fluctuations in the atmosphere (Lenschow et al., 2000). For this, data with high temporal resolution are needed in order to resolve the timescale of these fluctuations. The method based on the analysis of the autocovariance function was developed in Senff et al. (1994) and Wulfmeyer (1999a, b) and summarized in Lenschow et al. (2000). Recently, this technique was applied to WV DIAL (e.g., Muppa et al., 2014) and Doppler lidar data (Lenschow et al., 2012), elastic backscatter lidar data (Pal et al., 2010), and water vapor Raman lidar data (Wulfmeyer et al., 2010; Turner et al., 2014a, b). Recently it was applied for the first time to temperature lidar data by using measurements of the UHOH RRL during HOPE (Behrendt et al., 2014). The comparison between the errors derived with Poisson statistics and turbulence analysis confirms that the total statistical error is mainly due to photon shot noise.

2.2 Current setup

The rotational Raman lidar of UHOH aims at measurements in the atmospheric boundary layer and lower free troposphere during daytime. A scheme of the setup is shown in Fig. 3. As a laser source, an injection-seeded frequency-tripled Nd:YAG laser (Spectra Physics GCR290-50) is used. Only the third harmonic radiation at 354.83 nm is transmitted into the atmosphere. The human eye is less sensitive to UV wavelengths than to wavelengths in the visible spectrum, and therefore eye safety is achieved at even smaller distances for UV lidars than for lidars using, e.g., the second harmonic radiation of a Nd:YAG laser. In consequence, the first and second harmonics are separated in the transmitter from the third using a Pellin–Broca prism and directed into beam dumps. The separation by a prism is preferable to a beam splitter due

Table 1. Parameter of interference filters. AOI: angle of incidence; CWL: central wavelength; FHWM: full width at half maximum.

	IF0	IF1	IF2a	IF2b	IF3	IF4
AOI, deg	0	4.8	6.0	6.0	4.8/6.2	3.9
CWL, nm	355	354.8	354.15	354.15	353.30/353.05	407.7
FWHM	8.5	0.3	0.3	0.3	0.5	0.3
Peak transmission	0.56	0.62	0.53	0.65	0.52	0.75
Refl. at 354.8 nm		< 0.1				
Transm. at 354.8 nm	0.56	0.62	$< 1 \times 10^{-3}$	$< 1 \times 10^{-3}$	$< 1 \times 10^{-6}$	$< 1 \times 10^{-7}$

to the fact that the wavelengths are separated spatially, and definitely no radiation at the other wavelengths remains in the outgoing light. The transmitted power in the UV is around 10 W at a repetition rate of 50 Hz. The beam is expanded by a factor of 6.5 to prevent damage on the transmitting optics and to achieve eye safety in less than 400 m distance (taking also hotspots in the beam profile into account). The expanded beam is directed via three mirrors (10 cm diameter) onto the first mirror of the beam steering unit. Together with a second mirror, this setup enables us to direct the beam to any azimuth and elevation angle of interest. The whole system is mounted on a mobile platform which can be moved to different measurement sites for field campaigns.

The receiving telescope is of the Ritchey–Chrétien–Cassegrain type. Its primary mirror has a diameter of 40 cm. To reduce the daylight background, the focused signal passes a pinhole with selectable diameter. The diameter was set to 3 mm during the HOPE campaign, which results in a full field of view of 0.75 mrad. After passing this field stop, the light is parallelized with a lens and then split by a dichroic beam splitter: while light with wavelengths shorter than 375 nm is transmitted into the main receiver unit, longer wavelengths are reflected. The vibrational-rotational Raman signal of water vapor excited by 354.83 nm is shifted by wavenumbers around 3657.05 cm^{-1} (e.g., Avila et al., 2004) from the excitation wavelength and is thus around 407.7 nm and obtained from the light reflected by the beam splitter. The transmission efficiency of the beam splitter is 0.93 for 355 nm and 0.02 for 408 nm. Reflectivity at 408 nm is above 0.95.

The signal transmitted by the beam splitter passes a daylight-reducing filter IF0 (Eureca, peak transmission of 0.56) and enters the main part of the receiver for the detection of the elastic and rotational Raman signals, which are sequentially mounted (Behrendt and Reichardt, 2000; Behrendt et al., 2002, 2004; Radlach et al., 2008). In front of the first rotational Raman channel, two identical interference filters IF2a and IF2b are used to achieve sufficient suppression of the elastic signal. In Table 1 all filter parameters are listed. The first filter pair has a transmission of 0.34 for 354.15 nm and 10^{-6} for 354.83 nm. The passbands have a full width at half maximum of 0.3 nm. The filter for extracting the second rotational Raman signal can be toggled between two distinct angles to optimize the performance of temperature measurements during conditions of high and low background (see below). For the time being, the change of setting has to be done manually, but the angles are fixed by a pivot tooth. IF3 has a peak transmission of 0.52 and is 0.5 nm wide. The angles of incidence are 6° for the interference filters IF2a and b, and 4.8 and 6.2° for the second interference filter for the high- and low-background setting, respectively. The beam divergence in the polychromator is 0.45° as determined by experimentally validated ray tracing calculations. All narrowband interference filters were manufactured by Materion, Barr Precision Optics & Thin Film Coating. As photomultiplier for the rotational Raman signals, a Hamamatsu R1924P is used. The elastic channel is equipped with a neutral density filter (transmission of 0.2) to prevent saturation of the photomultiplier (Hamamatsu R7400-U02) in the presence of optically thick clouds in the near range.

In the branch of the water vapor channel, an interference filter (IF4) transmits the desired wavelength range; for details see Table 1. The total suppression has an efficiency better than 10^{-8} for 355 nm and 10^{-6} for other wavelengths. The signal is focused on a photomultiplier (Hamamatsu R1924P) for detection.

Presently, two data acquisition systems are used for the four lidar signals. First, there is a transient recorder (LICEL GmbH, Berlin) with three acquisition channels. Each photomultiplier signal is recorded in analog mode with 3.75 m range resolution and in photon-counting mode with range resolutions of 3.75 and 37.5 m (see Table 2). The measurement data shown here were all derived with the analog data. In standard operation, backscatter signals of 500 shots are averaged to get a profile with 10 s integration time. At the beginning of the HOPE campaign, the elastic signal and the two Raman channels were recorded with this system. From early May on, the water vapor detection channel signal was implemented and its signal was recorded with the LICEL data acquisition instead of the elastic signal. The strong elastic signal was then stored with a transient recorder system (Compuscope 14100 card of GaGe Applied Inc.) with resolutions of 3 m and 1 s, albeit only in analog mode.

The data analysis contains the following steps. First, the photon-counting data are corrected for photomultiplier dead-time effects. A dead time of 4.8 ns was determined by comparing the analog with the photon-counting signal and used

Table 2. Signal raw resolution.

Raw data	Temporal resolution	Spatial resolution	Recording mode
Elastic signal	10 s	3.75 m	Analog + photon-counting
	10 s	37.5 m	Photon-counting
	1 s	3 m	Analog
Rotational Raman 1	10 s	3.75 m	Analog + photon-counting
	10 s	37.5 m	Photon-counting
Rotational Raman 2	10 s	3.75 m	Analog + photon-counting
	10 s	37.5 m	Photon-counting
Vibrational water vapor	10 s	3.75 m	Analog + photon-counting

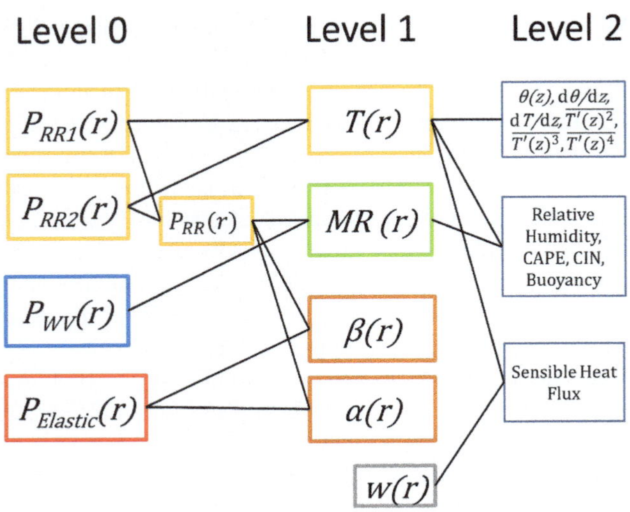

Figure 4. Diagram of the data products. Level 0 data are the background-corrected rotational Raman signals (yellow), the vibrational Raman signal of water vapor (blue), and the elastic backscatter signal (red). A temperature-independent reference signal P_{RR} is obtained from the temperature-dependent rotational Raman signals. Level 1 data products are temperature T, water vapor mixing ratio MR, particle backscatter coefficient β, and particle extinction coefficient α. Higher-level products derived from level 1 data: potential temperature θ, gradient of temperature and potential temperature, higher moments of turbulent temperature fluctuations T', relative humidity, buoyancy, CAPE, and CIN. For deriving the sensible heat flux, profiles of the vertical wind w, e.g., from a Doppler lidar, are used.

for the correction. It has to be noted that the dead time not only is influenced by the photomultiplier alone but also depends on the data acquisition system. Then, the signals are background-corrected and smoothed in range and time with gliding average lengths that depend on the noise conditions and application. For example, for turbulence analysis (Behrendt et al., 2014) high temporal resolution of 10 s is needed, while for studies of the temperature gradient a low statistical error is essential. How the statistical error is affected by the averaging can be seen in Eq. (5). Temperature and other data products are then determined with the analog and photon-counting signals separately and merged afterwards if needed. We found that the alternative approach of first merging the analog and photon-counting signals and then deriving the data products in a second step with the merged signals (Newsom et al., 2009) results in unstable temperature calibration functions because the merging procedure produces small erroneous fluctuations which influence the temperature measurements critically. To determine the statistical uncertainties of the data, the analog data are scaled to the photon-counting signals in order to attribute virtual count rates to the analog data (Whiteman et al., 2006). It turned out that this scaling results in accurate shot noise error estimates also for the analog data. In Behrendt et al. (2014) it is shown that the total statistical error depends mainly on the noise error. It is therefore possible to obtain a good error estimation by calculating the shot noise error through Poisson statistics.

From the combination of the two temperature-dependent signals, a temperature-independent molecular reference signal can also be obtained. The weighted sum of the signals with a weighting factor that depends on the system characteristics (Behrendt et al., 2002) can be used as a reference signal for the calculation of the water vapor MR, particle backscatter coefficient β, and extinction coefficient α (see also Fig. 4); no further vibrational nitrogen Raman signal like that used in other Raman lidar systems is needed. It should also be noted that the statistical uncertainty of the rotational Raman reference signals is lower than the uncertainty of the weaker nitrogen vibrational Raman signal.

The measured temperature profiles can be used to further derive, e.g., profiles of potential temperature θ, temperature gradients, variance, or other higher-order moments of turbulent temperature fluctuations (Behrendt et al., 2014). If profiles of vertical wind w are available with high temporal resolution, e.g., from a Doppler lidar, the sensible heat flux can be calculated. Also other products – like buoyancy (Corsmeier et al., 2011), CAPE (Convective Available Potential Energy), and CIN (Convective Inhibition) (Behrendt et al., 2011) – are possible.

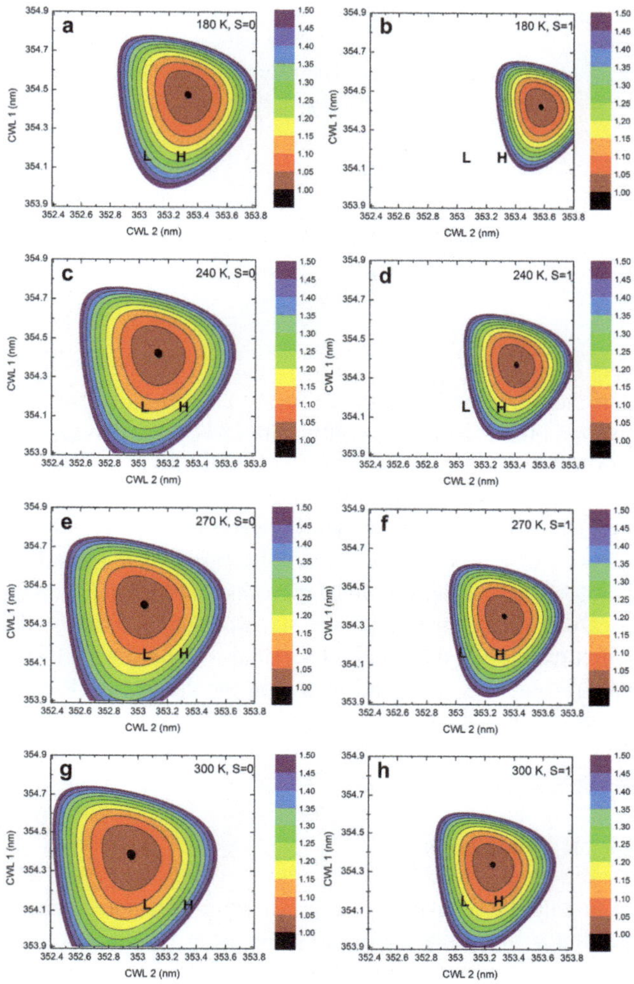

Figure 5. Statistical temperature uncertainty ΔT for different central wavelengths (CWL) of the RR filters. **(a)**, **(c)**, **(e)**, and **(g)** are without background ($S = 0$); **(b)**, **(d)**, **(f)**, and **(h)** are with background ($S = 1$). H denotes the configuration selected for high-background measurements, L the configuration for low-background measurements. The uncertainty values were normalized to the smallest value of each plot. The laser wavelength is 354.83 nm.

2.3 Determination of optimum configurations for low and high background

A switch for the central wavelength of the interference filter of the second rotational Raman channel was introduced during the HOPE campaign. By selecting lines for the second (high J) rotational Raman backscatter channel with larger spectral distance to the excitation wavelength, one can enhance the temperature sensitivity of the system. But while these lines are more temperature sensitive, they are also weaker in intensity. Consequently, there is a tradeoff between temperature dependence and signal intensity, which results in optimum settings which depend on the signal background relative to the Raman signal intensities (Radlach, 2009). In the following, refined simulation results on this problem of lowest-possible statistical uncertainty of the temperature measurements depending on the background are presented. While the passband of the second rotational Raman channel is changed, the passband of the first rotational Raman channel (low J) stays constant. The RR1 channel is already so close to the laser wavelength that further change to weaker signals would decrease the blocking of the elastically backscattered light in the signal to critical values.

Refined simulations to what had already been described by Radlach et al. (2008) were performed for the present laser wavelength, with the goal of finding an optimum setting for the central wavelength of the second rotational Raman channel for high and low background. In the following the central wavelength of the first rotational Raman signal detection channel is called CWL1 and the second CWL2. From Eq. (4), the 1σ statistical uncertainty of temperature measurements can be calculated for two close temperatures T_1 and T_2 with

$$\Delta T = \frac{\partial T}{\partial Q}\Delta Q \approx \frac{(T_1 - T_2)}{(Q_1 - Q_2)} Q \sqrt{\frac{P_{\mathrm{RR1}} + P_{\mathrm{B1}}}{P_{\mathrm{RR1}}^2} + \frac{P_{\mathrm{RR2}} + P_{\mathrm{B2}}}{P_{\mathrm{RR2}}^2}}. \quad (10)$$

A simulation of the spectrum at two temperatures T_1 and T_2 5 K apart combined with the filter transmission curves gives the ratio Q of the two channels. Scaling parameters of the background are the ratio of the background per 0.1 nm filter bandwidth and the intensity $P_J^{\max}$ of the strongest line of the rotational Raman spectrum (Radlach, 2009). This gives as a background

$$P_{\mathrm{B}} = S \frac{\Delta\lambda_{\mathrm{FWHM}}}{0.1\,\mathrm{nm}} P_J^{\max}. \quad (11)$$

For the scaling factor S we chose 1 in the high-background case and 0 in the low-background case. As the intensity of the Raman signal depends on height but the background is constant for all height bins of a profile, the scaling factor S changes for the different heights of a measured profile and, of course, with the solar background and thus latitude, time of the day, and season as well as the laser power and receiver efficiency of the lidar system. During daytime, S will be nearly 0 at low altitudes for a well-designed high-power lidar with high signal to background ratio, but S will increase quickly with altitude as the signal intensity decreases. With optimizations to $S = 0$ and $S = 1$, one is even prepared for high-background conditions, e.g., near cumulus clouds at noon. It was found that larger values for S do not change the optimum central wavelengths significantly compared to $S = 1$. But as detailed in the following, the differences between $S = 0$ and $S = 1$ are significant, which is the motivation for the new switch.

In Fig. 5 the results of the simulations are presented for both high and low background for temperature regimes of 180, 240, 270, and 300 K. The simulation was performed assuming a beam divergence of 0.45° on the interference filters in the receiver, which modifies the filter transmission curves accordingly. The beam divergence was chosen in agreement

Table 3. Relative statistical temperature error for the high- and low-background setting of the center wavelength CWL2 of the interference filter of the second (high J) rotational Raman channel (H and L, respectively). S is the scaling factor for the background level (see text for details). While $S = 0$ stands for no background, $S = 1$ characterizes high-background conditions. The resulting relative errors for the selected configurations are bold. With the optimized settings for CWL2 the statistical errors do not exceed 20 % higher values than the absolute optimum for atmospheric temperatures between 240 and 300 K.

	$S = 0$				$S = 1$			
CWL2	180 K	240 K	270 K	300 K	180 K	240 K	270 K	300 K
353.30 (H)	1.25	1.27	1.31	1.37	1.52	**1.17**	**1.13**	**1.14**
353.05 (L)	1.36	**1.13**	**1.10**	**1.10**	3.46	1.70	1.39	1.21

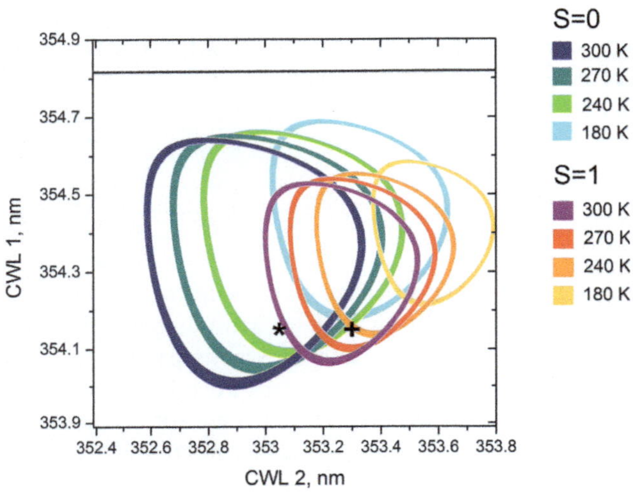

Figure 6. Overview of all cases shown in Fig. 5. The outer and inner borders of the rings mark 1.20 and 1.17 relative uncertainty, respectively. The selected configurations for high- and low-background conditions (marked with + and *, respectively) show lower relative uncertainties than 1.2 for all temperatures between 240 and 300 K (see also Table 3). The gray line marks the laser wavelength.

with the divergence in the polychromator in the current setup. For divergence values between 0.1 and 1.0°, the optimum wavelengths and iso-lines shift by 0.01 and 0.02 nm, respectively. The relative uncertainty ΔT is normalized to the smallest value in each case. The values of ΔT are of course lower during night- than during daytime, but this simulation is made to find an optimum setting within each regime and then compare the optimum settings. The central wavelengths of the first filter are limited to 354.2 nm and smaller to ensure that the elastic signal is blocked sufficiently for this channel.

The central wavelengths were 354.15 nm for CWL1 and 353.30 nm for CWL2 for high-background conditions for the measurements discussed here. With low background at night, this setting is not optimum, and therefore the pair of central wavelengths of 354.15 and 353.05 nm was used for low-background conditions. We use the corresponding angles of incidence mentioned in Table 1. In Fig. 6, an overview of all these temperature regimes is given. One can see that the measurements with our settings are not more than 20 % higher than the minimum uncertainty for all temperatures between 240 and 300 K. But this is only achieved by the two settings for CWL2.

To be able to use the advantages of both wavelengths, it is crucial to be able to toggle between the two settings in a short time period in order to avoid measurement gaps. Furthermore, the setting should be reproducible to avoid changes in calibration or overlap. This is realized by a filter holder which can be fixed by a pivot tooth at two positions. It was built in our workshop and tested on the campaign. Results are shown in Sect. 3.1.

3 Measurement examples

3.1 RR2 switch

To test the performance of the switch, a 24 h measurement was made including changes between the H and L setting. The switching time coincided with a radiosonde launch during nighttime. In general, "low-background" conditions are defined here as the time where the background is small compared to the rotational Raman signals in the altitude range of interest, i.e., between about half an hour after sunset and half an hour before sunrise. This was the case during the consecutive observation periods with a radiosonde launch at 20:00 UTC and early in the morning. All in all, there were three cases for the evaluation of the performance of the switch.

On average it took about 5 min to interrupt the data acquisition in a controlled way, change filter position, readjust the stray-light cover, and restart the data acquisition. When changing from high- to low-background setting, the signal intensity in the second Raman channel decreases, which results in a lower ratio Q (Fig. 7). But, as already discussed above, the relative sensitivity to temperature increases. The enhanced sensitivity is seen in Fig. 8, where the ratio between the two channels is normalized to the value at an arbitrary altitude (here the altitude was chosen in which the temperature profile of the radiosonde was 273 K, which was 2.6 km). In this visualization the different slope can be seen, which is larger for L than for H.

Finally, the Q vs. T curves from simulation and experiment (Fig. 9) are compared. For the experimental data, Q

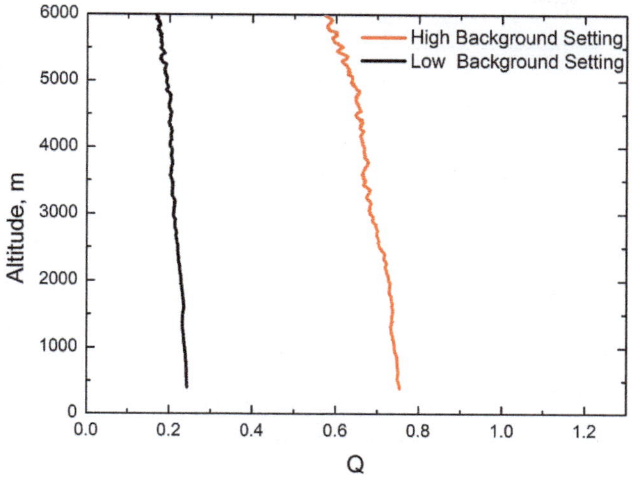

Figure 7. Measured ratio Q of the rotational Raman signals with the high- and low-background setting, H and L , respectively, at night. The data for the setting H were collected between 19:38 and 19:58 UTC on 18 May 2013, and data for the setting L just afterwards between 20:05 and 20:25 UTC. The intensity of the RR2 signal decreased by switching to L, hence the smaller ratio Q. The rotational signals were smoothed with a gliding average of 105 m height before the ratio was calculated.

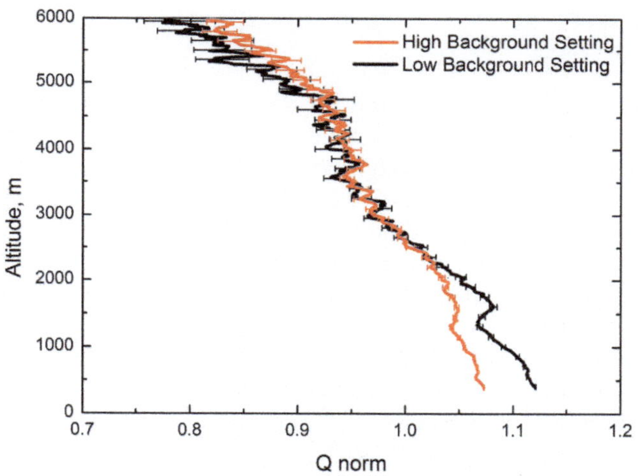

Figure 8. Same as Fig. 7 but normalized to 1 at a height of 2.6 km to illustrate the higher relative sensitivity of the low-background setting. Error bars show the statistical uncertainties.

values were assigned to T values of the radiosonde at the same height. The simulation calculates the resulting Q for temperatures from 240 to 300 K in 1 K steps for the actual filter curves including the central wavelengths. For the low-background setting, simulation and experiment agree very closely within the statistical error bars; for the high-background, a deviation of <1.5 K is found for Q values between 1.0 and 1.05. A comparison with Fig. 8 shows that this is the altitude region where an inversion layer was found. An inversion layer can lead to differences between

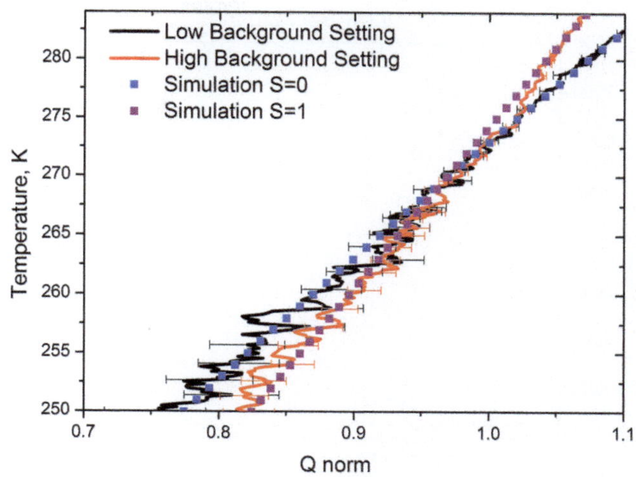

Figure 9. Same as Fig. 8 but against radiosonde temperature at the same height. In addition to the measurements, simulation data for the two settings are shown. In red and black are the measured values for Q with the statistical error, and in blue and violet the corresponding simulation.

radiosonde and lidar measurement due to averaging of the lidar data and sampling of a different air mass. The experimentally determined Q is linked to the temperature values of the radiosounding; a deviation of 1 to 2 K therefore can be attributed to the inversion. As there are for all other temperatures only deviations within the 1σ statistical uncertainties, we conclude that the agreement between experiment and simulation is good.

The statistical errors for different background values with the two settings were calculated. Nighttime measurements were used and, for the high-background case, background added according to a scaling factor of $S = 1$ at 1 km. In comparison with actually measured background values at noon with or without cloud coverage, the used background values are higher because the Raman signals of the UHOH RRL are quite strong. Thus, the comparison shows the performance in circumstances which can be considered a worst-case scenario.

Figure 10 shows the statistical errors with low background (measurement from 20:00 UTC, 20 min average) and both settings. Even during nighttime, there is a background of around 1 photon per bin for the integrated average over 500 laser pulses. Therefore S is not strictly equal to 0, but, in this case, S was less than 0.1 up to 2.1 km for the L setting and up to 3.2 km for the H setting. $S = 1$ was reached at 4.3 and 6.6 km. As expected, the setting L shows smaller errors for low altitudes and is favorable up to 7 km. The simulated error for high background is shown in Fig. 11. The absolute error values are in this case higher than in the case of low background. Below 1 km, S is less than 1, which results in a better performance of the low-background setting than the high-background setting. Above 1 km, S is larger than 1 and the

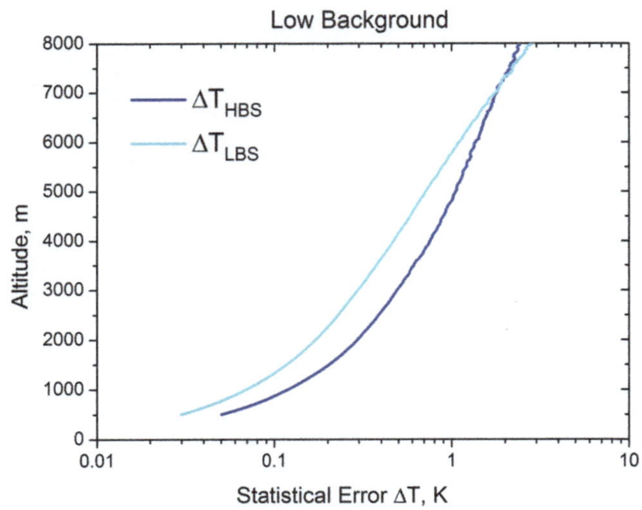

Figure 10. Comparison of the statistical error of the temperature measurements with both settings, ΔT_{HBS} and ΔT_{LBS}, at low-background conditions. The profiles were derived from data collected over about 20 min (55 000 laser shots) and with 105 m gliding average. Above 7 km altitude, the high-background configuration shows smaller errors due to the lower atmospheric temperatures at these heights and also the decrease in signal-to-noise ratio.

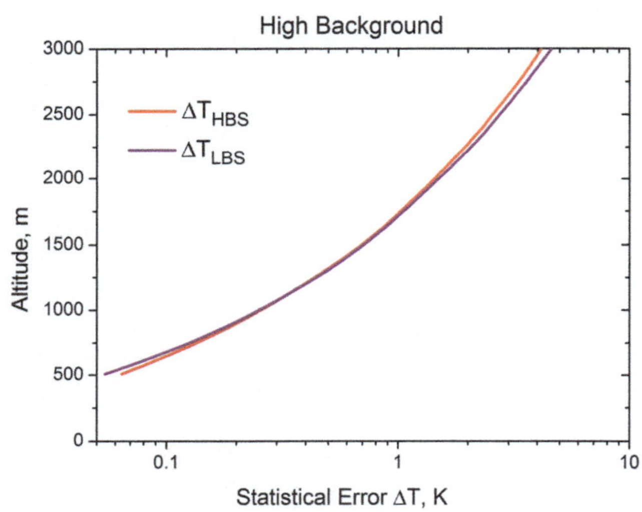

Figure 11. Same as Fig. 10 but for high-background conditions ($S = 1$ in 1 km altitude). The high-background setting is superior for measurements above 1 km altitude.

statistical errors are smaller with the high-background configuration.

To show the advantage of the new setting, the ratio between the error with high-background setting ΔT_{HBS} and the error with low-background setting ΔT_{LBS} was calculated for the different background conditions. The result is shown in Fig. 12. Using setting L is favorable during nighttime up to a height of 7 km, as already seen in Fig. 10. ΔT is reduced by 70 % for altitudes between 1 and 2 km. Above this altitude the advantage of L over H decreases. For high altitudes, H is still better than L even at night (see Fig. 6 for low temperatures < 240 K). This behavior is explained by the fact that the background factor S will be ≥ 1 at such altitudes. During daytime the signal intensity is high, such that $S < 1$ in low altitudes. Here both settings show similar statistical errors with advantages of up to 20 % (at 0.5 km) of the low-background configuration. With $S = 1$ at 1 km, the setting H is advantageous above this altitude.

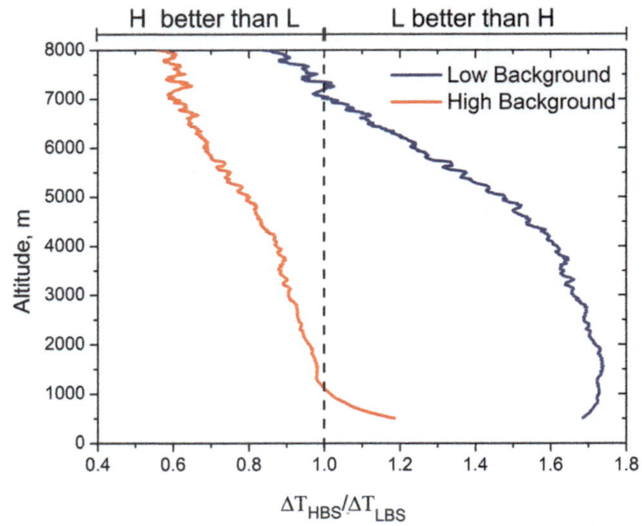

Figure 12. Ratio of high-background-setting uncertainties ΔT_{HBS} to low-background-setting uncertainties ΔT_{LBS}. A ratio above 1 shows better performance for the low-background setting compared to the high-background setting. In cases with no background, this advantage is clearly significant at lower altitudes but decreases with height. On the other hand, the high-background setting is superior already above 1 km in high-background conditions. But it should be noted that the altitude where the high-background setting becomes preferable depends on the signal-to-background ratio.

Because the background factor S assumed in the simulations is higher than in reality for the strong signals which are obtained with the UHOH RRL during cloud-free conditions, the altitude where the high-background setting shows an advantage lies above 1 km unless clouds cause higher background by scattering sunlight. By applying the setting H during daytime, a good to optimum measurement performance can be guaranteed even under these very high background conditions. It is interesting to note that also smaller laser power would result in higher values of S for all background levels. Thus this optimization approach can be used to adapt the optimum receiver passbands to the system parameters of other lidars.

3.2 Temperature gradient and humidity

Figure 13 shows the water vapor MR and the temperature gradient during a 25 h measurement period between 18 May 2013 at 15:00 UTC and 19 May 2013 at 16:00 UTC. The li-

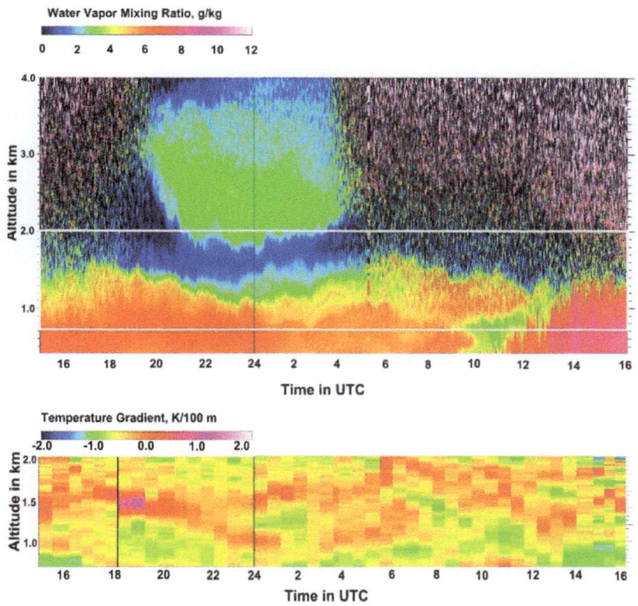

Figure 13. Time series of the water vapor mixing ratio (upper panel) and temperature gradient (lower panel) measured between 15:00 UTC on 18 May 2013 and 16:00 UTC on 19 May 2013. The white lines in the lower panel mark the height range of the data in the upper panel.

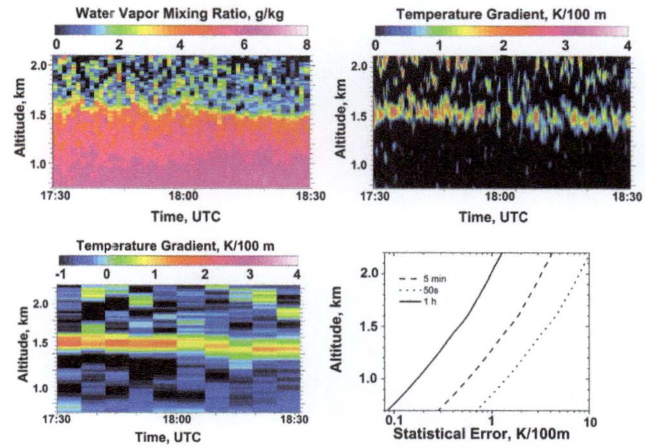

Figure 14. Detail from Fig. 13 of the time period from 17:30 to 18:30 UTC on 18 May. Upper left panel: water vapor mixing ratio with 1 min resolution; lower left panel: temperature gradient with 5 min; and upper left panel: 10 s profiles and 50 s gliding time average. Lower right panel: statistical error of temperature gradient measurements.

dar was pointing vertically during this time period. The setting of IF3 was changed at 20:00 UTC and at 05:00 UTC the next morning. MR data are shown with 1 min averaging. A 78 m gliding height average was applied to the data. The boundary layer top can be seen due to the gradient between the moist boundary layer and the lower values of MR in the free troposphere. The time series starts at 15:00 UTC, and the boundary layer is still convective at this time. The top of the boundary layer can be found around 1.6 km at 17:00 UTC. With sunset at 19:22 UTC there is a transition to a stable nocturnal boundary layer. Also a residual layer with a mixing ratio of 3 $g\,kg^{-1}$ can be observed above the boundary layer with around 6 $g\,kg^{-1}$ up to 1 km. Due to the lower solar background and the corresponding better measurement range a moist layer between 2 and 3 km can be observed at night. Sunrise is at 03:39 UTC, which can be seen by the increase of noise around this time. From 10:00 UTC on, the convective boundary layer starts to increase in altitude and shows higher values of mixing ratio (around 8 $g\,kg^{-1}$) than the day before. The height of the boundary layer top does not change significantly from 1.5 km after 14:00 UTC.

For the altitude region marked with the white box, the temperature gradient is shown in the lower panel. The resolution of temperature gradient data is 30 min and 105 m. These values have been chosen to monitor the mean boundary layer height continuously. A high positive gradient indicates a temperature inversion. Here one can see a positive gradient at 1.5 km with a value around 1.2 $K\,(100\,m)^{-1}$ between 17:30 and 19:00 UTC, which is remarkable to see in the late afternoon just before sunset. We would have not expected to find such high values in this altitude so late in the day. After sunset the gradient became weaker and split to the top of the nocturnal boundary layer and a residual layer at 1.7 km which was already observed with the water vapor mixing ratio time series. The residual layer stayed at this altitude during the whole measurement period. From 03:00 till 07:00 UTC another inversion can be seen between 1 and 1.3 km. A strong inversion was present on top of the growing convective boundary layer 10:00 till 15:00 UTC. This demonstrates that with the rotational Raman lidar temperature gradient layers can be identified and observed during longer time periods.

Figure 14 shows the time period from 17:30 to 18:30 UTC in more detail. The water vapor mixing ratio in the upper left panel has the same resolution as in Fig. 13. The temperature gradient is shown with two different temporal resolutions: 50 s gliding average over 10 s profiles on the upper right panel, 5 min on the lower left panel. The statistical uncertainty of the temperature gradient can be seen in the lower right panel. While the statistical error of the 5 min average is 1.1 $K\,(100\,m)^{-1}$ at 1.5 km altitude, the 50 s average shows an error of 4 $K\,(100\,m)^{-1}$. Therefore it can be used to determine qualitatively the altitude of the highest gradient, but not to identify absolute values. The altitude of the positive gradient and the top of the moist layer agree well for the shown time period even with this high temporal resolution.

The profiles of temperature, potential temperature, temperature gradient, and water vapor mixing ratio and their corresponding statistical uncertainties measured on 19 May 2013 between 13:00 and 13:30 UTC are shown in Fig. 15. This time period near local noon was chosen because the highest

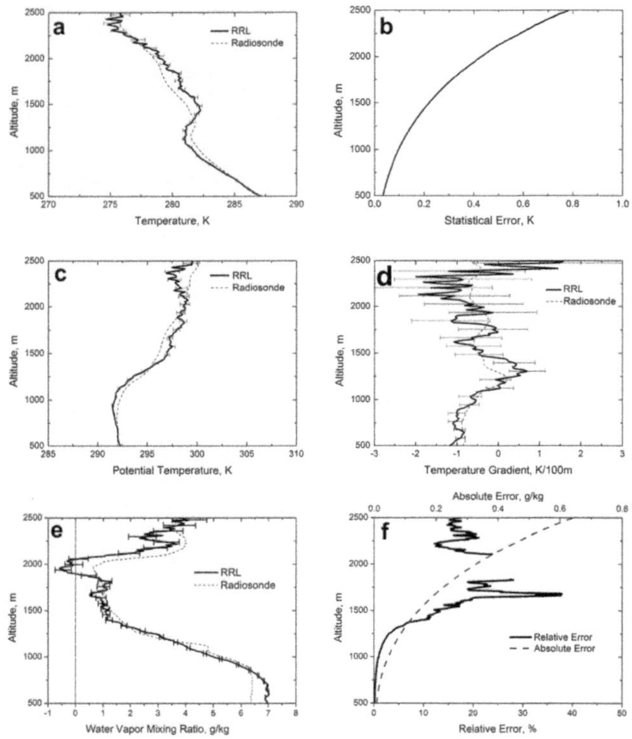

Figure 15. Measurements of the UHOH RRL between 13:00 and 13:30 UTC on 19 May 2013 with data of a local radiosonde launched at 13:00 UTC (dashed) for comparison. **(a)** Temperature, **(b)** statistical temperature error, **(c)** potential temperature **(d)**, temperature gradient **(e)**, water vapor mixing ratio, and **(f)** absolute and relative error of water vapor mixing ratio. Values of absolute error in the height range where the mixing ratio shows negative values are omitted. Error bars show the statistical errors of the lidar data. Altitudes below 500 m are affected by non-total overlap and are therefore not shown.

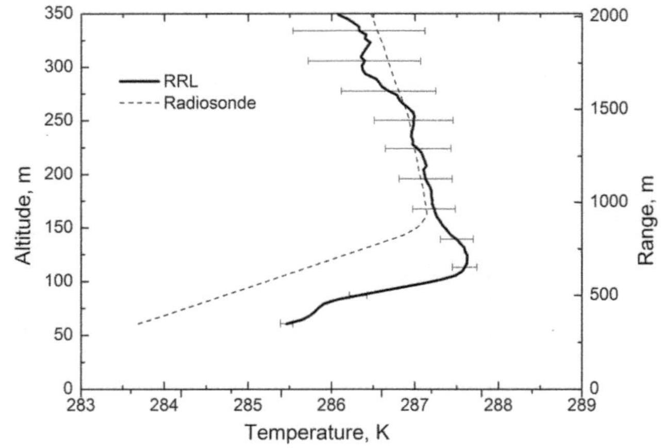

Figure 16. Results of low-elevation measurements: temperature measured with the UHOH RRL and a radiosonde (RS) against height a.g.l. The lidar data were collected from 02:40 to 02:57 UTC on 28 May 2013. The radiosonde measurement was started at 03:00 UTC. The altitude for the lidar profile is calculated from the range and corrected for topography. Lidar data at altitudes below 50 m are affected by overlap effects.

background values during the 25 h measurement period were found here; these examples thus illustrate the lower limit of the measurement performance, with all other periods having smaller statistical uncertainties. For the temperature profiles an average of 167 raw data profiles was used, and a gliding height average of 105 m was applied. A pressure profile measured by the radiosonde started at 13:00 UTC was used to calculate potential temperature. In the profiles of temperature, potential temperature, and temperature gradient one can see the characteristics of a well-mixed boundary layer up to about 1100 m. In the interfacial layer above, differences between the measurements of both instruments can be identified. As the lidar measurement is an average over half an hour and the radiosonde profile is just a snapshot, this is reasonable. For the water vapor profile measured with the lidar (Fig. 15e) a 154 m gliding average was applied. The moist boundary layer, a very dry layer just above, and a second moist layer above 2 km were found. In the dry layer, the uncertainty of the water vapor Raman lidar measurement gets larger in the absence of moisture due to the small water vapor Raman backscatter signal. In Fig. 15f the absolute and relative errors of the water vapor mixing ratio are shown. Of course, negative values of the water vapor mixing ratio are physically not possible. But in this case, they appear as measured values as a consequence of the small signal-to-noise ratio due to the small amount of water vapor at this height resulting in a signal which was hardly statistically distinguishable from the daylight background. Note that the 1σ statistical range indicates a probability of 68 % that the real value in found within. We do not want to shift these values to 0 because, in doing so, the mean of the data would be biased.

The profiles are shown with noise error bars derived by Eq. (4). A detailed error analysis including errors derived by turbulence analysis is discussed by Behrendt et al. (2014).

3.3 Low-elevation pointing

In vertical pointing mode low heights cannot be observed due to overlap effects. These effects can be corrected down to a certain altitude, but a minimum altitude of 300 m in the case of water vapor mixing ratio or temperature can unfortunately not be overcome with our single-telescope design. One solution for this problem is to perform measurements with small elevation angles. During the measurement shown in Fig. 16, the elevation angle was set to 10°. During HOPE, the measurement site was elevated above its surroundings by 10 to 15 m. This has been taken into account when transferring range to height above ground. The measurement geometry is illustrated in Fig. 17.

In Fig. 16 a comparison between the acquired lidar profile from 02:40 to 02:58 UTC and the radiosonde started at 03:00 UTC is shown. Raw data were treated like vertical ac-

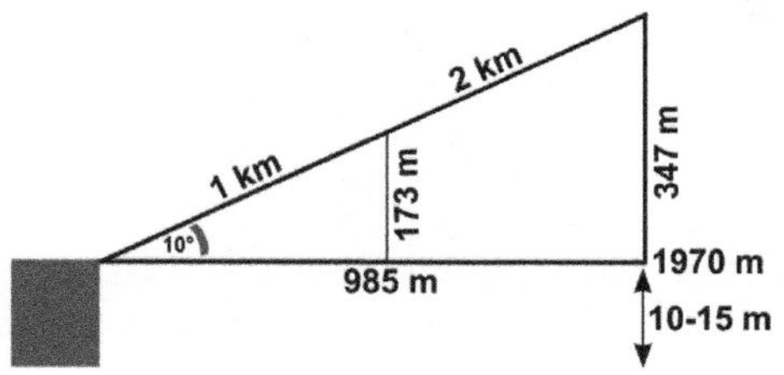

Figure 17. Illustration of the measurement geometry of Fig. 16.

quired profiles and projected to the vertical for the comparison. A positive gradient is present below 100 m in the lidar profile, whereas in the radiosonde profile it is shifted by 50 m to higher altitudes. The lidar profile is corrected for the altitude of the surrounding terrain. Whereas the altitudes of the inversion differ, the slope of the temperature profiles is identical. So we conclude that the observations indeed reveal differences of the temperature inversion profile of the nocturnal boundary layer.

4 Summary and outlook

In this paper the performance of the rotational Raman lidar of the University of Hohenheim during the HOPE campaign in April to May 2013 is presented.

To optimize the temperature measurements for both low and high background, a switch for the alignment of the interference filter for the second rotational Raman was implemented. It is possible to switch between the configurations in a short time period. The experimental data confirm the optimization simulations. The measurements proved the advantages of the new low-background configuration up to 7 km during night measurements, especially in altitudes below 4 km. An improvement of the statistical error up to 70 % can be found. Depending on the actual background level present during a measurement, the low-background configuration is also favorable in altitudes below 1 km during daytime. Due to the high signal intensities of the UHOH RRL, values of S less than 1 can be found in this altitude range. For larger background levels the improvement of the statistical error for using the H setting is up to 20 %. The advantage is not as large when changing from H to L, but one should keep in mind that a reduction of the statistical uncertainty by 20 % is equivalent to, e.g., an increase in laser power of 44 %.

The simulation was performed with the system parameters of the UHOH RRL. Therefore some points have to be taken into account when the results shown here are to be transferred to other systems. To get the exact values for another system, the simulation has to be repeated with the specific system parameters. However, the shifts from the excitation wavelength will be similar for other RRL systems operated in the UV. Additionally, the desired wavelength pairs depend on the temperature range being measured with the highest precision, including the background level in relation to the signal intensity. Bearing all these factors in mind, one can estimate suitable wavelength pairs from Fig. 6.

Furthermore, the UHOH RRL was equipped with an additional channel to detect the vibrational Raman lines of water vapor and now delivers, as a product, the atmospheric water vapor mixing ratio of the atmospheric boundary layer and lower free troposphere during day- and nighttime. As a molecular reference signal, a temperature-independent combination of the two rotational Raman signals is used – not a vibrational Raman signal of nitrogen as commonly used. The statistical error depends on humidity. During daytime the lower 2 km of the atmosphere can be observed with a time and height averaging of 20 min and 154 m gliding average to achieve an absolute noise error less than 0.5 $\mathrm{g\,kg^{-1}}$, which fulfills the requirements for boundary layer studies (Stull, 1988). Time series of water vapor mixing ratio and temperature gradient over a 25 h period were shown, and the diurnal changes in the boundary layer can be observed. Results from a low level measurement with capture of a strong inversion at 100 m show the capability to measure in a low-altitude range above ground with the beam steering unit.

Acknowledgements. The HOPE campaign was funded by the German Research Ministry under the project number 01LK1212 A. We thank KIT and the University of Cologne for performing the radiosonde launches.

Edited by: S. Buehler

References

Achtert, P., Khaplanov, M., Khosrawi, F., and Gumbel, J.: Pure rotational-Raman channels of the Esrange lidar for temperature and particle extinction measurements in the troposphere and lower stratosphere, Atmos. Meas. Tech., 6, 91–98, doi:10.5194/amt-6-91-2013, 2013.

Arshinov, J., Bobrovnikov, S., Serikov, I., Ansmann, A., Wandinger, U., Althausen, D., Mattis, I., and Müller, D.: Daytime operation of a pure rotational Raman lidar by use of a Fabry-Perot interferometer, Appl. Optics, 44, 17, 3593–3603, doi:10.1364/AO.44.003593, 2005.

Avila, G., Fernandez, J. M., Tejeda, G., and Montero, S.: The Raman Spectra and cross-sections of H_2O, D_2O, and HDO in the OH/OD-stretching regions, J. Mol. Spectrosc., 228, 38–65, 2004.

Balin, I., Serikov, I., Bobrovnikov, S., Simeonov, V., Calpini, B., Arshinov, Y., and van der Bergh, H.: Simultaneous measurement of atmospheric temperature, humidity, and aerosol extinction and backscatter coefficients by a combined vibrational-pure-rotational Raman lidar, Appl. Phys. B, 79, 775–782, 2004.

Behrendt, A.: Temperature Measurements with Lidar, Lidar: Range-Resolved Optical Remote Sensing of the Atmosphere, Springer, New York, 2005.

Behrendt, A. and Reichardt, J.: Atmospheric temperature profiling in the presence of clouds with a pure rotational Raman lidar by use of an interference-filter-based polychromator, Appl. Optics, 39, 1372–1378, 2000.

Behrendt, A., Nakamura, T., Onishi, M., Baumgart, R., and Tsuda, T.: Combined Raman lidar for the measurement of atmospheric temperature, water vapor, particle extinction coefficient, and backscatter coefficient, Appl. Optics, 36, 7657–7666, 2002.

Behrendt, A., Nakamura, T., and Tsuda, T.: Combined temperature lidar for measurements in troposphere, stratosphere and mesosphere, Appl. Optics, 14, 2930–2939, 2004.

Behrendt, A., Wulfmeyer, V., Riede, A., Wagner, G., Pal, S., Bauer, H., Radlach, M., Späth, F.: Three-dimensional observations of atmospheric humidity with a scanning differential absorption lidar, Proc. SPIE, 7475, 74750L, doi:10.1117/12.835143, 2009.

Behrendt, A., Pal, S., Aoshima, F., Bender, M., Blyth, A., Corsmeier, U., Cuesta, J., Dick, G., Dorninger, M., Flamant, C., Di Girolamo, P., Gorgas, T., Huang, Y., Kalthoff, N., Khodayar, S., Mannstein, H., Träumner, K., Wieser, A., and Wulfmeyer, V.: Observation of convection initiation processes with a suite of state-of-the-art research instruments during COPS IOP 8b, Q. J. Roy. Meteor. Soc., 137, 81–100, 2011.

Behrendt, A., Wulfmeyer, V., Hammann, E., Muppa, S. K., and Pal, S.: Profiles of second- to third-order moments of turbulent temperature fluctuations in the convective boundary layer: first measurements with Rotational Raman Lidar, Atmos. Chem. Phys. Discuss., 14, 29019–29055, doi:10.5194/acpd-14-29019-2014, 2014.

Bhawar, R., Di Girolamo, P., Summa, D., Flamant, C., Althausen, D., Behrendt, A., Kiemle, C., Bosser, P., Cacciani, M., Champollion, C., Di Iorio, T., Engelmann, R., Herold, C., Pal, S., Riede, A., Wirth, M., and Wulfmeyer, V.: The water vapour intercomparison effort in the framework of the convective and orographically-induced precipitation study: airborne-to-ground-based and airborne-to-airborne lidar systems, Q. J. Roy. Meteor. Soc., 137, 325–348, 2011.

Cooney, J.: Measurement of atmospheric temperature profiles by Raman Backscatter, J. Appl. Meteorol., 11, 108–112, 1972.

Corsmeier, U., Kalthoff, N., Barthlott, C., Aoshima, F., Behrendt, A., Di Girolamo, P., Dorninger, M., Handwerker, J., Kottmeier, C., Mahlke, H., Mobbs, S., Norton, E. G., Wickert, J., and Wulfmeyer, V.: Processes driving deep convection over complex terrain: a multi-scale analysis of observations from COPS IOP 9c, Q. J. Roy. Meteor. Soc., 137, 137–155, 2011.

Di Girolamo, P., Marchese, R., Whiteman, D. N., and Demoz, B. B.: Rotational Raman lidar measurements of athmospheric temperature in the UV, Geophys. Res. Lett., 31, L01106, doi:10.1029/2003GL018342, 2004.

Di Girolamo, P., Summa, D., and Ferretti, R.: Multiparameter Raman lidar measurements for the characterization of a dry stratospheric intrusion event, J. Atmos. Ocean. Tech., 26, 1742–1762, 2009.

Immler, F., Engelbart, D., and Schrems, O.: Fluorescence from atmospheric aerosol detected by a lidar indicates biogenic particles in the lowermost stratosphere, Atmos. Chem. Phys., 5, 345–355, doi:10.5194/acp-5-345-2005, 2005.

Kalthoff, N., Adler, B., Wieser, A., Kohler, M., Träumner, K., Handwerker, J., Corsmeier, U., Khodayar, S., Lambert, D., Kopmann, A., Kunka, N., Dick, G., Ramatschi, M., Wickert, J., and Kottmeier, C.: KITcube – a mobile observation platform for convection studies deployed during HyMeX, Meteorol. Z., 22, 633–647, 2013.

Leblanc, T. and McDermid, I. S.: Accuracy of Raman lidar water vapor calibration and its applicability to longterm measurements, Appl. Optics, 30, 5592–5603, 2008.

Lenschow, D. H., Wulfmeyer, V., and Senff, C.: Measuring second-though fourth-order moments in noisy data, J. Atmos. Ocean. Tech., 17, 1330–1346, 2000.

Lenschow, D. H., Lothon, M., Mayor, S. D., and Sullivan, P. P.: A comparison of higher-order vertical velocity moments in the convective boundary layer from lidar with in situ measurements and LES, Bound.-Lay. Meteorol., 143, 107–123, 2012.

Mao, J., Hua, D., Wang, Y., Gao, F., and Wang, L.: Accurate temperature profiling of the atmospheric boundary layer using an ultraviolet rotational Raman lidar, Opt. Commun., 282, 3113–3118, 2009.

Melfi, S. H., Lawrence, J. D., McCormick, M. P.: Observation of Raman scattering by water vapor in the atmosphere, Appl. Phys. Lett., 15, 295–297, 1969.

Muppa, S. K., Behrendt, A., Späth, F., Wulfmeyer, V., Metzendorf, S., and Riede, A.: Turbulent humidity fluctuations in the convective boundary layer: case studies using DIAL measurements, Bound. Lay. Meteorol., submitted, 2014.

Newsom, R. K., Turner, D. D., Mielke, B., Clayton, M., Ferrare, R., and Sivaraman, C.: Simultaneous analog and photon counting detection for Raman lidar, Appl. Optics, 20, 3903–3914, 2009.

Pal, S., Behrendt, A., and Wulfmeyer, V.: Elastic-backscatter-lidar-based characterization of the convective boundary layer and investigation of related statistics, Ann. Geophys., 28, 825–847, doi:10.5194/angeo-28-825-2010, 2010.

Radlach, M.: A scanning eye-safe rotational Raman lidar in the ultraviolet for measurements of tropospheric temperature fields, doctoral thesis, Stuttgart, 2009.

Radlach, M., Behrendt, A., and Wulfmeyer, V.: Scanning rotational Raman lidar at 355 nm for the measurement of tropospheric temperature fields, Atmos. Chem. Phys., 8, 159–169, doi:10.5194/acp-8-159-2008, 2008.

Schotland, R. M.: Errors in the lidar measurement of atmospheric gases by differential absorption, J. Appl. Meteorol., 13, 71–77, 1974.

Schwitalla, T. and Wulfmeyer, V.: Radar data assimilation experiments using the IPM WRF Rapid Update Cycle, Meteorol. Z. 1, 79–102, 2014.

Senff, C., Bösenberg, J., and Peters, G.: Measurements of water vapor flux profiles in the convective boundary layer with lidar and radar-RASS, J. Atmos. Ocean. Tech., 11, 85–93, 1994.

Sherlock, V., Garnier, A., Hauchecorne, A., and Keckhut, P.: Implementation and validation of a Raman lidar measurement of middle and upper tropospheric water vapor, Appl. Optics, 27, 5838–5850, 1999.

Stevens, B. and Bony, S.: What are climate models missing?, Science, 340, 1053–1054, 2013.

Stull, R. B.: An Introduction to Boundary Layer Meteorology, Springer, Heidelberg, New York, 1988.

Su, J., McCormick, M. P., Wu, Y., Lee III, R. B., Lei, L., Liu, Z., and Leavor, K. R.: Cloud temperature measurement using rotational Raman lidar, J. Quant. Spectrosc. Ra., 125, 45–50, 2013.

Turner, D. D. and Goldsmith, J. E. M.: Twenty-four-hour Raman lidar water vapor measurements during the atmospheric radiation measurement program's 1996 and 1997 water vapor intensive observation periods, J. Atmos. Ocean. Tech., 16, 1062–1076, 1999.

Turner, D. D., Ferrare, R. A., Wulfmeyer, V., and Scarino, A. J.: Aircraft evaluation of ground-based Raman lidar water vapor turbulence profiles in convective mixed layers, J. Atmos. Ocean. Tech., 31, 1078–1088, 2014a.

Turner, D. D., Wulfmeyer, V., Berg, L. K., and Schween, J. H.: Water vapor turbulence profiles in stationary continental convective mixed layers, J. Geophys. Res., 119, 11 pp., doi:10.1002/2014JD022202, 2014b.

Wagner, G., Behrendt, A., Wulfmeyer, V., Späth, F., and Schiller, M.: High-power Ti:sapphire laser at 820 nm for scanning ground-based water–vapor differential absorption lidar, Appl. Optics, 52, 11, 2454–2469, 2013.

Whiteman, D. N., Melfi, S. H., and Ferrare, R. A.: Raman lidar system for the measurement of water vapor and aersosols in the Earth's atmosphere, Appl. Optics, 31, 16, 3068–3082, 1992.

Whiteman, D. N., Demoz, B., di Girolamo, P., Comer, J., Veselovskii, I., Evans, K., Wang, Z., Sabatino, D., Schwemmer, G., Gentry, B., Lin, R-F., Behrendt, A., Wulfmeyer, V., Browell, E., Ferrare, R., Ismail, S. and Wang, J.:Raman Lidar Measurements during the International H2O Project. Part II: Case Studies, J. Atmos. Oceanic Technol., 23, 170-183, 2006.

Wulfmeyer, V.: Investigation of turbulent processes in the lower troposphere with water-vapor DIAL and Radar-RASS, J. Atmos. Sci., 56, 1055–1076, 1999a.

Wulfmeyer, V.: Investigations of humidity skewness and variance profiles in the convective boundary layer and comparison of the latter with large eddy simulation results, J. Atmos. Sci., 56, 1077–1087, 1999b.

Wulfmeyer, V. and Bösenberg, J.: Ground-based differential absorption lidar for water-vapor profiling: assessment of accuracy, resolution, and meteorological applications, Appl. Optics, 37, 3825–3844, 1998.

Wulfmeyer, V. and Feingold, G.: On the relationship between relative humidity and particle backscattering coefficient in the marine boundary layer determined with differential absorption lidar, J. Geophys. Res., 105, 4729–4741, 2000.

Wulfmeyer, V., Pal, S., Turner, D. D., and Wagner, E.: Can water vapour Raman lidar resolve profiles of turbulent variables in the convective boundary layer?, Bound.-Lay. Meteorol., 136, 253–284, 2010.

Zeyn, J., Lahmann, W., and Weitkamp, C.: Remote daytime measurements of tropospheric temperature profiles with a rotational Raman lidar, Opt. Lett., 21, 16, 1301–1303, 1996.

3

Trends of ozone total columns and vertical distribution from FTIR observations at eight NDACC stations around the globe

C. Vigouroux[1], T. Blumenstock[2], M. Coffey[3], Q. Errera[1], O. García[4], N. B. Jones[5], J. W. Hannigan[3], F. Hase[2], B. Liley[6], E. Mahieu[7], J. Mellqvist[8], J. Notholt[9], M. Palm[9], G. Persson[8], M. Schneider[4], C. Servais[7], D. Smale[6], L. Thölix[10], and M. De Mazière[1]

[1]Department of Atmospheric Composition, Belgian Institute for Space Aeronomy (BIRA-IASB), Brussels, Belgium
[2]Karlsruhe Institute of Technology (KIT), Institute for Meteorology and Climate Research (IMK-ASF), Karlsruhe, Germany
[3]Atmospheric Chemistry Division, National Center for Atmospheric Research (NCAR), Boulder, Colorado, USA
[4]Izaña Atmospheric Research Centre (IARC), Agencia Estatal de Meteorología (AEMET), Santa Cruz de Tenerife, Spain
[5]Centre for Atmospheric Chemistry, University of Wollongong, Wollongong, Australia
[6]Department of Atmosphere, National Institute of Water and Atmospheric Research Ltd (NIWA), Lauder, New Zealand
[7]Institute of Astrophysics and Geophysics, University of Liège (ULg), Liège, Belgium
[8]Department of Earth and Space Science, Chalmers University of Technology, Göteborg, Sweden
[9]Institute of Environmental Physics, University of Bremen, Bremen, Germany
[10]Climate Research, Finnish Meteorological Institute (FMI), Helsinki, Finland

Correspondence to: C. Vigouroux (corinne.vigouroux@aeronomie.be)

Abstract. Ground-based Fourier transform infrared (FTIR) measurements of solar absorption spectra can provide ozone total columns with a precision of 2 % but also independent partial column amounts in about four vertical layers, one in the troposphere and three in the stratosphere up to about 45 km, with a precision of 5–6 %. We use eight of the Network for the Detection of Atmospheric Composition Change (NDACC) stations having a long-term time series of FTIR ozone measurements to study the total and vertical ozone trends and variability, namely, Ny-Ålesund (79° N), Thule (77° N), Kiruna (68° N), Harestua (60° N), Jungfraujoch (47° N), Izaña (28° N), Wollongong (34° S) and Lauder (45° S). The length of the FTIR time series varies by station but is typically from about 1995 to present. We applied to the monthly means of the ozone total and four partial columns a stepwise multiple regression model including the following proxies: solar cycle, quasi-biennial oscillation (QBO), El Niño–Southern Oscillation (ENSO), Arctic and Antarctic Oscillation (AO/AAO), tropopause pressure (TP), equivalent latitude (EL), Eliassen–Palm flux (EPF), and volume of polar stratospheric clouds (VPSC).

At the Arctic stations, the trends are found mostly negative in the troposphere and lower stratosphere, very mixed in the middle stratosphere, positive in the upper stratosphere due to a large increase in the 1995–2003 period, and non-significant when considering the total columns. The trends for mid-latitude and subtropical stations are all non-significant, except at Lauder in the troposphere and upper stratosphere and at Wollongong for the total columns and the lower and middle stratospheric columns where they are found positive. At Jungfraujoch, the upper stratospheric trend is close to significance ($+0.9 \pm 1.0$ % decade^{-1}). Therefore, some signs of the onset of ozone mid-latitude recovery are observed only in the Southern Hemisphere, while a few more years seem to be needed to observe it at the northern mid-latitude station.

1 Introduction

While the past negative trend in the ozone layer has been successfully attributed to the increase of ozone depleting substances and reproduced by chemistry–climate models, understanding and predicting the current and future ozone layer, and especially attributing an ozone recovery to the positive effect of the Montreal Protocol and its Amendments and Adjustments, is still a challenge. This results from natural variability, observation uncertainties, and changes in dynamics and temperature induced by the increase of greenhouse gases (WMO, 2010). Long-term measurements of total and vertical ozone are required to understand the ozone response to different natural and anthropogenic forcings. Since the long-term satellite experiments ceased to operate (i.e., SAGE, HALOE), the satellite community is working on merging the past records to the new measurements performed by a number of satellite instruments launched since 2000 (e.g., Bodeker et al., 2013; Kyrölä et al., 2013; Sioris et al., 2014; Chehade et al., 2014). Reliable data from stable instruments are needed to validate these satellite-extended data sets, and to offer an alternative determination of ozone total and vertical changes. Ground-based (Dobson, Umkehr) and ozonesonde data are traditionally used for these studies, already reporting trends in the 1985 ozone report (WMO, 1985) and followed in 1998 by lidar and microwave measurements (WMO, 1998). Ground-based FTIR (Fourier transform infrared) measurements derived from high-resolution solar absorption spectra provide an additional ozone data set; they were first used for trend studies in Vigouroux et al. (2008) with 10 years of data (1995–2004) at several European stations and then were updated in the WMO (2010) report. Additional similar studies have been performed at individual stations (Mikuteit, 2008; García et al., 2012). These measurements have their own advantages. First, for atmospheric gases such as ozone, which have very narrow absorption lines, the ozone absorption signatures are self-calibrated with the reference being the surrounding continuum. Therefore, the derived absolute ozone columns depend mainly on the employed spectroscopic parameters which dominate the systematic uncertainty budget. Second, they can provide not only ozone total columns with a precision of 2 %, but also low vertical resolution profiles, obtained from the temperature and pressure dependence of the absorption line shapes. This leads to about four independent partial columns, one in the troposphere and three in the stratosphere up to about 45 km, with a precision of about 5–6 %. The instrumental line shape (ILS), which depends on the alignment of the spectrometer, impacts the absorption line shape on which the ozone profile retrievals are based. Hence, it is important to have accurate knowledge of the ILS in order to derive correct ozone profiles and trends.

The work discussed in this paper expands on the previous study of Vigouroux et al. (2008): it is based on longer time series, it includes FTIR data from stations outside of Europe, and it uses a stepwise multiple linear regression model including several explanatory variables for the trend evaluation. It is presented as follows: Sect. 2 provides information about the FTIR ozone observations (retrieval strategies, characterization of the vertical information, time series and seasonality). Section 3 describes the stepwise multiple linear regression model applied to the ozone time series. Section 4 presents and discusses the trend results as well as the explained part of ozone variability. Section 5 summarizes the conclusions.

2 Ground-based FTIR ozone observations

2.1 FTIR monitoring

Table 1 identifies the ground-based FTIR stations, all part of NDACC (Network for the Detection of Atmospheric Composition Change), that contribute to the present work. The latitudinal coverage is good: from 79° N to 45° S. These stations perform regular solar absorption measurements, under clear-sky conditions and over a wide spectral range (around 600–4500 cm^{-1}); the derived time series of total column abundances of many atmospheric species are available in the NDACC database (http://www.ndacc.org). While the stations are all currently active, they started their regular monitoring activities at different times. The period of measurement used for ozone trend analysis at each station is summarized in Table 1, together with the instrument manufacturer and type. Some of the stations performed measurements even earlier but these older spectra, taken with different spectrometers, have to be carefully reanalyzed first before being included in a trend study. The instruments currently used are the high-resolution spectrometers Bruker 120 M, 125 M, 120 HR, and 125 HR, which can achieve a spectral resolution of 0.0035 cm^{-1} or better. The Bomem DA8 used in the first years of Wollongong measurements has a spectral resolution of 0.004 cm^{-1}.

2.2 FTIR retrieval strategy

We refer to Vigouroux et al. (2008) for more details on the ozone FTIR inversion principles, which are based on the optimal estimation method (Rodgers, 2000). The effort of retrieval homogenization initiated in Vigouroux et al. (2008) has been pursued and we report in Table 2 the common retrieval parameters. The spectroscopic database has been updated to HITRAN 2008 (Rothman et al., 2009). All stations are employing the daily pressure and temperature profiles from NCEP (National Centers for Environmental Prediction). A common source for the ozone a priori profiles is used: the model WACCM4 (Whole Atmosphere Community Climate Mode; Garcia et al., 2007) calculated at each FTIR station, except at Harestua where a climatology based on ozonesondes and HALOE measurements is used. Finally, the interfering species fitted in the ozone retrievals, usually with a simple scaling of their a priori profile, are the same for all stations, except Harestua, namely, H_2O, CO_2, C_2H_4, and the ozone isotopologues $^{668}O_3$ and $^{686}O_3$. At Harestua, only H_2O and CO_2 are fitted.

Table 1. Characteristics of the FTIR stations that contribute to the present work: location and altitude (in km a.s.l.), time period covered by the ozone measurements used in the present trend analysis, and instrument type.

Station	Latitude	Longitude	Altitude (km)	Time period	Instrument
Ny-Ålesund	79° N	12° E	0.02	1995–2012	Bruker 120 HR
Thule	77° N	69° W	0.22	1999–2012	Bruker 120 M
Kiruna	68° N	20° E	0.42	1996–2007	Bruker 120 HR
				2007–2012	Bruker 125 HR
Harestua	60° N	11° E	0.60	1995–2009	Bruker 120 M
				2009–2012	Bruker 125 M
Jungfraujoch	47° N	8° E	3.58	1995–2012	Bruker 120 HR
Izaña	28° N	16° W	2.37	1999–2005	Bruker 120 M
				2005–2012	Bruker 125 HR
Wollongong	34° S	151° E	0.03	1996–2007	Bomem DA8
				2007–2012	Bruker 125 HR
Lauder	45° S	170° E	0.37	2001–2012	Bruker 120 HR

Some retrieval parameters still differ from station to station, either for historical reasons or for the inherent specificities of the different locations. They are also summarized in Table 2.

First, two different profile retrieval algorithms are widely used depending on each team's expertise: PROFFIT9 (Hase, 2000) at Kiruna and Izaña and SFIT2 (Pougatchev et al., 1995) at the six other stations. It has been demonstrated in Hase et al. (2004) that the profiles and total column amounts retrieved from these two different algorithms under identical conditions are in excellent agreement.

Second, the microwindow sets involve some common lines at all stations, which ensures that only a small bias is expected due to the different microwindow choices. Either some additional thin microwindows are used together with the 1000–1005 cm^{-1} or, at Kiruna and Izaña, a different choice was led by the priority given to avoid the more intense H_2O lines while still having a high DOFS (degrees of freedom for signal). All choices of microwindows lead to the required 4–5 DOFS, thanks to the numerous ozone lines with different intensities which give information both in the stratosphere and the troposphere. The test has been made at Kiruna and Ny-Ålesund to use the 1000–1005 cm^{-1} window only and, as expected, only little impact has been observed: except for Ny-Ålesund's tropospheric trends (1.4%/decade), we obtained small trend differences of between 0.0 and 0.8 % $decade^{-1}$, which is in all cases well below the uncertainty on the trends (see Section 4). However, it is planned, within the InfraRed Working Group of NDACC, to fix a common choice of microwindows for future improved homogenization.

Third, the main interfering species in this spectral region is water vapor, and it has been dealt with differently depending on the station: at the Wollongong and Lauder stations, the H_2O profile is retrieved simultaneously with the ozone profile, adding the microwindow of 896.4–896.6 cm^{-1} for a better H_2O determination. At Kiruna, Izaña and Jungfraujoch, the H_2O a priori profiles are only scaled in the ozone retrieval but these a priori profiles have been preliminarily retrieved in dedicated H_2O microwindows for each spectrum (Schneider et al. (2006) for Kiruna and Izaña; Sussmann et al. (2009) for Jungfraujoch). For the very dry Jungfraujoch site, it has been found that preliminary H_2O retrievals do not improve the quality of the ozone retrievals. At Ny-Ålesund and Thule, water vapor is treated as the other interfering species: only a scaling of a single a priori profile from WACCM4 is made. Therefore, except at the two latter stations, the H_2O profile variability has been well taken into account. This may be a future improvement to be done in Ny-Ålesund and Thule strategies. However, the random uncertainties due to the water vapor interference are not dominating the ozone error budget (see Sect. 2.3), and we expect a negligible impact on the ozone trends due to the H_2O treatment.

Fourth, the choice of the regularization (a priori covariance matrix, $\mathbf{S}_a$, and signal to noise ratio, SNR) cannot be easily homogenized because it depends on the real variability of ozone which is different at each station location and on the real SNR achieved by each spectrometer. In optimal estimation, the choice of the a priori covariance matrix $\mathbf{S}_a$ is an important parameter of the inversion process and, together with the measurement noise error covariance matrix $\mathbf{S}_\epsilon$, it will lead to the following averaging kernel matrix $\mathbf{A}$ (Rodgers, 2000):

$$\mathbf{A} = (\mathbf{K}^T\mathbf{S}_\epsilon^{-1}\mathbf{K} + \mathbf{S}_a^{-1})^{-1}\mathbf{K}^T\mathbf{S}_\epsilon^{-1}\mathbf{K}, \quad (1)$$

where $\mathbf{K}$ is the weighting function matrix that links the measurement vector $\boldsymbol{y}$ to the state vector $\boldsymbol{x}$: $\boldsymbol{y} = \mathbf{K}\boldsymbol{x} + \boldsymbol{\epsilon}$, with $\boldsymbol{\epsilon}$ representing the measurement error. In our retrievals, we assume $\mathbf{S}_\epsilon$ to be diagonal, in which case the diagonal elements are the inverse square of the SNR. The diagonal elements of $\mathbf{S}_a$ represent the assumed variability of the target gas volume mixing ratio (VMR) at a given altitude, and the non-diagonal elements represent the correlation between the VMR at different altitudes. We can see in Table 2 that, ex-

Table 2. Summary of the ozone retrieval parameters. All microwindow (mw) limits are given in per centimeter units. Ny: Ny-Ålesund; Th: Thule; Ha: Harestua; Ju: Jungfraujoch.

Parameters	Ny-Ålesund/Thule	Harestua/Jungfraujoch	Kiruna/Izaña	Wollongong/Lauder
Spectroscopic database	HITRAN 2008	HITRAN 2008	HITRAN 2008	HITRAN 2008
Pressure and temperature	NCEP	NCEP	NCEP	NCEP
Ozone a priori profiles	WACCM4	WACCM4 (Ju) climatology based on sondes and HALOE (Ha)	WACCM4	WACCM4
Retrieval code	SFIT2[a] v3.94	SFIT2[a] v3.94	PROFFIT9[b]	SFIT2[a] v3.94
Microwindows	1000–1005 782.56–782.86 (Ny) 788.85–789.37 (Ny) 993.3–993.8 (Ny)	1000–1005	991.25–993.80 1001.47–1003.04 1005.0–1006.9 1007.347–1009.003 1011.147–1013.553	1000–1005 782.56–782.86 788.85–789.37 993.3–993.8 896.4–896.6 (H_2O)
Interfering species	$H_2O, CO_2, C_2H_4,$ $^{668}O_3, ^{686}O_3$	$H_2O, CO_2, C_2H_4,$ $^{668}O_3, ^{686}O_3$ (Ju) H_2O, CO_2 (Ha)	$H_2O, CO_2, C_2H_4,$ $^{668}O_3, ^{686}O_3$	$H_2O, CO_2, C_2H_4,$ $^{668}O_3, ^{686}O_3$
H_2O treatment – a priori profile	One single profile (Ny) Preliminary retrievals in dedicated H_2O mws (Th)	One single profile (Ha) Preliminary retrievals in dedicated H_2O mws (Ju)	Preliminary retrievals in dedicated H_2O mws	One single profile
– fit in ozone mw	Scaling retrieval only	Scaling retrieval only	Scaling retrieval only	Profile retrieval
Regularization: – $\mathbf{S}_a$	Diagonal: 20 % (Ny) Diagonal: 30 % (Th) No interlayer correlation	Diagonal: 5–11 % (Ha) Diagonal: 10 % (Ju) No interlayer correlation (Ha) Interlayer correlation: Gaussian decay 4 km (Ju)	Tikhonov regularization L1	Diagonal: 10 % Interlayer correlation: exponential decay 4 km
– SNR	Real SNR (depending on each spectrum), except[c] regions at 1000.85–1001.45 1003.16–1004.5 set to SNR = 1 (Ny) Constant = 50 (Th)	Constant = 100 (Ju) Constant = 200 (Ha)	Depending on each spectrum	Constant = 150
Instrumental Line Shape	Fixed ideal (Ny) Fixed from LINEFIT (Th)	Fixed from LINEFIT (Ha) second-order polynomial fit of EAP (Ju)	Fixed ideal (Kiruna) Fixed from LINEFIT (Izaña)	Fixed ideal except Bomem spectra: fourth-order polynomial fit of EAP

[a] Pougatchev et al. (1995);
[b] Hase (2000);
[c] in order to mask strong H_2O absorptions.

cept at Harestua, Kiruna and Izaña, the stations are using an a priori covariance matrix with diagonal elements constant with altitude corresponding to 10, 20 or 30 % variability, the largest variability taking place at the high latitude stations Ny-Ålesund and Thule. At Harestua, the diagonal elements of $\mathbf{S}_a$ correspond to 11 % in the stratosphere, decreasing down to 6 % in the troposphere and to 5 % above 35 km. Except at Ny-Ålesund, the SNR value is not the real one coming from each individual spectrum but an effective SNR that is used as a regularization parameter. This effective SNR is smaller than the value derived from the inherent noise in the spectra, since the residuals in a spectral fit are not only coming from pure measurement noise but also from uncertainties in the model parameters. At Kiruna and Izaña, the regularization is made using the Tikhonov L1 constraint (Tikhonov, 1963). The regularization choice ($\mathbf{S_a}$ and SNR) is made at each station in order to obtain stable retrievals with reasonable DOFS. The regularization, via the **A** matrix, will impact, together with the real natural variability of ozone, the smoothing uncertainty which is the dominant source for the tropospheric and lower stratospheric columns. However, this is mainly a random uncertainty source and it as been shown at Izaña that using Tikhonov regularization or a $\mathbf{S_a}$ matrix obtained from ozone climatological measurements does not impact the ozone trends significantly (García et al., 2012).

The last important parameter is the instrumental line shape (ILS). As already mentioned, the ILS impacts the absorption line shape on which the ozone profile retrievals are based. Hence, if it is not properly included in the forward model or in the retrieval process, and if the alignment of the instrument is changing over time, this could impact the derived ozone trends (García et al., 2012). There are three options for considering the ILS and the choice is led by the type of spectrometer and the availability of cell measurements. A perfect alignment of the instrument would provide an "ideal" ILS: the modulation efficiency amplitude (also called the effective apodization parameter, EAP, at the maximum optical path difference) and the phase error remain equal to 1 and 0, respectively, along the optical path differences (OPDs). This perfect alignment can usually be achieved and maintained over time by the stable Bruker 120 or 125 HR. Even when those spectrometers are used, the alignment must be controlled by HBr or N_2O absorption measurements in a low-pressure gas cell and the use of the LINEFIT code, as described in Hase et al. (1999). In this approach, the loss of modulation efficiency and the phase error can be described (1) by 40 parameters (20 for each) at equidistant OPDs, or (2) simply by 2 parameters assuming a linear decline of the modulation efficiency with OPD and a constant phase error. At all stations using the 120 or 125 HR spectrometers, and where the cell measurements were available for the whole period and taken at least twice a year (Ny-Ålesund, Kiruna, Lauder, Wollongong from 2007), the ILS retrieved from LINEFIT was found good and stable: less than 2 % of loss in modulation efficiency at the maximum OPD. It has been therefore considered and fixed as ideal in the forward model. For the stations where the cell measurements were available and where the ILS could not be considered ideal, which was the case for the stations running a Bruker 120 M instrument, the ILS was fixed in the forward model to the parameters obtained by LINEFIT using either option (1) at Thule and Izaña or option (2) at Harestua. At Jungfraujoch up to the early 2000's, and at Wollongong, when the Bomem instrument was used, no cell measurements were performed; hence, it is not possible to use the LINEFIT results in the forward model. To take into account that the ILS may not be ideal, the modulation efficiency is retrieved simultaneously with the ozone profiles by using a polynomial fit of order 2 (Jungfraujoch) or 4 (Wollongong). The phase error has been neglected, i.e., it is treated as ideal. An argument against the use of the ozone absorption line shape to retrieve simultaneously the ozone profiles and the ILS is that a change on the ozone concentration at a given altitude may be interpreted wrongly as a change in the ILS. However, it was found that at Jungfraujoch the fitting of the ILS, instead of assuming that it is ideal, improved the agreement with correlative ozone profiles measurements (Barret et al., 2002), leading to the conclusions that there was enough information in the absorption line shapes to isolate correctly the ILS effect. We conducted the test at Ny-Ålesund to use a polynomial fit (order 2) of the modulation efficiency instead of a fixed ideal ILS. We found very small impact on the trends (less than 0.6% decade^{-1} for all layers). Of course the situation may differ for stations with worse alignment if this one cannot be reproduced by a polynomial fit of the modulation efficiency. Another solution to deal with periods without cell measurements would be to retrieve independently the ILS using N_2 and CO_2 lines in the historical solar spectra, since these gases have very well-known vertical profiles, and then fix the ILS to these preliminary derived values in the ozone retrievals.

2.3 Vertical information in FTIR retrievals

The vertical information contained in the FTIR retrievals can be characterized by the averaging kernel matrix **A** (Eq. 1), as described in detail in Vigouroux et al. (2008). It has been shown in this previous paper that the ozone retrievals provide 4–5 DOFS, depending on the station. Therefore, in addition to total column trends, we provide ozone trends in four independent partial column layers, corresponding to the vertical information. The layer limits have been chosen such that the DOFS is at least 1.0 in each associated partial column. The adopted layers are independent according to the resolution of the averaging kernels, as can be seen in Fig. 1 where the partial column averaging kernels of the four layers in the case of Jungfraujoch and Izaña are shown. Similar averaging kernels are obtained at each station (not shown). Also shown is the sensitivity which is, at each altitude k, the sum of the elements of the corresponding averaging kernel $\sum_i A_{ki}$ and represents roughly the fraction of the retrieval that comes from the measurement rather than from the a priori information. At Izaña, the sensitivity does not decrease towards 0 at about 50 km (Fig. 1) because of the use of Tikhonov regularization instead of optimal estimation (García et al., 2012). In the present work, small changes have been made in the partial column limits in comparison to Vigouroux et al. (2008): we avoid the tropopause region at each station, in order to have a better separation between the layer that we call the "tropospheric" layer and the lower stratospheric layer. Due to the high tropopause heights at Izaña (14.9 km) and Wollongong (13.8 km), compared to mid- and high-latitude stations (from 10.1 km at Ny-Ålesund to 11.8 km at Jungfraujoch), we use

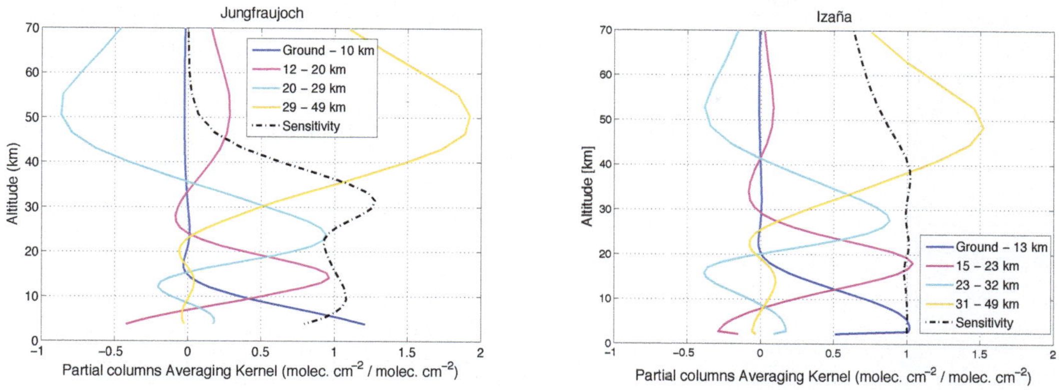

Figure 1. Partial column averaging kernels (molec cm^{-2} (molec cm^{-2})$^{-1}$) for ozone retrievals at Jungfraujoch (left) and Izaña (right) stations.

different partial column limits for these two stations. The upper limit of the upper layer is here 49 km, the altitude above which the DOFS is small (from about 0.01 to 0.04 depending on station), instead of 42 km in Vigouroux et al. (2008), chosen as the altitude above which the sensitivity was below 0.5. We still gain from 0.06 (Jungfraujoch) up to 0.23 (Lauder) DOFS in this 7 km wide range with poorer sensitivity. For Harestua, the chosen layer limits give a DOFS of only 0.9 and 0.75 in the ground–10 km and in the 29–49 km layers, respectively.

We provide in Table 3, for each station, the partial column limits of the four defined layers (Trop: troposphere; LowS: lower stratosphere; MidS: middle stratosphere; UppS: upper stratosphere). The detailed error budget for ozone FTIR retrievals has been described in Vigouroux et al. (2008) and more recently in García et al. (2012) for Izaña; we only just summarize in Table 3 the total random uncertainties obtained for the present choice of layers and for the total columns (TotC). As obtained in the two previous papers, and not shown here, the smoothing error is the dominant random error source for the tropospheric and lower stratospheric layer, while the temperature dominates the random error budget for the middle and upper stratospheric layers and for total columns. Also found in these two papers, and not repeated here, is the validation of the FTIR total and partial columns with correlative data (Dobson, Brewer, UV–Vis, ozonesondes, lidar).

2.4 FTIR ozone time series

Figure 2 displays the time series of ozone total columns at each ground-based FTIR station. Because we consider only solar absorption measurements, the time series at Ny-Ålesund, Thule, and Kiruna cover only the mid-March–September, late-February–mid-October and mid-January–mid-November periods, respectively. The seasonal variation is isolated in Fig. 3, which shows the monthly mean total columns over the periods of measurements. We clearly see the well-known seasonal cycle of ozone total column having a maximum in spring at all stations, and the higher amplitude of the seasonal variation at higher latitudes (Brasseur and Solomon, 1984).

Figure 3 shows also the monthly means of the four partial columns defined in the previous section (Table 3). In the upper stratospheric layer, the ozone maximum occurs in summer (early summer at high latitudes shifting to late summer with decreasing latitude), in agreement with higher photochemical production of ozone during this season. In the lower stratospheric layer, the ozone maximum is in late winter/early spring at all latitudes. The situation is more variable for the middle stratospheric layer: still late winter/early spring for Harestua, Jungfraujoch, Lauder and Wollongong, but the latter shows a second maximum in late summer, and a small amplitude of the seasonal cycle. For the three higher latitude stations – Ny-Ålesund, Thule and Kiruna – the maximum is still in spring, extending to May for the two latter stations. At Izaña, the maximum is in summer in the mid-

Table 3. Partial column (PC) limits for the four altitude layers containing at least one DOFS. The random uncertainties are given for each partial column. Trop: troposphere; LowS: lower stratosphere; MidS: middle stratosphere; UppS: upper stratosphere; TotC: total columns; Gd: ground; Err.: total random uncertainties.

Layers	Stations	PC limits	Err.
Trop	Izaña/Wollongong	Gd–13/12 km	6 %
	Other stations	Gd–9/10 km	5 %
LowS	Izaña/Wollongong	15–23 km	5 %
	Other stations	12–20 km	4 %
MidS	Izaña/Wollongong	23–32 km	5 %
	Other stations	20–29 km	5 %
UppS	Izaña/Wollongong	31–49 km	5 %
	Other stations	29–49 km	5 %
TotC	Izaña/Wollongong	–	2 %
	Other stations	–	2 %

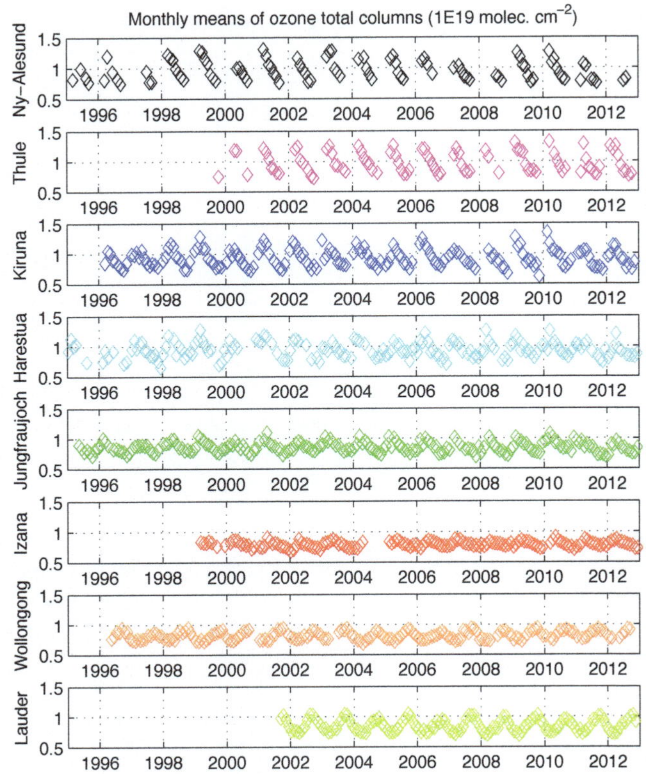

Figure 2. Time series of monthly means of ozone total columns at each station.

dle stratosphere. For the tropospheric column, we observe a maximum in spring at all stations, but at Jungfraujoch it extends also into summer.

3 Multiple regression model

The ozone FTIR total and partial column trends in Vigouroux et al. (2008); WMO (2010); García et al. (2012) were calculated with a bootstrap resampling method, applied to the daily means time series. In these studies, only the seasonal cycle and a linear trend were taken into account, the remaining natural ozone variability was then an additional noise in the ozone trend determination. To reduce the uncertainties on the trends and to better understand what drives ozone variability and trends, we use in the present study a multiple linear regression (MLR) model. To reduce the autocorrelation in the residuals, we use here the monthly means time series. Furthermore, to account for the still significant auto-correlation in the residuals, a Cochrane–Orcutt transformation (Cochrane and Orcutt, 1949) is applied to the final model. This gives more reliable confidence intervals for the regression parameters.

The following regression model is applied to the monthly means of ozone total and partial column time series $Y(t)$:

$$Y(t) = A_0 + A_1 \cdot \cos(2\pi t/12) + A_2 \cdot \sin(2\pi t/12) + A_3 \cdot \cos(4\pi t/12) + A_4 \cdot \sin(4\pi t/12) + A_5 \cdot t + \sum_{k=6}^{n} A_k \cdot X_k(t) + \epsilon(t), \quad (2)$$

where A_0 is the intercept, the A_1–A_4 parameters describe the ozone seasonal cycle, A_5 is the annual trend, $X_k(t)$ are the explanatory variables (proxies time series) and A_k their respective coefficient, and $\epsilon(t)$ represents the residuals.

To select the final regression model, we have included several proxies, which represent processes that are known to impact ozone, in a stepwise regression procedure that keeps or rejects each proxy: the initial model (seasonal cycle and trend) is fitted first. Second, iteratively, if any proxies, not already in the model, have p values smaller than an entrance tolerance (0.05), i.e., if it is unlikely that they would have a 0 coefficient if added to the model, then we add the one with the smallest p value. Otherwise, if any proxies in the model have p values greater than an exit tolerance (0.10), we remove the one with the largest p value and we repeat the whole process until no single step improves the model. Hence, the final set of parameters can vary with the station and with the partial columns concerned. In this paper, a proxy is called "non-significant" when it has not been retained by the stepwise procedure. This choice of not using a fixed model for all stations and partial columns avoids overfitting the data and is justified by the large latitudinal range of the stations (e.g., the VPSC or ENSO proxies will not impact the stations in the same way) and by the different processes driving ozone variability at different altitudes.

The proxies that have been tested in the stepwise regression procedure are summarized in Table 4. The two most common explanatory variables found in the literature are the solar radio flux at F10.7 cm (SOLAR) which represents the 11-year solar cycle (following e.g., Newchurch et al., 2003; Randel and Wu, 2007), and the zonal winds measured at Singapore at 30 and 10 hPa (following e.g., Brunner et al., 2006) which represent the quasi-biennial oscillation (QBO). The proxy used for the El Niño–Southern Oscillation (ENSO) is the multivariate ENSO index (MEI), following Randel et al. (2009). Different time lags (from 0 to 4 months) between ENSO and ozone time series have been tested. The other dynamical proxies that have been explored are the tropopause pressure (TP) at each station (following e.g., Appenzeller et al., 2000), the equivalent latitude (EL) at three altitude levels around each station, the Arctic Oscillation (AO) or the Antarctic Oscillation (AAO) indices depending on the station location (e.g., Appenzeller et al., 2000; Frossard et al., 2013), and the vertical component of the Eliassen–Palm flux (EPF) at 100 hPa averaged over 45–75° north and south, as a proxy for the Brewer–Dobson circulation (e.g., Brunner et al., 2006). These dynamical proxies are connected, e. g. the

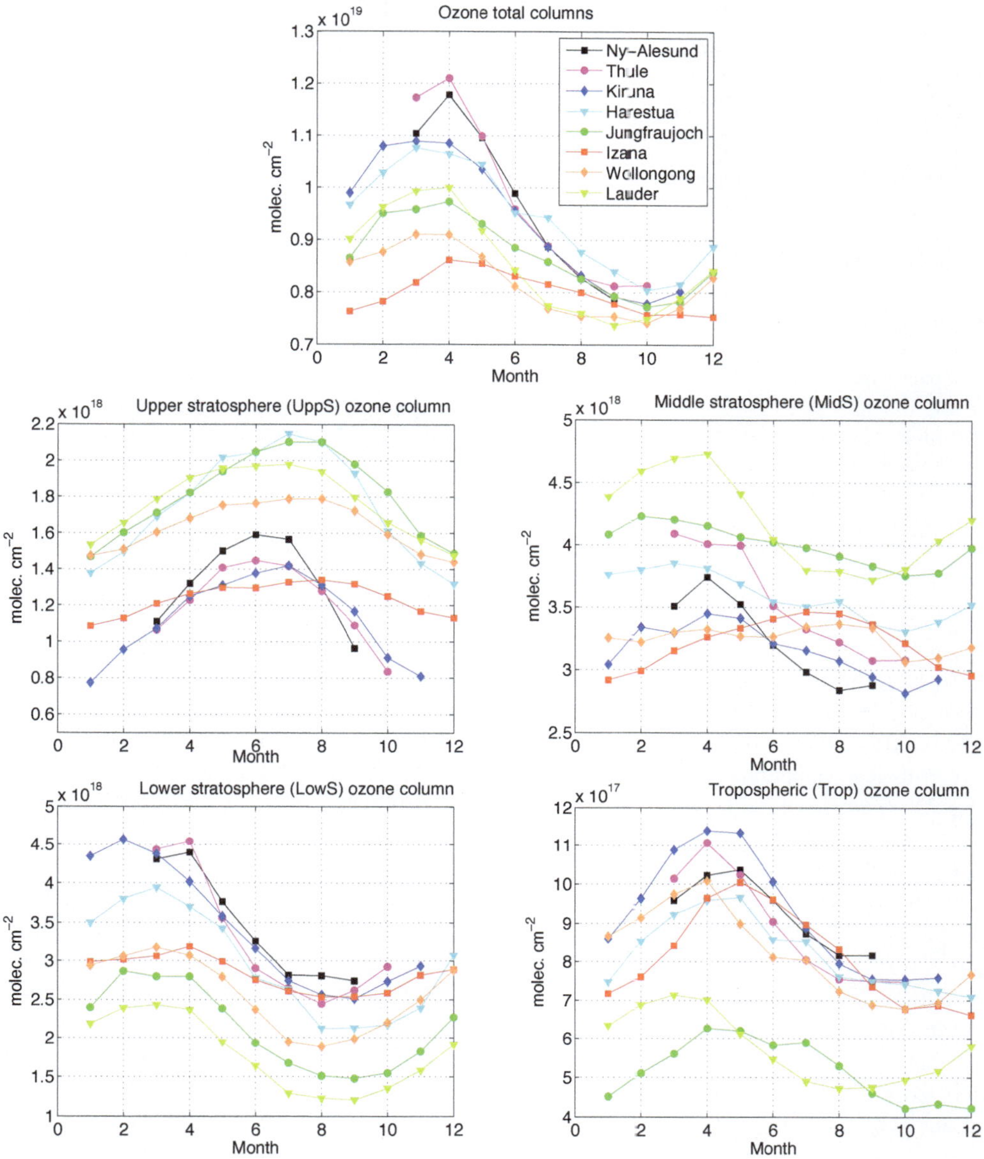

Figure 3. Monthly means of ozone total and partial columns for the whole periods of measurements. See Table 3 for the limits of the partial columns at each station. The seasonal cycle for Southern Hemisphere stations, Wollongong and Lauder, has been shifted by 6 months for better comparison.

NAO (North Atlantic Oscillation, closely related to AO) and the tropopause pressure (Appenzeller et al., 2000), the eddy heat flux (proportional to EPF) and the AO (Weber et al., 2011), but we let the stepwise regression model choose the most adapted proxy for each station and partial column. Concerning the equivalent latitude, we did not construct an integrated equivalent proxy valuable for the whole ozone "integrated" total column as in Wohltmann el al. (2005). Here, we simply use the equivalent latitude calculated from ERA Interim reanalysis (Dee et al., 2011) at three altitude levels corresponding approximately to the middle of our three stratospheric layers (ELL for LowS, ELM for MidS, and ELU for UppS; see Table 3 for the layer limits), namely at 370, 550, and 950 K, respectively, for all stations except Izaña and Wollongong (460, 700, and 1040 K, respectively).

Lastly, the volume of polar stratospheric clouds (VPSC) is used as a proxy for polar ozone loss (e.g., Brunner et al., 2006). The VPSC proxy has been multiplied by the effective equivalent stratospheric chlorine (EESC) time series calculated with a mean age of air of 5.5 years, in order to take into account the time for the ozone-depleting substances to reach the poles (http://acdb-ext.gsfc.nasa.gov/Data_services/automailer/index.html). To account for the cumulative effect over months of the EPF and the VPSC*EESC proxies on ozone, we have followed the approach of Brunner et al.

Table 4. Name, short description, and source of the proxies that have been tested in the stepwise regression model.

Name	Description	Source
SOLAR	solar radio flux at 10.7 cm	ftp://ftp.ngdc.noaa.gov/STP/SOLAR_DATA/SOLAR_RADIO /FLUX/Penticton_Adjusted/monthly/MONTHLY.ADJ
QBO30	zonal winds measured at Singapore at 30 hPa	http://www.geo.fu-berlin.de/en/met/ag/strat/produkte/qbo/index.html
QBO10	zonal winds measured at Singapore at 10 hPa	http://www.geo.fu-berlin.de/en/met/ag/strat/produkte/qbo/index.html
ENSO	multivariate ENSO index (MEI)	http://www.esrl.noaa.gov/psd/enso/mei/
AO/AAO	Arctic Oscillation	http://www.cpc.ncep.noaa.gov/products/precip/CWlink/daily_ao_index /monthly.ao.index.b50.current.ascii
	Antarctic Oscillation	http://www.cpc.ncep.noaa.gov/products/precip/CWlink/daily_ao_index /aao/monthly.aao.index.b79.current.ascii
TP	tropopause pressure	http://www.esrl.noaa.gov/psd/data/gridded /data.ncep.reanalysis.tropopause.html
EL(L/M/U)	equivalent latitude at three altitude levels: 370, 550, and 950 K: high-/mid-latitude stations 460, 700, and 1040 K: subtropical stations	calculated at BIRA from ERA interim reanalysis
EPF	vertical component of the EP flux	http://www.awi.de/en/research/research_divisions/climate_science /atmospheric_circulations/projects/candidoz/ep_flux_data/
VPSC	volume of polar stratospheric clouds	calculated at FMI

(2006) (see their Eq. 4), with the same dependence of their constant τ on season and latitude of the station.

For the two QBO proxies (30 and 10 hPa), if retained in the stepwise procedure, four seasonal parameters can be added to the model. The $A_k \cdot X_k(t)$ term of Eq. (2) is then replaced by

$$(A_k + A_{k+1} \cdot \cos(2\pi t/12) + A_{k+2} \cdot \sin(2\pi t/12) \\ + A_{k+3} \cdot \cos(4\pi t/12) + A_{k+4} \cdot \sin(4\pi t/12)) \cdot X_k(t). \quad (3)$$

Depending on the station and on the layer, none, one or both of the two proxies QBO30 and QBO10 will be retained in the model, with or without their additional seasonal parameters. We will call from here on "QBO contribution" the sum of all possible contributions of QBO30 and QBO10.

Since the time series involved in the present study start at the earliest in 1995, we do not include two commonly used explanatory variables: the aerosol optical thickness needed to represent the effect on ozone of the large volcanic eruptions of El Chichón (1982) and Mount Pinatubo (1991), and the EESC proxy which can be used as direct proxy for the halogen loading of the stratosphere instead of the piecewise linear trend (PWLT) with a turnaround in 1996/1997 often used in time series starting well before this turnaround point (WMO, 2010). Our linear-trend estimates are therefore better comparable to the studies which use the PWLT method. At polar stations, the turnaround is occurring a few years later, so that the use of the EESC proxy could be an alternative to the simple linear trend for these stations. However, we preferred to adopt the same approach for all the stations. It is possible that when the FTIR record is longer, one will be able to distinguish between the EESC impact on ozone and a possible additional trend due to processes that are not represented in the model.

4 Results and discussion

In Fig. 4, we show the individual contribution C_{frac} of each proxy retained by the stepwise procedure to the coefficient of determination $R^2 = \sum C_{frac}$, for each station and partial column. The individual contribution C_{frac} of a proxy is the product of the standardized regression coefficient of this proxy with the correlation coefficient between the proxy and the observations (Scherrer, 1984). In Fig. 4, the seasonal parameters' contribution (A_1– A_4 in Eq. 2), which gives in most cases the very dominant part of the explained variability, is not shown for better clarity of the other proxies' contribution. However, we give it for completeness in Table 5, together with R^2. In the following discussion, we will highlight some selected features which are visible in the ozone time series and which can be attributed to a specific proxy. The final MLR model is the sum of all the significant proxies and, therefore, the effect of a specific proxy can be visible in the plots in some years, but masked in other years.

In Table 6, we give the annual ozone trend at each station for each layer obtained with the stepwise multiple linear regression model. The uncertainties on the trends correspond to the 95 % confidence interval. A trend is considered significant if it is larger than the uncertainty.

4.1 High latitude stations

In addition to the three Arctic stations Ny-Ålesund, Thule and Kiruna, we will consider Harestua (60° N) as a high latitude station since, in terms of trends, Harestua appears to behave similarly to the Arctic stations.

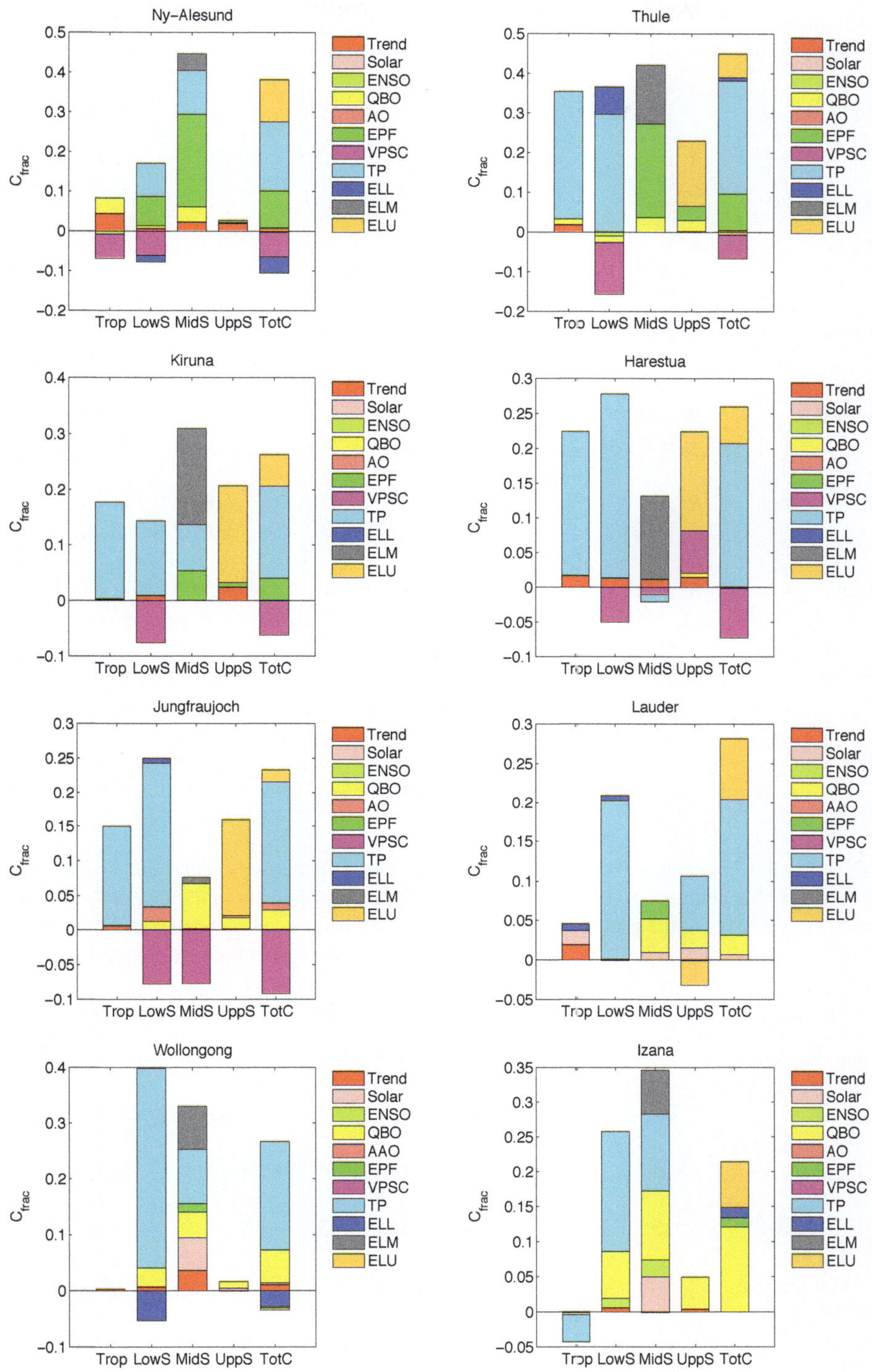

Figure 4. Individual contributions C_{fract} of the proxies to the coefficient of determination R^2. R^2 and the dominant contribution of the seasonal cycle C_{seas} are given in Table 5.

4.1.1 Tropospheric (Trop) columns

In the troposphere, the high latitude stations, except Kiruna, show negative significant ozone trends (Table 6). The spatial and temporal variability in the Arctic and the different sampling at the stations Thule/Ny-Ålesund due to polar night (see Fig. 2) makes it difficult to compare the trend results. We see in Fig. 5 that at Ny-Ålesund the negative trend occurs in the second part of the period (2004–2012), which is also observed at Thule (not shown). On the contrary, at Harestua, the

Table 5. Coefficient of determination R^2 and contribution of the seasonal cycle C_{seas} determined within the final model. See Table 3 for the limits of the layers, different for subtropical stations and mid-/high-latitude stations.

FTIR station	Trop	LowS	MidS	UppS	Total columns
Ny-Ålesund	$R^2=0.75$ $C_{seas}=0.73$	$R^2=0.92$ $C_{seas}=0.82$	$R^2=0.72$ $C_{seas}=0.27$	$R^2=0.74$ $C_{seas}=0.72$	$R^2=0.95$ $C_{seas}=0.68$
Thule	$R^2=0.86$ $C_{seas}=0.50$	$R^2=0.92$ $C_{seas}=0.71$	$R^2=0.83$ $C_{seas}=0.41$	$R^2=0.81$ $C_{seas}=0.58$	$R^2=0.96$ $C_{seas}=0.58$
Kiruna	$R^2=0.85$ $C_{seas}=0.67$	$R^2=0.89$ $C_{seas}=0.82$	$R^2=0.54$ $C_{seas}=0.23$	$R^2=0.78$ $C_{seas}=0.58$	$R^2=0.89$ $C_{seas}=0.69$
Harestua	$R^2=0.77$ $C_{seas}=0.54$	$R^2=0.74$ $C_{seas}=0.51$	$R^2=0.36$ $C_{seas}=0.25$	$R^2=0.67$ $C_{seas}=0.45$	$R^2=0.75$ $C_{seas}=0.56$
Jungfraujoch	$R^2=0.73$ $C_{seas}=0.58$	$R^2=0.83$ $C_{seas}=0.66$	$R^2=0.53$ $C_{seas}=0.53$	$R^2=0.93$ $C_{seas}=0.77$	$R^2=0.81$ $C_{seas}=0.67$
Izaña	$R^2=0.83$ $C_{seas}=0.87$	$R^2=0.72$ $C_{seas}=0.46$	$R^2=0.80$ $C_{seas}=0.45$	$R^2=0.69$ $C_{seas}=0.64$	$R^2=0.77$ $C_{seas}=0.56$
Wollongong	$R^2=0.69$ $C_{seas}=0.69$	$R^2=0.86$ $C_{seas}=0.52$	$R^2=0.42$ $C_{seas}=0.09$	$R^2=0.77$ $C_{seas}=0.75$	$R^2=0.87$ $C_{seas}=0.63$
Lauder	$R^2=0.89$ $C_{seas}=0.85$	$R^2=0.94$ $C_{seas}=0.73$	$R^2=0.78$ $C_{seas}=0.70$	$R^2=0.89$ $C_{seas}=0.82$	$R^2=0.95$ $C_{seas}=0.66$

Table 6. Annual trend (in % decade^{-1}) and their 95 % uncertainty ranges. Due to polar night, the measurements at Ny-Ålesund, Thule and Kiruna cover only the mid-March–September, late-February–mid-October, and mid-January–mid-November periods, respectively. All time series end in September/December 2012 for the present study. The time of start is repeated for each station. See Table 3 for the limits of the layers, different for subtropical stations and mid-/high-latitude stations. Trends indicated in bold are significant.

FTIR station	Trop	LowS	MidS	UppS	Total columns
Ny-Ålesund 1995	**−5.8 ± 3.2**	**−4.2 ± 3.1**	**−5.5 ± 3.8**	**+6.7 ± 5.3**	**−3.0 ± 1.5**
Thule 1999 (October)	**−5.3 ± 4.4**	−0.4 ± 6.3	+0.2 ± 4.4	−2.3 ± 6.5	−2.1 ± 2.6
Kiruna 1996	−0.9 ± 2.5	**−3.9 ± 2.6**	+0.4 ± 2.6	**+7.4 ± 3.4**	−0.3 ± 1.6
Harestua 1995	**-3.1 ± 2.0**	**-5.3 ± 4.6**	**+4.8 ± 4.3**	**+7.8 ± 5.5**	+1.0 ± 2.2
Jungfraujoch 1995	−2.5 ± 2.7	−0.5± 3.3	−0.6 ± 1.2	+0.9 ± 1.0	−0.4 ± 1.2
Izaña 1999	+0.7 ± 2.8	−1.7 ± 2.2	−0.1 ± 2.0	+1.6 ± 2.6	+0.5 ± 1.2
Wollongong 1996	−2.2 ± 2.8	**+3.1± 2.7**	**+4.0 ± 2.0**	+0.2 ± 1.6	**+1.9 ± 1.1**
Lauder with / without SOLAR 2001	**+7.7 ± 3.5** **+5.0 ± 4.4**	-3.8± 4.1 –	-0.2 ± 3.5 −1.1 ± 3.4	**+2.8± 2.4** +1.7± 2.4	−0.3 ± 1.8 -0.6 ± 1.9

negative trend is occurring in the 1999–2007 period (Fig. 5, lower left panel). The second line of Fig. 5 shows the partial columns where the seasonal cycle is removed to emphasize the interannual variability, and the effect of individual proxies showing interannual differences. We have added in the third line of Fig. 5 the VPSC signal, i.e., the VPSC proxy time series multiplied by the corresponding parameter obtained in the MLR process ($A_k \cdot X_k(t)$ in Eq. 2). We see that the particular low tropospheric values in 1995, 2005 and 2011 at Ny-Ålesund can be related to the VPSC proxy and, therefore, by the influence of lower stratospheric ozone variability on the tropospheric columns. At the three other stations, this VPSC impact was not found to be significant, and the main driver of tropospheric variability is found to be the tropopause pressure TP (Fig. 4). The larger VPSC value in 1996 does not lead to a larger decrease in tropospheric ozone because it is compensated by a positive QBO signal, while the small ozone value in 2004 is related to a negative QBO signal (not shown).

As expected, the large VPSC values in 1995, 2005 and 2011 have also a significant impact on the lower stratospheric (LowS) values at Ny-Ålesund (middle column of Fig. 5), as well as in 1996, since the negative effect is not compensated by the QBO signal as in Trop. We can note that the VPSC impact is 10 times larger in LowS than in Trop (different scales in Fig. 5).

4.1.2 Lower stratospheric (LowS) columns

The VPSC proxy is found significant at the four high latitude stations for the lower stratospheric columns, being the main driver of ozone variability after TP (Fig. 4). We give the example of Ny-Ålesund and Kiruna in Fig. 5, where the effect of large amount of VPSC in 1996, 2005, and 2011 is clearly visible in both monthly means and deseasonalized time series. We show in addition the EPF and TP signals at Ny-Ålesund and Kiruna, respectively, in the bottom panel. It can be seen that the TP signal at Kiruna in 2005 also contributed to even lower ozone that particular year. The larger LowS values at Ny-Ålesund in 1999 are due to a combination of the TP (not shown) and EPF signals.

In the lower stratosphere, at all high latitude stations, except Thule, we observe significant negative trends (Table 6). At Thule, the shorter time period associated with the high variability of this layer at high latitude gives a large uncertainty on the trend.

4.1.3 Middle stratospheric (MidS) columns

The results are mixed for the middle stratospheric layers, as noticed previously for the seasonal cycles. The trend is significantly negative at Ny-Ålesund and non-significant at Thule. The trend is non-significant at Kiruna, and significantly positive at Harestua. The EPF proxy explains about 25 % of the variability at Ny-Ålesund and Thule, and about 5 % at Kiruna (Fig. 4). This is illustrated in Fig. 6 for Ny-Ålesund and Thule, where we see nicely the same features at both stations in the middle stratospheric columns (e.g., higher columns in 2009, 2010; lower columns in 2011), associated with the EPF time series.

4.1.4 Upper stratospheric (UppS) columns

In the upper stratosphere, the three stations with similar time periods show a significant positive trend. In the three cases, the increase in ozone partial columns occurs in the 1995–2003 period, after which a leveling off is observed (Fig. 7). If we run the MLR model on the same time period as Thule (October 1999–2012), all the stations show non-significant trends. Since the EESCs were still increasing until about 2000 at polar regions (WMO, 2010), the significant positive trends obtained at high latitude stations in the upper stratosphere cannot be explained by the effect of the Montreal Protocol on ozone depleting substances. At present we do not have an explanation for this increase in ozone during the 1995–2003 period. The 11-year solar cycle might contribute to it, since the increase in solar activity from 1996 to its maximum in 2001–2002 is in phase with the ozone increase during the same period. The solar cycle signal at Ny-Ålesund, shown in Fig. 7 as an illustration, turns out to be non-significant after the Cochrane–Orcutt transformation is applied, so its contribution is small and not visible in Fig. 4. The solar cycle might be found non-significant at the other stations because the expected decrease of ozone during the declining phase of the solar cycle (2002–2009) is not observed. This could be a sign that this decrease is compensated by a positive linear trend, which could be due to the declining EESCs, but also to the increase of greenhouse gases (WMO, 2010). More years are needed to understand unequivocally the increase in 1995–2003, followed by a leveling off, and distinguish between the ozone responses due to solar cycle, EESCs and other possible proxies not included in the present study.

4.1.5 Total columns

We observe that the total column ozone trends are small and non-significant at all high latitude stations, except at Ny-Ålesund (-3.0 ± 1.5 % decade^{-1} or -10.8 ± 5.6 DU decade^{-1}). The negative trend at Ny-Ålesund occurs in the 2003–2012 period, as for the lowest altitude layers. At all stations, the dominant contributions to the total column variability are the TP, the VPSC, the ELU, and, except at Harestua, the EPF proxies. We see nicely in Table 5, how well the proxies explained the additional variability at the Arctic stations, e.g., at Ny-Ålesund $R^2 = 0.95$, compared to the contribution of the seasonal cycle $C_{seas} = 0.68$.

4.2 Mid-latitude and subtropical stations

In this study, we have two mid-latitude stations (Jungfraujoch at 47° N and Lauder at 45° S) and two subtropical stations (Izaña at 28° N and Wollongong 34° S).

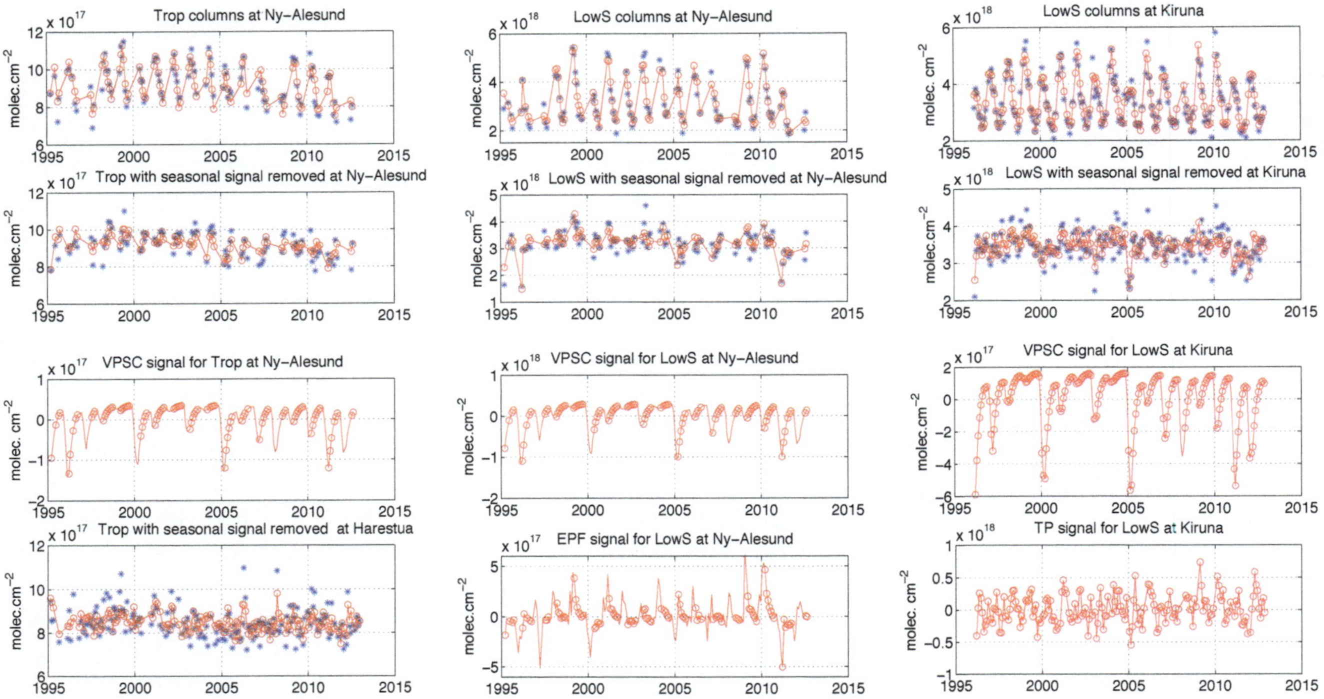

Figure 5. Left (top to bottom panels): (1) monthly means of the tropospheric columns (Trop) at Ny-Ålesund (blue: FTIR, red: MLR model); (2) same but with the seasonal signal removed; (3) the VPSC signal obtained from the MLR model for Trop at Ny-Ålesund, for each month of the period (red line), and at each FTIR observed month (red circle); (4) monthly means of Trop at Harestua with the seasonal cycle removed. Middle panels: (1–3) same as left panels but for the lower stratospheric columns (LowS) at Ny-Ålesund; (4) the EPF signal obtained for the LowS at Ny-Ålesund. Right panels: (1–3) same as middle panels but at Kiruna; (4) the tropopause pressure (TP) signal obtained for the LowS at Kiruna.

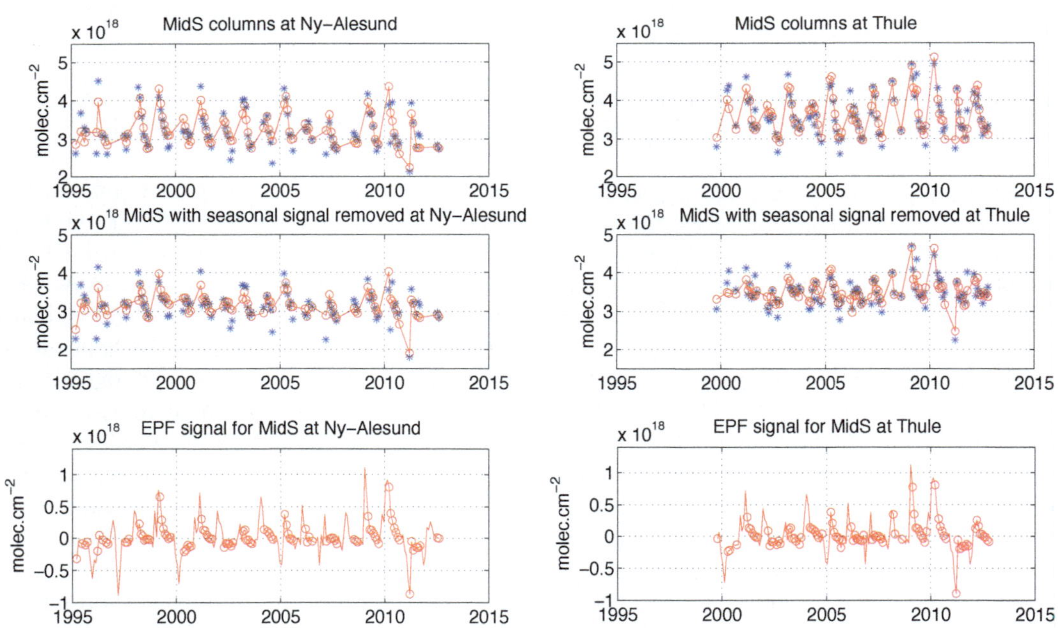

Figure 6. Top panels: monthly means of the middle stratospheric columns (MidS) at Ny-Ålesund (left) and Thule (right) (blue: FTIR, red: MLR model). Middle panels: same but with the seasonal signal removed. Bottom panels: the EPF signal obtained in each case from the MLR model, for each month of the period (red line), and at each FTIR observed month (red circle).

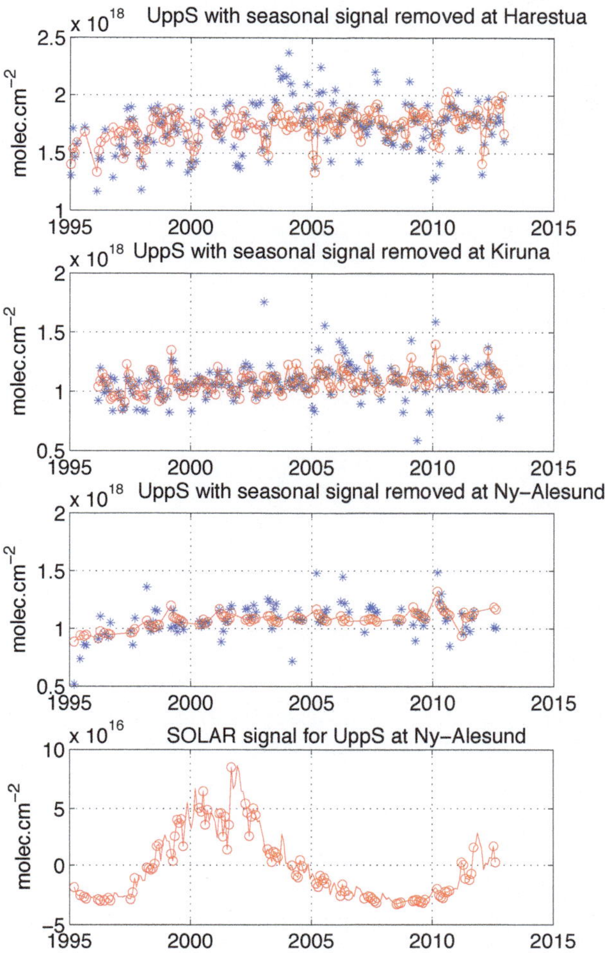

Figure 7. Monthly means of the upper stratospheric columns (UppS) with the seasonal cycle removed at, from top to bottom: Harestua, Kiruna and Ny-Ålesund (blue: FTIR, red: MLR model). Bottom panel: the solar cycle signal obtained at Ny-Ålesund from the MLR model, before the Cochrane–Orcutt transformation.

4.2.1 Tropospheric (Trop) columns

The tropospheric trends are non-significant at Jungfraujoch, Izaña and Wollongong, and significantly positive at Lauder. The trend at Jungfraujoch is $-2.5 \pm 2.7\,\%\,\text{decade}^{-1}$; however, we see in Fig. 8 that the tropospheric columns are increasing up to 1999 and then show a linear decrease, in agreement with aircraft and surface alpine sites in the study of Logan et al. (2012). If we limit our time period to the 1998–2008 period as in Logan et al. (2012), we also find a significant negative trend ($-6.3 \pm 4.9\,\%\,\text{decade}^{-1}$). However, this is largely due to the high ozone values 1998–1999, and for the period 2000–2012 we obtain still a non-significant trend of $-2.9 \pm 3.4\,\%\,\text{decade}^{-1}$. At Izaña, the tropospheric trends derived from ozonesondes were found non-significant in García et al. (2012), in agreement with our study, but the uncertainties were large. The situation is more mixed in the Southern Hemisphere: the tropospheric trend at Wollongong is not significant while it is significantly positive at Lauder ($+7.7 \pm 3.5\,\%\,\text{decade}^{-1}$). The trend at Lauder is in agreement with the study of Oltmans et al. (2013), who obtain about $+5\,\%\,\text{decade}^{-1}$ in the lower and middle troposphere with ozonesonde measurements at Lauder. We find a significant positive impact of the solar cycle at Lauder and it is clearly seen in Fig. 8. This is not in agreement with Chandra et al. (1999), in which the solar cycle shows a strong but negative impact on tropospheric columns for a non-polluted region. At Lauder, presently only a short time period (2001–2012) is available for trend studies, and we hope to have more clarification on this subject with more years of data. However, if we remove the solar cycle proxy from the MLR model, we still obtain a significant trend of $+5.0 \pm 4.4\,\%\,\text{decade}^{-1}$. For this short time series, we have added in Table 6 the trends that are obtained if the solar cycle is removed from the model.

4.2.2 Lower stratospheric (LowS) columns

The trends in the lower stratosphere are non-significant at Jungfraujoch, Izaña and Lauder, and significantly positive at Wollongong. The cause of the significant positive trend at Wollongong is not fully explained at present. A part of it is due to a small negative trend in the ELL proxy. If we remove this proxy from the MLR model, we observe a non-significant positive trend of $+2.4 \pm 2.8\,\%\,\text{decade}^{-1}$.

The dominant proxy is TP for all stations. At the Jungfraujoch station, the VPSC proxy, which in the case of Jungfraujoch corresponds to the transport of polar ozone loss to mid-latitudes, explains about 8 % of the variability (Fig. 4). The VPSC proxy is non-significant at the southern hemispheric station Lauder, in agreement with the more stable and isolated vortex in the Antarctic compared to the Arctic. The AO proxy is found significant at Jungfraujoch while the corresponding AAO proxy is non-significant at Lauder.

We show the time series of the lower stratospheric columns at Jungfraujoch in Fig. 9 together with the AO and QBO signals. We see that in 2010 ozone shows larger values and that this is explained by the combination of a very negative AO index (the corresponding parameter in the MLR is negative and gives the positive signal in 2010 shown in Fig. 9) and easterly phase of the QBO. This is in agreement with Nair et al. (2013), who applied a MLR model to the mean of ozone anomalies at Observatoire de Haute-Provence (OHP) from different instruments (lidar, ozonesondes and satellites). However, we did not find a significant contribution from the EPF proxy, which according to Nair et al. (2013) also contributed to the high ozone values in 2010. We can state that our vertical and total column ozone trends are in agreement with the Nair et al. (2013) results when taking the error bars into account, but the latter study found significant positive trends at OHP while our trends at Jungfraujoch are all non-significant.

As expected, the QBO contribution to ozone variability is more important at the subtropical station Izaña, which is

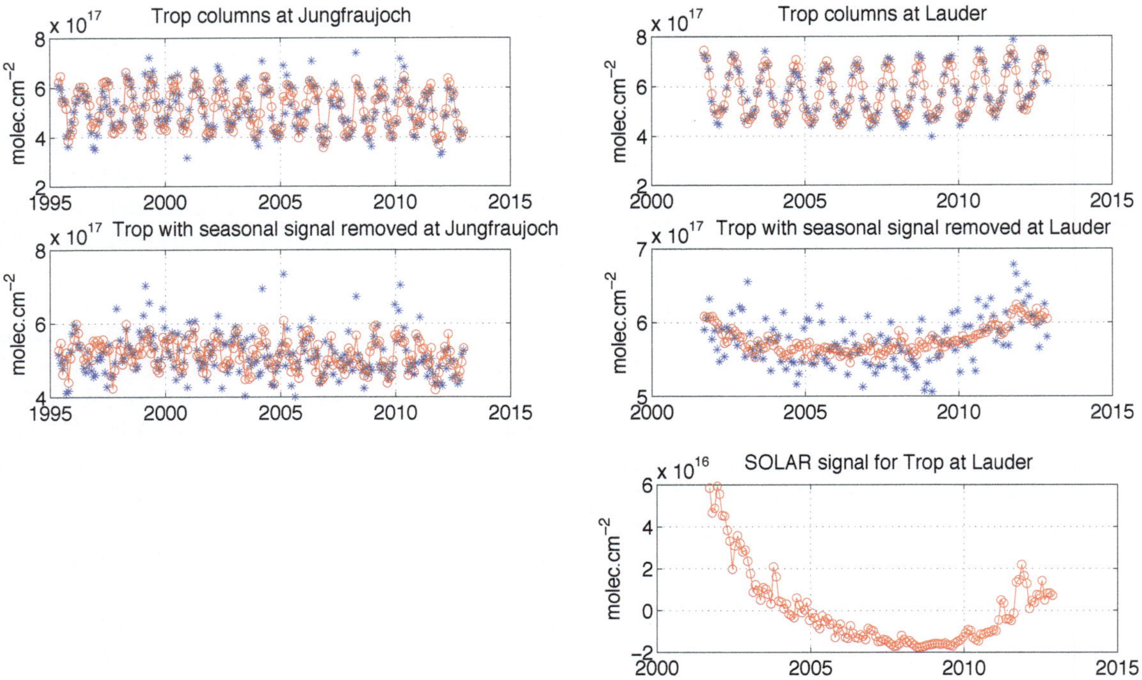

Figure 8. Top panels: monthly means of the tropospheric columns (Trop) at Jungfraujoch (left) and Lauder (right) (blue: FTIR, red: MLR model). Middle panels: same but with the seasonal signal removed. Bottom panel: the solar cycle signal obtained at Lauder from the MLR model.

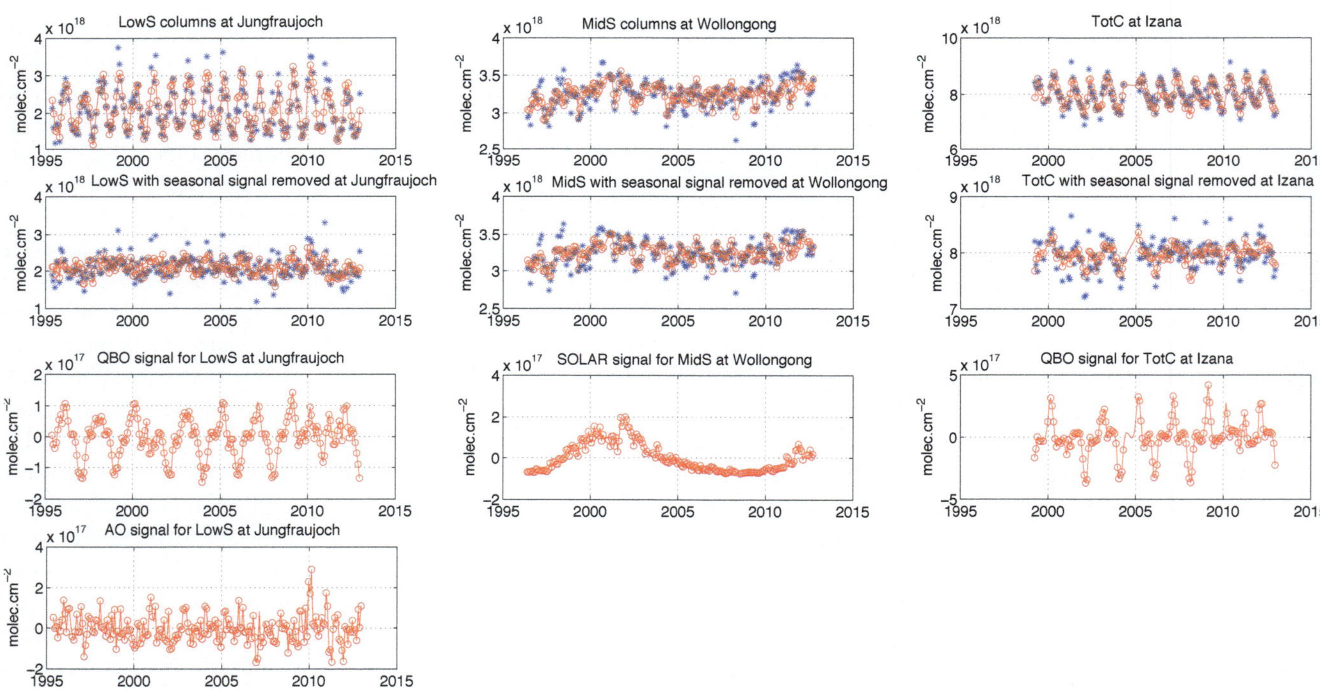

Figure 9. Top panels: monthly means of the lower stratospheric columns (LowS) at Jungfraujoch (left), middle stratospheric columns (MidS) at Wollongong (middle), and total columns (TotC) at Izaña (right) (blue: FTIR, red: MLR model). Middle panels: same but with the seasonal signal removed. Bottom panels: QBO and AO signals obtained from the MLR model at Jungfraujoch (left), SOLAR signal at Wollongong (middle), and QBO signal at Izaña (right).

also the only station where the ENSO proxy was found to make a significant, but small, contribution to the variability (Fig. 4). We illustrate the QBO effect at Izaña in Fig. 9 for total columns.

4.2.3 Middle stratospheric (MidS) columns

The situation for the middle stratosphere is very similar to that of the lower stratosphere: all trends are found non-significant except at Wollongong where it is positive. It is in this 23–32 km layer for subtropical stations that the solar cycle shows the most important contribution (Fig. 4). This is not what has been reported in Randel and Wu (2007) and Tourpali et al. (2007), where the ozone response to the solar cycle was maximum in the tropical lower and upper stratosphere, and minimum in the middle stratosphere. At Wollongong, the middle stratospheric ozone response is about 6 % between solar minimum and solar maximum (see Fig. 9) while values of 1 % have been reported (Sioris et al., 2014) at about 25 km. However, the recent work of Chiodo et al. (2014) shows that the apparent solar cycle signal in the tropical lower stratosphere for the period 1960–2004 is due to the two volcanic eruptions of El Chichón in 1982 and Mt. Pinatubo in 1991, and the authors find robust solar cycle signals only in the middle and upper stratosphere. In the upper stratospheric layer at Wollongong, the response to the solar cycle is indeed also significant and is about 2.5 % between solar minimum and solar maximum which is in agreement with previous studies (WMO, 2010). At Izaña, the solar contribution is found negative in the 23–32 km layer, which seems doubtful. Again, this concerned one of the shortest time series of the study (1999–2012) and could be corrected with future measurements.

4.2.4 Upper stratospheric (UppS) columns

The trends in the upper stratospheric layer are all positive in these latitudes but significant only at Lauder ($+2.8 \pm 2.4$ % decade^{-1}). Our trend at Jungfraujoch station ($+0.9 \pm 1.0$ % decade^{-1}) corresponds well to the observed trend ($+1.5$ % decade^{-1}) at OHP in Nair et al. (2013) in the 31–39 km range, although it is found significant in this latter study. The MLR model explains 93 % of the variability at Jungfraujoch ($R^2 = 0.93$), namely, 77 % of the variability comes from the seasonality and the remaining 16 % from the proxies, mainly the ELU and QBO (see Fig. 4). At Lauder, the trend in the 30–40 km range from lidar measurements is also found significantly positive for the period 2000–2012 with trend values (+2–3 % decade^{-1}) similar to FTIR (W. Steinbrecht, personal communication, 2013). If we remove the solar cycle signal in the MLR for the short time series of Lauder, the trend becomes smaller and non-significant ($+1.7 \pm 2.4$ % decade^{-1}). More years of data will improve the confidence in the solar cycle signal in the short time series.

4.2.5 Total columns

The total column trends are non-significant at the mid-latitude stations (-0.4 ± 1.2 % decade^{-1} or -1.4 ± 3.8 DU decade^{-1} at Jungfraujoch; -0.3 ± 1.8 % decade^{-1} or -1.1 ± 5.9 DU decade^{-1} at Lauder), non-significant at Izaña ($+0.5 \pm 1.2$ % decade^{-1} or $+1.4 \pm 3.6$ DU decade^{-1}), and significantly positive at Wollongong ($+1.9 \pm 1.1$ % decade^{-1} or $+5.8 \pm 3.5$ DU decade^{-1}). The total column trend at Jungfraujoch is in agreement within error bars with the result of Nair et al. (2013) at OHP when they use the PWLT method (-5.5 ± 3.3 DU decade^{-1}), but again the trend at OHP is found significantly positive. When the EESC proxy is used in their study a trend of $+4.2 \pm 0.8$ DU decade^{-1} is found. The same behavior is seen more globally in a recent study using merged satellite data from 1979 to 2012 (Chehade et al., 2014): for the latitude of Jungfraujoch, the trends are about +3–4 DU decade^{-1} for the 1997–2012 period and non-significant if the PWLT method is used, while significant when the EESC proxy is used, which decreases the uncertainty on the trends. It seems that at Jungfraujoch, our time series is still too short to observe this positive trend. At the latitude of Izaña, the merged satellite data set shows a +3–4 DU decade^{-1} for the 1997–2012 period, with the more recent SBUV/SBUV-2 MOD v8.6, non-significant using the PWLT (in agreement with our study) and significant using the EESC proxy. Since our time series start at best in 1995, the EESC proxy is not really "separable" from a linear trend study at our mid-latitude and subtropical stations. When more years of data are available, the same sensitivity study (PWLT vs. EESC) could be tested at least for polar stations where the turnaround point is expected around 2000.

It is also interesting to note that, using the PWLT method, at the latitude of Wollongong, Chehade et al. (2014) found a positive significant trend of about +3 DU decade^{-1}, while at the latitude of Lauder the trend is decreased to about +1 DU decade^{-1} (non-significant) in good agreement with what FTIR observed. When they use the EESC proxy, the trend is increasing with latitude so that at the Lauder latitude it reaches about 4–5 DU decade^{-1}.

Our non-significant trends at Jungfraujoch, Izaña and Lauder, and positive trend at Wollongong are also in agreement with the recent study of Coldewey-Egbers et al. (2014), which provides trends using a similar period (1995–2013) of merged satellite data sets. For Wollongong, since the total column positive trend is due to the ozone trends in the lower and middle stratosphere, it cannot be attributed unambiguously to the EESCs decline.

5 Conclusions

We have exploited the time series of ozone total and partial columns (Trop, LowS, MidS, UppS) at eight NDACC FTIR stations (Ny-Ålesund, 79° N; Thule, 77° N; Kiruna, 68° N;

Harestua, 60° N; Jungfraujoch, 47° N; Izaña, 28° N; Wollongong, 34° S; Lauder, 45° S) to derive vertically resolved trends, using a MLR model including the main proxies well-known for impacting the ozone variability.

After the seasonal variation, the TP proxy is the dominant driver of ozone variability at all stations, mainly for the troposphere, lower stratosphere and total columns, while the EL proxy is an important contributor to the middle and upper stratosphere, as well as to the total column variabilities. At the highest latitude stations (68–79° N), the EPF proxy contributes substantially to the middle stratospheric and total column variabilities. The VPSC proxy for polar ozone loss contributes to the lower stratosphere and total column variabilities at the Arctic stations but also at Jungfraujoch, while is it non-significant at the southern hemispheric station Lauder. At the mid-latitude and subtropical stations, the QBO proxy is a substantial contributor to ozone variability, especially at the lowest latitude station, Izaña. The AO/AAO and ENSO proxies are significant only at Jungfraujoch and Izaña, respectively. At Wollongong, the 2.5 % ozone response to solar cycle in the upper layer is in agreement with previous studies, but the response in the middle stratosphere (~ 6 %) is much larger than previously reported (~ 1 %). The 11-year solar cycle effect is still subject of debate (WMO, 2010; Chiodo et al., 2014), so that an additional decade of measurements would help in fixing its real impact on ozone. This is particularly true for our shortest time series of Lauder, Izaña and Thule.

The trends at the high latitude stations are negative in the troposphere, except at Kiruna where it is non-significant. Except at Thule, the high latitude stations show significant negative trends in the lower stratosphere. The situation is mixed in the middle stratosphere where the trend is significantly negative at Ny-Ålesund, non-significant at Thule and Kiruna, and significantly positive at Harestua. The trends of the three high latitude stations with a similar time period are all positive in the upper stratosphere, but this increase is taking place during the 1995–2003 period while the EESCs were still increasing until about 2000 in the polar region (WMO, 2010). However all four stations give non-significant trends in the upper stratosphere for the October 1999–2012 period, which could be the onset of the upper stratospheric ozone recovery at high latitude. The total column trends are non-significant at all high latitude stations, except at Ny-Ålesund where it is negative. This is in agreement (except at Ny-Ålesund) with model predictions that the Arctic March ozone recovery to 1980 levels will occur around 2026 (WMO, 2010). However, the high year-to-year total column variability at these latitudes, driven mainly by lower stratospheric variability due to the polar temperature variations, does not allow yet to draw conclusions from the current trends for Arctic total ozone in the coming few years.

The trends for mid-latitude and subtropical stations are all non-significant, except at Lauder in the troposphere and upper stratosphere and at Wollongong for the total columns and the lower and middle stratospheric columns. Some signs of the onset of ozone mid-latitude recovery are observed only in the Southern Hemisphere, while a few more years seem to be needed to observe it at the northern stations.

To conclude, among the numerous available satellite and ground-based data sets measuring vertical distributions of ozone that are useful for ozone trend evaluations (Hassler et al., 2014), the NDACC ground-based FTIR measurements have their particular assets. Indeed, several stations, well distributed around the globe, are now reaching almost 20 years of measurements and will continue measuring ozone in the future: to the eight stations of this work could be added, after homogenization of the retrieval analysis and/or few more years of data, Eureka (80°N), Rikubetsu (44°N), Bremen (53°N), Mauna Loa (20°N), and Arrival Heights (78°S). This provides long time series of ozone that are reliable over time, provided that the ILS is properly taken into account. This is also the only data set, with Umkehr measurements, that provides simultaneously total columns, tropospheric columns and three stratospheric columns that reach 40–45 km. This data set is suitable for an alternative determination of ozone vertical changes, as demonstrated in this study, but also for validation of the satellite-merged data sets and detection of possible drifts.

Acknowledgements. The National Center for Atmospheric Research is supported by the National Science Foundation. The observation program at Thule, Greenland, is supported under contract by the National Aeronautics and Space Administration and the site is also supported by the NSF Office of Polar Programs. We wish to thank the Danish Meteorological Institute for support at Thule. We would like to thank U. Raffalski and P. Völger for technical support at IRF Kiruna. The University of Liège team acknowledges the support of the F.R.S. – FNRS, the Fédération Wallonie-Bruxelles, the International Foundation High Altitude Research Stations Jungfraujoch and Gornergrat (HFSJG, Bern) and Meteoswiss. The NIWA Lauder MIR-FTIR project is core-funded through New Zealand's Ministry of Business, Innovation and Employment. We are grateful to the many colleagues having contributed to FTIR data acquisition at the various sites. We wish to thank S. Godin-Beekmann and M. Pastel (LATMOS) for useful discussions on equivalent latitude proxy, and R. Kivi (KMI) for discussion on ozonesonde trends in the Arctic. This study has been supported by the EU FP7 project NORS, the ESA PRODEX project A3C, as well as the AGACC-II project within the Science for a Sustainable Development research program funded by the Belgian Science Policy Office.

Edited by: M. Dameris

References

Appenzeller, C., Weiss, A. K., and Staehelin, J.: North Atlantic Oscillation modulates total ozone winter trends, Geophys. Res. Lett., 27, 1131–1134, 2000.

Barret, B., De Mazière, M., and Demoulin, P.: Retrieval and characterization of ozone profiles from solar infrared spectra at the Jungfraujoch, J. Geophys. Res., 107, 4788, doi:10.1029/2001JD001298, 2002.

Bodeker, G. E., Hassler, B., Young, P. J., and Portmann, R. W.: A vertically resolved, global, gap-free ozone database for assessing or constraining global climate model simulations, Earth Syst. Sci. Data, 5, 31–43, doi:10.5194/essd-5-31-2013, 2013.

Brasseur, G. and Solomon, S: Aeronomy of the Middle Atmosphere, 441 pp., D. Reidel Publishing Company, Dordrecht, Holland, 1984.

Brunner, D., Staehelin, J., Maeder, J. A., Wohltmann, I., and Bodeker, G. E.: Variability and trends in total and vertically resolved stratospheric ozone based on the CATO ozone data set, Atmos. Chem. Phys., 6, 4985–5008, doi:10.5194/acp-6-4985-2006, 2006.

Chandra, S., Ziemke, J. R., and Stewart, R. W.: An 11-year solar cycle in tropospheric ozone from TOMS measurements, Geophys. Res. Lett., 26, 185–188, 1999.

Chehade, W., Weber, M., and Burrows, J. P.: Total ozone trends and variability during 1979–2012 from merged data sets of various satellites, Atmos. Chem. Phys., 14, 7059–7074, doi:10.5194/acp-14-7059-2014, 2014.

Chiodo, G., Marsh, D. R., Garcia-Herrera, R., Calvo, N., and García, J. A.: On the detection of the solar signal in the tropical stratosphere, Atmos. Chem. Phys., 14, 5251–5269, doi:10.5194/acp-14-5251-2014, 2014.

Cochrane, D. and Orcutt, G. H.: Application of least squares regression to relationships containing auto-correlated error terms, J. Am. Stat. Assoc., 44, 32–61, 1949.

Coldewey-Egbers, M., Loyola, D. G., Braesicke, P., Dameris, M., van Roozendael, M., Lerot, C., and Zimmer, W.: A new health check of the ozone layer at global and regional scales, Geophys. Res. Lett., 41, 4363–4372, doi:10.1002/2014GL060212, 2014.

Dee, D. P., Uppala, S. M., Simmons, A. J., Berrisford, P., Poli, P., Kobayashi, S., Andrae, U., Balmaseda, M. A., Balsamo, G., Bauer, P., Bechtold, P., Beljaars, A. C. M., van de Berg, L., Bidlot, J., Bormann, N., Delsol, C., Dragani, R., Fuentes, M., Geer, A. J., Haimberger, L., Healy, S. B., Hersbach, H., Hólm, E. V., Isaksen, L., Kållberg, P., Köhler, M., Matricardi, M., McNally, A. P., Monge-Sanz, B. M., Morcrette, J.-J., Park, B.-K., Peubey, C., de Rosnay, P., Tavolato, C., Thépaut, J.-N., and Vitart, F.: The ERA-Interim reanalysis: configuration and performance of the data assimilation system, Q. J. Roy. Meteor. Soc., 137, 553–597, doi:10.1002/qj.828, 2011.

Frossard, L., Rieder, H. E., Ribatet, M., Staehelin, J., Maeder, J. A., Di Rocco, S., Davison, A. C., and Peter, T.: On the relationship between total ozone and atmospheric dynamics and chemistry at mid-latitudes – Part 1: Statistical models and spatial fingerprints of atmospheric dynamics and chemistry, Atmos. Chem. Phys., 13, 147–164, doi:10.5194/acp-13-147-2013, 2013.

García, O. E., Schneider, M., Redondas, A., González, Y., Hase, F., Blumenstock, T., and Sepúlveda, E.: Investigating the long-term evolution of subtropical ozone profiles applying ground-based FTIR spectrometry, Atmos. Meas. Tech., 5, 2917–2931, doi:10.5194/amt-5-2917-2012, 2012.

Garcia, R. R., Marsh, D. R., Kinnison, D. E., Boville, B. A., and Sassi, F.: Simulation of secular trends in the middle atmosphere, 1950–2003, J. Geophys. Res., 112, D09301, doi:10.1029/2006JD007485, 2007.

Hase, F., Blumenstock, T., and Paton-Walsh, C.:Analysis of the instrumental line shape of high-resolution Fourier transform IR spectrometers with gas cell measurements and new retrieval software, Appl. Optics, 38, 3417–3422, 1999.

Hase, F.: Inversion von Spurengasprofilen aus hochaufgelösten bodengebundenen FTIR-Messungen in Absorption, Dissertation, Wissenschaftliche Berichte Forschungszentrum Karlsruhe, FZKA 6512; ISSN 0947-8620, Forschungszentrum Karlsruhe, Karlsruhe, Germany, 2000.

Hase, F., Hannigan, J. W., Coffey, M. T., Goldman, A., Höpfner, M., Jones, N. B., Rinsland, C. P., and Wood, S. W.: Intercomparison of retrieval codes used for the analysis of high-resolution, ground-based FTIR measurements, J. Quant. Spectrosc. Ra., 87, 25–52, 2004.

Hassler, B., Petropavlovskikh, I., Staehelin, J., August, T., Bhartia, P. K., Clerbaux, C., Degenstein, D., Mazière, M. De, Dinelli, B. M., Dudhia, A., Dufour, G., Frith, S. M., Froidevaux, L., Godin-Beekmann, S., Granville, J., Harris, N. R. P., Hoppel, K., Hubert, D., Kasai, Y., Kurylo, M. J., Kyrölä, E., Lambert, J.-C., Levelt, P. F., McElroy, C. T., McPeters, R. D., Munro, R., Nakajima, H., Parrish, A., Raspollini, P., Remsberg, E. E., Rosenlof, K. H., Rozanov, A., Sano, T., Sasano, Y., Shiotani, M., Smit, H. G. J., Stiller, G., Tamminen, J., Tarasick, D. W., Urban, J., van der A, R. J., Veefkind, J. P., Vigouroux, C., von Clarmann, T., von Savigny, C., Walker, K. A., Weber, M., Wild, J., and Zawodny, J. M.: Past changes in the vertical distribution of ozone – Part 1: Measurement techniques, uncertainties and availability, Atmos. Meas. Tech., 7, 1395–1427, doi:10.5194/amt-7-1395-2014, 2014.

Kyrölä, E., Laine, M., Sofieva, V., Tamminen, J., Päivärinta, S.-M., Tukiainen, S., Zawodny, J., and Thomason, L.: Combined SAGE II–GOMOS ozone profile data set for 1984–2011 and trend analysis of the vertical distribution of ozone, Atmos. Chem. Phys., 13, 10645–10658, doi:10.5194/acp-13-10645-2013, 2013.

Logan, J. A., Staehelin, J., Megretskaia, I. A., Cammas, J.-P., Thouret, V., Claude, H., De Backer, H., Steinbacher, M., Scheel, H.-E., Stübi, R., Fröhlich, M., and Derwent, R.: Changes in ozone over Europe: Analysis of ozone measurements from sondes, regular aircraft (MOZAIC) and alpine surface sites, J. Geophys. Res., 117, D09301, doi:10.1029/2011JD016952, 2012.

Mikuteit, S.: Trendbestimmung stratosphärischer Spurengase mit Hilfe bodengebundener FTIR-Messungen, Dissertation, Forschungszentrum Karlsruhe, FZK Report No. 7385, Germany, 2008.

Nair, P. J., Godin-Beekmann, S., Kuttippurath, J., Ancellet, G., Goutail, F., Pazmiño, A., Froidevaux, L., Zawodny, J. M., Evans, R. D., Wang, H. J., Anderson, J., and Pastel, M.: Ozone trends derived from the total column and vertical profiles at a northern mid-latitude station, Atmos. Chem. Phys., 13, 10373–10384, doi:10.5194/acp-13-10373-2013, 2013.

Newchurch, M. J., Yang, E.-S., Cunnold, D. M.,Reinsel, G. C., Zawodny, J. M., and Russell III, J. M.: Evidence for slowdown in stratospheric ozone loss: First stage of ozone recovery, J. Geophys. Res., 108, 4507, doi:10.1029/2003JD003471, 2003.

Oltmans, S. J., Lefohn, A. S., Shadwick, D., Harris, J. M., Scheel, H. E., Galbally, I., Tarasick, D. W., Johnson, B. J., Brunke, E.-G., Claude, H., Zeng, G., Nichol, S., Schmidlin, F., Davies, J., Cuevas, E., Redondas, A., Naoe, H., Nakano, T., Kawasato, T.: Recent tropospheric ozone changes – a pattern dominated by slow or no growth, Atmos. Environ., 67, 331–351, 2013.

Pougatchev, N. S., Connor, B. J., and Rinsland, C. P.:Infrared measurements of the ozone vertical distribution above Kitt Peak, J. Geophys. Res., 100, 16689–16697, 1995.

Randel, W. J., and Wu, F.: A stratospheric ozone profile data set for 1979–2005: Variability, trends, and comparisons with column ozone data, J. Geophys. Res., 112, D06313, doi:10.1029/2006JD007339, 2007.

Randel, W. J., Garcia, R. R., Calvo, N., and Marsh, D.: ENSO influence on zonal mean temperature and ozone in the tropical lower stratosphere, Geophys. Res. Lett., 36, L15822, doi:10.1029/2009GL039343, 2009.

Rodgers, C. D.: Inverse methods for atmospheric sounding: Theory and Practice, Series on Atmospheric, Oceanic and Planetary Physics – Vol. 2, World Scientific Publishing Co., Singapore, 2000.

Rothman, L. S., Gordon, I. E., Barbe, A., Benner, D. C., Bernath, P. F., Birk, M., Boudon, V., Brown, L. R., Campargue, A., Champion, J.-P., Chance, K., Coudert, L. H., Danaj, V., Devi, V. M., Fally, S., Flaud, J.-M., Gamache, R. R., Goldmanm, A., Jacquemart, D., Kleiner, I., Lacome, N., Lafferty, W. J., Mandin, J.-Y., Massie, S. T., Mikhailenko, S. N., Miller, C. E., Moazzen-Ahmadi, N., Naumenko, O. V., Nikitin, A. V., Orphal, J., Perevalov, V. I., Perrin, A., Predoi-Cross, A., Rinsland, C. P., Rotger, M., Šimečková, M., Smith, M. A. H., Sung, K., Tashkun, S. A., Tennyson, J., Toth, R. A., Vandaele, A. C., and Vander Auwera, J.: The Hitran 2008 molecular spectroscopic database, J. Quant. Spectrosc. Ra., 110, 533–572, 2009.

Scherrer, B.: Biostatistique, Gaëtan Morin, Chicoutimi, 1984.

Schneider, M., Hase, F., and Blumenstock, T.: Ground-based remote sensing of HDO/H2O ratio profiles: introduction and validation of an innovative retrieval approach, Atmos. Chem. Phys., 6, 4705–4722, doi:10.5194/acp-6-4705-2006, 2006.

Sioris, C. E., McLinden, C. A., Fioletov, V. E., Adams, C., Zawodny, J. M., Bourassa, A. E., Roth, C. Z., and Degenstein, D. A.: Trend and variability in ozone in the tropical lower stratosphere over 2.5 solar cycles observed by SAGE II and OSIRIS, Atmos. Chem. Phys., 14, 3479–3496, doi:10.5194/acp-14-3479-2014, 2014.

Sussmann, R., Borsdorff, T., Rettinger, M., Camy-Peyret, C., Demoulin, P., Duchatelet, P., Mahieu, E., and Servais, C.: Technical Note: Harmonized retrieval of column-integrated atmospheric water vapor from the FTIR network – first examples for long-term records and station trends, Atmos. Chem. Phys., 9, 8987–8999, doi:10.5194/acp-9-8987-2009, 2009.

Tikhonov, A.: On the solution of incorrectly stated problems and a method of regularization, Dokl. Acad. Nauk SSSR, 151, 501–504, 1963.

Tourpali, K., Zerefos, C. S., Balis, D. S., and Bais, A. F.: The 11-year solar cycle in stratospheric ozone: comparison between Umkehr and SBUVv8 and effects on surface erythemal irradiance, J. Geophys. Res., 112, D12306, doi:10.1029/2006JD007760, 2007.

Vigouroux, C., De Mazière, M., Demoulin, P., Servais, C., Hase, F., Blumenstock, T., Kramer, I., Schneider, M., Mellqvist, J., Strandberg, A., Velazco, V., Notholt, J., Sussmann, R., Stremme, W., Rockmann, A., Gardiner, T., Coleman, M., and Woods, P.: Evaluation of tropospheric and stratospheric ozone trends over Western Europe from ground-based FTIR network observations, Atmos. Chem. Phys., 8, 6865–6886, 2008, http://www.atmos-chem-phys.net/8/6865/2008/.

Weber, M., Dikty, S., Burrows, J. P., Garny, H., Dameris, M., Kubin, A., Abalichin, J., and Langematz, U.: The Brewer-Dobson circulation and total ozone from seasonal to decadal time scales, Atmos. Chem. Phys., 11, 11221–11235, doi:10.5194/acp-11-11221-2011, 2011.

Wohltmann, I., Rex, M., Brunner, D., and Mäder: Integrated equivalent latitude as a proxy for dynamical changes in ozone column, Geophys. Res. Lett., 32, L09811, doi:10.1029/2005GL022497, 2005.

World Meteorological Organization: Atmospheric Ozone: 1985, Global Ozone Research and Monitoring Project – Report No. 16, Geneva, 1998.

World Meteorological Organization: Scientific Assessment of Ozone Depletion: 1998, Global Ozone Research and Monitoring Project – Report No. 44, Geneva, 1998.

World Meteorological Organization: Scientific Assessment of Ozone Depletion: 2010, Global Ozone Research and Monitoring Project – Report No. 52, Geneva, 2011.

4

Biomass burning related ozone damage on vegetation over the Amazon forest: a model sensitivity study

F. Pacifico[1], G. A. Folberth[2], S. Sitch[3], J. M. Haywood[1,2], L. V. Rizzo[4], F. F. Malavelle[1], and P. Artaxo[5]

[1]College of Engineering, Mathematics and Physical Sciences, University of Exeter, Exeter, UK
[2]Met Office Hadley Centre, Exeter, UK
[3]Geography, College of Life and Environmental Sciences, University of Exeter, Exeter, UK
[4]Department of Earth and Exact Sciences, Institute of Environmental, Chemical and Pharmaceutics Sciences, Federal University of Sao Paulo, Sao Paulo, Brazil
[5]Department of Applied Physics, Institute of Physics, University of Sao Paulo, Sao Paulo, Brazil

Correspondence to: F. Pacifico (f.m.pacifico@exeter.ac.uk)

Abstract. The HadGEM2 earth system climate model was used to assess the impact of biomass burning on surface ozone concentrations over the Amazon forest and its impact on vegetation, under present-day climate conditions. Here we consider biomass burning emissions from wildfires, deforestation fires, agricultural forest burning, and residential and commercial combustion. Simulated surface ozone concentration is evaluated against observations taken at two sites in the Brazilian Amazon forest for years 2010 to 2012. The model is able to reproduce the observed diurnal cycle of surface ozone mixing ratio at the two sites, but overestimates the magnitude of the monthly averaged hourly measurements by 5–15 ppb for each available month at one of the sites. We vary biomass burning emissions over South America by $\pm 20, 40, 60, 80$ and 100 % to quantify the modelled impact of biomass burning on surface ozone concentrations and ozone damage on vegetation productivity over the Amazon forest. We used the ozone damage scheme in the "high" sensitivity mode to give an upper limit for this effect. Decreasing South American biomass burning emissions by 100 % (i.e. to zero) reduces surface ozone concentrations (by about 15 ppb during the biomass burning season) and suggests a 15 % increase in monthly mean net primary productivity averaged over the Amazon forest, with local increases up to 60 %. The simulated impact of ozone damage from present-day biomass burning on vegetation productivity is about 230 $TgC\,yr^{-1}$. Taking into account that uncertainty in these estimates is substantial, this ozone damage impact over the Amazon forest is of the same order of magnitude as the release of carbon dioxide due to fire in South America; in effect it potentially doubles the impact of biomass burning on the carbon cycle.

1 Introduction

Biomass burning is a global source of aerosol and trace gases, including ozone (O_3) precursors, and can lead to local and regional O_3 pollution. Tropospheric O_3 is a greenhouse gas and, above background concentrations, an air pollutant: it is harmful to human health (e.g. Lippmann, 1993; Burnett et al., 1997) and it damages plants (e.g. Rich et al., 1964; Fiscus et al., 2005; Felzer et al., 2007; Ainsworth et al., 2012). Tropospheric O_3 is a product of photochemical reactions whose main precursors are nitrogen oxides (NO_x), carbon monoxide (CO), methane (CH_4) and volatile organic compounds (VOCs) (Seinfeld and Pandis, 1998). VOCs are particularly important in Amazonia because of the large natural biogenic and biomass burning emissions (Karl et al., 2007).

In the Amazon forest, biomass burning is mostly anthropogenic, and mainly occurs during the dry season (August to October). Biomass burning emissions drastically change the composition of the atmosphere, e.g. diurnal maximum mixing ratios of tropospheric O_3 vary from 12 ppb during the wet season to values as high as 100 ppb in the biomass burning affected dry season (Kirkman et al., 2002; Sigler et al., 2002; Artaxo et al., 2002, 2005; Rummel et al., 2007).

Surface O_3 mixing ratios over 40 ppb are known to produce visible leaf injury and damage to plants, reducing crop productivity and posing a threat to food security; nonetheless different climatic conditions (e.g. soil moisture and water stress) also play a role in determining leaf stomatal closure and hence there will be variable impacts of the same O_3 concentrations (Ashmore, 2005), e.g. tropical rainforest vegetation may be particularly sensitive to surface O_3, even at concentrations below 40 ppb (a threshold associated with extra-tropical vegetation), due to high stomatal conductances. Moreover, tropical vegetation evolved in low background O_3 concentrations and could be more sensitive to O_3. In leaves, cellular damage caused by O_3 not only reduces photosynthetic rates but also requires increased resource allocation to detoxify and repair leaves (Ainsworth et al., 2012). Ozone damage to vegetation reduces plant productivity, decreasing the amount of carbon absorbed by plants, and hence has an impact on climate via an indirect radiative forcing (Sitch et al., 2007).

Tropical rainforests play an important role in the global carbon budget, as they cover 12 % of the earth's land surface and contain around 40 % of the terrestrial biosphere's carbon (Ometto et al., 2005; Taylor & Lloyd, 1992). It has been estimated that they may account for as much as 50 % of the global net primary productivity (Grace et al., 2001). Depending on age, land use and large-scale meteorological conditions, tropical forest ecosystems can act as net carbon sources or sinks, or they can be in approximate balance (Lloyd et al., 2007; Gatti et al., 2013), but it is uncertain if global environmental changes are forcing these ecosystems outside their range of natural variation (Sierra et al., 2007). However, biomass burning may further reduce natural sinks in the neighbouring intact forest, via air pollution and O_3 damage on vegetation, and thus currently the effects of biomass burning on the carbon cycle (Le Quéré et al., 2009) may be underestimated. Biomass burning is also an important aerosol source: regional levels of particulate matter are very high in the dry season in Amazonia (Artaxo et al., 2013), and the increase in diffuse radiation due to changes in aerosol loadings can increase net ecosystem exchange (NEE) quite significantly (Oliveira et al., 2007; Cirino et al., 2014). After a certain level of aerosol optical depth, the decrease in radiation fluxes can significantly reduce NEE over Amazonia (Cirino et al., 2014). This study does not consider the effects of the changes in diffuse radiation due to biomass burning on photosynthesis, or the impact of aerosols on O_3 chemistry via changing photolysis rate. That will be the focus of a separate study. Our specific aim is to estimate the effect of O_3-induced changes on vegetation productivity due to biomass burning.

Importantly, Sitch et al. (2007) performed their assessment of the potential impact of O_3 on vegetation using an offline simulation where monthly mean O_3 concentrations derived with a global chemistry climate model were used in determining the impacts of O_3 damage. Here we use an online flux-gradient approach to quantify the impact of biomass burning on surface O_3 concentration and O_3 damage on vegetation over the Amazon forest (see model description below). The HadGEM2 (Hadley Centre Global Environment Model 2; Collins et al., 2011; The HadGEM2 Development Team, 2011) earth system climate model is used to study these interactions. We show results of the evaluation of surface O_3 simulated with HadGEM2 against observations in the Amazon forest and model experiments quantifying the impact of biomass burning on plant productivity.

2 Methods

We used HadGEM2 to simulate surface O_3 concentrations and O_3 damage on vegetation for present-day (2001–2009) climate conditions. Our version of HadGEM2 includes the O_3 damage scheme developed by Sitch et al. (2007). We evaluated simulated surface O_3 against observations taken at two sites in the Amazon forest: Porto Velho (Brazil; 8.69° S; 63.87° W), a site heavily impacted by biomass burning emissions, and site ZF2 in the Cuieiras forest reserve in Central Amazonia (Brazil; 2.59° S; 60.21° W). A description of the sites can be found in Artaxo et al. (2013). In a sensitivity study we varied biomass burning emissions over South America by $\pm 20, 40, 60, 80, 100$ % to quantify the potential impact of biomass burning on surface O_3 concentrations and O_3 damage over the Amazon forest.

3 Model description

HadGEM2 is a fully coupled earth system model (ESM; Collins et al., 2011). It is built around the HadGEM2 atmosphere-ocean general circulation model and includes a number of earth system components: the ocean biosphere model diat-HadOCC (Diatom-Hadley Centre Ocean Carbon Cycle, a development of the HadOCC model of Palmer and Totterdell, 2001), the Top-down Representation of Interactive Foliage and Flora Including Dynamics (TRIFFID) dynamic global vegetation model (Cox, 2001), the land-surface and carbon cycle model MOSES2 (Met Office Surface Exchange Scheme; Cox et al., 1998, 1999; Essery et al., 2003), the interactive Biogenic Volatile Organic Compounds (iBVOC) emission model (Pacifico et al., 2012), the United Kingdom Chemistry and Aerosol (UKCA) model (O'Connor et al., 2014) and an interactive scheme of O_3 damage on vegetation (Sitch et al., 2007; Clark et al., 2011).

The configuration used here is a version of HadGEM2-UKCA with extended tropospheric chemistry (N96L38); the resolution is 1.25° latitude $\times$ 1.875° longitude ($\sim 200 \times 140$ km) at the equator with 38 vertical levels extending up to 39 km altitude. The land-based anthropogenic, biomass burning, and shipping emissions are taken from Lamarque et al. (2010), and represent a decadal (1997–2006) mean centred on the year 2000. The use of an emission pattern from 1997–2006 can lead to an overestimation of O_3 concentrations by the model, since the emissions vary from year to year and are expected to be lower in recent years due to the reduction in Amazonian deforestation via burning, consequently reducing the amount of O_3 precursors. HadGEM2 runs at a 30 min time step with the exception of global radiation, which is updated every 3 h and provides radiative fluxes between those time steps via interpolation. This configuration is described and evaluated in O'Connor et al. (2014) with the exception of the Extended Tropospheric Chemistry (ExtTC) that has been applied in this work. The ExtTC mechanism has been designed to represent the key species and reactions in the troposphere in as much detail as is necessary to simulate atmospheric composition–climate couplings and feedbacks while retaining the capability to conduct decade-long climate simulations. UKCA-ExtTC simulates the spatial distribution and evolution in time of 89 chemical species, 63 of which are model tracers. The model includes emissions from anthropogenic, biogenic, soil and wildfire sources for 17 species: $NO_x(=NO+NO_2)$, CH_4, CO, hydrogen (H_2), methanol, formaldehyde, acetaldehyde and higher aldehydes, acetone, methyl ethyl ketone, ethane (C_2H_6), propane (C_3H_8), butanes and higher alkanes, ethene (C_2H_4), propene (C_3H_6), isoprene, (mono)terpenes, and a lumped species representing aromatics (toluene + xylene) from anthropogenic sources.

Emissions of biogenic species (isoprene, terpenes, methanol, acetone) are computed by iBVOC and provided to UKCA at every time step. The isoprene emission scheme is that of Pacifico et al. (2011). Terpenes, methanol and acetone emissions are simulated with the model described in Guenther et al. (1995). Anthropogenic and wildfire emissions are prescribed from monthly mean emission data sets prepared for the Coupled Model Intercomparison Project 5 (CMIP5, Taylor et al., 2012) using the historic scenario (Lamarque et al., 2010). Given the difficulty in prescribing a diurnal cycle for fire emissions, these monthly mean emissions are kept constant during the day. Wetland methane emissions are prescribed from data from Gedney et al. (2004). Soil-biogenic NO_x emissions are prescribed using the monthly distributions provided by the Global Emissions Inventory Activity (http://www.geiacenter.org/inventories/present.html), which are based on the global empirical model of soil-biogenic NO_x emissions of Yienger and Levy (1995). NO_x emissions from global lightning activity are parameterized based on the convective cloud top height following Price and Rind (1992, 1994) and are thus sensitive to the model climate. UKCA also includes a dry deposition scheme based on the resistance in-series approach as outlined in Wesely (1989). Physical removal of soluble species is parameterized as a first-order loss process based on convective and stratiform rainfall rates (Collins et al., 2011).

The TRIFFID vegetation module of HadGEM2 simulates the dynamics of five plant functional types (PFTs): broadleaf trees, needleleaf trees, shrubs, and C_3 and C_4 grass (i.e. grasses using the C_3 and C_4 photosynthetic pathway, respectively). Changes in the extent of croplands over time are not simulated but are prescribed from land use maps prepared for CMIP5. Here we use the historic (1850–2000; Hurtt et al., 2009) data sets, as described in Jones et al. (2011). A further four surface types (urban, inland water, bare soil and ice) are used in the land-surface scheme for the calculation of water and energy exchanges between the land and the atmosphere. Each model grid box can include varying proportions of several vegetation and/or surface types. The model does not include interactive deforestation due to fire.

The parameterization of O_3 damage on vegetation is that of Sitch et al. (2007). This scheme uses a flux-gradient approach to model O_3 damage, rather than empirical approaches based on the accumulated O_3 exposure above 40 ppb (e.g. Felzer, et al., 2005). The Sitch et al. (2007) parameterization assumes a suppression of net leaf photosynthesis by O_3 that varies proportionally to the O_3 flux through stomata above a specified critical O_3 deposition flux. The critical deposition flux depends on O_3 concentration near the leaves, but also on stomatal conductance. This scheme also includes a relationship between stomatal conductance and photosynthesis, determining a reduction in stomatal conductance through O_3 deposition. As the O_3 flux itself depends on the stomatal conductance, which in turn depends on the net rate of photosynthesis, the model requires a consistent solution for the net photosynthesis, stomatal conductance and O_3 deposition flux. This approach to modelling O_3 effects on photosynthesis accounts for the complex interaction between CO_2 and O_3 effects, and can be used to study future climate impacts. This scheme includes a "high" and "low" parameterization for each PFT to represent species more sensitive and less sensitive to O_3 effects; in our analysis we use the "high" sensitivity mode to establish the maximum response. The model was calibrated with data from temperate and boreal vegetation. Calibration data for other ecosystems, including tropical vegetation, are currently unavailable.

4 Description of the model experiments

All simulations use HadGEM2 in its atmosphere-only configuration, i.e. with all implemented couplings between atmosphere and land surface (including carbon cycle) active but without the atmosphere–ocean coupling. HadGEM2 was initialized with equilibrium concentrations of the major chemical components (O_3, CO, H_2, total reactive nitrogen (NO_y), biogenic volatile organic compounds (BVOCs)) taken from the CMIP5 simulation (see description of the simulations in Jones et al., 2011). Methane mixing ratios were prescribed as specified by CMIP5, with values of 1750 ppb for the present day. The decade-mean CO_2 atmospheric mixing ratio was 368 ppm.

Monthly means of sea surface temperature and sea ice cover were prescribed using climatologies derived from the appropriate decade of the Hadley Centre CMIP5 transient climate run (Jones et al., 2011). The vegetation distribution for each of our simulations was prescribed using the simulated vegetation averaged for the same decade from this transient climate run, on which we superimposed crop area as given in the CMIP5 historic and future land use maps (Hurtt et al., 2009; Riahi et al., 2007).

We performed a 9-year (2001–2009) control simulation for present-day climate conditions initialized from a centennial transient climate simulation with ocean couplings (Jones et al., 2011). We analysed the last 8 years of the simulation, as the first year of simulation was used as spin-up. A single year is considered sufficient for spin-up because one year is around five times longer than the lifetime of the longest lived atmospheric species (with the exception of methane) involved in O_3 chemistry. The control simulation was driven by anthropogenic and wildfire emissions of trace gases and aerosols via historical scenarios (Global Fire Emissions Database GFEDv2; Lamarque et al., 2010; van der Werf et al., 2006) of anthropogenic and wildfire emissions.

HadGEM2 is able to reproduce the main spatial distribution of surface temperature (Fig. S1 in the Supplement) and precipitation (Fig. S2). Surface temperature simulated with HadGEM2 exhibits a bias in the region of up to 2 °C colder than in the observations over the Amazon forest. Simulated precipitation rate is in reasonable agreement with observations. The model is able to reproduce the main features of the seasonal cycle of precipitation, but tends to simulate less precipitation in September and November than the observations (Fig. S3).

Simulated HadGEM2 net primary productivity (NPP) is compared against a meta-analysis of field data from the Ecosystem Model Data Model Intercomparison (EMDI) project (Olson et al., 2001). Measurements from the 81 "class A" ("well documented and intensively studied") sites, representative of all major global biomes, are compared against our simulations. Traditionally, global vegetation models underestimate NPP in tropical ecosystems, and tend towards an asymptote of $\sim 1000\,\mathrm{g\,C\,m^{-2}}$ (Prentice et al., 2007). HadGEM2 is able to reproduce the main geographical variations of NPP globally (Fig. S4), especially in the Northern Hemisphere, where more plentiful observations are available. In addition HadGEM2 is able to better simulate higher tropical NPP.

Ozone concentration simulated with HadGEM2–UKCA–ExtTC agrees better with observations at higher altitudes and higher latitudes (Fig. S5). The model performs more poorly than the ACCENT mean over tropical areas, especially closer to the surface. Comparison with a selection of observed profiles of O_3 concentration shows that the model overestimates O_3 for some locations but is in extremely good agreement for others. Over the tropics the agreement is better in the few continental profiles than in the marine environment (Fig. S6). Some differences may be expected given that the observations are from campaigns with specific meteorological conditions, while the model simulations represent a multi-year mean from the model. Comparison with a selection of surface O_3 observations (Fig. S7) confirms again how the model shows a better agreement with observations taken at higher latitudes.

We also perform 10 experiments that differ from the control simulation in terms of assumed biomass burning emissions, i.e. biomass burning emissions over South America are either increased or decreased by $\pm 20, 40, 60, 80, 100\,\%$, while emissions over the rest of the world are kept unchanged. The vegetation distribution was not adjusted for loss of vegetation due to fire. We define biomass burning emissions as those from wildfires, deforestation fires, agricultural forest burning, and residential and commercial combustion, including fuel wood burning, charcoal production and biofuel combustion for cooking and heating (Lamarque et al., 2010). The dominant fire types in South America are from deforestation and degradation fires in an arc around Amazonia, with some regional hotspots of agricultural burning (see Fig. 13 in van der Werf et al., 2010). Between 2001 and 2009 the percentage contribution to annual fire emissions from fire types (deforestation and degradation, grassland and savanna, woodland, forest, agriculture) are (59, 22, 10, 8, 2 %) over Southern Hemisphere South America (Fig. 13 van der Werf et al., 2010), with minor differences in this region between this data set (Global Fire Emissions Database GFEFv3) and the earlier GFEDv2 used in this study (see Fig. 16 in van der Werf et al., 2010). The residential and commercial combustion contribution accounts for 1 and 8 % of the total annual biomass burning emissions of CO and NO_x respectively.

This set of experiments allows us to simulate the impact of biomass burning on surface O_3 and vegetation productivity. The control simulation was also used to evaluate surface O_3 mixing ratios against measurements over the Amazon forest.

5 Model site-level evaluation

Over the data-sparse Amazonian region, comprehensive spatial data sets of surface O_3 concentration are extremely limited. We evaluated simulated surface O_3 against observations from two sites that have full annual analyses of O_3 concentration: Porto Velho (Brazil; 8.69° S; 63.87° W) and site ZF2 in the Cuieiras forest reserve (Brazil; 2.59° S; 60.21° W). O_3 mixing ratios were measured with a UV absorption analyser (Thermo 49i, USA). Observations from both sites have an estimated 4 % uncertainty, considering zero noise, zero and span drifts reported in the instrument manual, and the frequency of zero and span checks performed along the experiments.

The Porto Velho sampling site is located in a forest reserve about 5 km NE (generally upwind) from the city of Porto Velho. Large land use change and regional biomass burning makes its atmospheric conditions characteristic of those of the Amazon forest with significant human interference (Brito

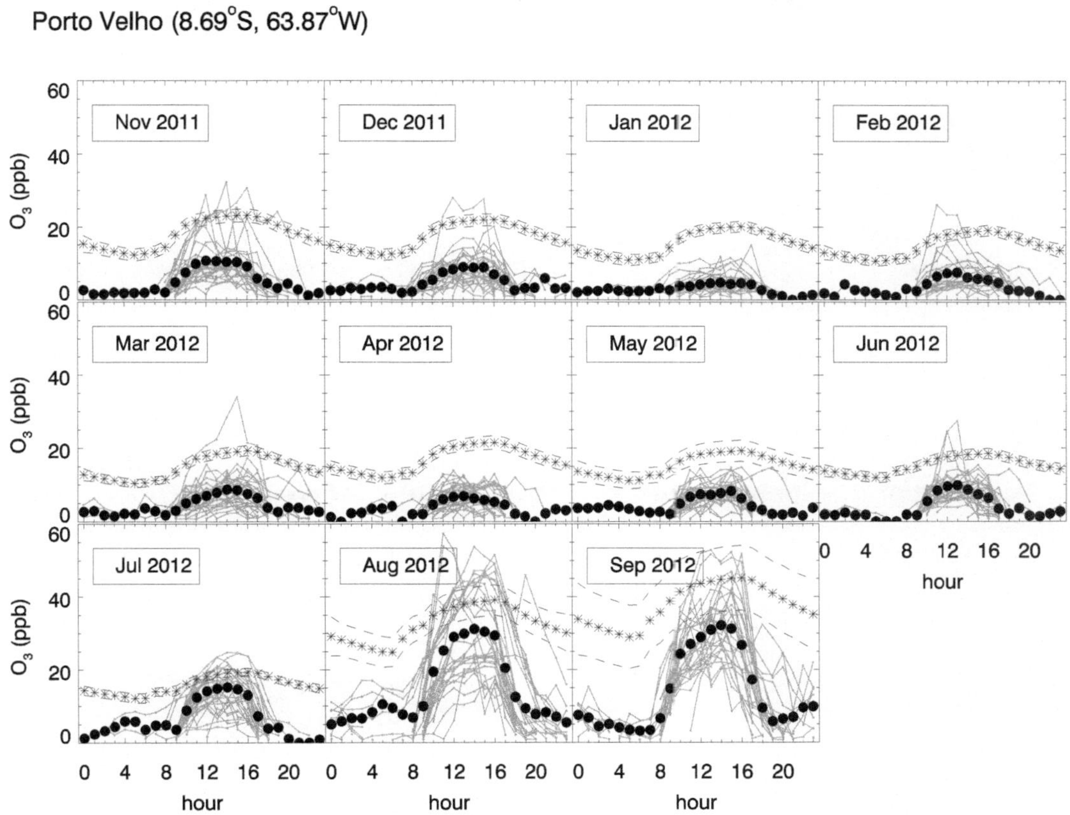

Figure 1. Comparison of measured (dots) and simulated (stars) monthly averaged diurnal cycle of surface O_3 mixing ratios at the Porto Velho site, including measured day-to-day variability (grey lines) and standard deviation (dashed lines) for the model results. The measurements have an uncertainty of 4 %.

et al., 2014). The whole region of Porto Velho has been subject to land use change since the 1980s. In Porto Velho, the dry season is from June to October and the wet season from November to May. Measurements of surface O_3 mixing ratios were taken from November 2011 to October 2012 in a forest clearance, at 5 m a.g.l.

The Cuieiras forest reserve in Central Amazonia encloses 380 km^2 of pristine tropical rainforest. The reserve is located in the central Amazon Basin, 60 km NNW of downtown Manaus and 40 km from the metropolis margins. This site is relatively undisturbed, as no biomass burning occurs in the forest reserve. Here rain showers are frequent with a short dry season from July to October. Measurements were taken at 39 m a.g.l. at the TT34 tower. The forest canopy height near the tower varied between 30 and 35 m, and the site is described in Martin et al. (2010), Rizzo et al. (2013) and Artaxo et al. (2013). Most of the time, the prevailing trade winds blow over 2000 km of the intact tropical forest before reaching the measurement tower. However, the site was also affected by regional transport of pollutants, from either biomass burning or urban sources (Rizzo et al., 2013). Measurements of surface O_3 mixing ratios were taken from April 2010 to June 2012, with the exception of a few months due to instrument maintenance.

We compared simulated (averaged over 8 years of simulations) against observed average O_3 diurnal cycles at each site for each available month. The model overestimates observed monthly averaged hourly O_3 mixing ratios at the surface by about 5–15 ppb for all months at the Porto Velho site, but it reproduces the diurnal and seasonal cycle, including the months affected by biomass burning, i.e. August and September, at the Porto Velho site (Fig. 1). The model is able to reproduce the diurnal cycle, including magnitude, at the ZF2 site for about 8 months out of 24. The model overestimates surface monthly averaged hourly O_3 mixing ratios by about 5–10 ppb for the rest of the months, which are also the months with lower surface O_3 mixing ratios (Fig. 2).

6 Results

Our analysis is focused on the region enclosed in the red rectangle in Fig. 3: this is a highly vegetated region with homogeneous topography, and it includes the two sites used for the model evaluation (Porto Velho and ZF2 in the Cuieiras forest reserve). This region of analysis is covered by two PFTs in HadGEM2: broadleaf trees, which is the predominant one, and C_3 grass (Fig. 3).

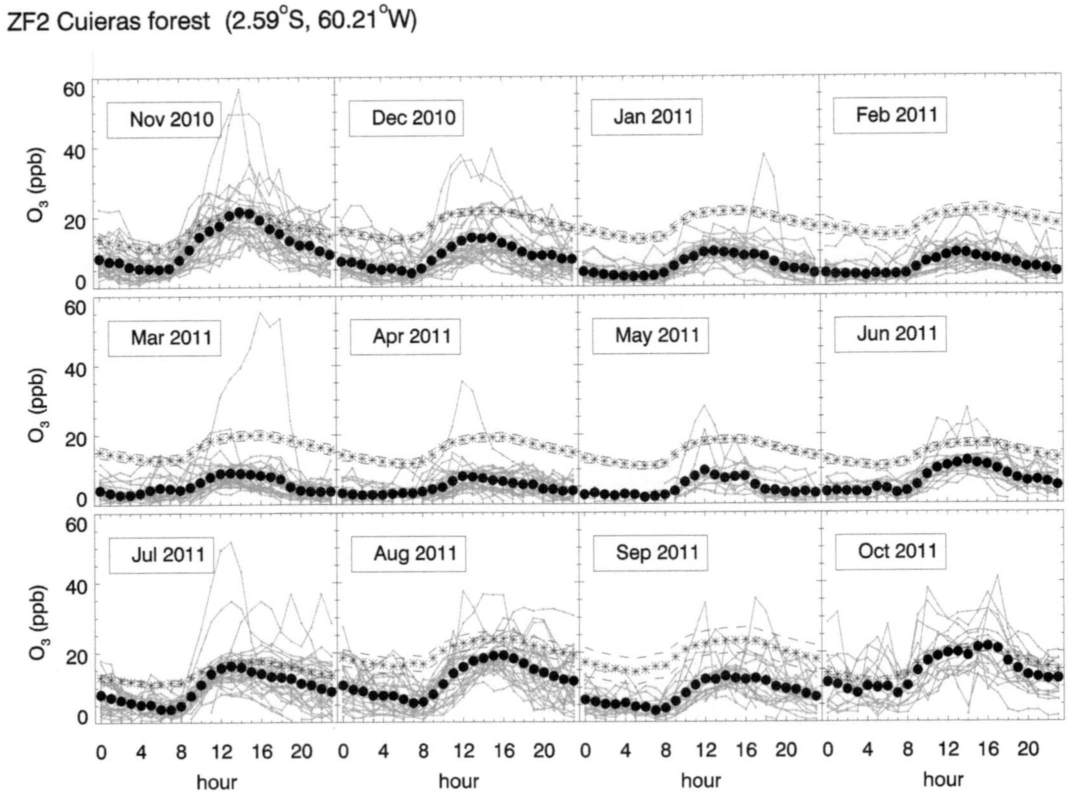

Figure 2. Comparison of measured (dots) and simulated (stars) monthly averaged diurnal cycle of surface O_3 mixing ratios at the ZF2 site in the Cuieiras forest reserve, including measured day-to-day variability (grey lines) and standard deviation (dashed lines) for the model results. The measurements have an uncertainty of 4 %. We show one of the two available years of observations.

Surface O_3 mixing ratios simulated with HadGEM2 are higher during the months of August, September and October over the Amazon forest, and in particular over our region of analysis, because of the higher biomass burning emissions in the model during these months. Monthly average surface O_3 mixing ratios in our control simulation peak at 55–60 ppb in this region (Fig. 4), while the average over the region of analysis peaks at about 30 ppb in August and September, less in October (Fig. 5a, black line).

Monthly total NPP in our control simulation reaches its minimum during the months of August and September (Fig. 5b, black line), at about 300 TgC month^{-1}, corresponding to the end of the dry season.

Decreasing biomass burning emissions over South America by −20, −40, −60, −80, −100 % decreases surface O_3 mixing ratios and increases net productivity. Conversely, increasing biomass burning emissions over South America by +20, +40, +60, +80, +100 % increases surface O_3 mixing ratios over the region of analysis and subsequently reduces net productivity because of O_3 damage on vegetation (Fig. 5c).

These sensitivity tests suggest that decreasing biomass burning emissions by 100 % over South America brings monthly mean surface O_3 mixing ratios averaged over the region of analysis to about the observed 15 ppb for each month (Fig. 5a, dark blue line), even during the dry season, with no values over 35 ppb for any grid-cell (Fig. 6). Increasing biomass burning emissions by 100 % suggests that monthly mean mixing ratios of surface O_3 averaged over the region of analysis reach 40 ppb in August (Fig. 5a), with peaks of about 65–70 ppb in some grid-cells (Fig. 6, left). For both increases and decreases of between 20 and 80 % in South American biomass burning the model simulates almost linear changes in surface O_3 mixing ratios (Fig. 6; the figure shows increases and reductions by 40, 60 and 100 %).

Suppressing biomass burning emissions (i.e. decreasing biomass burning emissions by 100 %) over South America increases total NPP over the region of analysis by about 15 %, to about 350–370 TgC month^{-1}, with peak increases of 60 % for a few grid-cells, in August and September (Fig. 6b): this quantifies the impact of present-day biomass burning on vegetation productivity. When increasing biomass burning emissions over South America by 100 %, monthly total NPP over the region of analysis is reduced by about 10 %, i.e. to about 250 TgC month^{-1}, in August and September (Fig. 5b), with peak values of 50–60 % reductions for few grid-cells (Fig. 6c). For reductions by 20 to 80 % in South American biomass burning the model varies NPP almost linearly

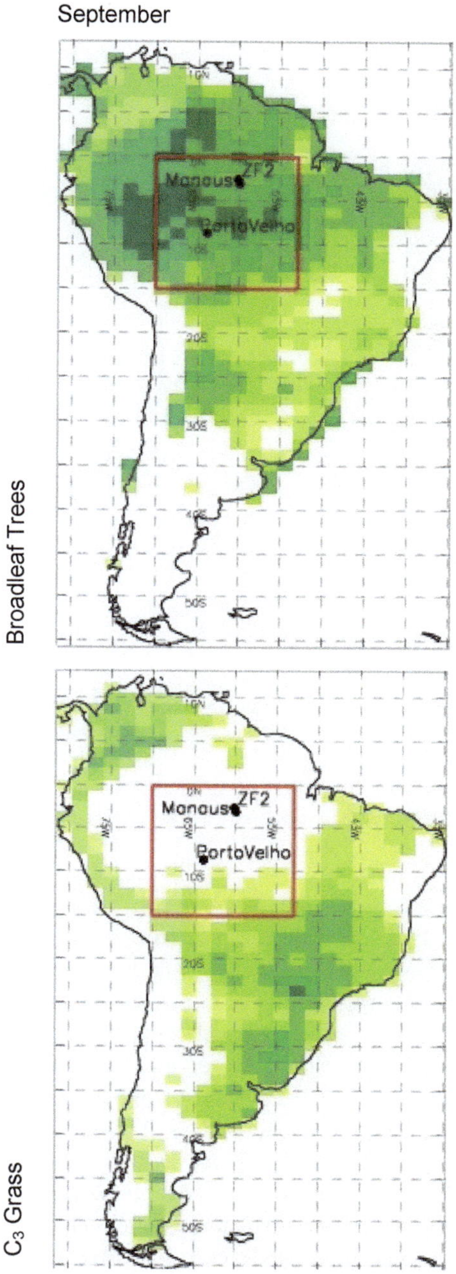

Figure 3. Vegetation cover in HadGEM2 for the month of September. The red rectangle is our region of analysis. The two sites used in the model evaluation (the sites of Porto Velho and ZF2 site in the Cueiras forest reserve) are also marked.

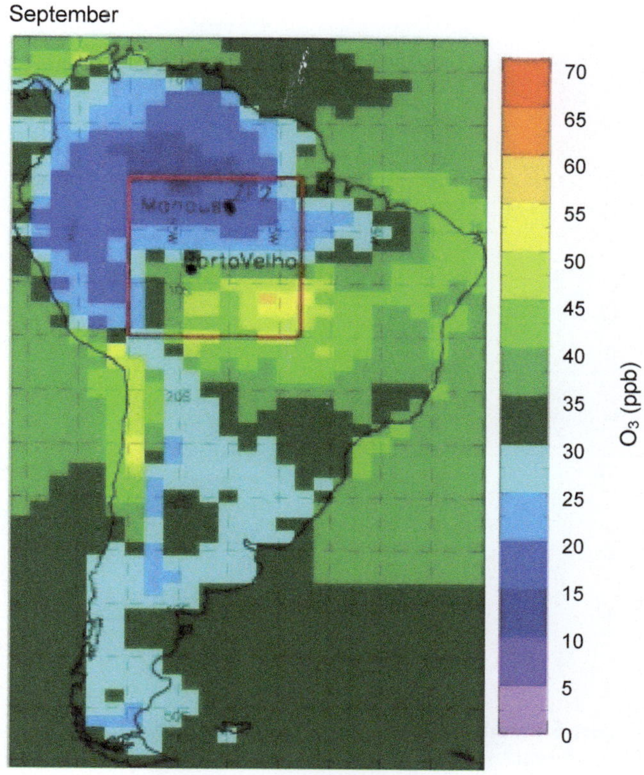

Figure 4. Monthly average surface O_3 mixing ratio simulated with HadGEM2 for the month of September (average over 8 years of simulations).

(Fig. 5c). However, the increase in South American biomass burning by 20 to 80 % determines a very similar decrease in NPP, e.g. between a 7 and 10 % decrease in August (Fig. 5c). Both increasing and reducing South American biomass burning between 20 and 80 % increases the number of grid-cells where a significant variation of NPP takes place (Fig. 6b). The percentages given above are significant against interannual variability in the control simulation, i.e. we only take account of the variations above one standard deviation in the control simulation. We also exclude from our analysis the grid-cells with low productivity, i.e. where NPP in the control simulation is below $50\,g\,C\,m^{-2}\,month^{-1}$ (i.e. we focus on high productivity regions, e.g. forests).

7 Discussion and conclusions

The HadGEM2 model overestimates the magnitude of the O_3 diurnal cycle at the two sites used in the evaluation. Overestimation of simulated O_3 in the Amazonian boundary layer has been observed in other modelling studies, especially in clean air conditions (Bela et al., 2015). Nonetheless, our model reproduces the main features of the diurnal and seasonal cycle. In particular, the model is able to reproduce the increase in surface O_3 during the biomass burning season.

As stated in the model description section, biomass burning emissions are prescribed as monthly mean and kept constant during the day, and this can have an impact on the hourly and day-to-day variation of surface O_3. For example, O_3 production will respond differently depending on whether biomass burning emissions occur during the day or at night, affecting simulated surface O_3 mixing ratios. These issues can be improved by modelling fire and biomass burn-

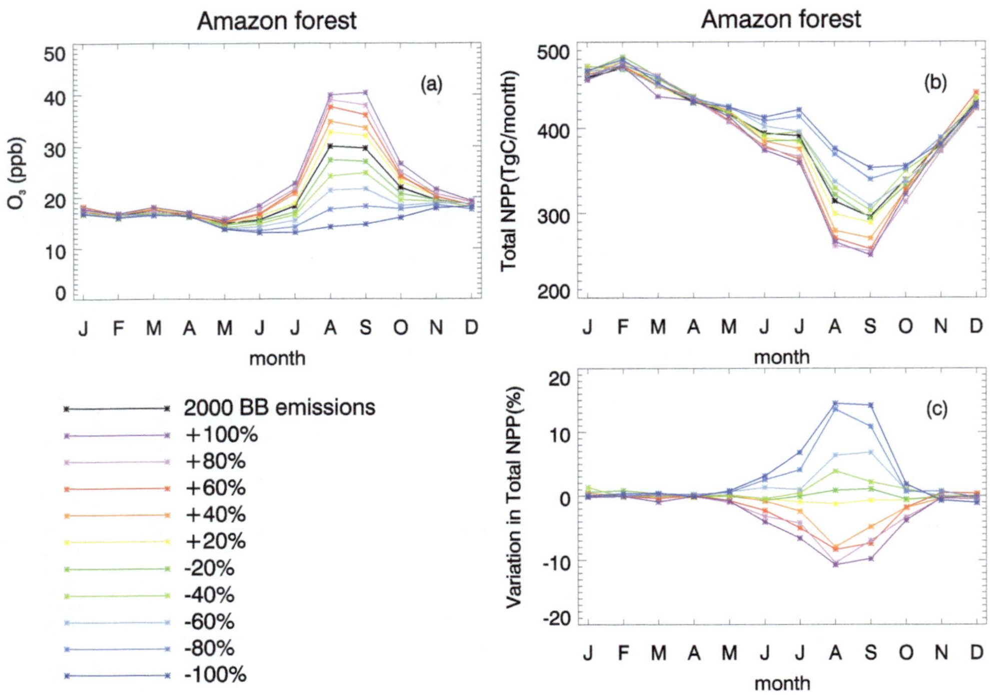

Figure 5. (a) Simulated monthly surface O_3 mixing ratios; **(b)** simulated monthly total NPP; **(c)** simulated monthly variation in total NPP. The plots show the results for the control simulation (i.e. using the decadal mean biomass burning emissions from Lamarque et al. (2010) centred on year 2000; 2000 biomass burning (BB) emissions) and the various experiments with increased (+) or decreased (−) biomass burning emissions over South America by 20, 40, 60, 80 and 100 %. All data are averaged over the region of analysis for 8 years of simulations.

ing emissions interactively. The inclusion of an interactive fire model in HadGEM is currently under development.

The model overestimates surface O_3 mixing ratios by 5–15 ppb for several months at the ZF2 site in the Cuieiras forest reserve and for all available months at the Porto Velho site. The reasons for these systematic biases in surface O_3 mixing ratio are likely manifold. In a complex, highly coupled system such as the HadGEM2 ESM it is not always easy to disentangle all processes and attribute model biases to specific components.

We attribute the systematic biases in the surface O_3 mixing ratio to the following most likely reasons:

1. Model resolution in both the horizontal and the vertical dimension

2. Uncertainties in emissions, in terms of magnitude, seasonality and location

3. Uncertainties in the O_3 dry deposition at the surface

Other factors such as uncertainties in the chemical mechanism, the photolysis rates, lightning NO_x production over the area and transport of O_3 and precursors will certainly contribute. We will briefly discuss the three most important (in our opinion) factors that contribute to the systematic biases.

The relatively coarse resolution of a global ESM simulates mixing ratios of trace species (both trace gases and aerosols) that represent averages over large areas. This issue has been discussed previously in the literature, mostly in relation to air quality modelling (see, e.g., Valari and Menut, 2008; Tie et al., 2010; Appel et al., 2011; Thompson and Selin, 2012). In our case one grid box equals approximately 30 000 km^2 (i.e. $200 \times 150\,km^2$ in longitude and latitude). The implicit averaging pertains to both emission and concentration fields; the predominant consequence is a dilution in each grid-cell. Depending on the chemical regime, this can lead to reduced or enhanced net O_3 production. Additionally, HadGEM2-ES has a relatively coarse vertical resolution. HadGEM2-ES has a lowest model layer depth of 40 m (global average) and the vertical profile of O_3 will undoubtedly show a gradient as the loss mechanism for O_3 is dominated by the surface (e.g. Colbeck and Harrison, 1967). The measurement level may explain part of the model overestimation, since it is well known that O_3 mixing ratios strongly decrease with height due to deposition within the canopy. The lowest layer of the model has a midpoint height 20 metres above the displacement height for the particular grid box (generally approximated as 2/3 of the average height of the obstacle, in this case the canopy), while measurements were taken at 5 m and 39 m a.g.l., respectively, at Porto Velho and ZF2 which are located either in or just above canopy level. Rummel et al. (2007) reports a 5–15 ppb O_3 decrease from 52 to 11 m a.g.l. in a forest site in Amazonia. This steep gradient near the surface is due to surface deposition but also due to in-canopy chemical pro-

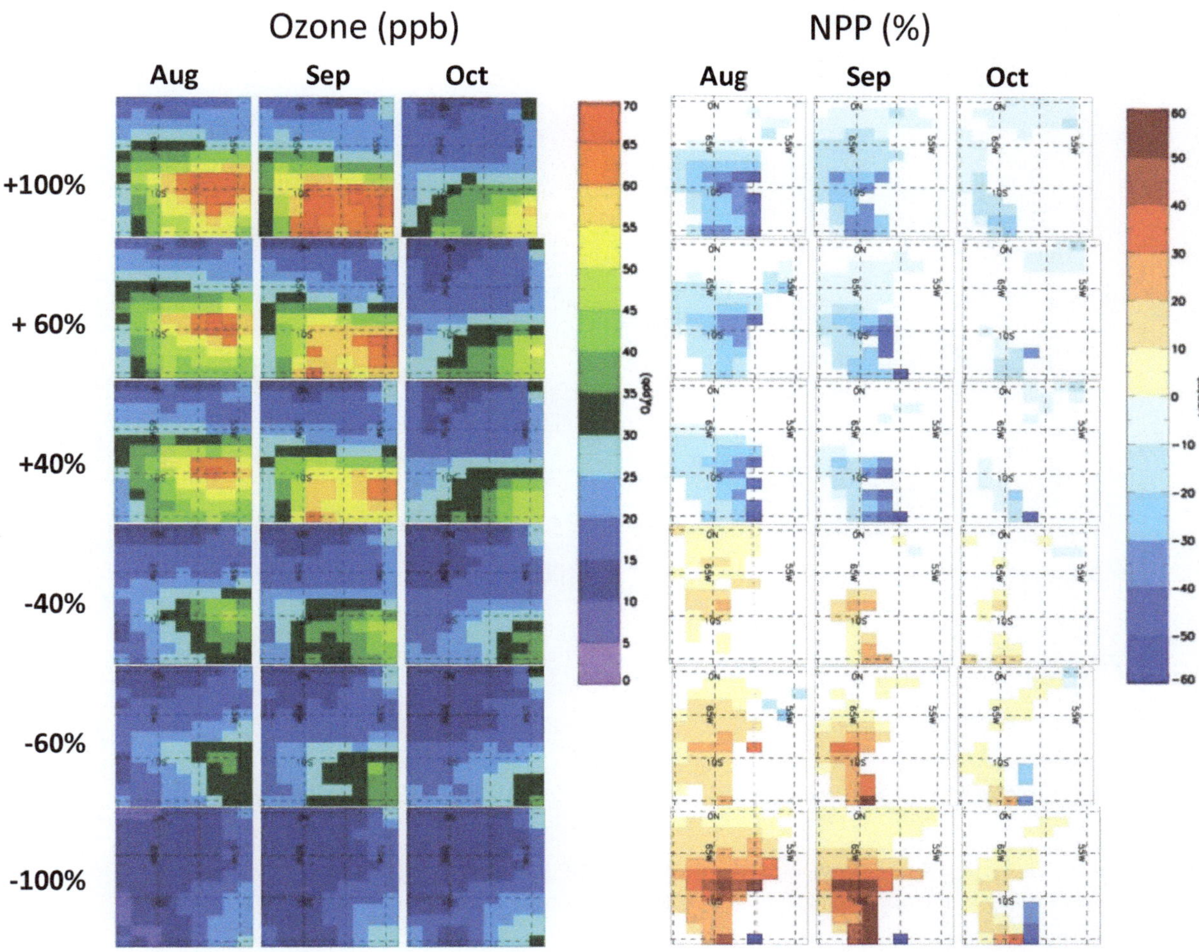

Figure 6. From the left: simulated variation in surface O_3 mixing ratios and NPP over the region of analysis for the months of August, September and October.

cessing (cf., e.g., Stroud et al., 2005; Gordon et al., 2014). The latter is not represented in HadGEM2-ES.

The remote environment of the Amazon forest is dominated by relatively high concentrations of VOCs, particularly of biogenic origin, and low concentrations of NO_x. It is an NO_x-limited environment. In such an environment O_3 is destroyed by reactions with BVOCs (mainly isoprene and (mono-)terpenes). This destruction is more pronounced the higher the BVOC concentration becomes. Consequently, conditions in the global model are likely to differ from that of a measurement at a specific point such as those we compare to in Figs. 1 and 2. It is a known problem in model evaluation.

Another issue related to model resolution, when comparing global models to point-like observations, is the uncertainty in global emission inventories with respect to both magnitude and location. In particular the latter will result in discrepancies between modelled concentrations of O_3 and its precursors and point-like observations. But the uncertainties in emission magnitude are also substantial and can reach a factor of 2 or more in the case of biogenic VOCs (e.g. Guenther et al., 2006; Arneth et al., 2008, 2011; Pacifico et al., 2011, 2012).

Third, and again related to model resolution, is the representation of O_3 dry deposition at the surface. Its magnitude and diurnal cycle will depend on boundary layer turbulence, surface roughness, land surface type, vegetation type, soil moisture, photosynthetic activity and more. In a recent sensitivity study by Folberth et al. (2015), O_3 surface concentrations showed the largest sensitivity to perturbations in O_3 surface dry deposition fluxes. Underestimating O_3 surface dry deposition, in particular during the night, preventing a complete flush of the planetary boundary layer with respect to O_3, will lead to systematic biases.

A comparison with Rummel et al. (2007) indicates that O_3 dry deposition velocities on average compare favourably with observations. Rummel at al. (2007) reported day-time velocities of up to $2.25\,cm\,s^{-1}$ and night-time velocities of typically around $0.5\,cm\,s^{-1}$ during the wet season; during the dry season reported day-time velocities are between $0.25\,cm\,s^{-1}$ and $1.0\,cm\,s^{-1}$, and night-time velocities be-

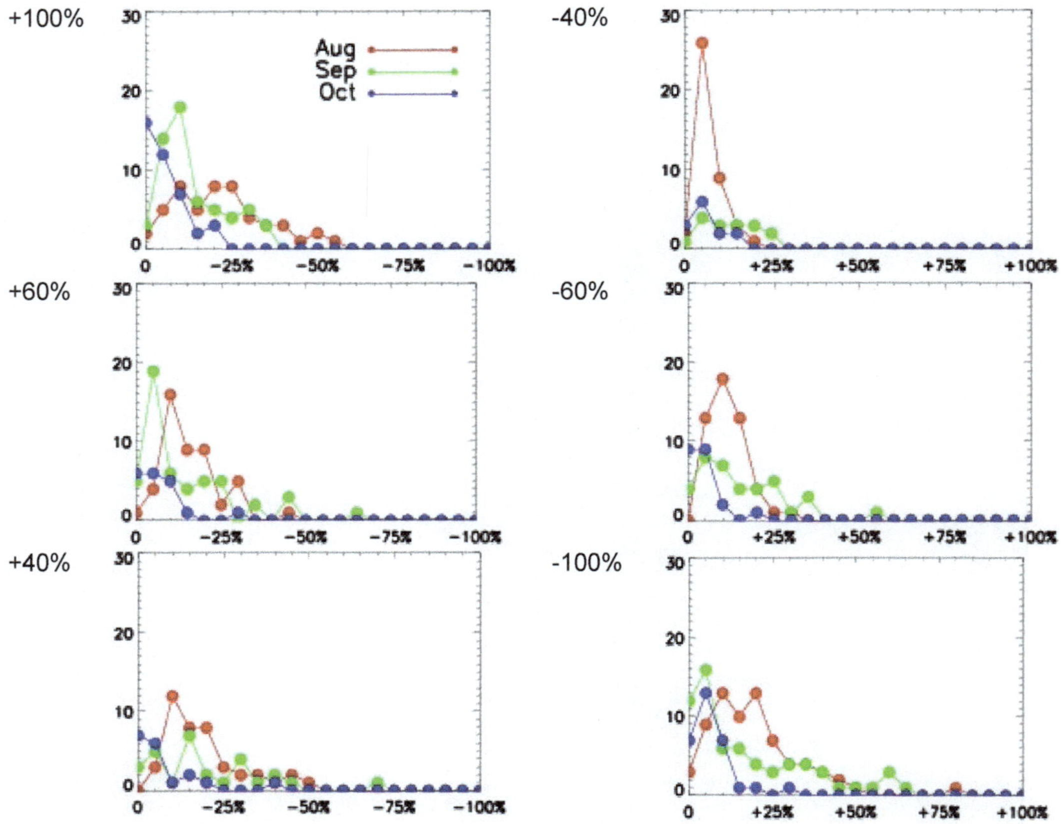

Figure 7. Probability density function (histogram) of the variation in NPP for the months of August, September and October. The plots show the variation between the experiments with South American biomass burning increased/decreased by 40, 60 and 100 % and the control simulation.

tween $0.2\,cm\,s^{-1}$ and $1.6\,cm\,s^{-1}$ for one site in the Amazon region. HadGEM2-ES predicts annual mean O_3 deposition velocities of 0.5 to $0.6\,cm\,s^{-1}$ (see Fig. S8), in fair agreement with the observations. Furthermore, the model is able to capture well the variability between the wet season and the dry season. More data are needed to conduct a robust evaluation, but this admittedly crude comparison is sufficient to demonstrate a basic capability of HadGEM2-ES to reproduce observed ozone deposition velocities in the Amazon region to a reasonable degree.

Interestingly, however, the latter process may also represent a redeeming feature of the model. According to our model of O_3 plant damage it is the total O_3 flux into the plant that determines the amount of damage caused to the photosynthetic activity and, hence, carbon assimilation. However, the total O_3 flux (or dose) is a function of both O_3 surface concentrations and dry deposition, i.e. for plants there is a compensation effect when concentrations are overestimated while deposition velocities are underestimated. Underestimating the O_3 dry deposition flux implies reduced O_3 plant uptake, and consequently an underestimation of the plant damage and productivity losses. However, it also leads to higher O_3 concentrations, which subsequently act to increase plant O_3 uptake and damage, compensating for the initial effects on productivity. Still, a detailed assessment and quantification of this interdependence of O_3 concentration and dry deposition fluxes is beyond the scope of this study and must be referred to future research.

August, September and October are the months when biomass burning and surface O_3 concentrations are higher over the Amazon forest, but also the months when plant productivity is at its lowest, which will tend to suppress the impact of O_3 damage on plant productivity. This is because stomatal conductance is reduced due to water limitations (also accounted for in the model) during the dry season, thus reducing the flux of both carbon dioxide and O_3 into the leaves, and consequently reducing O_3 plant damage.

Ashmore (2005) noted how O_3 exposure is poorly correlated with flux into leaves and also the potential for damagingly high O_3 fluxes in leaves at concentrations significantly below 40 ppb at maximum stomatal conductance. Consequently, global vegetation models as used in this study have adopted flux-based parameterizations to represent O_3 impacts on vegetation, moving away from application of the earlier exposure-based metrics, e.g. accumulated O_3 exposure above a threshold of 40 ppb, AOT40.

The parameterization of O_3 damage used in this study is calibrated for high-latitude vegetation. Unfortunately data for calibrating this O_3 damage scheme for tropical vegetation are currently not available and observations of O_3 damage in the Amazon forest are very limited. Observations of O_3 damage on tropical forests are urgently needed, including observations at moderate (e.g. 20–30 ppb) and high surface O_3 mixing ratios.

The simulated impact of present-day biomass burning on vegetation productivity over our area of analysis is about 230 TgC yr^{-1} (i.e. the difference between the dark blue line and the black line in Fig. 5b) using the "high" sensitivity mode in the O_3 damage scheme. Taking into account that the uncertainty in these estimates is substantial, this O_3 damage impact over the Amazon forest is of the same order of magnitude as the release of CO_2 due to land fire in South America, as quantified in van der Werf et al., 2010 – 293 TgC yr^{-1} from Table 7 of that paper – in effect potentially doubling the impact of biomass burning on the CO_2 fluxes. This highlights the urgent need for more tropical data on plant O_3 damage to better constrain estimates.

Despite overestimating surface O_3 mixing ratios, our model simulates only a moderate reduction in NPP associated with elevated O_3 due to biomass burning emissions (Fig. 7). Given that our model systematically overestimates O_3 mixing ratio, assuming accurate dry deposition, and that we use our model in the high sensitivity mode, our simulations where we increase biomass burning emissions by 100 % suggest a maximum 10 % average reduction in monthly plant productivity, and peak reductions of 50–60 % in a few gridcells. This is because, despite the increase in biomass burning, monthly average surface O_3 mixing ratios do not exceed a moderate 40 ppb. Moreover, our model does not include deforestation due to fire, which would reduce vegetation cover when increasing biomass burning emissions in our sensitivity experiments, reducing NPP and BVOC emissions further. However, local and daily/hourly impact of O_3 damage on plant productivity can be higher.

Estimates of the magnitude of the reduction in plant productivity due to O_3 damage can be improved with additional field studies and improving the representation of tropospheric O_3 in ESMs (sources, chemistry and sinks). Nevertheless, considering these processes in a coupled system can provide an improvement in robustness of conclusions, as, for example, it can treat processes with a specific diurnal cycle, such as photosynthesis and surface O_3, interactively on a short timescale (e.g. half an hour in our model).

Acknowledgements. This work was funded by the Natural Environment Research Council (NERC) South AMerican Biomass Burning Analysis (SAMBBA) project grant code NE/J010057/1. The UK Met Office contribution to this project was funded by the DECC under the Hadley Centre Climate Programme contract (GA01101). The Brazilian contribution was funded by Fundacao de Amparo a Pesquisa do Estado de Sao Paulo (FAPESP, projects 08/58100-2 and 12/14437-9). We thank INPA (Instituto Nacional de Pesquisas da Amazonia) for the coordination work of the LBA Experiment. We thank USP technicians for the support on data sampling: Alcides Ribeiro, Ana Lucia Loureiro, Fernando Morais and Fabio Jorge.

Edited by: M. O. Andreae

References

Ainsworth, E. A., Yendrek, C. R., Sitch, S., Collins, W. J., and Emberson, L. D.: The Effects of Tropospheric Ozone on Net Primary Productivity and Implications for Climate Change, Annu. Rev. Plant Biol., 63, 637–661 , 2012.

Appel, K. W., Foley, K. M., Bash, J. O., Pinder, R. W., Dennis, R. L., Allen, D. J., and Pickering, K.: A multi-resolution assessment of the Community Multiscale Air Quality (CMAQ) model v4.7 wet deposition estimates for 2002–2006, Geosci. Model Dev., 4, 357–371, doi:10.5194/gmd-4-357-2011, 2011.

Arneth, A., Monson, R. K., Schurgers, G., Niinemets, Ü., and Palmer, P. I.: Why are estimates of global terrestrial isoprene emissions so similar (and why is this not so for monoterpenes)?, Atmos. Chem. Phys., 8, 4605–4620, doi:10.5194/acp-8-4605-2008, 2008.

Arneth, A., Schurgers, G., Lathiere, J., Duhl, T., Beerling, D. J., Hewitt, C. N., Martin, M., and Guenther, A.: Global terrestrial isoprene emission models: sensitivity to variability in climate and vegetation, Atmos. Chem. Phys., 11, 8037–8052, doi:10.5194/acp-11-8037-2011, 2011.

Artaxo, P. Martins, J. V., Yamasoe, M. A., Procópio, A. S., Pauliquevis, T. M., Andreae, M. O., Guyon, P., Gatti, L. V., and Leal., A. M. C.: Physical and chemical properties of aerosols in the wet and dry season in Rondônia, Amazonia, J. Geophys. Res., 107, 8081–8095, 2002.

Artaxo, P., Gatti, L V., Leal, A. M. C, Longo, K. M., de Freitas, S. R., Lara, L. L., Pauliquevis, T. M., Procópio, A. S., and Rizzo, L. V.: Atmospheric Chemistry in Amazonia: The forest and the biomass burning emissions controlling the composition of the Amazonian atmosphere, Acta Amazonica, 35, 185–196, 2005.

Artaxo, P., Rizzo, L. V., Brito, J. F., Barbosa, H. M. J., Arana, A., Sena, E. T., Cirino, G. G., Bastos, W., Martin, S. T., and Andreae, M. O.: Atmospheric aerosols in Amazonia and land use change: from natural biogenic to biomass burning conditions, Faraday Discuss., 165, 203–235, 2013.

Ashmore, M. R: Assessing the future global impacts of ozone on vegetation, Plant Cell Environ., 28, 949–964, 2005.

Bela, M. M., Longo, K. M., Freitas, S. R., Moreira, D. S., Beck, V., Wofsy, S. C., Gerbig, C., Wiedemann, K., Andreae, M. O., and Artaxo, P.: Ozone production and transport over the Amazon Basin during the dry-to-wet and wet-to-dry transition seasons, Atmos. Chem. Phys., 15, 757–782, doi:10.5194/acp-15-757-2015, 2015.

Brito, J., Rizzo, L. V., Morgan, W. T., Coe, H., Johnson, B., Haywood, J., Longo, K., Freitas, S., Andreae, M. O., and Artaxo, P.: Ground-based aerosol characterization during the South American Biomass Burning Analysis (SAMBBA) field experiment, Atmos. Chem. Phys., 14, 12069–12083, doi:10.5194/acp-14-12069-2014, 2014.

Burnett, R. T., Brook, J. R., Yung, W. T., Dales, R. E., and Krewski, D.: Association between Ozone and Hospitalization for Respiratory Diseases in 16 Canadian Cities, Environ. Res., 72, 24–31, 1997.

Cirino, G. G., Souza, R. A. F., Adams, D. K., and Artaxo, P.: The effect of atmospheric aerosol particles and clouds on net ecosystem exchange in the Amazon, Atmos. Chem. Phys., 14, 6523–6543, doi:10.5194/acp-14-6523-2014, 2014.

Clark, D. B., Mercado, L. M., Sitch, S., Jones, C. D., Gedney, N., Best, M. J., Pryor, M., Rooney, G. G., Essery, R. L. H., Blyth, E., Boucher, O., Harding, R. J., Huntingford, C., and Cox, P. M.: The Joint UK Land Environment Simulator (JULES), model description – Part 2: Carbon fluxes and vegetation dynamics, Geosci. Model Dev., 4, 701–722, doi:10.5194/gmd-4-701-2011, 2011.

Colbeck, I. and Harrison, R. M.: Dry deposition of ozone: some measurements of deposition velocity and of vertical profiles to 100 metres, Atm. Environ., 19, 11, 1807–1818, 1967

Collins, W. J., Bellouin, N., Doutriaux-Boucher, M., Gedney, N., Halloran, P., Hinton, T., Hughes, J., Jones, C. D., Joshi, M., Liddicoat, S., Martin, G., O'Connor, F., Rae, J., Senior, C., Sitch, S., Totterdell, I., Wiltshire, A., and Woodward, S.: Development and evaluation of an Earth-System model – HadGEM2, Geosci. Model Dev., 4, 1051–1075, doi:10.5194/gmd-4-1051-2011, 2011.

Cox, P. M., Huntingford, C., and Harding, R. J.: A canopy conductance and photosynthesis model for use in a GCM land surface scheme, J. Hydrol., 212, 79–94, 1998.

Cox, P. M., Betts, R. A., Bunton, C. B., Essery, R. L. H., Rowntree, P. R., and Smith, J.: The impact of new land surface physics on the GCM simulation of climate and climate sensitivity, Clim. Dyn., 15, 183–203, 1999.

Cox, P. M.: Description of the "TRIFFID" Dynamic Global Vegetation Model, Tech. Note 24, 17 pp., Met Off. Hadley Cent., Exeter, UK, 2001.

Emmons, L., Hauglustaine, D., Muller, J., Carroll, M., Brasseur, G., Brunner, D., Staehelin, J., Thouret, V., and Marenco, A.: Data composites of airborne observations of tropospheric ozone and its precursors, J. Geophys. Res., 105, 20497–20538, 2000.

Essery, R. L. H., Best, M. J., Betts, R. A., Cox, P. M., and Taylor, C. M.: Explicit representation of subgrid heterogeneity in a GCM Land Surface Scheme, J. Hydrometeorol., 4, 530–543, 2003.

Felzer, B., Reilly, J., Melillo, J., Kicklighter, D., Sarofim, M., Wang, C., Prinn, R., and Zhuang, Q.: Future effects of ozone on carbon sequestration and climate change policy using a global biogeochemical model, Clim. Change 73, 345–373, 2005.

Felzer, B. S., Cronin, T., Reilly, J. M., Melillo, J. M., and Wang, X.: Impacts of ozone on trees and crops, C. R. Geosci., 339, 784–798, 2007.

Fiscus, E. L., Booker, F. L., and Burkey, K. O.: Crop responses to ozone: uptake, modes of action, carbon assimilation and partitioning, Plant Cell Environ, 28, 997–1011, 2005.

Fishman, J., Hoell, J., Bendura, R., McNeil, R., and Kirchhoff, V.: NASA GTE TRACE A experiment (September October 1992): Overview, J. Geophys. Res., 101, 23865–23879, doi:10.1029/96JD00123, 1996.

Folberth, G. A., Abraham, N. L., Dalvi, M., Johnson, C. E., Morgenstern, O., O'Connor, F. M., Pacifico, F., Young, P. A., Collins, W. J., and Pyle, J. A.: Evaluation of the new UKCA climate-composition model. Part IV. Extension to Tropospheric Chemistry and Biogeochemical Coupling between Atmosphere and Biosphere, to be summitted to Geosci. Model Dev., 2015.

Gatti, L.V., Gloor, M., Miller, J. B., Doughty, C. E., Malhi Y., Domingues, L. G., Basso, L. S., Martinewski, A., Correia, C. S., Borges, V. F., Freitas, S., Braz, R., Anderson, L. O., Rocha, H., Grace, J., Phillips, O. L., and Lloyd, J.: Drought sensitivity of Amazonian carbon balance revealed by atmospheric measurements, Nature, 506, 76–80, 2014.

Gedney, N., Cox, P. M., and Huntingford, C.: Climate feedback from wetland methane emissions, Geophys. Res. Lett., 31, L20503, doi:10.1029/ 2004GL020919, 2004.

Gordon, M., Vlasenko, A., Staebler, R. M., Stroud, C., Makar, P. A., Liggio, J., Li, S.-M., and Brown, S.: Uptake and emission of VOCs near ground level below a mixed forest at Borden, Ontario, Atmos. Chem. Phys., 14, 9087–9097, doi:10.5194/acp-14-9087-2014, 2014.

Grace, J., Mahli, Y., Higuchi, N., and Meir, P.: Productivity and carbon fluxes of tropical rain forest, in: Terrestrial Global Productivity, edited by: Roy, J., Mooney, H. A., and Saugier, B., Academic Press, San Diego, 401–426 , 2001.

Guenther, A., Hewitt, C. N., Erickson, D., Fall, R., Geron, C., Graedel, T., Harley, P., Klinger, L., Lerdau, M., Mckay, W. A., Pierce, T., Scholes, B., Steinbrecher, R., Tallamraju, R., Taylor, J., and Zimmerman, P.: A global model of natural volatile organic compound emissions, J. Geophys. Res., 100, 8873–8892, 1995.

Guenther, A., Karl, T., Harley, P., Wiedinmyer, C., Palmer, P. I., and Geron, C.: Estimates of global terrestrial isoprene emissions using MEGAN (Model of Emissions of Gases and Aerosols from Nature), Atmos. Chem. Phys., 6, 3181–3210, doi:10.5194/acp-6-3181-2006, 2006.

Harris, I., Jones, P. D., Osborn, T. J., Lister, D. H.: Updated high-resolution grids of monthly climatic observations – the CRU TS3.10 Dataset, Int. J. Climatol., 34, 632–642, 2014.

Hurtt, G. C., Chini, L. P., Frolking, S., Betts, R., Feedema, J., Fischer, G., Klein Goldewijk, K., Hibbard, K., Janetos, A., Jones, C., Kindermann, G., Kinoshita, T., Riahi, K., Shevliakova, E., Smith, S., Stehfest, E., Thomson, A., Thorton, P., van Vuuren, D., and Wang, Y.: Harmonization of global land-use scenarios for the period 1500–2100 for IPCC-AR5, iLEAPS Newsl., 7, 6–8, 2009.

Jones, C. D., Hughes, J. K., Bellouin, N., Hardiman, S. C., Jones, G. S., Knight, J., Liddicoat, S., O'Connor, F. M., Andres, R. J., Bell, C., Boo, K.-O., Bozzo, A., Butchart, N., Cadule, P., Corbin, K. D., Doutriaux-Boucher, M., Friedlingstein, P., Gornall, J., Gray, L., Halloran, P. R., Hurtt, G., Ingram, W. J., Lamarque, J.-F., Law, R. M., Meinshausen, M., Osprey, S., Palin, E. J., Parsons Chini, L., Raddatz, T., Sanderson, M. G., Sellar, A. A., Schurer, A., Valdes, P., Wood, N., Woodward, S., Yoshioka, M., and Zerroukat, M.: The HadGEM2-ES implementation of CMIP5 centennial simulations, Geosci. Model Dev., 4, 543-570, doi:10.5194/gmd-4-543-2011, 2011.

Karl, T., Yokelson, R., Guenther, A., Greenberg, J., Blake, D., and Artaxo, P.: TROFFEE (TROpical Forest and Fire Emissions Experiment): Investigating Emission, Chemistry, and Transport of Biogenic Volatile Organic Compounds in the Lower Atmosphere over Amazonia, J. Geophys. Res., 112, D18302, doi:10.1029/2007JD008539, 2007.

Kirkman, G. A., Gut, A., Ammann, C., Gatti, L. V, Cordova, A. M., Moura, M. A. L., and Meixner, F. X.: Surface exchange of nitric oxide, nitrogen dioxide, and ozone at a cattle pasture in Rondônia, Brazil, J. Geophys. Res., 107, 8083, doi:10.1029/2001JD000523, 2002.

Lamarque, J.-F., Bond, T. C., Eyring, V., Granier, C., Heil, A., Klimont, Z., Lee, D., Liousse, C., Mieville, A., Owen, B., Schultz, M. G., Shindell, D., Smith, S. J., Stehfest, E., Van Aardenne, J., Cooper, O. R., Kainuma, M., Mahowald, N., McConnell, J. R., Naik, V., Riahi, K., and van Vuuren, D. P.: Historical (1850–2000) gridded anthropogenic and biomass burning emissions of reactive gases and aerosols: methodology and application, Atmos. Chem. Phys., 10, 7017–7039, doi:10.5194/acp-10-7017-2010, 2010.

Le Quéré, C., Raupach, M. R., Canadell, J. G., Marland, G., Bopp, L., Ciais, P., Conway, T. J., Doney, S. C., Feely, R. A., Foster, P. N., Friedlingstein, P., Gurney, K., Houghton, R. A., House, J. I., Huntingford, C., Levy, P. E., Lomas, M. R., Majkut, J., Metzl, N., Ometto, J. P., Peters, G. P., Prentice, I. C., Randerson, J. T., Running, S. W., Sarmiento, J. L., Schuster, U., Sitch, S., Takahashi, T., Viovy, N., van der Werf, G. R., and Woodward, F. I.: Trends in the sources and sinks of carbon dioxide, Nature Geosciences, 2, 831–836 , 2009.

Lippmann, M.: Health effects of tropospheric ozone: review of recent research findings and their implications to ambient air quality standards, J. Exp. An. Environ. Epid., 3, 103–129, 1993.

Lloyd, J., Kolle, O., Fritsch, H., de Freitas, S. R., Silva Dias, M. A. F., Artaxo, P., Nobre, A. D., de Araújo, A. C., Kruijt, B., Sogacheva, L., Fisch, G., Thielmann, A., Kuhn, U., and Andreae, M. O.: An airborne regional carbon balance for Central Amazonia, Biogeosciences, 4, 759–768, doi:10.5194/bg-4-759-2007, 2007.

Logan, J., Megretskaia, I., Miller, A., Tiao, G., Choi, D., Zhang, L., Stolarski, R., Labow, G., Hollandsworth, S., Bodeker, G., Claude, H., De Muer, D., Kerr, J., Tarasick, D., Oltmans, S., Johnson, B., Schmidlin, F., Staehelin, J., Viatte, P., and Uchino, O.: Trends in the vertical distribution of ozone: A comparison of two analyses of ozonesonde data, J. Geophys. Res., 104, 26373–26399, doi:10.1029/1999JD900300, 1999.

Martin, S. T., Andreae, M. O., Althausen, D., Artaxo, P., Baars, H., Borrmann, S., Chen, Q., Farmer, D. K., Guenther, A., Gunthe, S. S., Jimenez, J. L., Karl, T., Longo, K., Manzi, A., Müller, T., Pauliquevis, T., Petters, M. D., Prenni, A. J., Pöschl, U., Rizzo, L. V., Schneider, J., Smith, J. N., Swietlicki, E., Tota, J., Wang, J., Wiedensohler, A., and Zorn, S. R.: An overview of the Amazonian Aerosol Characterization Experiment 2008 (AMAZE-08), Atmos. Chem. Phys., 10, 11415–11438, doi:10.5194/acp-10-11415-2010, 2010.

O'Connor, F. M., Johnson, C. E., Morgenstern, O., Abraham, N. L., Braesicke, P., Dalvi, M., Folberth, G. A., Sanderson, M. G., Telford, P. J., Voulgarakis, A., Young, P. J., Zeng, G., Collins, W. J., and Pyle, J. A.: Evaluation of the new UKCA climate-composition model – Part 2: The Troposphere, Geosci. Model Dev., 7, 41–91, doi:10.5194/gmd-7-41-2014, 2014.

Oliveira, P. H. F., Artaxo, P., Pires Jr, C., de Lucca, S., Procópio, A., Holben, B., Schafer, J., Cardoso, L. F., Wofsy, S. C., and Rocha, H. R.: The effects of biomass burning aerosols and clouds on the CO_2 flux in Amazonia, Tellus Series B-Chemical and Physical Meteorology, 59B, 338–349, 2007.

Olson, R. J., Scurlock, J. M. O., Prince, S. D., Zheng, D. L., and Johnson, K. R.(Eds.): NPP Multi-Biome: NPP and Driver Data for Ecosystem Model-Data Intercomparison, Data set available at: http://www.daac.ornl.gov (last access: 6 March 2015) from the Oak Ridge National Laboratory Distributed Active Archive Center, Oak Ridge, Tennessee, USA, 2001.

Ometto, J. P., Nobre, A. D., Rocha, H., Artaxo, P., and Martinelli, L.: Amazônia and the Modern Carbon Cycle: Lessons Learned. Oecologia, 143, 483–500, 2005.

Pacifico, F., Harrison, S. P., Jones, C. D., Arneth, A., Sitch, S., Weedon, G. P., Barkley, M. P., Palmer, P. I., Serça, D., Potosnak, M., Fu, T.-M., Goldstein, A., Bai, J., and Schurgers, G.: Evaluation of a photosynthesis-based biogenic isoprene emission scheme in JULES and simulation of isoprene emissions under present-day climate conditions, Atmos. Chem. Phys., 11, 4371–4389, doi:10.5194/acp-11-4371-2011, 2011.

Pacifico, F., Folberth, G. A., Jones, C. D., Harrison, S. P., and Collins, W. J.: Sensitivity of biogenic isoprene emissions to past, present, and future environmental conditions and implications for atmospheric chemistry, J. Geophys. Res., 117, D22302, doi:10.1029/2012JD018276, 2012.

Palmer, J. R., and Totterdell, I. J.: Production and export in a Global Ocean Ecosystem Model, Deep Sea Res., Part I, 48, 1169–1198, 2001.

Prentice, I. C. , Bondeau A., Cramer W., Harrison, S. P., Hickler, T., Lucht, W., Sitch, S., Smith, B., and Sykes, M. T.: Dynamic global vegetation modeling: quantifying terrestrial ecosystem responses to large-scale environmental change, in: Terrestrial ecosystems in a changing world, edited by: Canadell, J. G., Pataki, D. E., Pitelka, L. F., IGBP Series, Berlin, Springer, 175–192, 2007.

Price, C. and Rind, D.: A simple lightning parameterization for calculating global lightning distributions, J. Geophys. Res., 97, 9919–9933, 1992.

Price, C. and Rind, D.: Modeling global lightning distributions in a general circulation model, Mon. Weather Rev., 122, 1930–1939 , 1994.

Riahi, K., Gruebler, A., and Nakicenovic, N.: Scenarios of long-term socio-economic and environmental development under climate stabilization, Technol. Forecast. Soc. Change, 74, 887–935, 2007.

Rich, S.: Ozone damage to plants, Ann. Rev. Phytopathol., 2, 253–266, 1964.

Rizzo, L. V., Artaxo, P., Müller, T., Wiedensohler, A., Paixão, M., Cirino, G. G., Arana, A., Swietlicki, E., Roldin, P., Fors, E. O., Wiedemann, K. T., Leal, L. S. M., and Kulmala, M.: Long term measurements of aerosol optical properties at a primary forest site in Amazonia, Atmos. Chem. Phys., 13, 2391–2413, doi:10.5194/acp-13-2391-2013, 2013.

Rummel, U., Ammann, C., Kirkman, G. A., Moura, M. A. L., Foken, T., Andreae, M. O., and Meixner, F. X.: Seasonal variation of ozone deposition to a tropical rain forest in southwest Amazonia, Atmos. Chem. Phys., 7, 5415–5435, doi:10.5194/acp-7-5415-2007, 2007.

Seinfeld, J. H. and Pandis, S. N.: Atmospheric Chemistry and Physics: from Air Pollution to Climate Change, Chapter 6 , J. Wiley, New York, 1998.

Sierra, C. A., Harmon, M. E., Moreno, F. H., Orrego, S. A., and Del Valle, J. I.: Spatial and temporal variability of net ecosystem production in a tropical forest: testing the hypothesis of a significant carbon sink, Glob. Change Biol., 13, 838–853, 2007.

Sigler, J. M., Fuentes, J. D., Heitz, R. C., Garstang, M., and Fisch, G.: Ozone dynamics and deposition processes at a deforested site in the Amazon basin, Ambio, 31, 21–27, 2002.

Sitch, S., Cox, P. M., Collins, W. J., and Huntingford, C.: Indirect radiative forcing of climate change through ozone effects on the land-carbon sink, Nature, 448, 791–794, 2007.

Stroud, C., Makar, P., Karl, T., Guenther, A., Geron, C., Turnipseed, A., Nemitz, E., Baker, B., Potosnak, M., and Fuentes, J. D.: Role of canopy-scale photochemistry in modifying biogenic-atmosphere exchange of reactive terpene species: Results from the CELTIC field study, J. Geophys. Res., 110, D17303, doi:10.1029/2005JD005775, 2005.

Taylor, J. A. and Lloyd, J.: Sources and sinks of atmospheric CO_2, Australian Journal of Botany, 40, 407–418, 1992.

Taylor, K. E., Stouffer, R., J., and Meehl, G. A.: An Overview of CMIP5 and the Experiment Design, B. Am. Meteorol. Soc., 93, 485–498 , 2012.

The HadGEM2 Development Team: G. M. Martin, Bellouin, N., Collins, W. J., Culverwell, I. D., Halloran, P. R., Hardiman, S. C., Hinton, T. J., Jones, C. D., McDonald, R. E., McLaren, A. J., O'Connor, F. M., Roberts, M. J., Rodriguez, J. M., Woodward, S., Best, M. J., Brooks, M. E., Brown, A. R., Butchart, N., Dearden, C., Derbyshire, S. H., Dharssi, I., Doutriaux-Boucher, M., Edwards, J. M., Falloon, P. D., Gedney, N., Gray, L. J., Hewitt, H. T., Hobson, M., Huddleston, M. R., Hughes, J., Ineson, S., Ingram, W. J., James, P. M., Johns, T. C., Johnson, C. E., Jones, A., Jones, C. P., Joshi, M. M., Keen, A. B., Liddicoat, S., Lock, A. P., Maidens, A. V., Manners, J. C., Milton, S. F., Rae, J. G. L., Ridley, J. K., Sellar, A., Senior, C. A., Totterdell, I. J., Verhoef, A., Vidale, P. L., and Wiltshire, A.: The HadGEM2 family of Met Office Unified Model climate configurations, Geosci. Model Dev., 4, 723–757, doi:10.5194/gmd-4-723-2011, 2011.

Thompson, A., Witte, J., McPeters, R., Oltmans, S., Schmidlin, F., Logan, J., Fujiwara, M., Kirchhoff, V., Posny, F., Coetzee, G., Hoegger, B., Kawakami, S., Ogawa, T., Johnson, B., Vomel, H., and Labow, G.: Southern Hemisphere Additional Ozonesondes (SHADOZ) 1998–2000 tropical ozone climatology – 1. Comparison with Total Ozone Mapping Spectrometer (TOMS) and ground-based measurements, J. Geophys. Res., 108, 8238, doi:10.1029/2001JD000967, 2003a.

Thompson, A., Witte, J., Oltmans, S., Schmidlin, F., Logan, J., Fujiwara, M., Kirchhoff, V., Posny, F., Coetzee, G., Hoegger, B., Kawakami, S., Ogawa, T., Fortuin, J., and Kelder, H.: Southern Hemisphere Additional Ozonesondes (SHADOZ) 1998–2000 tropical ozone climatology – 2. Tropospheric variability and the zonal wave-one, J. Geophys. Res., 108, 8241, doi:10.1029/2002JD002241, 2003b.

Thompson, T. M. and Selin, N. E.: Influence of air quality model resolution on uncertainty associated with health impacts, Atmos. Chem. Phys., 12, 9753–9762, doi:10.5194/acp-12-9753-2012, 2012.

Tie, X., Brasseur, G., and Ying, Z.: Impact of model resolution on chemical ozone formation in Mexico City: application of the WRF-Chem model, Atmos. Chem. Phys., 10, 8983–8995, doi:10.5194/acp-10-8983-2010, 2010.

van der Werf, G. R., Randerson, J. T., Giglio, L., Collatz, G. J., Kasibhatla, P. S., and Arellano Jr., A. F.: Interannual variability in global biomass burning emissions from 1997 to 2004, Atmos. Chem. Phys., 6, 3423–3441, doi:10.5194/acp-6-3423-2006, 2006.

van der Werf, G. R., Randerson, J. T., Giglio, L., Collatz, G. J., Mu, M., Kasibhatla, P. S., Morton, D. C., DeFries, R. S., Jin, Y., and van Leeuwen, T. T.: Global fire emissions and the contribution of deforestation, savanna, forest, agricultural, and peat fires (1997–2009), Atmos. Chem. Phys., 10, 11707–11735, doi:10.5194/acp-10-11707-2010, 2010.

Valari, M. and Menut, L.: Does an increase in air quality Models' resolution bring surface ozone concentrations closer to reality?, J. Atm. Oceanic Tech., 25, 1955–1968, 2008.

Wesely, M. L.: Parameterization of surface resistances to gaseous dry deposition in regional-scale numerical models, Atmos. Environ., 23, 1293–1304, 1989.

Yienger, J. J. and Levy II, H.: Global inventory of soil-biogenic NO_x emissions, J. Geophys. Res., 100, 11447–11464, 1995.

5

Impact of pollution controls in Beijing on atmospheric oxygenated volatile organic compounds (OVOCs) during the 2008 Olympic Games: observation and modeling implications

Y. Liu[1,2], B. Yuan[1,*], X. Li[1,], M. Shao[1], S. Lu[1], Y. Li[1], C.-C. Chang[3], Z. Wang[1], W. Hu[1], X. Huang[4], L. He[4], L. Zeng[1], M. Hu[1], and T. Zhu[1]**

[1]State Joint Key Laboratory of Environmental Simulation and Pollution Control, College of Environmental Sciences and Engineering, Peking University, Beijing, China
[2]Chinese Research Academy of Environmental Sciences, Beijing, China
[3]Research Center for Environmental Changes, Academia Sinica, Taipei, Taiwan
[4]School of Environment and Energy, Peking University Shenzhen Graduate School, Shenzhen, China
[*]now at: Earth System Research Laboratory, National Oceanic and Atmospheric Administration (NOAA), Boulder, Colorado, USA
[**]now at: Institut für Energie- und Klimaforschung Troposhäre (IEK-8), Forschungszentrum Jülich, Jülich, Germany

Correspondence to: M. Shao (mshao@pku.edu.cn)

Abstract. Oxygenated volatile organic compounds (OVOCs) are important products of the photo-oxidation of hydrocarbons. They influence the oxidizing capacity and the ozone-forming potential of the atmosphere. In the summer of 2008, 2 months of emission restrictions were enforced in Beijing to improve air quality during the Olympic Games. Observational evidence reported in related studies that these control measures were efficient in reducing the concentrations of primary anthropogenic pollutants (CO, NO_x and non-methane hydrocarbons, i.e., NMHCs) by 30–40 %. In this study, the influence of the emission restrictions on ambient levels of OVOCs was explored using a neural network analysis with consideration of meteorological conditions. Statistically significant reductions in formaldehyde (HCHO), acetaldehyde (CH_3CHO), methyl ethyl ketone (MEK) and methanol were found to be 12.9, 15.8, 17.1 and 19.6 %, respectively, when the restrictions were in place. The effect of emission controls on acetone was not detected in neural network simulations, probably due to pollution transport from surrounding areas outside Beijing. Although the ambient levels of most NMHCs were reduced by ~ 35 % during the full control period, the emission ratios of reactive alkenes and aromatics closely related to automobile sources did not present much difference (< 30 %). A zero-dimensional box model based on the Master Chemical Mechanism version 3.2 (MCM3.2) was applied to evaluate how OVOC production responds to the reduced precursors during the emissions control period. On average, secondary HCHO was produced from the oxidation of anthropogenic alkenes (54 %), isoprene (30 %) and aromatics (15 %). The importance of biogenic sources for the total HCHO formation was almost on par with that of anthropogenic alkenes during the daytime. Anthropogenic alkenes and alkanes dominated the photochemical production of other OVOCs such as acetaldehyde, acetone and MEK. The relative changes of modeled HCHO, CH_3CHO, methyl vinyl ketone and methacrolein (MVK + MACR) before and during the pollution controlled period were comparable to the estimated reductions in the neural network, reflecting that current mechanisms can largely explain secondary production of those species under urban conditions. However, it is worth noting that the box model overestimated the measured concentrations of aldehydes by a factor of 1.4–1.7 without consideration of loss of aldehydes on aerosols, and simulated MEK was in good agreement with the measurements when primary sources were taken into consideration. These results suggest that the understanding of the OVOCs budget in the box model remains incomplete, and that there is still considerable uncertainty in particular missing sinks (unknown chemical and physical processes) for aldehydes and absence of direct emissions for ketones.

1 Introduction

Oxygenated volatile organic compounds (OVOCs), such as aldehydes, ketones and alcohols, are mostly produced in the atmosphere by the oxidation of biogenic, anthropogenic hydrocarbons and other organic species (Finlayson-Pitts and Pitts, 2000), and also directly emitted by vegetation (Park et al., 2013), biomass burning (Yokelson et al., 2009; Mason et al., 2001; Andreae and Merlet, 2001), fossil fuel combustion (Schauer et al., 1999, 2002) and industries (Singh et al., 1994). They are lost through oxidation by OH, photolysis, and deposition/surface uptake. Aircraft measurements showed that the vertical distribution (0–12 km) of the total concentrations of oxygenated organics was 2–5 times as abundant as the sum of C_2–C_8 non-methane hydrocarbons (NMHCs) at all altitudes in the remote Pacific troposphere (Singh et al., 2001). Formaldehyde (HCHO) and other oxygenated hydrocarbons dominated the total OH loss with VOCs (volatile organic compounds) in clean air masses (Goldan et al., 2004). In polluted atmospheres the interaction of primary and secondary VOCs is closely coupled with the formation of ozone and secondary organic aerosols (SOA). And usually OVOCs serve as intermediate products in these oxidation processes, which helps to estimate the formation potential of ozone (Shao et al., 2011) and other products. And OVOCs also affect the oxidizing capacity of the atmosphere. Some previous studies found that carbonyls are one of the most important radical sources, especially for the wintertime in polluted urban environments (Emmerson et al., 2005). Accordingly, it is essential to understand the sources, sinks and chemistry of OVOCs quantitatively, while the role of OVOCs in heavily polluted regions remains largely unexplored partly due to the lack of reliable observations of them.

Beijing, the capital of China and one of the most populous megacities in the world, is located in northern China. Accompanied with the rapid population growth and economic expansion, severe air pollution in Beijing has attracted global attention. The city has been known for its increasing ozone concentration (Parrish and Zhu, 2009; Shao et al., 2009) and hazy skies (Chan and Yao, 2008). In preparation for the 2008 Summer Olympics and to clean up the city's air, Beijing imposed a number of pollution control measures during the games. Thus, coordinated observations on such a restricted occasion create a valuable opportunity for studying how OVOCs respond to the emission reductions, testing and refining the current knowledge of formation pathways of secondary OVOCs under polluted conditions.

After Beijing was selected as the host of the 2008 Summer Olympics, the Beijing government started to implement a series of long-term emission control regulations, such as the closure of heavy polluting industries (chemical factories, cement plants, steel manufacturing, etc.) in the southeast of the city, relocating nearly 200 factories out of the Beijing area; also tightening the vehicle emission and fuel quality standards, and accelerating the retirement of old vehicles. More stringent, short-term measures were put into effect from 1 July to 20 September 2008 such as temporarily halting industrial production, suspending construction and evaporative emissions. Nearly 2 million vehicles were banned from the roads step by step (UNEP, 2009): (1) a total of 350 000 yellow-labeled vehicles with high emissions were not allowed to enter the city starting from 1 July; (2) 50 % of privately owned vehicles were restricted to run on alternate days (based on their license plate numbers) in the metropolis and 70 % of government-owned vehicles were halted from operating from 21 July to 20 September, which is a so-called "full control period". As a result, the traffic emissions on VOCs, CO, NO_x , and particulate matter (PM) were estimated to be reduced by 50 % (Wang et al., 2010b).

The 2 months of emission restrictions had a noticeable impact on major pollutants directly emitted from fuel combustions, which have been validated by on-road measurements (Wang et al., 2009a, b; Wang and Xie, 2009), ground-level monitoring (Wang et al., 2010a, c; Chou et al., 2011; Zhang et al., 2009) and satellite-based observations (Witte et al., 2009; Worden et al., 2012; Lyapustin et al., 2011). Satellite measurements over Beijing showed a 43 % reduction of tropospheric column NO_2 during the games (Witte et al., 2009), and a 32 % reduction in CO for 2008 with respect to 2007 was estimated from MOPITT (Measurement of Pollution in the Troposphere) satellite retrievals (Worden et al., 2012). Wang et al. (2009a) found a dramatic decrease of primary pollutants from on-road measurements, by up to 54 % for CO, 41 % for NO_x and 66 % for aromatics (benzene, toluene, ethylbenzene and xylenes, BTEX). And Wang et al. (2010a) reported the averaged mixing ratios of NMHCs near the Olympic Stadium during the games were reduced by 35 %, compared with the concentrations measured in June (before the restrictions). The monthly concentrations of NO, NO_2 and CO were reduced by 76.8, 29.7 and 27.8 % in August 2008 compared to previous years for the same month (Chou et al., 2011). In contrast, the observed mixing ratio of O_3 increased by 16 % during the full control, compared to the period prior to 20 July (Wang et al., 2010c). The higher levels of ozone were influenced by the declined O_3–NO titration and the shift in different O_3 formation regimes (Chou et al., 2011), changes in weather conditions, or/and by the transport of photochemical plumes from the surrounding areas to Beijing (Wang et al., 2010c). So, the influence of control measures on secondary pollutants still needs more investigation.

The limited OVOCs data sets available in Beijing suggest that the ambient levels of carbonyls were observed to be 3–5 times higher than the levels in Hong Kong, about 35 % of those in Mexico City (Pang and Mu, 2006), and comparable to those in Rio de Janeiro, Brazil (Grosjean et al., 2002). In summer, about 60–70 % of aldehydes in Beijing were associated with the photochemical oxidation of anthropogenic and biogenic VOCs, and 10–15 % of them were attributed to the primary emission from anthropogenic sources (Liu et al., 2009). For ketones and alcohols, anthropogenic primary emissions were shown to be the largest contributor (Liu et al., 2009; Yuan et al., 2012). The 2008 summer provides the opportunity to estimate the impact of reduced emissions on ambient OVOCs. In this study, firstly a neural network approach was used to link VOC mixing ratios to meteorological conditions, which helps to ascertain that the observed changes in VOCs are significant due to the emission restrictions, rather than variations caused by different meteorological conditions. Secondly, a box model using the Master Chemical Mechanism (MCM) was employed to investigate secondary formation of OVOCs during the full control period, in particular, of two most abundant aldehydes (formaldehyde, acetaldehyde), two ketones (acetone and MEK) and unique oxidation products of isoprene chemistry (MVK + MACR). Moreover, this study also aims to quantify the changes of secondary OVOCs by the existing mechanisms, responding to the reduced precursors (NMHCs) and other pollutants (CO, NO, NO_2, O_3), and to analyze the similarity or discrepancy between the measured and modeled changes in OVOCs before and during the full control.

2 Methodologies

2.1 Monitoring sites and experimental methods

As part of the CAREBEIJING 2008 campaign, the measurements of VOCs and other air pollutants (CO, NO_x , O_3, etc) were conducted at an intensive observation site on the campus of Peking University (PKU) from 3 July to 27 August 2008. The PKU site (39.99° N, 116.31° E) was located in the northwest of downtown Beijing, which has been considered to be representative of a typical urban environments with different mixtures of high density traffic, commercial, residential, and electronic companies (Liu et al., 2009; Cheng et al., 2008; Song et al., 2007). The air inlets for the instruments were set up on the top of a six-story building (~ 25 m a.g.l). Three online techniques were used to quantify NMHCs and OVOCs during the campaign, including a proton-transfer-reaction mass spectrometer (PTR-MS), a gas chromatograph with mass spectrometer and flame ionization detector (GC-MS/FID) and a Hantzsch fluorimetric monitor.

2.1.1 Online measurements for NMHCs and OVOCs

A high-sensitivity PTR-MS (Ionicon Analytik, Innsbruck, Austria) was used for ambient measurements of selected NMHCs and OVOCs. This technique ionizes VOC molecules by transferring protons (H^+) from protonated water ions (H_3O^+) in a drift tube, followed by a quadrupole mass spectrometer detection of the product ions (Lindinger et al., 1998; de Gouw and Warneke, 2007). The air sample was drawn through an 8 m long perfluoroalkoxy (PFA) tube (0.25 in. OD × 0.156 in. ID) into the gas-handling system by a diaphragm pump at the flow rate of 2–3 slpm. A small flow of 220–240 sccm was branched from the main flow to the PTR-MS. A 2.0 μm pore-size 47 mm PTFE (polytetrafluorethylene) filter at the inlet was used to remove particles from the sample flow.

The operating parameters of PTR-MS used herein are similar to those described previously in Yuan et al. (2010a). The instrument was operated at 2.3 mbar drift tube pressure and 600 V drift voltage, 35 masses were sampled with the time resolution of 160 s. Instrument background signals were measured every 35 cycles (~ 1.5 h) by switching the sample air to an activated charcoal trap for 5 cycles (~ 13 min). Target ions with the dwell time of 5 s for each cycle included OVOCs such as methanol ($m/z = 33$), acetaldehyde ($m/z = 45$), acetone ($m/z = 59$), MVK + MACR ($m/z = 71$), methyl ethyl ketone (MEK, $m/z = 73$); aromatics such as benzene ($m/z = 79$), toluene ($m/z = 93$), styrene ($m/z = 105$), C8 aromatics ($m/z = 107$) and C9 aromatics ($m/z = 121$); isoprene ($m/z = 69$) and acetonitrile ($m/z = 42$). The PTR-MS was calibrated every week using a commercial TO15 gas standard (Air Environmental Inc. Denver, USA) at seven concentration levels ranging from 1 to 15 ppb. For most species, the detection limits were between 23 and 60 pptv, except methanol (141 pptv), acetone (148 pptv) and MEK (110 pptv). The relative errors of the PTR-MS measurements decrease with the mixing ratios of target species, as shown in Fig. S1, and the precisions of all detected species were below 5 % at the level of 1 ppbv.

An automated GC-MS/FID system with two columns and two detectors was deployed by the Research Center for Environmental Changes (RCEC), Academia Sinica, Taiwan, which was used to measure C_2–C_4 hydrocarbons and C_5–C_{11} hydrocarbons simultaneously. Technical details for this instrument were presented in Chou et al. (2011). A total of 65 NMHCs were quantified hourly during the full control period (24 July–27 August), the precision for most of the species was within 2 % and the limits of detection were below 30 pptv.

The comparison between the PTR-MS and GC-MS data showed a good agreement for aromatics (benzene, toluene, C8 aromatics and C9 aromatics) during the campaign, as shown in Fig. S2a, with the slopes varying between 0.86 and 1.21 and the correlation coefficients larger than 0.89. The isoprene concentrations measured by PTR-MS were systematically higher than those measured by GC-MS, particularly for the mixing ratios lower than 1 ppbv, shown in Fig. S2b, possibly because signals at m/z 69 detected by PTR-MS were influenced by some interferences from pentanal, methylbutanal, pentenol (de Gouw et al., 2003) and cycloalkanes (Yuan et al., 2014). It should be noted that the artifact of PTR-MS measurements on isoprene leads to an additional background in anthropogenic-emission-dominated areas. So, the isoprene data from the GC-MS is used in the following model calculations.

HCHO was measured by an online Hantzsch fluorimetric monitor at the time resolution of 10 min. The instrument is described in previous publications (Dasgupta et al., 2005; Li et al., 2010) and was based on sensitive wet chemical fluorimetric detection of HCHO. A diffusion scrubber was designed to strip and collect HCHO from the gas phase into the liquid phase. The detection limit for HCHO was within 100 pptv, and the precision was below 10 %.

2.1.2 Measurements for other air pollutants and parameters

A comprehensive set of instruments was installed at the PKU site to monitor the air quality. CO was measured by a nondispersive infrared sensor (NDIR) with an integration time of 1 min (48 CTLE, Thermo Environmental Instruments, TEI, USA). Zero calibrations were done every 2 h and span calibrations were performed at midnight (00:00–1:00 LT) each day during the campaign. Ozone was measured by a UV absorption detector (Model 49i, TEI). NO was measured using a NO/O_3 chemiluminescence analyzer (Model 42iTL, TEI), with the instrument sequentially measuring NO_2 through a photolytic converter.

HONO was measured using an online stripping coil sampler/ion chromatography (IC) system at the time resolution of 15 min. The instrument has a detection limit of 8 ppt with an uncertainty of 7 %, other technical details were presented in Cheng et al. (2013).

For aerosol measurements at the site, particle size distributions from 3 to 900 nm (mobility diameter) were measured by a TDMPS (twin differential mobility particle sizer) system with 10 min time resolution, as presented in Wang et al. (2011). An Aerodyne HR-TOF-AMS (high-resolution time-of-flight aerosol mass spectrometer) was deployed to measure chemical compositions of PM1. A detailed description of the system was shown in Huang et al. (2010).

A meteorological station was operated simultaneously to monitor meteorological parameters (temperature, RH, pressure, wind speed and wind direction) on the campus, which was about 400 m away from the air quality intensive site. The photolysis frequencies of O^1D and NO_2 were measured at the PKU site by two filter radiometers (Bohn et al., 2008) with a time resolution of 5 s. The uncertainties of J values were within 10 %.

2.2 Methodology for neutral network

The temporal variation of VOC concentrations at a receptor site reflects a combination of changes in emission location and strength, meteorological conditions, chemical loss and secondary formation. Previous studies indicated that meteorological conditions, especially local wind speed and direction, played an important role in shaping the air quality of Beijing (Wu et al., 2008). We cannot quantify the influence of emission restrictions on ambient VOCs by simply comparing their levels during the control period with those before and after. Therefore, following the method introduced by Cermak and Knutti (2009), a feed-forward neural network was used to establish a statistical relationship between VOC mixing ratios and meteorological parameters from the reference data (i.e., before the full control) and then to predict the concentrations given the meteorological conditions during the games, if there were no restriction measures taken. Thus, the difference between the observed and predicted VOC concentrations is helpful to determine the effective changes due to the air quality measures.

The multilayer perceptron (MLP) network in SPSS17.0 was used in this work. As the concentrations of VOCs displayed near lognormal distributions, the natural logarithm of VOC concentration, ln(VOC), was selected as the "dependent variable" of inputs in the MLP network, and the meteorological parameters as "covariates" of inputs, including wind vector (u, v, previous day u, previous day v), precipitation, relative humidity, temperature and pressure. The observed data from 3 July to 20 July (before the full control) were used as the reference data ($n = 1611$) to set up a relationship between ln(VOC) and meteorological parameters. Of the reference data 70 % was chosen randomly as training sample to train the neural network, 20 % of the data as testing sample used to track errors during training in order to prevent overtraining, and the rest of the data (10 %) as the holdout sample used to assess the final neural network for validation. The network contained one hidden layer, sigmoid functions were chosen as the activation functions for hidden layer and output layer. Given that there was no statistical difference between the prediction and the corresponding observation for validation data, the same approach was used for predicting the concentrations of VOCs from 21 July to 27 August.

2.3 Model description for OVOC simulation

A zero-dimensional box model, which is similar to the one used by Lu et al. (2012) and Li et al. (2014), was employed in this work to examine the change of OVOC production under the influence of emission restrictions. The box model uses a subset of the MCM3.2 (available online at http://mcm.leeds.ac.uk/MCM) which contains fully explicit chemical mechanisms for 57 kinds of VOCs (including 7 oxygenated hydrocarbons) plus CH_4 and CO. The model calculations were constrained by measured C_2–C_{11} NMHCs,

CO, CH_4, O_3, NO, NO_2, HONO, and physical parameters (i.e., photolysis frequencies, water vapor concentration, temperature, and pressure). The time step of the model calculation was set to 30 min. Here, 57 VOCs were measured by online GC-MS/FID and PTR-MS including 20 C_2–C_{11} alkanes, 11 C_2–C_5 alkenes, 15 aromatics, acetylene, isoprene, pinenes, formaldehyde, acetaldehyde, methanol, acetone, MEK and MVK + MACR. The CH_4 concentration was assumed to be 2.5 ppm, taken from previous observations in Beijing (Su, 2003). The photolysis rates for O^1D and NO_2 were constrained with the measured values at the site. The photolysis rates for HONO, aldehydes and ketones were calculated by the model for clear sky conditions (Saunders et al., 2003) and then scaled by the measured $J(NO_2)$. Dry deposition rate for all modeled species was set to 1.2 cm s^{-1} by assuming a well-mixed boundary layer with a height of 1000 m, corresponding to a lifetime (τ_D) of 24 h. Several additional model scenarios (listed in Table 3) were constructed to test the sensitivity of simulated OVOC concentrations with assumed deposition rates and boundary layer evolution. The model was operated for the full control period (26 July–27 August) with 2 days spin-up time to reach steady state. The relative changes of the model results responding to the controlled emissions were compared with the observed changes of OVOCs before and during the full control (Sect. 4).

The model uncertainty is mainly caused by the uncertainty of all input parameters (VOCs, trace gas, meteorological parameters, etc.) and reaction rates used in the model. Here, the total uncertainty was estimated by the error propagation from the errors of all considered parameters, as described in Li et al. (2014). The modeled concentrations of OVOCs in the base model had an uncertainty of 40–50 %.

3 Results and discussion based on observations

3.1 Diurnal variations of VOCs before and during the control

Table 1 summarizes the averaged meteorological parameters before (3–20 July) and during the full control period (21 July–27 August). Meteorological conditions in the two time periods were similar, except that the average ambient T and UVA before the control were slightly higher than respective values during the full control, which would affect the emission of biogenic VOCs (dominated by isoprene). As an overview of the changes in VOC concentrations between the two periods, the 10 min-average diurnal variations of the mixing ratios of 12 VOC species were compared in Fig. 1. Aromatics, as a group of primary anthropogenic NMHCs, mainly come from vehicle exhaust and solvent usage in Beijing (Liu et al., 2005; Yuan et al., 2010b). They dropped to the minimum at 14:00–15:00 for both of the two time periods (Fig. 1a–f). Compared to the period prior to 20 July, an obvious decrease (50–60 %) in aromatics occurred after 21 July, especially at rush hours, which may be caused by the traffic restrictions. Oxygenated species at the PKU site, from both secondary photochemical production and primary emissions, reached high concentrations at noon and in the late evening (Fig. 1h–l). Some differences of ketones and aldehydes between the two time periods were found, but not as prominent as aromatics. It is difficult to assess the effect of control measures on OVOC species merely based on the changes in their absolute concentrations before and during the control. In the following section a neural network analysis is used to subtract meteorological effects and quantify the influence of emission reductions on ambient levels of NMHCs and OVOCs.

3.2 Effect of emission control measures on mixing ratios of VOCs

Assuming that no emission restrictions had been in place from 21 July to 27 August, the "uncontrolled" VOCs corresponding to the meteorological conditions encountered at that time were predicted by the MLP network using the relationship between VOCs and meteorological parameters from the reference data (before 21 July). Figures 2 and 3 present the probability distributions of the observed (red) and predicted (blue) ln(VOC) before and during the control period for NMHCs and OVOCs, respectively. For most aromatics, the observation and prediction values from the uncontrolled period matched well with each other, but the observed data from the controlled period clearly shifted towards lower values compared to the prediction. This suggests that the influence of the traffic restriction on aromatics was effective. Acetonitrile (CH_3CN), unlike other NMHCs, did not show any difference between the observed and predicted values for both the controlled and uncontrolled periods. As acetonitrile is usually considered as a tracer for biomass burning and is seldom detected in automobile emissions (Wang et al., 2007), its concentration should be hardly affected by the traffic control. The measured median value of acetonitrile before 21 July was slightly higher than that after 21 July. This discrepancy (15 %) in the absolute concentrations between the two time periods could be explained by the changes in meteorological conditions or pollution transport. The reductions of most OVOCs were also observed relative to what was to be expected without emission controls, although the deviations in OVOCs were not as notable as aromatics (Fig. 3).

The predicted and observed results of NMHCs and OVOCs in the controlled period are compared in Table 2, and the results for the uncontrolled period are listed in Table S1. The median concentrations of aromatics dropped by 32–47 % compared to the levels expected under the same meteorological conditions without traffic controls. The median values of formaldehyde, acetaldehyde, MEK and methanol were decreased by 12.9, 15.8, 17.1 and 19.6 %, respectively. Isoprene, unlike aromatics, exhibited a broad peak from the early morning to the afternoon (Fig. 1g), which followed the

Table 1. Summary of meteorological parameters at the PKU site before and during the full control (average $\pm 1\sigma$ SD).

Time period		T (°)	RH (%)	Precipitation (mm)	WS ($m\,s^{-1}$)	UVA ($W\,m^{-2}$)
Before the full control	3–20 July	28.9 ± 4.5	66.9 ± 20.0	0.04 ± 0.37	1.1 ± 0.9	6.6 ± 8.7
During the full control	21 July–27 August	28.0 ± 4.1	67.9 ± 17.5	0.05 ± 0.51	0.9 ± 0.8	5.3 ± 7.1

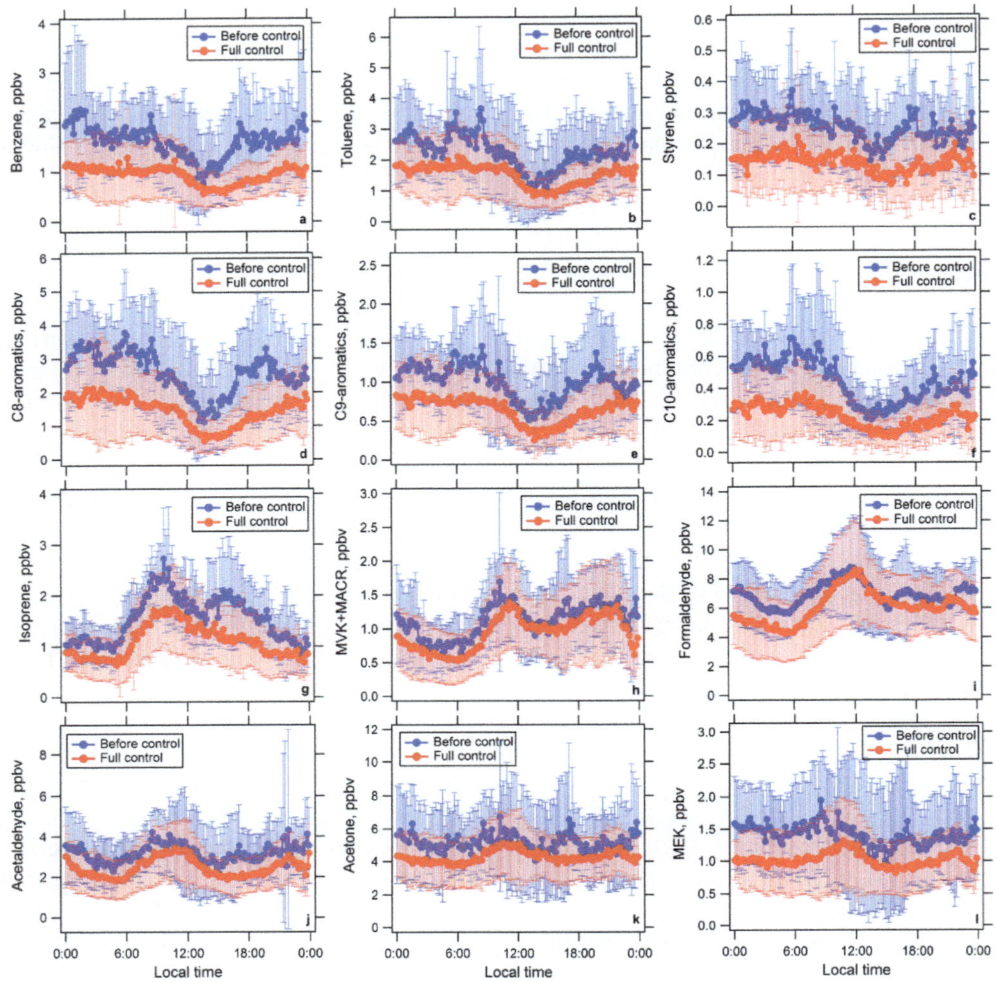

Figure 1. The 10 min average diurnal variations of several NMHCs and OVOCs measured by PTR-MS at the PKU site before (1–20 July, blue) and during the full control (21 July–27 August, red), respectively. **(a)** benzene, **(b)** toluene, **(c)** styrene, **(d)** C8 aromatics, **(e)** C9 aromatics, **(f)** C10 aromatics, **(g)** isoprene, **(h)** MVK + MACR, **(i)** formaldehyde, **(j)** acetaldehyde, **(k)** acetone, **(l)** MEK. The error bars represent standard deviations (SDs) for those compounds before (blue) and during the full control (red).

solar radiation and temperature cycles and showed the characteristics of local biogenic emissions in the daytime. In addition, their low concentrations at night likely indicate small local emissions from vehicles near the site. Compared to the values before the control isoprene and MVK + MACR were estimated to be reduced by 26 and 11 % from 21 July to 27 August, respectively, which was caused by a combined effect of the lower temperature, solar radiation and the control measures.

The relative difference between predicted and measured acetone was as small as 1 %, implying that the emission controls had little or no effect on acetone. The unchanged acetone levels were probably influenced by its high regional background, which was around 1.7–2 ppb at the PKU site accounting for 43–47 % of acetone concentration (Liu et al., 2009; Yuan et al., 2012). The acetone background was closely related to transport of photochemical plumes from the surrounding areas (such as Hebei Province and Tianjin) to Beijing. As the emission reductions were implemented

Table 2. Comparison of the observed and predicted values of median ln(VOC) and their corresponding concentrations during the full control period, and the deviations between observed and predicted concentrations, together with the outputs P(t) from Student's t test for the two data sets, where P(t) < 0.05 implies that the difference in the two data sets is statistically significant at the confidence level of 95 %.

	Median ln(VOC)			Concentration (ppbv)		Deviation of concentrations	
	observation	prediction	P(t)	observation	prediction	(%)	R
Benzene	−0.23	0.25	0.00	0.79	1.28	−38.2	0.48
Toluene	0.24	0.63	0.00	1.27	1.88	−32.5	0.41
C8 aromatics	0.12	0.73	0.00	1.12	2.08	−46.0	0.44
C9 aromatics	−0.63	−0.17	0.00	0.53	0.84	−37.0	0.26
C10 aromatics	−1.70	−1.05	0.00	0.18	0.35	−47.5	0.32
Styrene	−2.11	−1.66	0.00	0.12	0.19	−36.5	0.22
Acetonitrile	−1.45	−1.46	0.72	0.23	0.23	0.4	0.36
HCHO	1.72	1.85	0.00	5.56	6.38	−12.9	0.45
Acetaldehyde	0.78	0.95	0.00	2.18	2.59	−15.8	0.52
Acetone	1.39	1.40	0.36	4.02	4.06	−1.0	0.36
MEK	−0.11	0.08	0.00	0.90	1.08	−17.0	0.64
Methanol	1.95	2.17	0.00	7.06	8.78	−19.6	0.31
Isoprene	−0.07	0.24	0.00	0.94	1.27	−26.6	0.55
MVK + MACR	−0.27	−0.15	0.00	0.76	0.86	−11.3	0.72

within the Beijing area, i.e., in a range of about 150 km radius, the regional background of acetone should hardly be affected by the emission controls. So, it seems to be difficult to reduce the level of acetone in Beijing if relying solely on the control of local emissions.

With the exception of acetonitrile and acetone, the deviation for all other species between the full control period and the uncontrolled days is statistically significant from Student's t test at the 95 % level (P(t)<0.05). By contrast, the reference data from the uncontrolled period were well predicted by the MLP simulations (as shown in Fig. S3), with correlation coefficients ranging from 0.79 to 0.94 and successful t tests for all the species (presented in Table S1). Thus, the emission restrictions implemented in Beijing had significant effects on the NMHCs dominated by vehicular emissions, which is consistent with those results in Wang et al. (2010a). The reductions in OVOC species are moderate in comparison to their precursors (e.g., NMHCs), which are subject to the combined influence of controlled direct emissions, local production from reduced precursors and regional formation processes (transport). In subsequent model analyses (Sect. 4), the changes of OVOCs produced from local NMHC oxidation will be discussed.

3.3 Emission ratios for anthropogenic NMHCs

As NMHCs play significant roles on the formation of secondary OVOCs, emission ratios (ERs) of NMHCs before and during the full control were compared in this section to determine whether the emission restrictions also altered the relative emission of OVOC precursors (i.e., source pattern of NMHCs), in addition to reducing their total emission amounts. According to previous studies on NMHC source apportionment in Beijing (Liu et al., 2005; Song et al., 2007; Lu et al., 2007; Wang et al., 2010a), vehicle exhaust is the largest contributor to ambient NMHCs with 40–58 %, followed by solvent usage and painting processes (18–30 %), gasoline evaporation (7–13 %) and chemical plants (3–15 %).

The emission ratio for VOCs is defined as the ratio of VOC species relative to a reference compound in fresh emissions without undergoing photochemical processes. It can be determined using a photochemical-age-based method presented in de Gouw et al. (2005), Warneke et al. (2007) and Borbon et al. (2013). The degradation of NMHCs by their reactions with OH is described as Eq. (1), where ER_{NMHC} represents the emission ratio of NMHC species relative to CO, and k_{NMHC} and k_{CO} are the OH rate coefficients for NMHC and CO, respectively. The ERs reflect the comprehensive effect of all emission sources on ambient NMHCs at the site. Here, the photochemical age of the sampled air masses was calculated by the measured ratio of toluene / benzene (T / B) using Eq. (2), and the initial T / B was set to 2.8 ± 0.7. The 24 h averaged concentration of OH was taken as 2.5×10^6 molecules cm^{-3} which comes from OH measurements at a suburban site near Beijing during the CAREBEIJING 2006 campaign (Lu et al., 2013). The ratio of NMHCs (e.g., toluene, m,p-xylene) to CO decreases with the photochemical age (Fig. 4). The yellow line in Fig. 4a is the linear fit between ln([toluene]/[CO]) and the photochemical age. The emission ratio of toluene relative to CO ($ER_{toluene}$) is determined by the intercept of the linear fit. It is noted that a constant CO background of 120 ppbv was subtracted from measured CO when plotting VOC/CO versus photohemical age, the CO background was determined from the minimum concentrations when prevailing winds

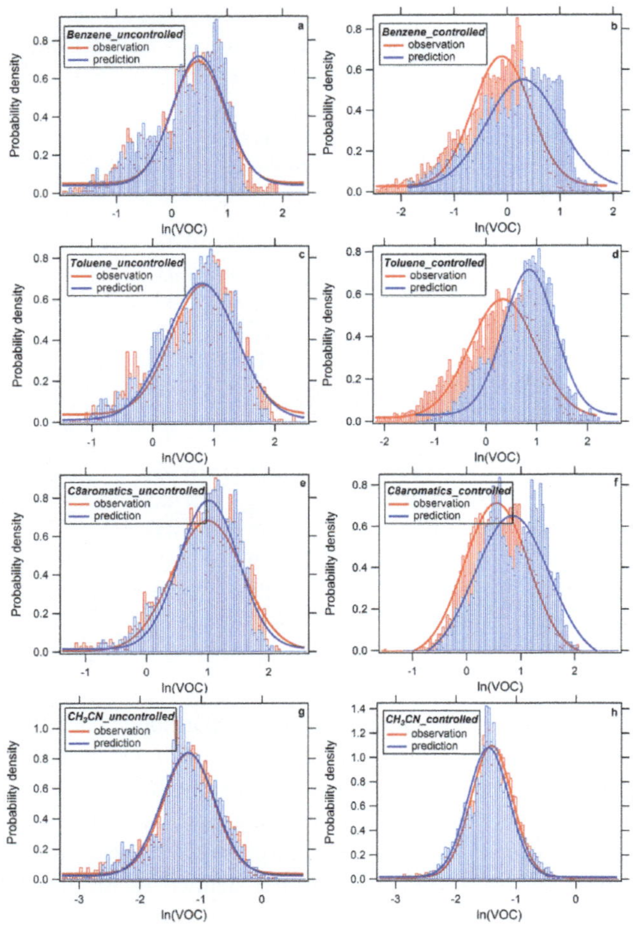

Figure 2. Probability density distribution of the observed ln(VOC) (red) and the predicted values (blue) before the full control (left column) and after the full control (right column) for NMHC species: **(a, b)** benzene, **(c, d)** toluene, **(e, f)** C8 aromatics, and **(g, h)** acetonitrile.

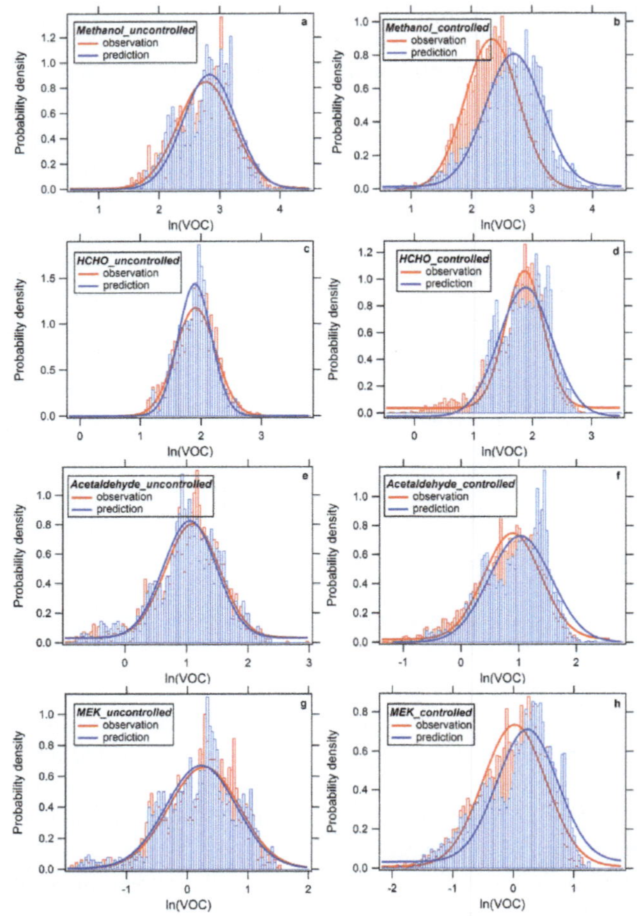

Figure 3. Probability density distribution of the observed ln(VOC) (red) and the predicted values (blue) before the full control (left column) and after the full control (right column) for OVOC species: **(a, b)** methanol, **(c, d)** HCHO, **(e, f)** acetaldehyde, and **(g, h)** MEK.

were from the northeast and relatively clean air was measured.

$$\frac{[\mathrm{NMHC}]}{[\mathrm{CO}]} = \frac{[\mathrm{NMHC}]}{\Delta\mathrm{CO}} = \mathrm{ER}_{\mathrm{NMHC}} \times \exp[-(k_{\mathrm{NMHC}} - k_{\mathrm{CO}})[\mathrm{OH}]\Delta t] \quad (1)$$

$$\Delta t = \frac{1}{[\mathrm{OH}](k_{\mathrm{toluene}} - k_{\mathrm{benzene}})} \times \left[\ln\left(\frac{[\mathrm{toluene}]}{[\mathrm{benzene}]}\bigg|_{t=0} \right) - \ln\left(\frac{[\mathrm{toluene}]}{[\mathrm{benzene}]}\bigg|_{t=t} \right) \right] \quad (2)$$

As discussed in Warneke et al. (2007), the error of emission ratio arises from the uncertainty of measured benzene and toluene and the estimation of the initial T / B ratio in Eq. (2). The total ER error was then calculated conservatively by linear addition of the above two errors. The measurement uncertainty of benzene and toluene by PTR-MS or GC-MS is about 10–15 %. As initial T / B ratio from the same source category varies with different operating conditions, the error from selecting initial ratios can be regarded as the uncertainty of source profiles, which is usually in the range of 15–20 %. So, the total uncertainty of ER estimation is around 30–35 %.

Since the GC-MS/FID data was only available for the full control period of 2008, VOC data measured at the PKU site in August 2005 (Liu et al., 2009) were used as a reference for uncontrolled situations. The emission ratios for all measured hydrocarbons versus CO in Beijing for August 2008 and August 2005 are compared in Fig. 5, and the values are tabulated in Table S2 together with some ER results from two US cities in previous work (Warneke et al., 2007; Borbon et al., 2013). In general, the ERs for all groups of NMHCs in the 2 years agree with each other within a factor of 1.5 (indicated by the shaded area in Fig. 5). For traffic-related NMHCs including most alkenes, acetylene, benzene, toluene and ethylbenzene, the difference of the ERs between 2005 and 2008 ranged from ±6.5 to ±29 %, within the range of combined error (30 %) for ER calculations. This suggests that the pollution control measures did not significantly change the emission composition of the species closely related to vehicu-

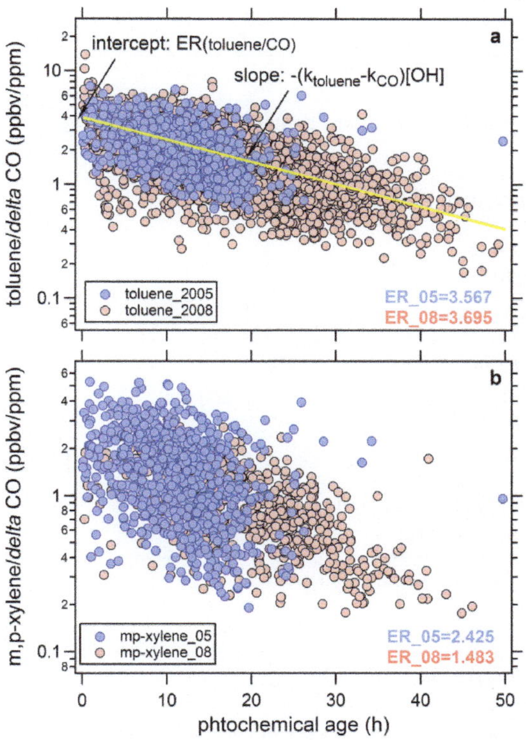

Figure 4. The ratios of toluene/CO **(a)** and m,p-xylene/CO **(b)** vs. the photochemical age of the sampled air mass in Beijing for August 2008 (pink) and August 2005 (blue).

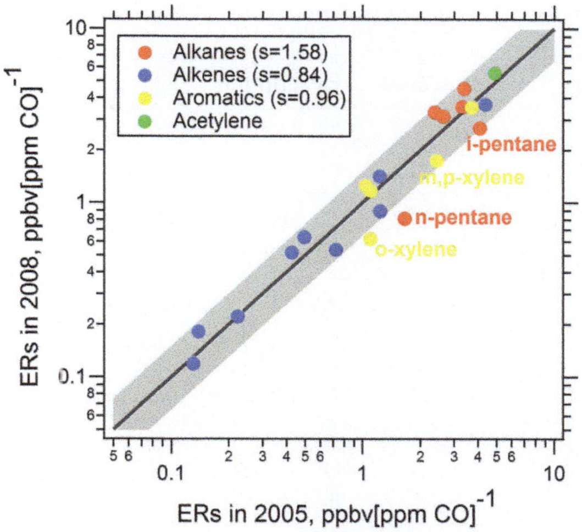

Figure 5. Comparison of the emission ratios (ERs) of NMHCs relative to CO in the 2008 and 2005 summers at the PKU site, where each point represents one compound and different colors indicate the classes of NMHCs. The black line represents the 1 : 1 line and the shaded area shows the agreements within a factor of 1.5.

lar emissions, although the absolute concentrations of those species did drop by half (e.g., toluene/CO in Fig. 4a). A recent work by Wang et al. (2014) also reported a good agreement of emission ratios for most NMHCs at the PKU site in the summertime of 2008, 2010 and 2011. These results reflect the similarity of NMHC compositions for typical urban emissions of Beijing.

Some alkanes (ethane, *n*-butane, *i*-butane, *n*-pentane) and aromatics (*m*,*p*-xylene and *o*-xylene, labeled in Fig. 5) showed a larger variability (35–50 %) in the two ER data sets for 2005 and 2008. For example, the emission ratio for *m*,*p*-xylene in August 2008 was observed to be 38 % lower than the value in August 2005, shown in Fig. 4b. Besides an additional error (5–10 %) from subtracting a constant CO background in ER estimation, one possible reason for this larger difference is that the abovementioned species also come from "other" emission sources in addition to vehicles. C_8 aromatics were observed as the main components from solvent usage in Beijing (Yuan et al., 2010b). The 2008 control measures included temporary closures of chemical plants, painting processes and constructions involving evaporation emissions. If the influence of the control measures on emissions of solvent usage and of automobile source occurred to different degrees, the averaged emission profiles of the city would be changed for the characteristic species (i.e., xylenes) of solvent usage.

The ERs of hydrocarbons related to vehicle emissions (acetylene, ethylene, propene, benzene and toluene) in Beijing generally agree with those in two US cities (in Table S2), indicating that emission patterns of automobile sources in different cities showed a similarity. Greater deviations of lighter alkanes, some C_4 and C_5 alkenes and higher aromatics were found probably because of the unique characteristics of emission sources (e.g., fuel types, vehicle fleet ages, solvent and paint types, industrial procedures, etc.) for each city.

4 Modeling analysis for the changes of OVOCs

To better understand the relationship between reduced precursors (NMHCs) and OVOC concentrations during the CAREBEIJING 2008 experiment, a box model using the recent version of MCM was applied in this study. According to the source apportionments of OVOCs in Beijing (Liu et al., 2009; Yuan et al., 2012), 60–70 % of aldehydes are attributed to hydrocarbon oxidation (secondary formation), the contributions from primary sources are more important for ketones (38–80 %) and alcohols (48–74 %), and the contribution of secondary formation was not detected in methanol. Therefore, the following sections will focus on the formation of aldehydes and ketones. The primary emissions for aldehydes and ketones are not included in the base model but added to subsequent model scenarios for comparison.

Table 3. Model scenarios performed in OVOC simulations during the CAREBEIJING 2008.

Simulation	Description
M0	MCM3.2 mechanism, $\tau_D = 24$ h, the boundary layer height (BLH) was assumed to be 1000 m, i.e., $v_{depo} = 1.2\,cm\,s^{-1}$
M1	As M0, the variation of BLH was taken into consideration, $\tau_D = v_{depo}/BLH$, and v_{depo} was set to $1.2\,cm\,s^{-1}$ as what was in M0
M2	As M1, $\tau_D = v_{depo}/BLH$, but v_{depo} were different for calculated species, vertical dilution rates ($\lambda_{dil,vert}$) were added, $\lambda_{dil,vert} = \max(0, \frac{1}{H(t)}\frac{dH(t)}{dt})^{(1)}$
M3	As M2, uptake of formaldehyde and acetaldehyde by aerosol were included, i.e., $\frac{dC}{dt} = -\frac{\gamma \times S_{RH} \times v \times C}{4}$, γ, $S_{RH}^{(2)}$, and v represent uptake coefficient, RH corrected aerosol surface concentration, and molecular velocity, respectively; here we assumed $\gamma = 10^{-3}$ for HCHO and CH_3CHO
M4	As M3, using emission ratios of OVOCs relative to CO (ER_{OVOC}) and measured CO, the primary aldehydes and ketones were added to the model run
M5	As M4, the concentrations of NMHCs, CO, NO_x, O_3 were increased in order to simulate the scenario without traffic restrictions

(1) Vertical dilution rates were determined by the method used in Riemer et al. (2009).
(2) S_{RH} is the RH-corrected aerosol surface concentration, it was estimated by an empirical function:
i.e., $S_{RH} = f(RH) \times S_{dry} = (1 + a \times RH^b) \times S_{dry}$.
Here ^{a} and ^{b} were set to 4.34 and 6.72, respectively, taken from Liu et al. (2013); S_{dry} was from the SMPS measurement during the campaign.

4.1 Description of model scenarios and OH simulation results

Six model scenarios for OVOC simulations are summarized in Table 3. M0 is the base case model with the settings described in Sect. 2.3. In M1, the variation of the boundary layer height (BLH) at a suburban site near Beijing taken from Liu et al. (2011) was included in the model run. The dry deposition rate in M1 was set to $1.2\,cm\,s^{-1}$, as used in the model scenario M0. In M2, different dry deposition rates were applied to the simulated species. The values were taken from Emmerson et al. (2007) for HNO_3 ($2\,cm\,s^{-1}$), organic peroxides ($0.5\,cm\,s^{-1}$), H_2O_2 and organic nitrates ($1.1\,cm\,s^{-1}$), from Li et al. (2014) for HCHO and other aldehydes ($1\,cm\,s^{-1}$). In addition, the vertical dilution rates that represent entrainment of a growing boundary layer were introduced to all modeled species in M2. These are estimated from the growth rate of boundary layer heights (Riemer et al., 2009), described by the equation given in Table 3. In M3, the loss of two aldehydes on particle surface was estimated by using the uptake coefficient of 10^{-3} form the literature (Jayne et al., 1996; Li et al., 2011); the equation (Jayne et al., 1996) for calculating uptake rate is shown in Table 3. In M4, the primary sources for formaldehyde, acetaldehyde, acetone and MEK were included by employing the emission ratios of OVOCs relative to CO and measured CO concentration.

OH concentrations were not directly measured in August 2008. As strong relations between observed OH and ozone photolysis frequency ($J(O^1D)$) were found in field studies (Rohrer et al., 2014; Rohrer and Berresheim, 2006), it makes sense to estimate OH radicals using the empirical OH-$J(O^1D)$ relations when lacking direct measurements of OH. Here, the modeled OH by the box model was compared with the calculated OH by an observed dependence of OH on $J(O^1D)$ that was obtained from OH measurements in CAREBEIJING 2006 (Lu et al., 2013). Figure 6a shows the average diurnal profiles of modeled and calculated OH, indicating that the modeled OH in the daytime shows a broader peak and it was overestimated by $\sim 22\,\%$ compared to the calculations from the empirical relation. Relatively higher nighttime OH was determined from the calculation, due to the intercept of the empirical OH-$J(D^1D)$ function. This intercept includes all processes that are light independent (Rohrer and Berresheim, 2006), implying that unknown OH production probably occurred at night (Lu et al., 2014). As the uncertainties of OH measurement and $J(O^1D)$ measurement are 20 and 10 %, respectively, from Lu et al. (2013), in addition to a fitting error between OH and $J(O^1D)$, the difference between the modeled and calculated OH is acceptable. This hints that the box model employed here provides a reasonable explanation of radical chemistry in the urban environments.

4.2 OVOCs simulation

4.2.1 HCHO and CH_3 CHO

Unlike OH, a large difference with a factor of 2–4 was found between the measured and modeled concentrations of aldehydes by the base model (M0), as summarized in Table S3.

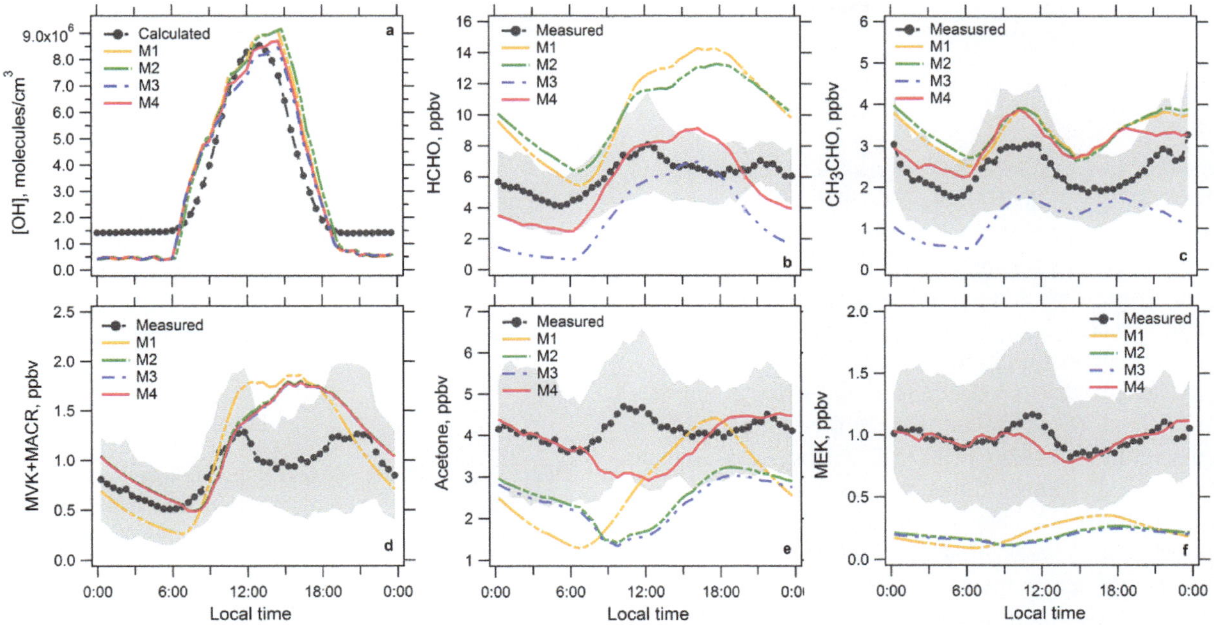

Figure 6. The 30 min average modeled and measured (or calculated) concentrations of **(a)** OH radical, **(b)** HCHO, **(c)** acetaldehyde, **(d)** MVK + MACR, **(e)** acetone and **(f)** MEK plotted as diurnal patterns from 26 July to 27 August. The dashed black lines with solid dots indicate measured (or calculated) data, grey areas show one standard deviation from the averaged values. The dash-dotted yellow lines, dashed green lines, dash-dotted blue lines and solid magenta lines represent the modeled concentrations by simulations M1, M2, M3 and M4, respectively.

After modifying deposition rates or/and vertical dilution, the averaged concentrations of HCHO and acetaldehyde in M1 and M2 were reduced by 30–37 % compared to the base case. But the modeled concentrations of the two aldehydes are still higher (> 1.5 times) than the measured values. The considerations of BLH variation, different deposition and vertical dilution can help to explain 40–50 % of the large discrepancy between M0 and observations. These results are consistent with a recent study on modeling of formaldehyde and glyoxal in the Pearl River Delta (PRD) (Li et al., 2014).

Li et al. (2014) estimated that the uptake of formaldehyde and glyoxal by aerosols had the largest contribution (~ 50 %) to aldehyde sinks in the presence of acidic aerosols. The H^+_{aer} presented in the particle phase was calculated from inorganic ions (NH_4^+, SO_4^{2-}, NO_3^- and Cl^-) by HR-TOF-AMS using the method in Zhang et al. (2007). For most days of campaign the average acidity of aerosols was close to neutral due to high ammonia in Beijing. However, for several days (such as 31 July, 1, 4, 11 and 14 August) the averaged H^+_{aer} in the daytime was up to 0.01448 mol L^{-1}, corresponding to a pH value of 1.84, which indicates high aerosol acidity occasionally occurred during the campaign. Jayne et al. (1996) reported a large uptake of formaldehyde by aqueous surface at low temperature and high aerosol acidity. Some lab experiments (Li et al., 2011) and field studies also showed loss of HCHO on aerosols is possible and driven by the liquid water content of the aerosol phase (Toda et al., 2014). In this study, the loss of aldehydes on aerosols through heterogeneous uptake processes was included in M3 and M4 by using the uptake coefficient of 10^{-3} for two aldehydes. On average, the modeled HCHO and CH_3CHO by M3 were decreased by 64 and 58 %, respectively, compared to M2. Therefore, the loss of aldehydes on aerosol particles might be important in the polluted areas with high production rates of aerosols. Further research on sinks of aldehydes, particularly for heterogeneous uptake processes, is still needed in future studies.

The model scenarios M0–M3 only estimated photochemical production of OVOCs (secondary OVOCs) and did not include their primary emissions from sources. As OVOC source apportionment showed that Beijing is characterized by high anthropogenic emissions of OVOCs (primary OVOCs) (Yuan et al., 2012; Chen et al., 2014), primary aldehydes were added to model simulation M4 by using the ERs of OVOCs relative to CO (HCHO/CO = 2.74, CH_3CHO/CO = 2.82) and measured CO. The ERs of OVOCs were calculated from the multivariate regression results for OVOC data at the PKU site in Yuan et al. (2012). Compared with M0–M3, the average of modeled HCHO and CH_3CHO by M4 show a better agreement with the observations. As shown in Fig. 6b and c, the maximum of modeled HCHO in M4 occurred at 15:00–16:00 LT, which was delayed by about 3 h from the observations. The diurnal pattern of simulated acetaldehyde was well matched with the observed variation most of the time. The calculated primary HCHO and CH_3CHO by M4 contributed 41 and 58 % of the total modeled concentrations, respectively, higher than the corresponding results (33 % for HCHO and 42 % for CH_3CHO) from the PMF model (Chen et al., 2014).

4.2.2 MACR

The model overpredicted the overall concentrations of MVK + MACR by 25 % (Table S3). The diurnal cycle of modeled MVK + MACR is similar in shape and magnitude to the measurements for early morning and nighttime, while the model overestimated the midday peak by a factor of 1.5 in the afternoon (Fig. 6d). This suggests that the box model is able to represent nocturnal sinks of MVK + MACR, but cannot fully explain their productions or/and sinks in the daytime. Karl et al. (2010) found that the observed deposition velocities for MVK + MACR in tropical ecosystems were up to 2.4 cm s^{-1}, 3–4 times higher than as used in our model runs (0.6–0.8 cm s^{-1}). They also reported that the uptake of MVK + MACR by vegetation followed an exponential increase with leaf temperature, and presented a light dependency as well. So, more MVK + MACR possibly deposits on leaves in the afternoon, which gives evidence of the gap between modeled and observed MVK+MACR during that period.

4.2.3 Acetone and MEK

The modeled concentrations of MEK in M0 underestimated the measurements by 56 % (given in Table S3), which is probably caused by the absence of direct anthropogenic emissions of MEK in the model. Sommariva et al. (2011) also found the underprediction of ketones using the box model in the ICARTT (International Consortium for Atmospheric Research on Transport and Transformation) campaign. Similar to HCHO and CH_3CHO, the primary ketones were added in M4 by using measured CO and emission ratios of ketones (acetone / CO = 2.23, MEK / CO = 1.22). The modeled ketones in M4 were comparable to the measurements, as listed in Table S3. For diurnal variations, the agreement between modeled and measured MEK concentrations is generally good (shown in Fig. 6f). Figure 7 indicates the simulated contributions of primary and secondary sources to MEK at the PKU site. Primary sources dominated the ambient MEK concentration (~ 80 %) on a 24 h basis, and the contribution of secondary MEK increased to 25 % at noon and early afternoon.

The model cannot reproduce the observed variations of acetone. The acetone peak at noon was probably attributed to some temperature-related emissions (e.g., solvent evaporation). As direct emissions of acetone have not always kept pace with CO, the primary acetone derived from the ER expressed by the acetone / CO ratio cannot explain it. Besides combustion sources, in urban areas acetone also directly comes from solvent usage and evaporation, particularly in chemical processing procedures. So, it needs to find a slightly reactive or inert species to work as an indicator

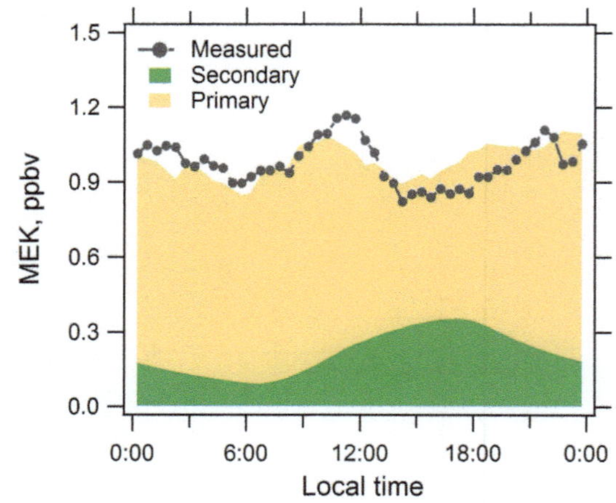

Figure 7. Partitioning of MEK to primary emissions (yellow area) and secondary chemical formation (green area) at the PKU site, and the dashed black lines with solid dots represent measured concentrations.

of non-combustion sources of acetone. CO and acetylene are frequently used as reference compounds in ERs, while they come mostly from automobile exhaust and fuel combustion, they were seldom observed in solvent usage. Other relatively inert VOCs such as ethane and propane show more variable emission ratios for different sources, so they are also not suitable to be a unique tracer for non-combustion sources. Thus, more detailed measurements on emission characteristics of acetone sources need to be investigated in further studies.

In summary, the box model overestimated the concentrations of aldehydes by a factor of 1.4–1.7 without considering the uptake by aerosols. This discrepancy is mainly attributed to missing sinks, such as vertical dilution, transport and heterogeneous uptake on aerosols. In the presence of fresh emissions, the box model predicted the concentrations and variations of MEK well, but it cannot explain observed acetone and is likely to be affected by unidentified primary emissions or high backgrounds. Despite the existence of drawbacks on estimation of sinks, the box model can still give an overview of the relative importance of different precursors on the production of secondary OVOCs. It can also be used as a tool to quantify the relative changes in OVOC production due to the reduced precursors and other pollutants during the emission control period.

4.3 Secondary production of OVOCs

Figure 8a–c illustrate averaged diurnal variations of the production of formaldehyde, acetaldehyde and acetone from different groups of NMHCs in M2 during the full control period. As an average, the photooxidation of alkenes contributed most to secondary HCHO production with a percentage of 54 %, followed by the oxidation of isoprene (30 %) and aromatics (15 %), the contributions of pinenes and alkanes were negligible (< 1 %). The contributions of C_2–C_4

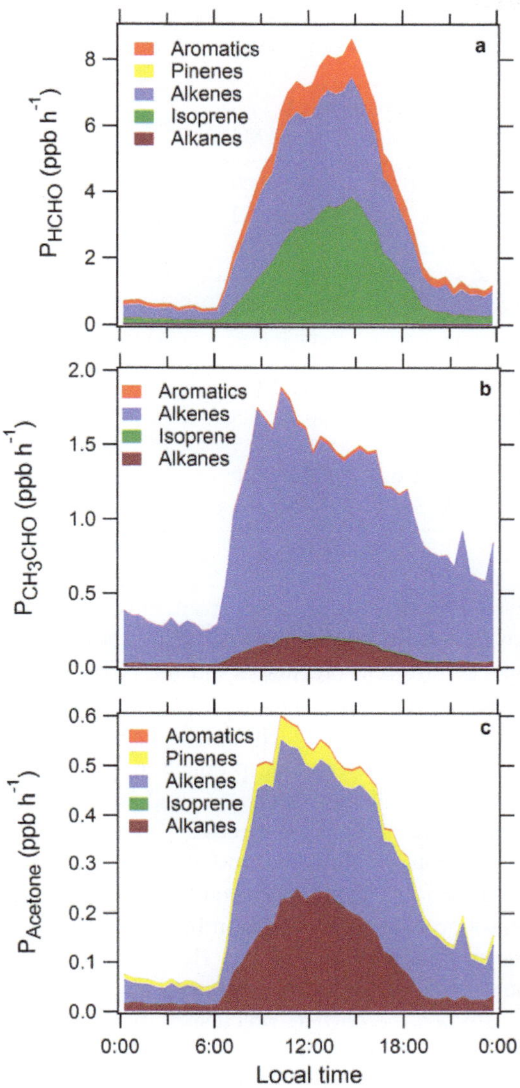

Figure 8. Averaged production of **(a)** formaldehyde, **(b)** acetaldehyde and **(c)** acetone during the full control period in the simulation of M4.

alkenes and isoprene were well matched, accounting for 39.7 and 37.5 % of the total of averaged HCHO production, respectively (listed in Table S4).

Acetaldehyde production was found to be dominated by alkene oxidation (91 %) and the contributions of alkanes and aromatics were minor (7 and 1 %, respectively). Usually the dominant secondary CH_3CHO source is from alkane oxidation (ethane, *n*-butane, *i*-butane) via the reaction of the C2H5O2 peroxy radical (Carter, 1990; Sommariva et al., 2008). But Sommariva et al. (2011) found that in urban plumes in the northeastern US, propene and other alkenes significantly contributed to the formation of acetaldehyde via the reaction of the HYPROPO alkoxy radical, especially in more fresh air masses during the first 2 days, when their contribution to acetaldehyde production is more important (up to 25 %). Then the role of C3–C5 alkenes decreases very quickly because of their high reactivity. Compared to urban plumes in the USA, air masses in Beijing are much fresher and closer to the emission sources. In Fig. 4 we can see that the majority of photochemical ages of air masses during the campaign are shorter than 40 h, i.e., within 2 days. In this case, alkene oxidations become the most important processes for CH3CHO formation.

For acetone, 60 % of its production was from alkenes, 29 % from alkanes, 10 % from pinenes, and the remaining 1 % was attributed to aromatics and isoprene. Secondary MEK was predominantly produced from the oxidation of alkanes (>95 %). The averaged results show that the oxidation of anthropogenic precursors is the dominant production pathway for secondary OVOCs in Beijing.

The NMHC oxidation for all OVOC compounds displays an obvious diurnal pattern with a peak at noon. Table 4 summarizes the averaged production rates and corresponding percentages of HCHO from different NMHC groups. For the full control period (M4), the pathway of isoprene oxidation accounted for up to 40 % of the total production in the daytime, comparable to that of anthropogenic alkenes (46 %). The remaining (13 %) of HCHO production was attributed to aromatics. The contribution of biogenic sources decreased to 23 % during the night and, accordingly, anthropogenic sources became the dominant contributor to HCHO production with a combined percentage of 76 % (alkenes + aromatics). The differences in contributions between day and night are closely related to different diurnal patterns of anthropogenic and biogenic NMHCs.

4.4 Changes in OVOCs responding to the control measures

As indicated above, ambient concentrations of NMHCs were reduced by 30–45 % due to the decrease in the city's emissions from 21 July to 27 August. Here, the effect of the reduced precursors together with other gas pollutants (CO, NO, NO_2 and O_3) on secondary OVOCs has been estimated using the box model.

Since the whole group of NMHCs were not measured by online GC-MS before the full control (3–20 July), the model scenario M5 ran with increased concentrations of alkanes, alkenes, aromatics and isoprene by 30, 35, 40 and 15 %, respectively, according to the results from this work and a previous study on NMHCs by Wang et al. (2010a), in order to simulate the situation of the uncontrolled period. The concentrations of NO, NO_2 and CO in M5 were increased by 35, 25 and 17 %, while O_3 was decreased by 16 %, taken from the observed changes in Wang et al. (2010c) and Chou et al. (2011). Then, the relative changes of secondary OVOCs responding to the control measures were determined by the difference between the model scenario of M5 (the uncontrolled period) and M4 (the full control period).

Table 4. The average of production rates and corresponding percentages of formaldehyde from different precursors in the model simulations for the full control period (M4) and the scenario without the control measures (M5). All the values are presented as average ± SD.

	09:00–17:00 LT		18:00–08:00 LT	
	M4	M5	M4	M5
$P_{alkanes}$ ($ppbh^{-1}$)	0.01 ± 0.00	0.01 ± 0.00	0.00 ± 0.00	0.00 ± 0.00
$P_{alkenes}$ ($ppbh^{-1}$)	3.08 ± 1.65	3.53 ± 1.93	0.75 ± 0.72	0.87 ± 0.82
$P_{aromatics}$ ($ppbh^{-1}$)	0.88 ± 0.50	1.00 ± 0.58	0.20 ± 0.18	0.23 ± 0.20
$P_{isoprene}$ ($ppbh^{-1}$)	2.77 ± 1.76	2.95 ± 1.97	0.35 ± 0.45	0.37 ± 0.48
$P_{pinenes}$ ($ppbh^{-1}$)	0.01 ± 0.01	0.01 ± 0.01	0.01 ± 0.01	0.01 ± 0.01
P_{total} ($ppbh^{-1}$)	6.75 ± 3.68	7.50 ± 4.22	1.31 ± 1.32	1.47 ± 1.46
$P_{alkanes}$ (%)	0.1 ± 0.1	0.1 ± 0.1	0.2 ± 0.2	0.2 ± 0.2
$P_{alkenes}$ (%)	46.5 ± 8.1	48.0 ± 8.3	59.2 ± 5.5	60.3 ± 5.5
$P_{aromatics}$ (%)	13.2 ± 1.9	13.3 ± 2.0	16.8 ± 2.6	16.7 ± 2.7
$P_{isoprene}$ (%)	40.0 ± 8.5	38.4 ± 8.6	23.1 ± 6.2	22.1 ± 6.0
$P_{pinenes}$ (%)	0.2 ± 0.1	0.1 ± 0.1	0.8 ± 0.6	0.6 ± 0.5

Table 5. The average of relative changes in OVOC concentrations from the scenario without pollution control measures (M5) to the control period (M4), compared to the corresponding values from the MLP network. Modelled change (%) = (M4–M5)/M5 × 100.0.

Compounds	Reduction from MLP %	Modeled changes (%) average ± SD
Formaldehyde	−12.9	−11.1 ± 2.4
Acetaldehyde	−15.8	−15.1 ± 1.9
MVK + MACR	−11.3	−12.0 ± 3.6
Acetone	−1.0	−12.8 ± 1.4
MEK	−17.0	−13.3 ± 0.1
OH		11.2 ± 5.5
HO_2		22.3 ± 8.1

The modeled reductions in HCHO, CH_3CHO and MVK + MACR from M5 to M4 agree quite well with the corresponding results from the MLP network (given in Table 5), suggesting the measurable changes of those three species can be reasonably represented by the photochemical production from the oxidation of reduced precursors in the box model. For MEK, the relative change from the box model is 3.7 % lower than the MLP results, which is likely related to the uncertainty of primary emissions. However, we found a large discrepancy of acetone changes in the two methods. The modeled changes (12.8 %) of acetone were much larger than the MLP result (1 %). As discussed in Sect. 3.2, there is no significant change in measured acetone during the full control period probably due to high backgrounds in the region. The change (12 %) of secondary production of acetone estimated by the model is relatively less important compared to transport in Beijing, it was covered by the high levels of acetone from a neighboring region. It seems difficult for the box model to evaluate such effects of background or transport.

The HCHO production rates from most of the precursors have been lowered during the full control period, but the ranks of different precursors did not change much (details listed in Table S4). Due to the emission restrictions the total production rate of HCHO decreased from 7.50 to 6.75 $ppbh^{-1}$, and the daytime productions of HCHO from the oxidation of alkenes, aromatics and isoprene were reduced by 13, 12 and 6 %, respectively (from Table 4). The relative contribution of alkene oxidations in the daytime was decreased by 1.5 % from the uncontrolled days (M5) to the full control period (M4) and, accordingly, the contribution of isoprene was increased by 1.7 %. Here, it should be noted that the relative importance of isoprene chemistry has been enhanced when the NMHC precursors from anthropogenic sources were reduced by 30–40 %. This would affect the relative potentials of reactive NMHCs for the formation of secondary pollutants (such as OVOCs and ozone).

5 Conclusions

In the summer of 2008 the pollution control measures implemented in Beijing provide a unique opportunity for studying how the primary and secondary air pollutants react to the reductions of anthropogenic emissions. Using a neural network analysis, we concluded that the emission controls taken in Beijing had a notable effect on reducing the ambient concentrations of formaldehyde, acetaldehyde, MEK and methanol. The influence of emission restrictions on OVOCs (except acetone) was found to be statistically significant when the variations due to meteorological conditions were excluded by the MLP network analysis. However, the effects of local emission controls on acetone seem to have been submerged in high background levels and no obvious changes were detected in acetone during the full control period. This highlights the complexity of secondary air pollutants (like

OVOCs and O_3), which needs to be treated as a regional issue.

The chemical production and degradation of OVOCs in Beijing under the emission control conditions have been determined quantitatively using a box model constrained with measurements of OVOCs precursors, gas pollutants and other physical parameters. The most important precursor of HCHO is isoprene, accounting for 30 % of the total HCHO production. However, as a whole, the anthropogenic source is still the main contributor to secondary HCHO formation, with alkene oxidation accounting for 54 % and aromatics for 15 %. Unlike HCHO, anthropogenic NMHCs were found to play predominant roles on the chemical production of acetaldehyde, acetone and MEK. Approximately 90 % of acetaldehyde production can be attributed to the oxidation of alkenes. Anthropogenic alkenes and alkanes contributed 60 and 29 % of acetone formation, respectively, and biogenic precursors (pinenes) contributed the other 10 %. The relative changes in the modeled aldehydes and MVK + MACR due to the emission restrictions are shown to be consistent with the estimated results from the neural network, reflecting that the current chemical mechanisms can largely represent the realistic formation processes of aldehydes and MVK+MACR in the high NO_x urban conditions.

Compared to measurements of OVOCs, the box model overestimated the measured aldehydes by a factor of 1.4–1.7 without taking into consideration their loss on aerosols, and it was able to roughly explain diurnal variations of acetaldehyde most of time. The model–measurement discrepancy for aldehydes is mostly caused by missing sinks in the model, including physical dilution, transport and heterogeneous processes on the surface of aerosols. After adding the primary concentrations of ketones to the box model, the concentrations and diurnal cycles of MEK can be well represented by the model simulations. But the modeled changes of acetone become unimportant due to the high background and transport of acetone in Beijing and its surrounding areas

The analysis of the summer 2008 situation in Beijing improves our understanding of the complicated relationship between anthropogenic emissions and pollution levels of primary and secondary VOCs. The concentrations of NMHCs were reduced by 30–40 % due to the emission controls. As vehicle emissions are the most important sources of VOCs in the city, the emission ratios of the traffic-related species (C_2 and C_3 alkenes and light aromatics) in the controlled days are found to be fairly similar to the values before the controls and in other urban areas. The relatively similar emission ratios for the above hydrocarbons in cities will help in providing a general idea of the photochemical evolution of gas-phase organic carbons in urban plumes based on model simulations. As China is currently in the midst of a sustained effort to improve the air quality in its megacities, the results presented in this study can serve as a case for further understanding of the atmospheric chemistry not only for Beijing but also for large regions and as an important aid for policymakers considering ways to reduce pollution more efficiently in the long term.

Acknowledgements. The authors wish to thank the entire CARE-BEIJING 2008 team for their excellent support and collaboration, especially Chunsheng Zhao's group for processing metrological data at the PKU site. We also gratefully acknowledge Joost A. deGouw's group from the NOAA Earth System Research Laboratory for the provision of VOC data measured at the PKU site in 2005. This study was funded by the Natural Science Foundation for Outstanding Young Scholars (grant no. 41125018) and Natural Science Foundation key project (grant no. 41330635). This work was also supported by the National Natural Science Foundation of China (grant no. 41005067 and 41175112).

Edited by: A. B. Guenther

References

Andreae, M. O. and Merlet, P.: Emission of trace gases and aerosols from biomass burning, Global Biogeochem. Cy., 15, 955–966, 2001.

Bohn, B., Corlett, G. K., Gillmann, M., Sanghavi, S., Stange, G., Tensing, E., Vrekoussis, M., Bloss, W. J., Clapp, L. J., Kortner, M., Dorn, H.-P., Monks, P. S., Platt, U., Plass-Dülmer, C., Mihalopoulos, N., Heard, D. E., Clemitshaw, K. C., Meixner, F. X., Prevot, A. S. H., and Schmitt, R.: Photolysis frequency measurement techniques: results of a comparison within the ACCENT project, Atmos. Chem. Phys., 8, 5373–5391, doi:10.5194/acp-8-5373-2008, 2008.

Borbon, A., Gilman, J. B., Kuster, W. C., Grand, N., Chevaillier, S., Colomb, A., Dolgorouky, C., Gros, V., Lopez, M., Sarda-Esteve, R., Holloway, J., Stutz, J., Petetin, H., McKeen, S., Beekmann, M., Warneke, C., Parrish, D. D., and de Gouw, J. A.: Emission ratios of anthropogenic volatile organic compounds in northern mid-latitude megacities: observations versus emission inventories in Los Angeles and Paris, J. Geophys. Res.-Atmos., 118, 2041–2057, doi:10.1002/Jgrd.50059, 2013.

Carter, W. P. L.: A detailed mechanism for the gas-phase atmospheric reactions of organic compounds, Atmos. Environ. Part A, 24, 481–518, 1990.

Cermak, J. and Knutti, R.: Beijing Olympics as an aerosol field experiment, Geophys. Res. Lett., 36, L10806, doi:10.1029/2009gl038572, 2009.

Chan, C. K. and Yao, X.: Air pollution in mega cities in China, Atmos. Environ., 42, 1–42, 2008.

Chen, W. T., Shao, M., Lu, S. H., Wang, M., Zeng, L. M., Yuan, B., and Liu, Y.: Understanding primary and secondary sources of ambient carbonyl compounds in Beijing using the PMF model, Atmos. Chem. Phys., 14, 3047–3062, doi:10.5194/acp-14-3047-2014, 2014.

Cheng, P., Cheng, Y. F., Lu, K. D., Su, H., Yang, Q., Zou, Y. K., Zhao, Y. R., Dong, H. B., Zeng, L. M., and Zhang, Y.: An online monitoring system for atmospheric nitrous acid (HONO) based on stripping coil and ion chromatography, J. Environ. Sci.-China, 25, 895–907, doi:10.1016/S1001-0742(12)60251-4, 2013.

Cheng, Y. F., Heintzenberg, J., Wehner, B., Wu, Z. J., Su, H., Hu, M., and Mao, J. T.: Traffic restrictions in Beijing during the Sino-African Summit 2006: aerosol size distribution and visibility compared to long-term in situ observations, Atmos. Chem. Phys., 8, 7583–7594, doi:10.5194/acp-8-7583-2008, 2008.

Chou, C. C.-K., Tsai, C.-Y., Chang, C.-C., Lin, P.-H., Liu, S. C., and Zhu, T.: Photochemical production of ozone in Beijing during the 2008 Olympic Games, Atmos. Chem. Phys., 11, 9825–9837, doi:10.5194/acp-11-9825-2011, 2011.

Dasgupta, P. K., Li, J. Z., Zhang, G. F., Luke, W. T., McClenny, W. A., Stutz, J., and Fried, A.: Summertime ambient formaldehyde in five US metropolitan areas: Nashville, Atlanta, Houston, Philadelphia, and Tampa, Environ. Sci. Technol., 39, 4767–4783, doi:10.1021/Es048327d, 2005.

de Gouw, J. and Warneke, C.: Measurements of volatile organic compounds in the earths atmosphere using proton-transfer-reaction mass spectrometry, Mass Spectrom. Rev., 26, 223–257, 2007.

de Gouw, J. A., Goldan, P. D., Warneke, C., Kuster, W. C., Roberts, J. M., Marchewka, M., Bertman, S. B., Pszenny, A. A. P., and Keene, W. C.: Validation of proton transfer reaction-mass spectrometry (PTR-MS) measurements of gas-phase organic compounds in the atmosphere during the New England Air Quality Study (NEAQS) in 2002, J. Geophys. Res.-Atmos., 108, 4682, doi:10.1029/2003JD003863, 2003.

de Gouw, J. A., Middlebrook, A. M., Warneke, C., Goldan, P. D., Kuster, W. C., Roberts, J. M., Fehsenfeld, F. C., Worsnop, D. R., Canagaratna, M. R., Pszenny, A. A. P., Keene, W. C., Marchewka, M., Bertman, S. B., and Bates, T. S.: Budget of organic carbon in a polluted atmosphere: Results from the New England Air Quality Study in 2002, J. Geophys Res.-Atmos, 110, D16305, doi:10.1029/2004JD005623, 2005.

Emmerson, K. M., Carslaw, N., and Pilling, M. J.: Urban atmospheric chemistry during the PUMA campaign 2: Radical budgets for OH, HO_2 and RO_2, J. Atmos. Chem., 52, 165–183, doi:10.1007/s10874-005-1323-2, 2005.

Emmerson, K. M., Carslaw, N., Carslaw, D. C., Lee, J. D., McFiggans, G., Bloss, W. J., Gravestock, T., Heard, D. E., Hopkins, J., Ingham, T., Pilling, M. J., Smith, S. C., Jacob, M., and Monks, P. S.: Free radical modelling studies during the UK TORCH Campaign in Summer 2003, Atmos. Chem. Phys., 7, 167–181, doi:10.5194/acp-7-167-2007, 2007.

Finlayson-Pitts, B. J. and Pitts, J. N.: Chemistry of the Upper and Lower Atmosphere: Theory, Experiments, and Application, Academic Press, San Diego, California, 2000.

Goldan, P. D., Kuster, W. C., Williams, E., Murphy, P. C., Fehsenfeld, F. C., and Meagher, J.: Nonmethane hydrocarbon and oxy hydrocarbon measurements during the 2002 New England Air Quality Study, J. Geophys. Res.-Atmos., 109, D21309, doi:10.1029/2003JD004455, 2004.

Grosjean, D., Grosjean, E., and Moreira, L. F. R.: Speciated ambient carbonyls in Rio de Janeiro, Brazil, Environ. Sci. Technol., 36, 1389–1395, 2002.

Huang, X. F., He, L. Y., Hu, M., Canagaratna, M. R., Sun, Y., Zhang, Q., Zhu, T., Xue, L., Zeng, L. W., Liu, X. G., Zhang, Y. H., Jayne, J. T., Ng, N. L., and Worsnop, D. R.: Highly time-resolved chemical characterization of atmospheric submicron particles during 2008 Beijing Olympic Games using an Aerodyne High-Resolution Aerosol Mass Spectrometer, Atmos. Chem. Phys., 10, 8933–8945, doi:10.5194/acp-10-8933-2010, 2010.

Jayne, J. T., Worsnop, D. R., Kolb, C. E., Swartz, E., and Davidovits, P.: Uptake of gas-phase formaldehyde by aqueous acid surfaces, J. Phys. Chem., 100, 8015–8022, doi:10.1021/jp953196b, 1996.

Karl, T., Harley, P., Emmons, L., Thornton, B., Guenther, A., Basu, C., Turnipseed, A., and Jardine, K.: Efficient Atmospheric Cleansing of Oxidized Organic Trace Gases by Vegetation, Science, 330, 816–819, doi:10.1126/science.1192534, 2010.

Li, X., Rohrer, F., Brauers, T., Hofzumahaus, A., Lu, K., Shao, M., Zhang, Y. H., and Wahner, A.: Modeling of HCHO and CHOCHO at a semi-rural site in southern China during the PRIDE-PRD2006 campaign, Atmos. Chem. Phys., 14, 12291–12305, doi:10.5194/acp-14-12291-2014, 2014.

Li, Y., Shao, M., Lu, S. H., Chang, C. C., and Dasgupta, P. K.: Variations and sources of ambient formaldehyde for the 2008 Beijing Olympic games, Atmos. Environ., 44, 2632–2639, doi:10.1016/j.atmosenv.2010.03.045, 2010.

Lindinger, W., Hansel, A., and Jordan, A.: On-line monitoring of volatile organic compounds at pptv levels by means of proton-transfer-reaction mass spectrometry (PTR-MS) – medical applications, food control and environmental research, Int. J. Mass Spectrom., 173, 191–241, 1998.

Liu, P. F., Zhao, C. S., Göbel, T., Hallbauer, E., Nowak, A., Ran, L., Xu, W. Y., Deng, Z. Z., Ma, N., Mildenberger, K., Henning, S., Stratmann, F., and Wiedensohler, A.: Hygroscopic properties of aerosol particles at high relative humidity and their diurnal variations in the North China Plain, Atmos. Chem. Phys., 11, 3479–3494, doi:10.5194/acp-11-3479-2011, 2011.

Liu, P. F., Zhao, C. S., Göbel, T., Hallbauer, E., Nowak, A., Ran, L., Xu, W. Y., Deng, Z. Z., Ma, N., Mildenberger, K., Henning, S., Stratmann, F., and Wiedensohler, A.: Hygroscopic properties of aerosol particles at high relative humidity and their diurnal variations in the North China Plain, Atmos. Chem. Phys., 11, 3479–3494, doi:10.5194/acp-11-3479-2011, 2011.

Liu, X. G., Li, J., Qu, Y., Han, T., Hou, L., Gu, J., Chen, C., Yang, Y., Liu, X., Yang, T., Zhang, Y., Tian, H., and Hu, M.: Formation and evolution mechanism of regional haze: a case study in the megacity Beijing, China, Atmos. Chem. Phys., 13, 4501–4514, doi:10.5194/acp-13-4501-2013, 2013.

Liu, Y., Shao, M., Zhang, J., Fu, L. L., and Lu, S. H.: Distributions and source apportionment of ambient volatile organic compounds in Beijing city, China, J. Environ. Sci. Heal. A, 40, 1843–1860, 2005.

Liu, Y., Shao, M., Kuster, W. C., Goldan, P. D., Li, X. H., Lu, S. H., and de Gouw, J. A.: Source identification of reactive hydrocarbons and oxygenated VOCs in the summertime in Beijing, Environ. Sci. Technol., 43, 75–81, 2009.

Lu, K. D., Rohrer, F., Holland, F., Fuchs, H., Bohn, B., Brauers, T., Chang, C. C., Häseler, R., Hu, M., Kita, K., Kondo, Y., Li, X., Lou, S. R., Nehr, S., Shao, M., Zeng, L. M., Wahner, A., Zhang, Y. H., and Hofzumahaus, A.: Observation and modelling of OH and HO_2 concentrations in the Pearl River Delta 2006:

a missing OH source in a VOC rich atmosphere, Atmos. Chem. Phys., 12, 1541–1569, doi:10.5194/acp-12-1541-2012, 2012.

Lu, K. D., Hofzumahaus, A., Holland, F., Bohn, B., Brauers, T., Fuchs, H., Hu, M., Häseler, R., Kita, K., Kondo, Y., Li, X., Lou, S. R., Oebel, A., Shao, M., Zeng, L. M., Wahner, A., Zhu, T., Zhang, Y. H., and Rohrer, F.: Missing OH source in a suburban environment near Beijing: observed and modelled OH and HO_2 concentrations in summer 2006, Atmos. Chem. Phys., 13, 1057–1080, doi:10.5194/acp-13-1057-2013, 2013.

Lu, K. D., Rohrer, F., Holland, F., Fuchs, H., Brauers, T., Oebel, A., Dlugi, R., Hu, M., Li, X., Lou, S. R., Shao, M., Zhu, T., Wahner, A., Zhang, Y. H., and Hofzumahaus, A.: Nighttime observation and chemistry of HO_x in the Pearl River Delta and Beijing in summer 2006, Atmos. Chem. Phys., 14, 4979–4999, doi:10.5194/acp-14-4979-2014, 2014.

Lu, S., Liu, Y., Shao, M., and Huang, S.: Chemical speciation and anthropogenic sources of ambient volatile organic compounds during summer in Beijing City, Frontiers of Environmental Science & Engineering, 1, 1–6, 2007.

Lyapustin, A., Smirnov, A., Holben, B., Chin, M., Streets, D. G., Lu, Z., Kahn, R., Slutsker, I., Laszlo, I., Kondragunta, S., Tanre, D., Dubovik, O., Goloub, P., Chen, H. B., Sinyuk, A., Wang, Y., and Korkin, S.: Reduction of aerosol absorption in Beijing since 2007 from MODIS and AERONET, Geophys. Res. Lett., 38, doi:10.1029/2011gl047306, 2011.

Mason, S. A., Field, R. J., Yokelson, R. J., Kochivar, M. A., Tinsley, M. R., Ward, D. E., and Hao, W. M.: Complex effects arising in smoke plume simulations due to inclusion of direct emissions of oxygenated organic species from biomass combustion, J. Geophys. Res.-Atmos., 106, 12527–12539, 2001.

Pang, X. B. and Mu, Y. J.: Seasonal and diurnal variations of carbonyl compounds in Beijing ambient air, Atmos. Environ., 40, 6313–6320, 2006.

Park, J.-H., Goldstein, A. H., Timkovsky, J., Fares, S., Weber, R., Karlik, J., and Holzinger, R.: Active atmosphere–ecosystem exchange of the vast majority of detected volatile organic compounds, Science, 341, 643–647, doi:10.1126/science.1235053, 2013.

Parrish, D. D. and Zhu, T.: Clean air for megacities, Science, 326, 674–675, doi:10.1126/science.1176064, 2009.

Riemer, N., West, M., Zaveri, R. A., and Easter, R. C.: Simulating the evolution of soot mixing state with a particle-resolved aerosol model, J. Geophys. Res.-Atmos., 114, D09202, doi:10.1029/2008jd011073, 2009.

Rohrer, F. and Berresheim, H.: Strong correlation between levels of tropospheric hydroxyl radicals and solar ultraviolet radiation, Nature, 442, 184–187, doi:10.1038/nature04924, 2006.

Rohrer, F., Lu, K., Hofzumahaus, A., Bohn, B., Brauers, T., Chang, C.-C., Fuchs, H., Haseler, R., Holland, F., Hu, M., Kita, K., Kondo, Y., Li, X., Lou, S., Oebel, A., Shao, M., Zeng, L., Zhu, T., Zhang, Y., and Wahner, A.: Maximum efficiency in the hydroxyl-radical-based self-cleansing of the troposphere, Nat. Geosci., 7, 559–563, doi:10.1038/ngeo2199, 2014.

Saunders, S. M., Jenkin, M. E., Derwent, R. G., and Pilling, M. J.: Protocol for the development of the Master Chemical Mechanism, MCM v3 (Part A): tropospheric degradation of non-aromatic volatile organic compounds, Atmos. Chem. Phys., 3, 161–180, doi:10.5194/acp-3-161-2003, 2003.

Schauer, J. J., Kleeman, M. J., Cass, G. R., and Simoneit, B. R. T.: Measurement of emissions from air pollution sources. 2. C-1 through C-30 organic compounds from medium duty diesel trucks, Environ. Sci. Technol., 33, 1578–1587, 1999.

Schauer, J. J., Kleeman, M. J., Cass, G. R., and Simoneit, B. R. T.: Measurement of emissions from air pollution sources. 5. C-1–C-32 organic compounds from gasoline-powered motor vehicles, Environ. Sci. Technol., 36, 1169–1180, 2002.

Shao, M., Wang, B., Lu, S. H., Liu, S. C., and Chang, C. C.: Trends in summertime non-methane hydrocarbons in Beijing city, 2004–2009, IGACtivities Newsletter, 42, 18–25, 2009.

Shao, M., Wang, B., Lu, S. H., Yuan, B., and Wang, M.: Effects of Beijing Olympics control measures on reducing reactive hydrocarbon species, Environ. Sci. Technol., 45, 514–519, doi:10.1021/es102357t, 2011.

Singh, H., Chen, Y., Staudt, A., Jacob, D., Blake, D., Heikes, B., and Snow, J.: Evidence from the Pacific troposphere for large global sources of oxygenated organic compounds, Nature, 410, 1078-1081, 2001.

Singh, H., O'Hara, D., Herlth, D., Sachse, W., Blake, D. R., Bradshaw, J. D., Kanakidou, M., and Crutzen, P. J.: Acetone in the atmosphere: distribution, sources, and sinks, J. Geophys. Res.-Atmos., 99, 1805–1819, 1994.

Sommariva, R., Trainer, M., de Gouw, J. A., Roberts, J. M., Warneke, C., Atlas, E., Flocke, F., Goldan, P. D., Kuster, W. C., Swanson, A. L., and Fehsenfeld, F. C.: A study of organic nitrates formation in an urban plume using a Master Chemical Mechanism, Atmos. Environ., 42, 5771–5786, doi:10.1016/j.atmosenv.2007.12.031, 2008.

Sommariva, R., de Gouw, J. A., Trainer, M., Atlas, E., Goldan, P. D., Kuster, W. C., Warneke, C., and Fehsenfeld, F. C.: Emissions and photochemistry of oxygenated VOCs in urban plumes in the Northeastern United States, Atmos. Chem. Phys., 11, 7081–7096, doi:10.5194/acp-11-7081-2011, 2011.

Song, Y., Shao, M., Liu, Y., Lu, S. H., Kuster, W., Goldan, P., and Xie, S. D.: Source apportionment of ambient volatile organic compounds in Beijing, Environ. Sci. Technol., 41, 4348–4353, 2007.

Su, F.: Study on the Variation and Sources of Volatile Oragnic Compounds in Beijing, Ph.D. thesis, Peking University, 2003 (in Chinese).

Toda, K., Yunoki, S., Yanaga, A., Takeuchi, M., Ohira, S.-I., and Dasgupta, P. K.: Formaldehyde Content of Atmospheric Aerosol, Environ. Sci. Technol., 48, 6636–6643, doi:10.1021/es500590e, 2014.

UNEP: Independent Envrironmental Assessment: Beijing 2008 Olympic Games, UNEP, Nairobi, Kenya, 2009.

Wang, Q., Shao, M., Liu, Y., Kuster, W., Goldan, P., Li, X., Liu, Y., and Lu, S.: Impact of biomass burning on urban air quality estimated by organic tracers: Guangzhou and Beijing as cases, Atmos. Environ., 41, 8380–8390, 2007.

Wang, T. and Xie, S. D.: Assessment of traffic-related air pollution in the urban streets before and during the 2008 Beijing Olympic Games traffic control period, Atmos. Environ., 43, 5682–5690, doi:10.1016/j.atmosenv.2009.07.034, 2009.

Wang, M., Zhu, T., Zheng, J., Zhang, R. Y., Zhang, S. Q., Xie, X. X., Han, Y. Q., and Li, Y.: Use of a mobile laboratory to evaluate changes in on-road air pollutants during the Beijing

2008 Summer Olympics, Atmos. Chem. Phys., 9, 8247–8263, doi:10.5194/acp-9-8247-2009, 2009a.

Wang, X., Westerdahl, D., Chen, L. C., Wu, Y., Hao, J. M., Pan, X. C., Guo, X. B., and Zhang, K. M.: Evaluating the air quality impacts of the 2008 Beijing Olympic Games: on-road emission factors and black carbon profiles, Atmos. Environ., 43, 4535–4543, doi:10.1016/j.atmosenv.2009.06.054, 2009b.

Wang, B., Shao, M., Lu, S. H., Yuan, B., Zhao, Y., Wang, M., Zhang, S. Q., and Wu, D.: Variation of ambient non-methane hydrocarbons in Beijing city in summer 2008, Atmos. Chem. Phys., 10, 5911–5923, doi:10.5194/acp-10-5911-2010, 2010a.

Wang, S. X., Zhao, M., Xing, J., Wu, Y., Zhou, Y., Lei, Y., He, K. B., Fu, L. X., and Hao, J. M.: Quantifying the air pollutants emission reduction during the 2008 Olympic Games in Beijing, Environ. Sci. Technol., 44, 2490–2496, doi:10.1021/Es9028167, 2010b.

Wang, T., Nie, W., Gao, J., Xue, L. K., Gao, X. M., Wang, X. F., Qiu, J., Poon, C. N., Meinardi, S., Blake, D., Wang, S. L., Ding, A. J., Chai, F. H., Zhang, Q. Z., and Wang, W. X.: Air quality during the 2008 Beijing Olympics: secondary pollutants and regional impact, Atmos. Chem. Phys., 10, 7603–7615, doi:10.5194/acp-10-7603-2010, 2010c.

Wang, M., Shao, M., Chen, W., Yuan, B., Lu, S., Zhang, Q., Zeng, L., and Wang, Q.: A temporally and spatially resolved validation of emission inventories by measurements of ambient volatile organic compounds in Beijing, China, Atmos. Chem. Phys., 14, 5871–5891, doi:10.5194/acp-14-5871-2014, 2014.

Wang, Z. B., Hu, M., Yue, D. L., Zheng, J., Zhang, R. Y., Wiedensohler, A., Wu, Z. J., Nieminen, T., and Boy, M.: Evaluation on the role of sulfuric acid in the mechanisms of new particle formation for Beijing case, Atmos. Chem. Phys., 11, 12663–12671, doi:10.5194/acp-11-12663-2011, 2011.

Warneke, C., McKeen, S. A., de Gouw, J. A., Goldan, P. D., Kuster, W. C., Holloway, J. S., Williams, E. J., Lerner, B. M., Parrish, D. D., Trainer, M., Fehsenfeld, F. C., Kato, S., Atlas, E. L., Baker, A., and Blake, D. R.: Determination of urban volatile organic compound emission ratios and comparison with an emissions database, J. Geophys. Res.-Atmos., 112, D10S47, doi:10.1029/2006JD007930, 2007.

Witte, J. C., Schoeberl, M. R., Douglass, A. R., Gleason, J. F., Krotkov, N. A., Gille, J. C., Pickering, K. E., and Livesey, N.: Satellite observations of changes in air quality during the 2008 Beijing Olympics and Paralympics, Geophys. Res. Lett., 36, L17803, doi:10.1029/2009gl039236, 2009.

Worden, H. M., Cheng, Y. F., Pfister, G., Carmichael, G. R., Zhang, Q., Streets, D. G., Deeter, M., Edwards, D. P., Gille, J. C., and Worden, J. R.: Satellite-based estimates of reduced CO and CO_2 emissions due to traffic restrictions during the 2008 Beijing Olympics, Geophys. Res. Lett., 39, L14802, doi:10.1029/2012gl052395, 2012.

Wu, Z. J., Hu, M., Lin, P., Liu, S., Wehner, B., and Wiedensohler, A.: Particle number size distribution in the urban atmosphere of Beijing, China, Atmos. Environ., 42, 7967–7980, doi:10.1016/j.atmosenv.2008.06.022, 2008.

Yokelson, R. J., Crounse, J. D., DeCarlo, P. F., Karl, T., Urbanski, S., Atlas, E., Campos, T., Shinozuka, Y., Kapustin, V., Clarke, A. D., Weinheimer, A., Knapp, D. J., Montzka, D. D., Holloway, J., Weibring, P., Flocke, F., Zheng, W., Toohey, D., Wennberg, P. O., Wiedinmyer, C., Mauldin, L., Fried, A., Richter, D., Walega, J., Jimenez, J. L., Adachi, K., Buseck, P. R., Hall, S. R., and Shetter, R.: Emissions from biomass burning in the Yucatan, Atmos. Chem. Phys., 9, 5785–5812, doi:10.5194/acp-9-5785-2009, 2009.

Yuan, B., Liu, Y., Shao, M., Lu, S., and Streets, D. G.: Biomass burning contributions to ambient VOCs species at a receptor site in the Pearl River Delta (PRD), China, Environ. Sci. Technol., 44, 4577–4582, doi:10.1021/es1003389, 2010a.

Yuan, B., Shao, M., Lu, S. H., and Wang, B.: Source profiles of volatile organic compounds associated with solvent use in Beijing, China, Atmos. Environ., 44, 1919–1926, doi:10.1016/j.atmosenv.2010.02.014, 2010b.

Yuan, B., Shao, M., de Gouw, J., Parrish, D. D., Lu, S., Wang, M., Zeng, L., Zhang, Q., Song, Y., Zhang, J., and Hu, M.: Volatile organic compounds (VOCs) in urban air: How chemistry affects the interpretation of positive matrix factorization (PMF) analysis, J. Geophys. Res.-Atmos., 117, D24302, doi:10.1029/2012jd018236, 2012.

Yuan, B., Warneke, C., Shao, M., and de Gouw, J. A.: Interpretation of volatile organic compound measurements by proton-transfer-reaction mass spectrometry over the deepwater horizon oil spill, Int. J. Mass. Spectrom., 358, 43–48, doi:10.1016/j.ijms.2013.11.006, 2014.

Zhang, Q., Jimenez, J. L., Worsnop, D. R., and Canagaratna, M.: A case study of urban particle acidity and its influence on secondary organic aerosol, Environ. Sci. Technol., 41, 3213–3219, doi:10.1021/Es061812j, 2007.

Zhang, X. Y., Wang, Y. Q., Lin, W. L., Zhang, Y. M., Zhang, X. C., Gong, S., Zhao, P., Yang, Y. Q., Wang, J. Z., Hou, Q., Zhang, X. L., Che, H. Z., Guo, J. P., and Li, Y.: Changes of atmospheric composition and optical properties over beijing 2008 Olympic Monitoring Campaign, Bull. Am. Meteorol. Soc., 90, 1633–1651, doi:10.1175/2009bams2804.1, 2009.

6

Two years of near real-time chemical composition of submicron aerosols in the region of Paris using an Aerosol Chemical Speciation Monitor (ACSM) and a multi-wavelength Aethalometer

J.-E. Petit[1,2,*], O. Favez[1], J. Sciare[2], V. Crenn[2], R. Sarda-Estève[2], N. Bonnaire[2], G. Močnik[3], J.-C. Dupont[4], M. Haeffelin[4], and E. Leoz-Garziandia[1]

[1]Institut National de l'Environnement Industriel et des Risques, Verneuil-en-Halatte, France
[2]Laboratoire des Sciences du Climat et de l'Environnement (CNRS-CEA-UVSQ), CEA Orme des Merisiers, Gif-sur-Yvette, France
[3]Aerosol d.o.o., Ljubljana, Slovenia
[4]Laboratoire de Météorologie Dynamique, Institut Pierre Simon Laplace, Ecole Polytechnique, Palaiseau, France
[*]now at: Air Lorraine, 20 rue Pierre Simon Laplace 57000 Metz, France

Correspondence to: J.-E. Petit (je.petit@air-lorraine.org)

Abstract. Aerosol mass spectrometer (AMS) measurements have been successfully used towards a better understanding of non-refractory submicron (PM_1) aerosol chemical properties based on short-term campaigns. The recently developed Aerosol Chemical Speciation Monitor (ACSM) has been designed to deliver quite similar artifact-free chemical information but for low cost, and to perform robust monitoring over long-term periods. When deployed in parallel with real-time black carbon (BC) measurements, the combined data set allows for a quasi-comprehensive description of the whole PM_1 fraction in near real time. Here we present 2-year long ACSM and BC data sets, between mid-2011 and mid-2013, obtained at the French atmospheric SIRTA supersite that is representative of background PM levels of the region of Paris. This large data set shows intense and time-limited (a few hours) pollution events observed during wintertime in the region of Paris, pointing to local carbonaceous emissions (mainly combustion sources). A non-parametric wind regression analysis was performed on this 2-year data set for the major PM_1 constituents (organic matter, nitrate, sulfate and source apportioned BC) and ammonia in order to better refine their geographical origins and assess local/regional/advected contributions whose information is mandatory for efficient mitigation strategies. While ammonium sulfate typically shows a clear advected pattern, ammonium nitrate partially displays a similar feature, but, less expectedly, it also exhibits a significant contribution of regional and local emissions. The contribution of regional background organic aerosols (OA) is significant in spring and summer, while a more pronounced local origin is evidenced during wintertime, whose pattern is also observed for BC originating from domestic wood burning. Using time-resolved ACSM and BC information, seasonally differentiated weekly diurnal profiles of these constituents were investigated and helped to identify the main parameters controlling their temporal variations (sources, meteorological parameters). Finally, a careful investigation of all the major pollution episodes observed over the region of Paris between 2011 and 2013 was performed and classified in terms of chemical composition and the BC-to-sulfate ratio used here as a proxy of the local/regional/advected contribution of PM. In conclusion, these first 2-year quality-controlled measurements of ACSM clearly demonstrate their great potential to monitor on a long-term basis aerosol sources and their geographical origin and provide strategic information in near real time during pollution episodes. They also support the capacity of the ACSM to be proposed as a robust and credible alternative to filter-based sampling techniques for long-term monitoring strategies.

1 Introduction

The understanding of the formation and fate of atmospheric particulate pollution in urban areas represents sanitary, scientific, economic, societal and political challenges, greatly amplified by increasing media coverage of pollution episodes all around the world. Growing evidence of adverse health effects of atmospheric pollutants (Chow et al., 2006; Pope and Dockery, 2006; Ramgolam et al., 2009) is illustrated by the fact that ambient air pollution has been characterized as carcinogenic since December 2013 by the International Agency for Research on Cancer (IARC, 2013). However, the "aerosol cocktail effect", directly linked to the complexity of the chemical composition and sources of the particulate phase, remains poorly understood.

In an effort to fill these lacks of knowledge, worldwide coordinated networking activities (such as Global Atmosphere Watch, European Monitoring and Evaluation Programme, and AErosol RObotic NETwork) have been documenting, for decades, the chemical, physical and optical properties of aerosol pollution in various environments. At a European level, this effort is also supported by the Aerosols, Clouds and Traces gases Research InfraStructure network (ACTRIS) program that aims at pooling high-quality data from state-of-the-art instrumentation such as the Aerosol Chemical Speciation Monitor (ACSM, Aerodyne Research Inc., Billerica, MA, USA).

The ACSM has recently been developed with the aim of robust and easy-to-use near real-time and artifact-free measurements of the major chemical composition of non-refractory submicron aerosol (organic matter (OM), NO_3^-, SO_4^{2-}, NH_4^+ and Cl^-) on a long-term basis (Ng et al., 2011). In parallel, a growing interest is also dedicated worldwide to the monitoring of black carbon (BC), considered as an adequate indicator of potential anthropogenic emissions having sanitary impacts (Janssen et al., 2011). In particular, the use of a seven-wavelength Aethalometer (Magee Scientific, USA) allows furthermore for BC source apportionment (Sandradewi et al., 2008), proven as robust over long-term periods (Herich et al., 2011). The combination of measurements from both instruments may thus constitute an efficient and relatively low-cost tool for the monitoring of submicron aerosol chemistry and a better knowledge of their phenomenology. Such a strategy may be particularly useful for documenting aerosol sources and their geographical origin large urban areas that are characterized by a complex mixture of gaseous and particulate pollutions. Paris (France) is one of the largest European megacities and is rather isolated from other major urban environments. With ~ 11 million inhabitants, the Paris region accounts for 20 % of the total French population distributed over only 2 % of its territory, leading to enhanced exposure to various types of pollution. Moreover, the flat orography of the Paris region favors pollution transport, making it representative of northwestern European aerosol pollution. Airparif, the regional air quality monitoring network, recently estimated that, since 2007, about 2 million people per year have been exposed to poor air quality (referring to the daily PM_{10} concentration for European limit values; AIRPARIF, 2014) in this region. Over the past 7 years, annual $PM_{2.5}$ concentrations in Paris have remained quite stable, although no continuous monitoring of the chemical composition of the particulate phase is available to investigate any trends in the major sources of fine aerosols.

A recent research program, based on a 1-year (2009–2010) daily filter sampling carried out at five various sites (traffic, urban, suburban and regional backgrounds; Ghersi et al., 2010), was a unique opportunity to give insight into the seasonal variations, sources and geographical origins of aerosol pollution in the region of Paris (Bressi et al., 2013, 2014; Petetin et al., 2014). However, long-term monitoring strategies based on the chemical analysis of aerosols sampled on filters are subject to various sampling and analytical artifacts (Appel et al., 1984; Turpin et al., 1994; Pathak et al., 2004; Cheng et al., 2009) and assumptions (an OC-to-OM ratio, for instance): they involve laborious laboratory analyses, they cannot capture processes governing diurnal variations of atmospheric pollutants, and fail to provide rapid diagnostics during pollution events.

In this context, Aerosol Mass Spectrometer (AMS) techniques have provided extremely valuable information on the artifact-free real-time chemical composition of submicron aerosols in urban areas over the past 10 years (Zhang et al., 2004, 2007; Jimenez et al., 2009). In Europe, OM and ammonium nitrate are generally the two main constituents of PM_1 (Zhang et al., 2007), showing, however, significant discrepancies during pollution episodes in terms of chemical composition. Real-time AMS data have improved the understanding of the physical and chemical (trans)formation pathways of both fractions, through the characterization of pollution dynamics and source apportionment analyses. Intensive field campaigns involving AMS measurements were performed during the 2009 summer and 2010 winter seasons in the framework of the European MEGAPOLI (Megacities: Emissions, urban, regional and Global Atmospheric POLlution and climate effects, and Integrated tools for assessment and mitigation) research program. They greatly improved the understanding of the sources and transformation processes of Paris aerosols, and especially their submicron organic fraction (Crippa et al., 2013a, b and c; Freutel et al., 2013; Healy et al., 2012, 2013; Laborde et al., 2013; Zhang et al., 2013). However, AMS techniques' cost, size and intensive control requirements make them impractical for unattended monitoring. Nevertheless, they may still represent the best strategy for investigating specific trends in aerosol sources, especially in the context of elevated and stable PM concentrations as observed over the region of Paris during the past few years. From that perspective, the recently commercialized ACSM may represent an interesting alternative and may ultimately represent the best strategy to deploy for long-term monitoring of submicron aerosol sources and geographical origins.

As part of the ACTRIS project, a new in situ atmospheric station was implemented in 2011 at a background site of the region of Paris, allowing the chemical, physical and optical characterization of submicron aerosol pollution on a regional scale. The key aim of the present paper is to describe and discuss one of the first long-term data sets obtained with the ACSM, offering opportunities for the evaluation of the scientific relevance of a new experimental strategy for long-term monitoring of near real-time chemical composition of PM_1. Seasonal trends, wind sector analysis, diurnal variations and pollution episodes retrieved from 2-year real-time measurement ACSM and BC data sets are presented and interpreted in order to refine the origins and parameters controlling the (trans)formation of particulate pollution over the region of Paris.

2 Material and methods

2.1 Sampling site and instrumentation

Long-term in situ observations of the chemical, optical and physical properties of atmospheric aerosols have been initiated at SIRTA (Site Instrumental de Recherche par Télédétection Atmosphérique, http://sirta.ipsl.fr) since June 2011 within the EU-FP7 ACTRIS (Aerosols, Clouds, and Traces gases Research InfraStructure Network, http://www.actris.net) program. Located 20 km southwest of Paris (2.15° E, 48.71° N, 150 m above sea level) in a semi-rural area, this atmospheric supersite is representative of the regional background pollution over the region of Paris (Haeffelin et al., 2005; Crippa et al., 2013a).

The chemical composition of non-refractory submicron aerosol has been continuously monitored using a Quadripole Aerosol Chemical Speciation Monitor (Aerodyne Research Inc.), which has been described in detail by Ng et al. (2011). Briefly, $PM_{2.5}$ aerosols are sampled at 3 L min^{-1} (from a $PM_{2.5}$ cyclone inlet) and then sub-sampled at 85 mL min^{-1} (volumetric flow) through an aerodynamic lens, focusing submicron particles (40–1000 nm aerodynamic diameter, A.D.) onto a 600 °C-heated conical tungsten vaporizer where non-refractory material is flash-vaporized and quasi-instantaneously ionized by electron impact at 70 eV. Fragments are detected following their mass-to-charge ratio by a quadrupole mass spectrometer. Briefly, the instrument calibration has been performed following the recommendation of Jayne et al. (2000) and Ng et al. (2011), where generated mono-disperse 300 nm A.D. ammonium nitrate particles are injected into both ACSM and a condensation particle counter (CPC) at different concentrations. Throughout the measuring period, three response factor (RF) calibrations and one $(NH_4)_2SO_4$ calibration were performed, summarized in Table 1. The low drift of the obtained slopes allowed the use of an average response factor of 2.72×10^{-11} (with a standard deviation of ±13 %), and relative ion efficiencies (RIE) of 5.9, 1.2 and 1.4 for ammonium, sulfate and organic matter, respectively, were used for the whole data set. Collection efficiencies were corrected using algorithms proposed by Middlebrook et al. (2012), and data were finally cross-validated using collocated PM_1 as well as $PM_{2.5}$ urban background measurements, retrieved from the regional association of air quality monitoring (AIRPARIF, http://www.airparif.asso.fr). The PM_1 and $PM_{2.5}$ data sets were obtained using tapered element oscillating microbalances (TEOM) equipped filter dynamic measurement systems (FDMS) as described by Grover (2005). A comprehensive determination of the overall uncertainty (as well as PM_1 components) associated with ACSM-derived measurements was carried out in November 2013 through an inter-comparison exercise (Crenn et al., submitted; Fröhlich et al., 2015). Here, the consistency of ACSM measurements has been assessed from the comparison with co-located measurements, as described in Sect. 3.

Table 1. Response factors obtained through IE calibrations from June 2011 to May 2013.

Date	Response factor	RIE_{NH_4}	RIE_{SO_4}
16/11/2011	2.31×10^{-11}	6	–
09/10/2012	2.98×10^{-11}	4.8	–
15/05/2013	2.84×10^{-11}	6.84	1.25
Average	2.72×10^{-11}	5.88	–
Standard deviation	13 %	17 %	–

A total number of ~26 000 ACSM data points (with a temporal resolution of 30 min) were collected from June 2011 to June 2013 covering on average 92 ± 9 % of each month over this 2-year period (it is to be noted that September–October 2012 and February–March 2013 were not taken into account within the latter calculation because the instrument was used for short-term intensive campaigns at other locations).

Aerosol light absorption coefficients b_{abs} were retrieved every 5 min from a seven-wavelength (370, 470, 520, 590, 660, 880 and 950 nm) AE31 Aethalometer from June 2011 to February 2013, and from a seven-wavelength (370, 470, 520, 590, 660, 880 and 950 nm) AE33 Aethalometer from February 2013 to May 2013. In both cases, instruments sampled aerosols with a $PM_{2.5}$ cut-off inlet, operating at 5 L min^{-1}. Filter-based absorption measurements need to be compensated for by multiple scattering in the filter matrix and for loading effects, using mathematical algorithms (Collaud Coen et al., 2010). While AE31 data were compensated for using the corrections of Weingartner et al. (2003) as described in Sciare et al. (2011), the use of the Dual-Spot Technology® in the AE33 avoids the need for manual post-processing to compensate the data (Drinovec et al., 2014). Both instruments performed absorption measurements simultaneously during 7 days in February 2013 (Fig. S1a in the Supplement). Absorption coefficients at 880 nm showed a slope of 0.93 and a very satisfactory r^2 (0.96, $n = 3023$

of 5 min data points). Black carbon concentrations for the whole (2-year) data set were then calculated from the absorption coefficient at 880 nm, with a mass absorption cross section (MAC) of 8.8 $m^2 g^{-1}$ (Fig. S1b), determined from the comparison with collocated filter measurements of elemental carbon (EUSAAR2 thermo-optical protocol, Cavalli et al., 2010). This value is close to the default input value implemented in the AE33 at 880 nm (7.77 $m^2 g^{-1}$). Although still under discussion (Bond and Bergstrom, 2006; Cappa et al., 2012), such relatively high MAC values might be related to a possible encapsulation of soot particles by organic/inorganic compounds at our regional background site, and to the presence of BC from wood-burning emissions during wintertime, both leading to an increase in BC mass absorption efficiency (Liousse et al., 1993; Bond and Bergstrom, 2006; Lack et al., 2008). A total number of ~280 000 BC data points (~133 000 5 min points from AE31 and ~147 000 1 min points from AE33) were collected from June 2011 to June 2013.

Ammonia measurements during selected periods (mainly during the spring, winter and summer seasons) were carried out using an AiRRmonia (Mechatronics Instruments BV, the Netherlands). Based on the conductimetric detection of ammonium, gaseous ammonia is sampled at 1 L min^{-1} through a sampling block equipped with an ammonia-permeable membrane; a water counter-flow allows ammonia to solubilize in ammonium. A second purification step is applied by adding 0.5 mM sodium hydroxide, leading to the detection of ammonium in the detector block. The instrument has been calibrated regularly using solutions of 0 and 500 ppb of ammonium. Two sets of sampling syringes ensure a constant flow throughout the instrument, but also create a temporal shift, estimated at 20 to 40 min by different studies (Cowen et al., 2004; Zechmeister-Boltenstern, 2010). In our case, this shift was set at 30 min.

Pre-fired 47 mm diameter quartz filters were sampled in $PM_{2.5}$ at the same location using a low-volume (1 $m^3 h^{-1}$) sampler (Partisol Plus, Thermo Environment) equipped with a volatile organic compound active charcoal denuder. Four-hour filters and 24 h filters were discontinuously sampled, respectively, from 10 February to 2 March 2012 and during the period from August 2012 to April 2013. These filters were analyzed for their water-soluble inorganic (anions and cations) and elemental/organic carbon contents using, respectively, an ion chromatography and Sunset OC / EC analyzer (EUSAAR2 thermal protocol), according to Sciare et al. (2008) and Cavalli et al. (2010).

Finally, standard meteorological parameters (temperature, relative humidity, wind speed and direction) were obtained from continuous measurements at Ecole Polytechnique, located 4 km east of our station, with an A100R Campbell Scientific cup anemometer for wind speed and a W200P weather vane for wind direction, at 10 m above ground level (m a.g.l.). Additionally, the boundary layer height (BLH) was derived from Pal et al. (2013) methodology. The attribution of the BLH was processed in combining a diagnostic of the surface stability from high-frequency sonic anemometer measurements and LIght Detection and Ranging (LIDAR) attenuated backscatter gradients from aerosols and clouds.

All measurements presented here are expressed in Coordinated Universal Time (UTC). Seasons are differentiated upon seasonal equinoxes.

2.2 Urban background $PM_{2.5}$ measurements

Within the framework of mandatory air quality monitoring, urban background measurements are continuously being carried out in the region of Paris. Hourly $PM_{2.5}$ data from TEOM-FDMS measurements were retrieved from the three stations representative of the Paris urban background (namely, Bobigny, Gennevilliers and Vitry-sur-Seine). Data sets are available online upon request at http://www.airparif.asso.fr.

2.3 Backtrajectories and non-parametric wind regression

To illustrate air mass origin during specific pollution episodes, 72 h backtrajectories were calculated every 3 h from the PC-based version of Hysplit (Draxler, 1999) with GDAS meteorological field data. Backtrajectories were set to end at SIRTA coordinates (48.71° N, 2.21° E) at 100 m a.g.l.

Non-parametric wind regression (NWR) is a smoothing algorithm (Henry et al., 2009) to alternatively display pollution roses, and has already been successfully applied to various atmospheric pollutants and pollution sources (Yu et al., 2004; Pancras et al., 2011; Olson et al., 2012). The objective is to estimate the concentration of a pollutant given any (θ, υ) couple (wind direction and speed, respectively), from measured wind speed and direction, and concentration.

$$E(\theta \mid \vartheta) = \frac{\sum_{i=1}^{N} K_1\left(\frac{\theta - W_i}{\sigma}\right) \cdot K_2\left(\frac{\vartheta - Y_i}{h}\right) \cdot C_i}{\sum_{i=1}^{N} K_1\left(\frac{\theta - W_i}{\sigma}\right) \cdot K_2\left(\frac{\vartheta - Y_i}{h}\right)}, \tag{1}$$

where E is the concentration estimate at a wind direction θ and speed υ; W_i, Υ_i and C_i the wind direction, speed and atmospheric concentrations, respectively, measured at t_i; σ and h the smoothing factors; and K_1 and K_2 two kernel smoothing functions defined as

$$K_1(x) = \frac{1}{\sqrt{2\pi}} \cdot e^{-0.5 \cdot x^2}, \; -\infty < x < \infty. \tag{2}$$

$$K_2(x) = 0.75 \cdot \left(1 - x^2\right), \; -1 < x < 1 = 0.$$

The choice of the two smoothing factors σ and h can be carried out using statistical calculations, although its empirical determination stays feasible, as the final interpretation should not be changed. Here, σ and h were set to 7 and 1.5, similarly

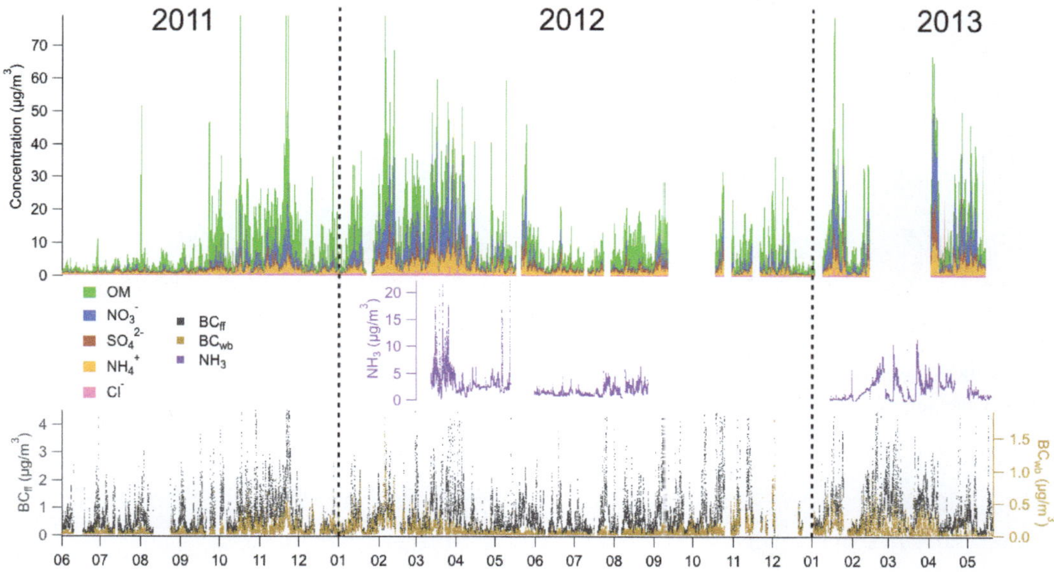

Figure 1. Time series of the major 30 min non-refractory (top, concentrations are aggregated) and 5 min refractory (bottom, concentrations are dissociated) PM_1 chemical constituents, and 5 min ammonia at SIRTA from June 2011 to May 2013. The two large data gaps in October 2012 and March 2013 correspond to two field intensive campaigns during which the ACSM was deployed elsewhere.

to Petit et al. (2014). Finally, the equivalent of the wind rose is calculated from the probability density

$$f(\theta,\vartheta)=\frac{1}{N\sigma h}\cdot\sum_{i=1}^{N}K_1\left(\frac{\theta-W_i}{\sigma}\right)K_2\left(\frac{\vartheta-Y_i}{h}\right), \qquad (3)$$

where N the is the total number of points.

Due to higher measurement uncertainties in wind direction at low speeds, data associated with wind speeds lower than $1\,\mathrm{m\,s^{-1}}$ were discarded, potentially inducing an underestimation of very local pollution events.

2.4 Source apportionment of carbonaceous aerosols

The measurement of aerosol absorption at multiple wavelengths is allowing for BC source apportionment. Organic molecules, especially polycyclic aromatic hydrocarbons and humic-like substances, strongly absorb in the UV and blue parts of the light spectrum. Based on the fact that these compounds are primarily related to biomass combustion, the deconvolution of BC into two contributions, fuel fossil and wood burning (BC_{ff} and BC_{wb}, respectively), can be carried out (Sandradewi et al., 2008). Such source apportionment has already been successfully performed during intensive field campaigns as well as for long-term monitoring periods, frequently enlightening the significant contribution of wood burning to ambient BC concentrations during wintertime (Favez et al., 2009, 2010; Sciare et al., 2011, Herich et al., 2011; Crippa et al. 2013a). Here, the 470 and 880 nm channels were used, with an absorption Ångström exponent of 2.1 and 1.0 for pure wood burning and traffic, respectively, similarly to previous work focusing on the February–March 2012 period of the same data set (Petit et al., 2014).

The source apportionment of our organic aerosol data is not presented here, although positive matrix factorization applied to the AMS or ACSM database is an efficient tool for the identification of organic aerosol primary sources and secondary formation processes (see, for instance, Lanz et al., 2007; Jimenez et al., 2009; El Haddad et al., 2013; Carbone et al., 2013; Bougiatioti et al., 2014; Petit et al. 2014). Such a work will be reported elsewhere (Crenn et al., submitted), as important issues related to the seasonal variation of specific organic aerosol factor profiles have to be addressed in many details, with a lot of sensitivity tests that are beyond the objectives of the present study.

3 Cross-validation of particulate chemical species concentrations

Figure 1 illustrates the temporal variations of chemical species concentrations used for the present study from June 2011 to May 2013. This extended duration highlights the robustness of used instruments, and in particular the ACSM, which did not undergo any major failures over this 2-year period. The consistency of the concentrations of each chemical constituent retrieved from the ACSM has been checked via comparisons with filter measurements (Fig. 2) as well as a chemical mass closure of PM_1 (Fig. 3).

ACSM nitrate is very consistent with filter measurements, the slope of the linear regression being close to 1 ($r^2=0.85$, $N=147$). No overestimation of ACSM nitrate is observed at high concentrations, which suggests the ability of the Middlebrook algorithm to properly correct our ACSM collec-

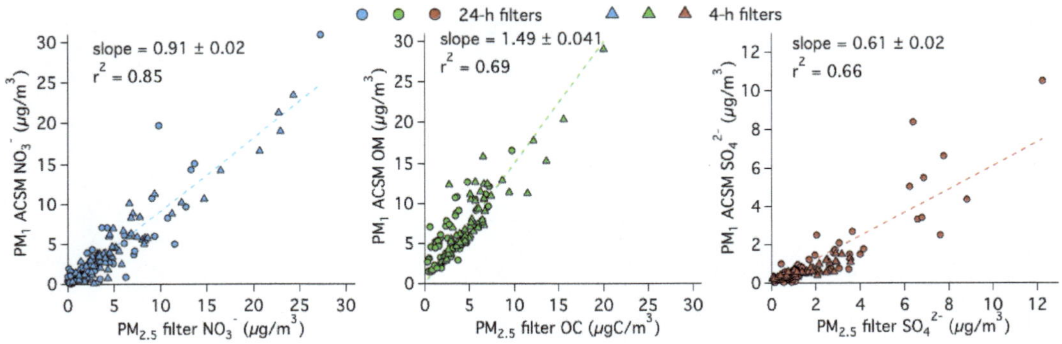

Figure 2. Scatterplot of chemically speciated ACSM measurements versus filter analyses for nitrate, organic matter (compared to OC filter-based measurements) and sulfate.

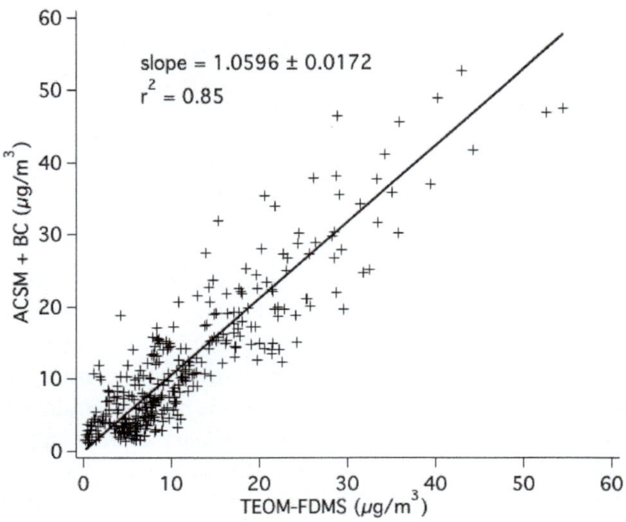

Figure 3. Mass closure exercise between daily averaged reconstructed PM_1 (ACSM + BC) and measured PM_1 by TEOM-FDMS.

tion efficiencies. Higher discrepancies are observed for sulfate. This feature has already been mentioned in previous studies for ACSM (Ng et al., 2011; Budisulistiorini et al., 2014) and AMS instruments (Takegawa et al., 2005). This could be partly related to the size distribution of sulfate, as fine ($PM_{2.5}$) sulfate can partially be associated with submicron sea salt and/or dust particles. Fine ammonium sulfate aerosols originating from secondary processes and long-range transport (Sciare et al., 2010; Freutel et al., 2013) may also present a larger size mode extending above 1 μm and partially not sampled by the ACSM. A sulfate ion efficiency calibration was also performed in May 2013 to investigate possible changes in RIE, but no significant discrepancy from the default value of 1.2 was found.

The OM-to-OC ratio obtained from the comparison between ACSM and filter-based measurements exhibits a mean value of approximately 1.5, which is lower than the value recommended for urban areas (1.6 ± 0.2, Turpin and Lim, 2001) and 33 % lower than and/or equal to values used in the Paris metropolitan area in previous studies (~ 2 in Bressi et al., 2013; 1.6 in Sciare et al., 2010). Although this ratio is subject to caution, by virtue of potential geographical and temporal discrepancies, the relatively low value observed here might be explained by the presence of organic material between 1 and 2.5 μm as well as filter sampling artifacts.

A chemical mass closure exercise, where the combination of validated ACSM and Aethalometer data is compared to co-located PM_1 TEOM-FDMS measurements, was used to assess the capacity of the two former instruments to correctly describe the PM_1 fraction over long-term periods. For this purpose, the reconstructed PM_1 (PM_{chem}) introduced here corresponds to the sum of all non-refractory species measured by the ACSM (OM, NO_3^-, SO_4^{2-}, NH_4^+ and Cl^-) and black carbon measured by Aethalometer, and is assumed to quasi-exhaustively account for submicron aerosols (Putaud et al., 2004). PM_{chem} daily averages were compared to the TEOM-FDMS data set, since the latter instrument is considered to be equivalent to the gravimetric reference method on this temporal scale. From June 2011 to May 2013, the 341-point (this number being due to the combined availability of ACSM, BC and PM data) scatterplot shows a very satisfactory correlation coefficient ($r^2 = 0.85$) with a slope of 1.06.

4 Representativeness of our 2-year observation period

Monthly mean atmospheric conditions were compared to standard meteorological parameters in order to investigate any anomalies over the 2011–2013 period (Fig. 4). Temperature, rainfall and sun exposure representative of the region of Paris were retrieved from monthly weather reports available at https://donneespubliques.meteofrance.fr, and are calculated from a 30-year period (1981–2010) (Arguez and Vose, 2011). A similar study was also performed for particulate matter concentrations, with representative $PM_{2.5}$ defined as the average $PM_{2.5}$ concentrations calculated from 2007 to 2014 at the three historical Airparif urban background stations.

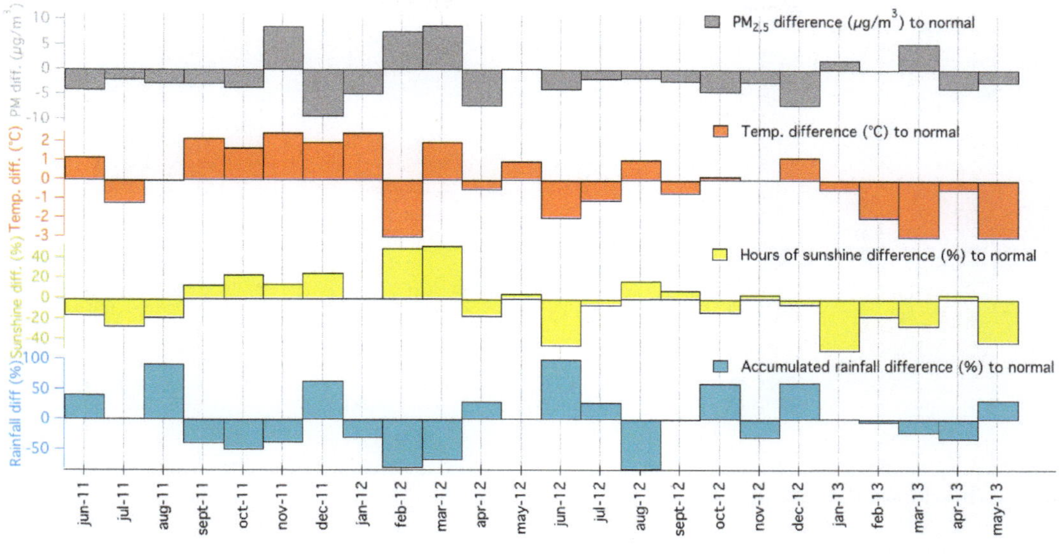

Figure 4. Comparison between observed and average $PM_{2.5}$, temperature, hours of sunshine and accumulated rainfall in the region of Paris.

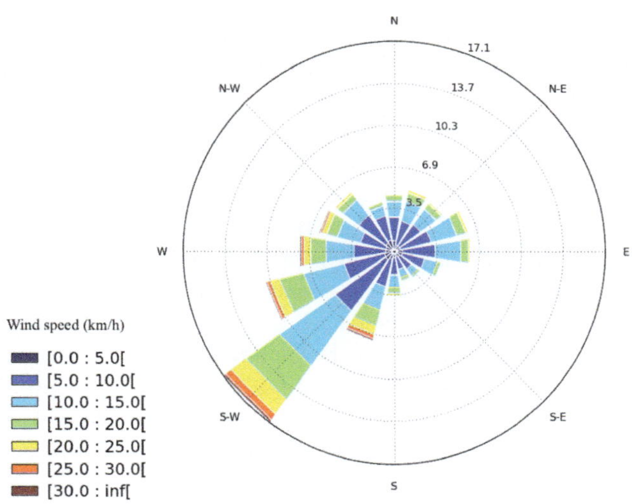

Figure 5. Average wind rose during June 2011 and June 2013; the radial axis represents the wind occurrence (in %).

Briefly, autumn 2011 was relatively mild, $PM_{2.5}$ levels being close to representative concentrations for the period. The end of winter 2011–2012 and early spring 2012 were particularly dry and sunny, enabling enhanced photochemical transformation, and exhibited unusually high $PM_{2.5}$ concentrations in February and March 2012. The summer of 2012 was chilly and rainy, especially in June 2012, leading to lower $PM_{2.5}$ levels (Yiou and Cattiaux, 2013). Finally, the first two months of 2013 were unusually cold, whereas March 2013 was remarkably representative of wintertime conditions. The highest observed discrepancies occur with the highest measured mass, which may highlight an intensification of pollution episodes. From a broader perspective, this feature is also observed through inter-annual variability of urban background $PM_{2.5}$ concentrations (Fig. S2). This underlines the need for continuous monitoring over several year periods. Interestingly, no direct link can be drawn between meteorological anomalies and unusually high $PM_{2.5}$ concentrations. Indeed, while the high $PM_{2.5}$ levels observed in February and March 2012 may be linked to unusually low temperatures, exceptionally high temperatures can also be associated with high $PM_{2.5}$ concentrations. This has to be related to the seasonal variability of sources, origins and (trans)formation pathways, and is investigated within the following sections, taking advantage of long-term trend analysis, wind regression, diurnal variations, and the analysis of pollution episodes.

Finally, it should be underlined that the Paris region is mostly influenced by winds coming from the southwestern (Fig. 5) sector. This sector is characterized by clean air masses from the Atlantic Ocean with high wind speeds, and is usually associated with low PM concentrations. The northeastern wind sector exhibits a smaller occurrence than previously observed between September 2009 and September 2010 (Supplementary information of Bressi et al., 2013).

5 Long-term trend and general features

Two-year temporal variations of the chemical composition of submicron aerosols (OM, NO_3^-, SO_4^{2-}, NH_4^+, Cl^-, BC_{ff} and BC_{wb}) and ammonia (NH_3) are presented in Fig. 1. Similarly to Bressi et al. (2013), a clear seasonal pattern is observed here, with the highest concentrations observed during winter and early spring, while summer periods exhibit the lowest pollution levels (Fig. 6a), which is also consistent with general patterns observed in northern Europe (Barmpadimos et al., 2012; Waked et al., 2014). Regardless of the season, OM dominates the PM_1 chemical composition, followed by ammonium nitrate, whose contribution is highest during

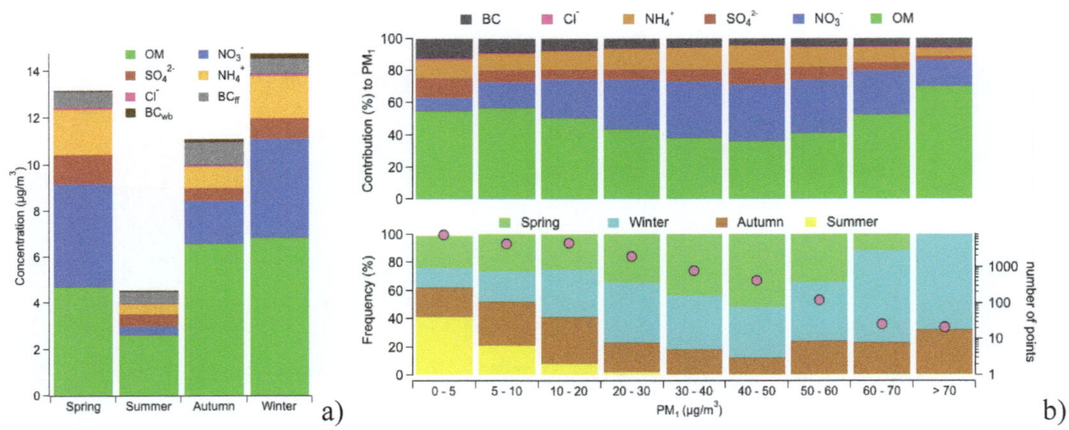

Figure 6. **(a)** PM_1 chemical composition for different mass classes (top), with the seasonal occurrence frequency and number of points in each bin (bottom) and **(b)** seasonal PM_1 chemical composition.

spring, a feature that is generally observed for European urban areas (Zhang et al., 2007; Putaud et al., 2010). Figure 6b presents the binned major chemical composition and the frequency per season of data points as a function of PM_1 concentration levels. The contribution of secondary inorganic aerosols (SIA, mostly NO_3^-, SO_4^{2-} and NH_4^+) increases with the increases in PM_1 mass until $50\,\mu g\,m^{-3}$, highlighting the role of inorganic secondary pollution during spring months (Fig. 6b). This well-documented pattern that has already been reported for the region of Paris in several studies (see, for instance, Sciare et al., 2010 and Bressi et al., 2013). Very interestingly, above $50\,\mu g\,m^{-3}$, organic contribution, as well as wintertime frequency, increases to dominate the chemical composition of the highest measured PM_1 concentrations with an associated increase in BC, a feature that has not been seen during the Airparif–Particules projects, essentially due to highly time resolved measurements, nor investigated during the MEGAPOLI project. There are well-defined occurrences of high concentrations (~ 150 data points of 30 min), suggesting sharp pollution events with a limited temporal duration, contrary to the 20–$50\,\mu g\,m^{-3}$ mass class presenting many more data points that highlight either a higher frequency of sharp events and/or pollution episodes with a longer temporal duration.

We have used here the BC / SO_4 ratio to assess potential transport of pollution. Sulfate mainly forms through heterogeneous processes with a slow kinetic rate and spreads over large scales (Putaud et al., 2004). For that reason, it can be considered as a good indicator of long-range transport assuming minor local SO_2 sources (background annual SO_2 concentrations of about $2\,\mu g\,m^{-3}$ in the region of Paris; AIRPARIF, 2014). On the contrary, black carbon in the region of Paris shows an important gradient from the city center to the regional background (Bressi et al., 2013) and can be used to better infer local (Paris city) influence at our background station. Although in situ sulfate formation may occur (for instance, during fog episodes; Healy et al., 2012) and long-range transport of BC may be observed over the region of Paris (Healy et al., 2012 and 2013), as a whole, the use of the BC / SO_4 ratio may support our study on local/regional/advected pollution. As shown in Fig. 7, the BC / SO_4 ratio decreases along with the increase in PM_1 (and thus secondary ion mass fraction), suggesting potential regional and/or trans-boundary transport, and large-scale pollution episodes, as previously reported by several studies in northern France (Bessagnet et al., 2005; Sciare et al., 2010; Bressi et al., 2013; Waked et al., 2014; Freutel et al., 2013). Very interestingly, this BC / SO_4 ratio dramatically increases for the highest concentrations, where the concomitant increases in the Ångström exponent, of the contribution of BC_{wb} relative to BC, along with the increase in OM and wintertime frequency (Fig. 7), suggest intense local and/or regional wood-burning pollution episodes during winter. Moreover, except for the single wood-burning episode observed on 5 February 2012 (described in Petit et al., 2014), all these intense PM pollution peaks ($PM_1 > 60\,\mu g\,m^{-3}$) also occurred at most of the rural/suburban/urban Airparif monitoring stations. This pattern underlines homogeneous meteorological conditions over the region of Paris, with "local" emissions being measured on a regional scale (within a distance of at least 50 km from the city center).

6 Seasonality and insights into geographical origins

Figure 8 displays the wind regression analysis plots for species of interest, namely, OM, NO_3^-, SO_4^{2-}, NH_3, BC_{ff} and BC_{wb}.

Overall, OM concentrations do not exhibit a particular dependence on wind direction, the regional background always staying at a significant contribution throughout seasons (~ 3–$6\,\mu g\,m^{-3}$). However, higher OM concentrations occurred in autumn and winter and are associated with very low wind speeds, suggesting higher local influence together with higher local wood-burning emissions (as previously

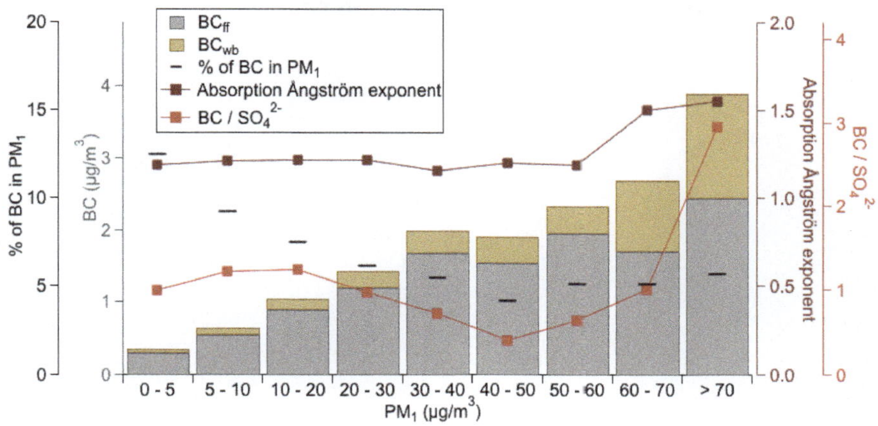

Figure 7. Source contribution to BC, absorption Ångström exponent, BC / SO_4 ratio (ACSM sulfate), and contribution of BC to PM_1, depending on PM_1 mass.

suggested from Figs. 6b and 7). During summer, OM concentrations are lower (by a factor of ~ 2.2) and show a more homogeneous distribution (e.g., with lower local influences).

As expected, semi-volatile nitrate concentrations are higher during the coldest months (in spring and winter). They are associated with relatively high wind speeds ($\sim 20\,\mathrm{km\,h^{-1}}$) coming from the northerly and northeasterly directions, suggesting significant medium- to long-range transport of ammonium nitrate during these seasons, which is consistent with similar observations reported for the region of Paris (Bressi et al., 2013; Freutel et al., 2013; Petetin et al., 2014). However, the significant nitrate concentrations observed for all the range of wind speeds from the N–NE direction suggest, at least for the lowest wind speed, a significant contribution of the region of Paris. Possible impacts of industrial activities in the Seine estuary (i.e., Rouen, Le Havre), especially during spring, may also be responsible for the noticeable nitrate hotspot observed in the northwestern sector. In autumn, nitrate concentrations are higher at low wind speeds, in agreement with the fact that traffic emissions are slightly higher in September and October than the rest of the year in Paris (V-Trafic report, 2014), and that BC_{ff} concentrations are also highest during these months. This is also consistent with a relatively fast nitrate formation mechanism from local NO_x emissions as reported by Petetin et al. (2014).

Sulfate features different behavior than nitrate, where the non-local origin is much more pronounced. High concentrations are associated with high wind speeds originating from the NNE, leading to the same conclusions as those reported in the literature on the major role of the long-range transport of this compound (Pay et al., 2012; Bressi et al., 2014; Petetin et al., 2014; Waked et al., 2014). Petrochemical and shipping activities may explain the observed hotspot in the marine northwestern sector, especially noticeable in spring, which may be linked to meteorological conditions enhancing ammonium sulfate formation and transport.

The region of Brittany, located less than 300 km west of the region of Paris, is the principal emitter of ammonia in France through intense agricultural activities (http://prtr.ec.europa.eu/DiffuseSourcesAir.aspx). However, no clear contribution from this region is observed from our wind regression analysis. This may be partly related to very few occurrences of air masses passing over Brittany and reaching the region of Paris. Despite hotspots from the northeast/east in spring, or from the north/northeast in winter, no clear wind sector is directly responsible for high NH_3 concentrations at our station, suggesting a diffuse regional source for this compound.

In Europe, BC_{ff} is assumed to be an excellent tracer of traffic emissions in urban areas (Herich et al., 2011, for instance). Although long-range transported BC_{ff} may not be excluded, as shown by Healy et al. (2012, 2013), here, wind regression analyses show that high BC_{ff} concentrations occurred at low wind speeds, highlighting the importance of local/regional traffic emissions in the Paris region, especially during the autumn and winter seasons. In spring, a clear distribution over a large range of wind speeds is noticeable in the NNE wind sector. This is consistent with the fact that Paris is located NNE of our station (e.g., a higher contribution of the Paris city plume to measured BC_{ff} concentrations at SIRTA). This is also related to a higher occurrence of this wind sector during spring.

Black carbon from biomass burning combustion (BC_{wb}) presents a clear seasonal trend similar to OM, with the highest concentrations during cold seasons at low wind speeds, suggesting increasing local influence in wood-burning emissions. The lowest boundary layer heights (BLHs) observed during wintertime favoring the accumulation of pollutants at ground level together with the large contribution of individual (domestic) wood-burning sources homogeneously spread over the region of Paris may explain the significant contribution of regional emissions observed during winter.

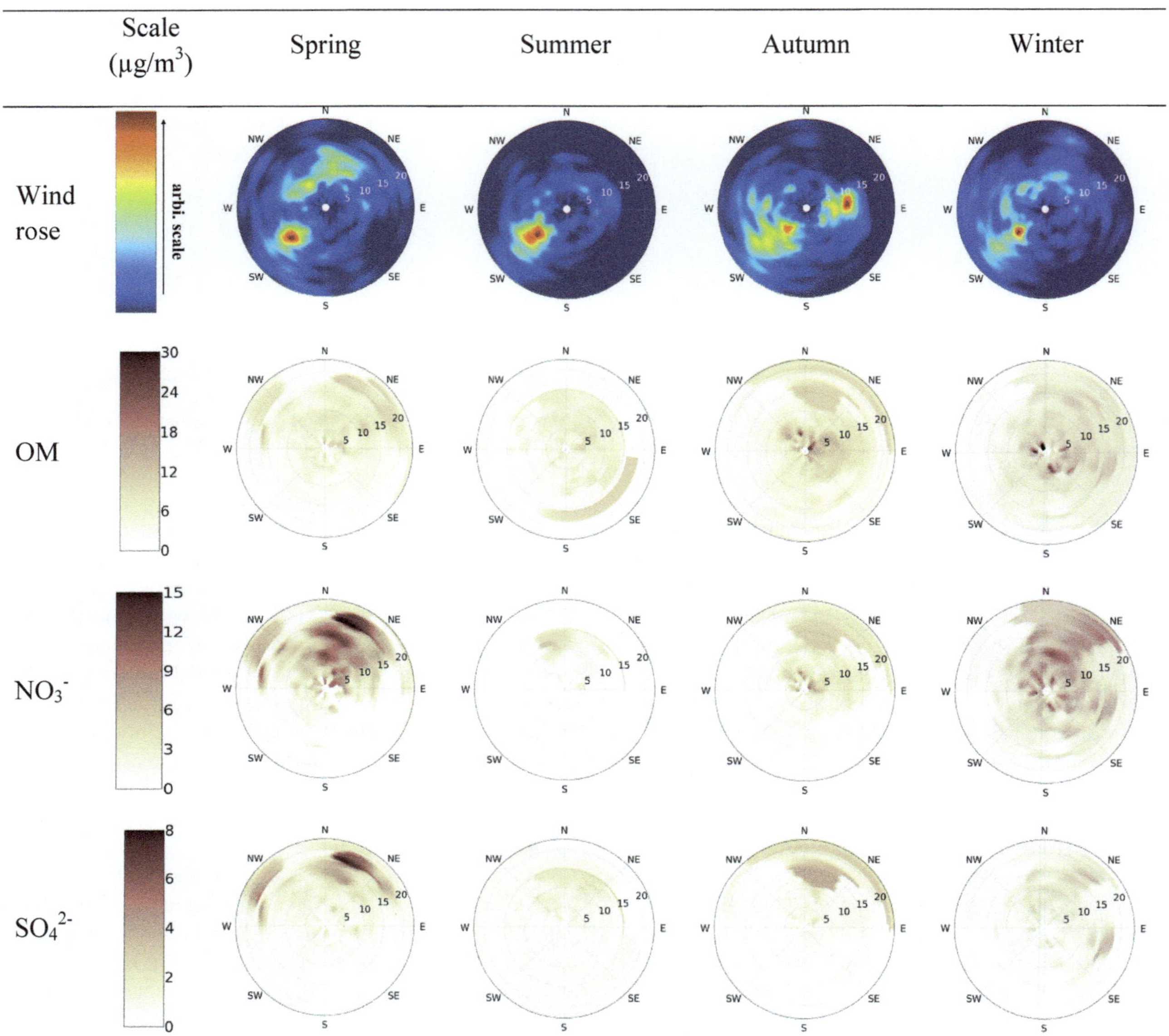

Figure 8.

Finally, it should be noted that the geographical origin of each investigated chemical constituent remains globally unchanged throughout the year, with a well-defined sectorized location. While SIA and BC_{ff} fractions are mainly associated with the NNE sector (coming from Paris and/or farther away), the highest OM and BC_{wb} concentrations exhibit strong local NW and SE sector origins. Various sources of organic matter also contribute to a significant contribution of the (unsectorized) regional background.

7 Weekly diurnal profiles and insight into sources and processes

Near real-time observations over long-term periods offer a unique opportunity to provide robust diurnal profiles for each season. First, Fig. 9 shows the average diurnal profiles of ambient temperatures (Fig. 9a) and BLH (Fig. 9b) across seasons. Weekly diurnal profiles for OM, NO_3^-, NH_4^+, NH_3, BC_{ff} and BC_{wb} are presented for different seasons from hourly averages (Fig. 10). Sulfate variations are not presented and discussed here because they lead to poor daily variations (average of $0.75\,\mu g\,m^{-3} \pm 2\,\%$), which are consistent with its mid- to long-range transport origin.

Clear weekly and diurnal patterns can be observed for carbonaceous aerosols. Independently of the investigated season, BC_{ff} presents a well-marked bimodal diurnal profile, with maxima in the morning (starting at 06:00 UTC) and the evening (starting at 17:00). This reflects the proximity of the traffic source (with daily commuting) and dilution in the boundary layer during daytime (Fig. 9b). With an average

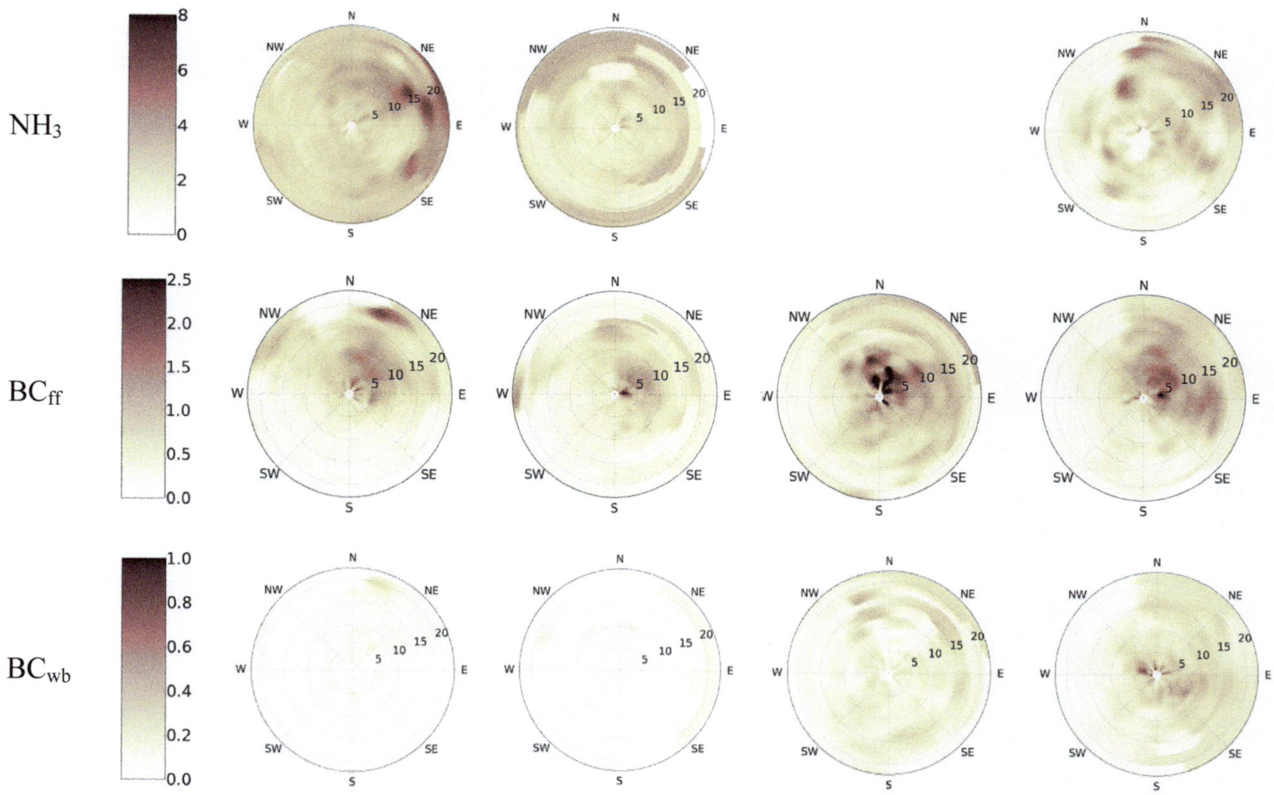

Figure 8. Seasonal NWR plots for the major components of PM_1 and gaseous ammonia. Radial and tangential axes represent the wind direction and speed ($km\,h^{-1}$), respectively.

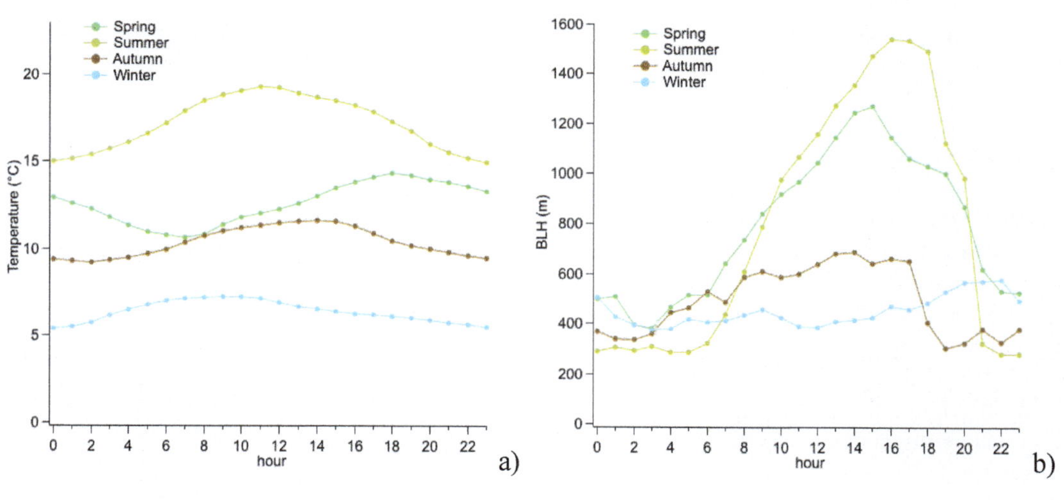

Figure 9. Average diurnal variations by seasons of temperature (**a**) and BLH (**b**).

of 0.61 $\mu g\,m^{-3}$, weekdays exhibit slightly higher concentrations than weekends (0.51 $\mu g\,m^{-3}$ on average). By comparison, the diurnal variability of BC_{wb} is revealed only in autumn and winter, with the combination of enhanced wood-burning emissions, low temperatures and BLH (Fig. 9), leading to a unimodal pattern with increasing concentrations after 18:00 UTC. Although individual wood-burning stoves only represent around 5 % of the means of heating in the region of Paris, they contribute almost 90 % of PM_{10} residential emissions in the region of Paris (Airparif emission inventory for the year 2010; Airparif, 2013) and are likely to represent the major contributor to BC_{wb}.

For OM, the highest variations (in terms of concentration amplitude) are observed during autumn and winter, with a growing influence of wood-burning heating, as OM concentrations nicely follow BC_{wb} ones. Levels of both com-

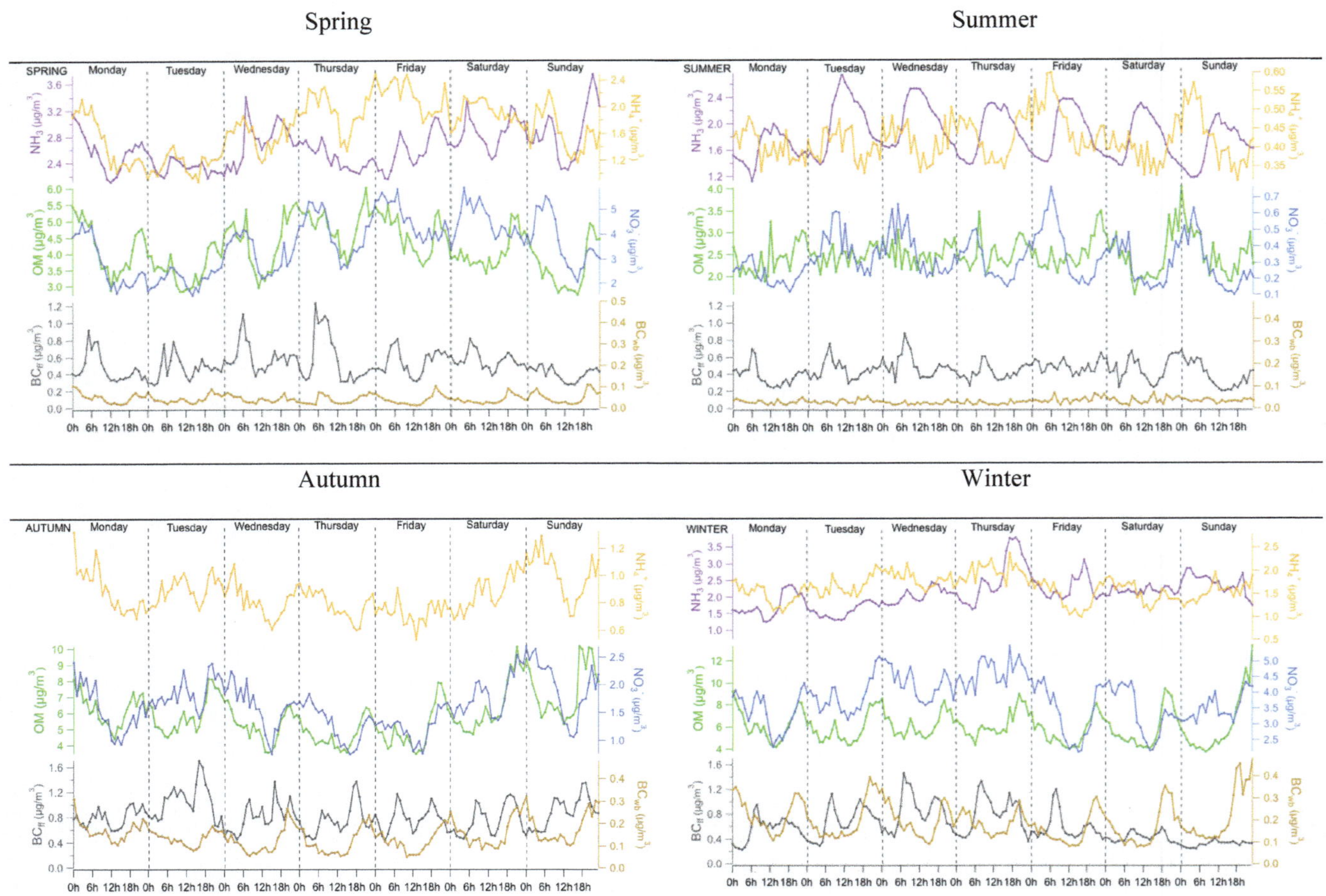

Figure 10. Seasonal weekly diurnal variations of OM (green), NO_3^- (blue), NH_4^+ (dark yellow), NH_3 (purple), BC_{ff} (black) and BC_{wb} (brown).

pounds during the evening are approximately 20 % higher during weekends than during weekdays. More specifically, low BLHs in winter (Fig. 9b) increase measured concentrations, leading, for example, to morning OM peaks that should be linked to traffic emissions. By contrast, the diurnal profile is rather flat, with poor temporal variations in summer, and is in accordance with the homogeneous geographical distribution from the NWR calculation for this season. The lack of a decrease in the afternoon during weekdays suggests rapid formation of secondary organic aerosols (SOA) from diverse anthropogenic (traffic for instance, as underlined by Platt et al., 2013 and Nordin et al., 2013) and biogenic sources (Carlton et al., 2009). During spring, OM globally follows the variations of nitrate, highlighting fast displacements of gas-particle equilibriums of semi-volatile material due to meteorological conditions. Some peaks are observed some days during the night, which could underline the residual contribution of wood-burning emissions in March and April.

For SIA, nitrate and ammonium display very similar diurnal and weekly profiles, illustrating the importance of ammonium nitrate by comparison with ammonium sulfate. Both compounds display well-marked diurnal profiles with max-

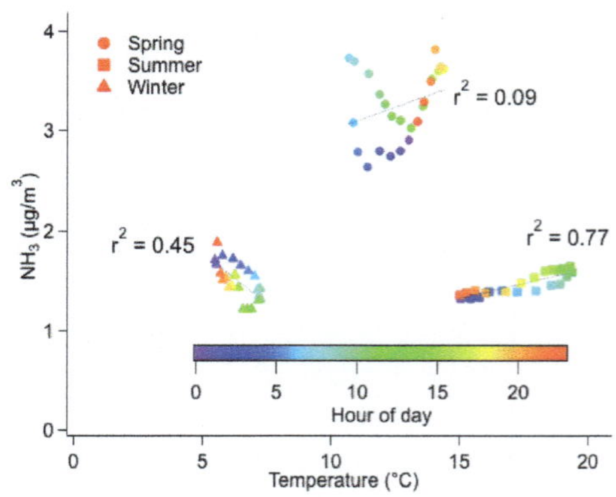

Figure 11. Correlation between ammonia and temperature in spring (circles), summer (squares) and winter (triangles) colored as a function of the hour of the day.

ima at night (especially in autumn and winter) and/or early morning (especially in spring and summer), which has to be related to the enhancement of ammonium nitrate formation at low temperatures and/or high relative humidity. The temporal variations of the two compounds can also be linked to the one of ammonia. For instance, during summertime, ammonia presents unimodal diurnal profiles, with the highest values around noon, and nicely follows temperature (Fig. 9a), in good agreement with previous studies (Bari et al., 2003; Lin et al., 2006). This phenomenon is exactly opposite to the variations of ammonium and nitrate exhibiting unimodal patterns with the highest concentrations during the night. Meteorological conditions can then fully explain the formation/partitioning of SIA as well as ammonia concentrations during summer.

Interestingly however, ammonia shows different profiles as a function of the season. In particular, during springtime, this compound displays a clear bimodal profile, with a morning and an evening peak, concomitant with traffic emissions and that come over elevated regional background levels due to the use of nitrogen-containing fertilizers in this period of the year. However, this bimodal pattern is not observed during the summer and winter seasons, where traffic also occurs. Although traffic-related ammonia has already been reported in urban environments (Edgerton et al., 2007; Pandolfi et al., 2012; Saylor et al., 2010) and several studies raise concerns about uncontrolled ammonia emissions from De–NO_x systems (Baum et al., 2001; Heeb et al., 2006 and 2012, for instance), this spring bimodal profile may also be related to parameters other than traffic emissions. Indeed, as already described by Bussink et al. (1996), emission of ammonia can occur during the evaporation of the morning dew, especially when soils are loaded with fertilizers. The morning decrease observed for ammonia in spring can then be associated with the growing of the mixing depth layer (Fig. 9b) while, in the afternoon, ammonia increases may be partly explained by temperature-driven gas-phase partitioning changes in ammonium nitrate.

8 PM_1 pollution episodes over the region of Paris

An in-depth characterization of each pollution episode over the region of Paris is particularly important in the context of mitigation policies, which are usually taken on a local scale during these episodes. Such an investigation should provide useful information regarding PM (trans)formation processes and help to identify parameters influencing the temporality of their chemical composition.

Statistical representativeness of pollution episodes (duration and intensity) may be addressed using our long-term data sets. Based on our 2-year data set, the highest 1 % of observed PM_1 concentrations ($q_{99} \sim 49\,\mu g\,m^{-3}$, representing around 200 data points of 30 min, i.e., approximately 100 h) mostly occur during February, April and November, while persistent pollution episodes ($PM_1 > 20\,\mu g\,m^{-3}$ during at least 3 consecutive days) mostly occur in early spring. More interestingly, the majority of the highest PM_1 concentrations fall within these persistent pollution episodes. As previously suggested from higher BC / SO_4 ratios (Sect. 5 and Fig. 7), the highest PM_1 concentration peaks are associated with rather local emissions. This result clearly points to the contribution of local/regional emissions during persistent pollution episodes. A more detailed analysis (episode-by-episode) is performed in the following to better characterize the local/regional versus advected PM pollution during persistent pollution episodes.

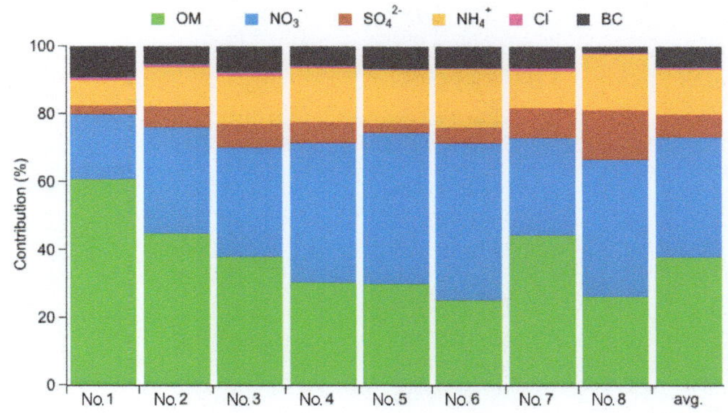

Figure 12. PM_1 chemical composition of the eight pollution episodes.

Eight persistent pollution episodes ($PM_1 > 20\,\mu g\,m^{-3}$ during at least 3 consecutive days) were detected between mid-2011 and mid-2013, and are displayed in Table 2 and Figs. 12 and 13. Figure 12 shows the averaged PM_1 chemical composition (in $\mu g\,m^{-3}$) for each episode, chronologically numbered, from 1 to 8. Table 2 summarizes key information for each episode. Figure 13 shows air mass origins, a wind rose and temporal variations of the chemical composition of each episode. As a general pattern for each episode, the chemical composition of PM_1 is dominated by OM and/or ammonium nitrate. Sulfate presents the highest variability (concentration standard deviation of 53 % over all episodes) compared to OM and nitrate (~ 30 %), possibly suggesting various contributions of advected pollution.

The following provides a thorough description of each episode.

Episode 1 (19–24 November 2011): while winds come from the NW and E sectors, 72 h backtrajectories originate from SSE and exhibit a recirculation over a part of northern France. Moreover, along with the BC / SO_4 ratio (3.56; i.e., the highest of all episodes) and a low BLH with no significant variations, the chemical composition is largely dominated by OM (60.8 % of PM_1), suggesting significant local influence. The contribution of BC_{wb} remains insignificant compared to BC_{ff}, which could underline the accumulation in the atmosphere of fossil-fuelled combustion sources (notably illustrated by the very low altitude of the air masses ending on the 21 and 23 November).

Episode 2 (5–13 February 2012): this episode presents two distinct phases. At the beginning, air masses come from the southeast, but originate from the east at low altitudes, along with very low temperatures (below 0 °C all day), high OM and BC concentrations and BC / SO_4 ratios (average

Table 2. Essential parameters describing the eight pollution episodes, such as the start and end dates, average temperature and relative humidity, fraction dominating the chemical composition (SIA stands for secondary inorganic aerosols), BC-to-SO_4 ratio and main geographical contribution.

Episode no.	Start–end date	Temp. (°C)	RH (%)	Chemical composition	BC / SO_4	Geographical contribution
1	19–24/11/2011	8.5	93	OM	3.56	Regional
2	05–13/02/2012	−4.7	71	OM then SIA	0.91	Strong local, then regional and advected
3	29/02–03/03/2012	8.2	95	SIA	1.12	Strong regional, low advected
4	12–17/03/2012	10.7	78	SIA	0.95	Advected and regional
5	23–26/03/2012	15	48	SIA	2.37	Strong advected, low regional
6	28–31/03/2012	12.3	62	SIA	1.42	Strong advected and regional
7	16–21/01/2012	−3	93	OM and SIA	0.72	Strong regional and advected
8	01–08/04/2013	4.2	64	SIA	0.12	Advected

of 22.6 and 0.6 $\mu g\,m^{-3}$, and 2.7, respectively, from 5 to 8 February). This is related to an intense local wood-burning episode already thoroughly described in Petit et al. (2014). Then, from 8 February, winds and air masses originate from NNE, and secondary inorganic ions, especially ammonium nitrate, dominate the chemical composition. The associated wind speed may underline mid- to long-range transport, although the impact of the Paris plume cannot be excluded here.

Episode 3 (29 February–3 March 2012): along with this pollution episode, trajectories have rapidly changed in origin but have remained low in altitude. The RH remained very high, reaching 100 % most of the time. Very interestingly, concentrations dropped on 1 and 3 March during the beginning of the day, coinciding with two stratus lowering fog events. These two fog events occurred during the second half of the night, and evaporated as the sun rose. The influence of fogs regarding the chemical transformation of PM_1 is notably highlighted by higher sulfate concentrations just after the evaporation of the first fog (and also when trajectories flew over the English Channel and Belgium), which could suggest transported SO_2 and oxidation over the region of Paris enhanced by fast fog processing (Kai et al., 2007; Rengarajan et al., 2011).

Episode 4 (12–17 March 2012): winds have originated from all directions (but mostly from NNE), suggesting anticyclonic conditions. The first half of the period exhibits a rather stable chemical composition (dominated by ammonium nitrate) and clear diurnal variations of RH, T and BHL. Then, after 15 March, daily amplitudes of the following three meteorological parameters increased: T reached 20 °C, RH 30 % and BHL 1000 m, compared to the first half, where they reached 15 °C, 50 % and 600 m, respectively. This caused rapid decreases in concentrations, due to higher temperature amplitudes, enhancing the gas partitioning of semi-volatile material, and an increase in BLH, allowing the dilution of atmospheric pollutants.

Episode 5 (23 March 2012–26 March 2013): air masses originated from the northeast to the east, and winds from the north to the northeast. This episode is characterized by the strong diurnal variation of OM and ammonium nitrate, due the high amplitude of the BLH and temperatures going above 15 °C, similarly to the previous episode. The high average BC / SO_4 ratio (2.37) is not representative of its temporality; the highest values are observed for lowest PM concentrations (26 March, afternoon). With this exception, low BC / SO_4 values (< 1) and the chemical composition dominated by ammonium nitrate suggest mid- and/or long-range transport.

Episode 6 (28–31 March 2012): it exhibits the same behavior as episode 5, with a clear medium- to long-range origin pattern (wind speed $\sim$ 10 $km\,h^{-1}$, chemical composition dominated by ammonium nitrate), but with backtrajectories coming from the northwest/northeast. Low altitudes of backtrajectories illustrate the accumulation of pollutants along the trajectory of the air masses. However, the BC peak on the morning of 30 March (the high BC_{ff} fraction suggests traffic emissions) could underline an influence of the Paris plume.

Episode 7 (16–21 January 2013): air masses display a coiling pattern around northern France. The BC / SO_4 ratio, remaining lower than 1, suggests advected pollution. However, the strong variability of BC_{wb} illustrates a significant influence of wood-burning emissions. No BHL data are available during this episode, but the altitude of backtrajectories may underline a more important dilution of the pollution.

Episode 8 (1–8 April 2013): this episode actually started on 22 March, but no ACSM data were available at that time; however, meteorological conditions from 22 March to 1 April were very similar, notably in terms of wind speeds and direction. It is characterized by air masses originating from the northeast and a very low BC / SO_4 ratio, illustrating a typical case of advected secondary pollution clearly dominated by ammonium nitrate and sulfate.

Overall, the observed variability, in terms of meteorological conditions, air mass origins, and chemical composition illustrates the variety of persistent pollution episodes, in terms of PM sources and different geographical origins. The BC / SO_4 ratio has been shown to represent a useful tool for assessing the local/regional/advected dimension of a specific pollution episode. Indeed, high ratios ($\geq$ 2) are usually associated with accumulation of local and/or regional emissions, while very low ratios ($\leq$ 0.5) are more representative

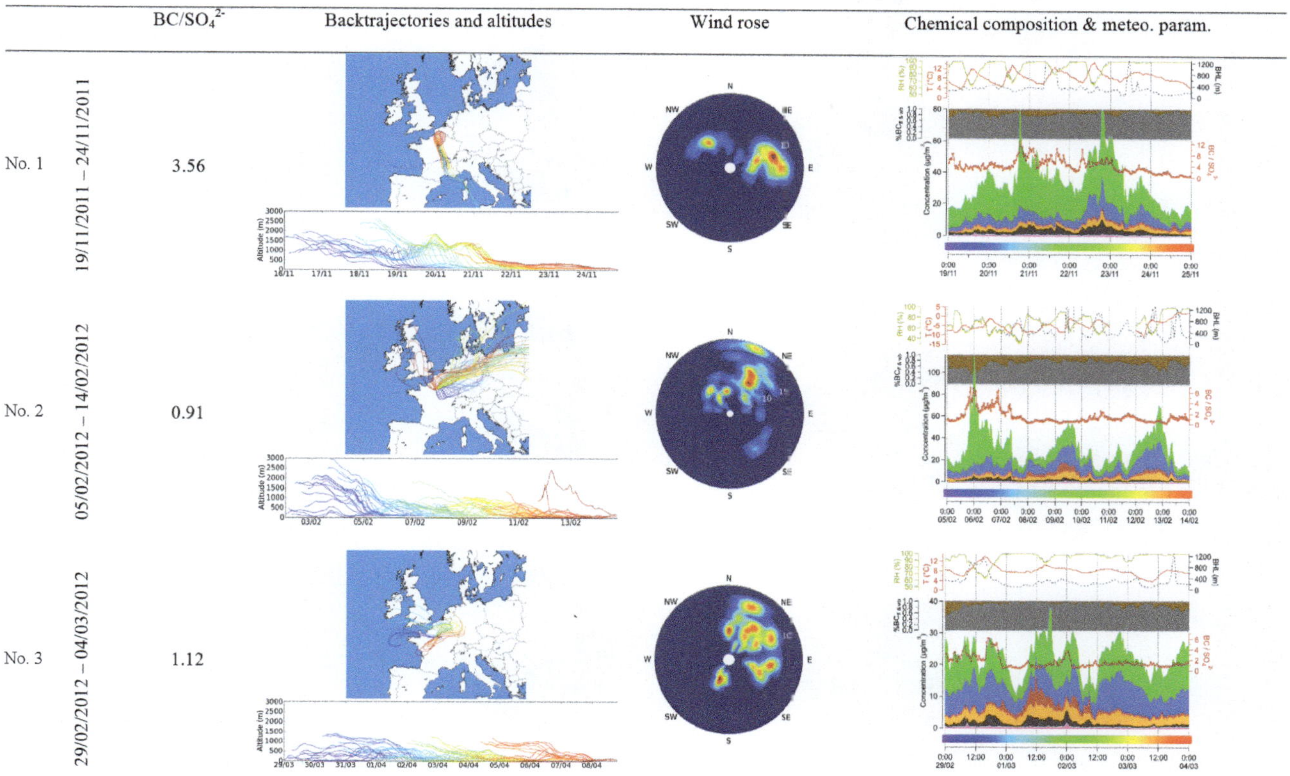

Figure 13.

of secondary advected pollutants. Ratios within this range should then be associated with a combined influence of regional and advected pollution. Finally, artifact-free ACSM data have been shown to be adequate for documenting semi-volatile aerosols (ammonium nitrate and a fraction of OM), which strongly contribute to PM_1 during persistent pollution episodes, and real-time measurements allow one to illustrate the close interactions between the chemical composition and meteorological parameters influencing its temporality.

9 Conclusions

The chemical composition of submicron (PM_1) aerosols was continuously monitored in near real time at a regional background site of the region of Paris between June 2011 and May 2013 using a combination of an ACSM and an Aethalometer. The obtained 2-year data set allows an appraisal of the robustness of ACSM measurements over several month periods, as well as Aethalometer measurements and BC source apportionment.

Non-parametric wind regression calculations have been performed for each season and provided useful information regarding the geographical origin of PM_1 chemical constituents. SIA, in particular ammonium sulfate, show a clear advected pattern, leading to a uniform signal over large scales. Ammonium nitrate also exhibits a significant contribution of regional and local emissions. The highest concentrations of OM were identified as having a major local origin, while regional background OM concentrations remain significant, especially in spring and summer. The region of Brittany (western France), the major hotspot of ammonia in France, seems to have little influence on the concentrations of this species at our station in the region of Paris; overall regional background concentrations of ammonia dominate, especially in spring. Similarly to OM, wintertime BC_{wb} concentrations are mainly from local emissions from domestic heating, although a noticeable regional background is still observed for this tracer of wood burning. As expected, BC_{ff} shows a clear local (nearby) origin, as well as a contribution from the Paris city plume, and remains fairly constant throughout seasons, due to its regional traffic origin.

Such near real-time observations over long-term periods offer a unique opportunity to provide robust diurnal profiles for each season. For instance, diurnal profiles of semi-volatile nitrate aerosols were observed in different seasons with temperatures favoring its partitioning into the particulate phase in the morning and in the gas phase in the afternoon. No clear contribution of traffic could be proven regarding ammonia variability, and the regional background seems to prevail.

All the persistent pollution episodes ($PM_1 > 20\,\mu g\,m^{-3}$ during at least 3 consecutive days) that occurred between 2011 and 2013 were carefully examined showing different meteorological conditions, sources and geographical ori-

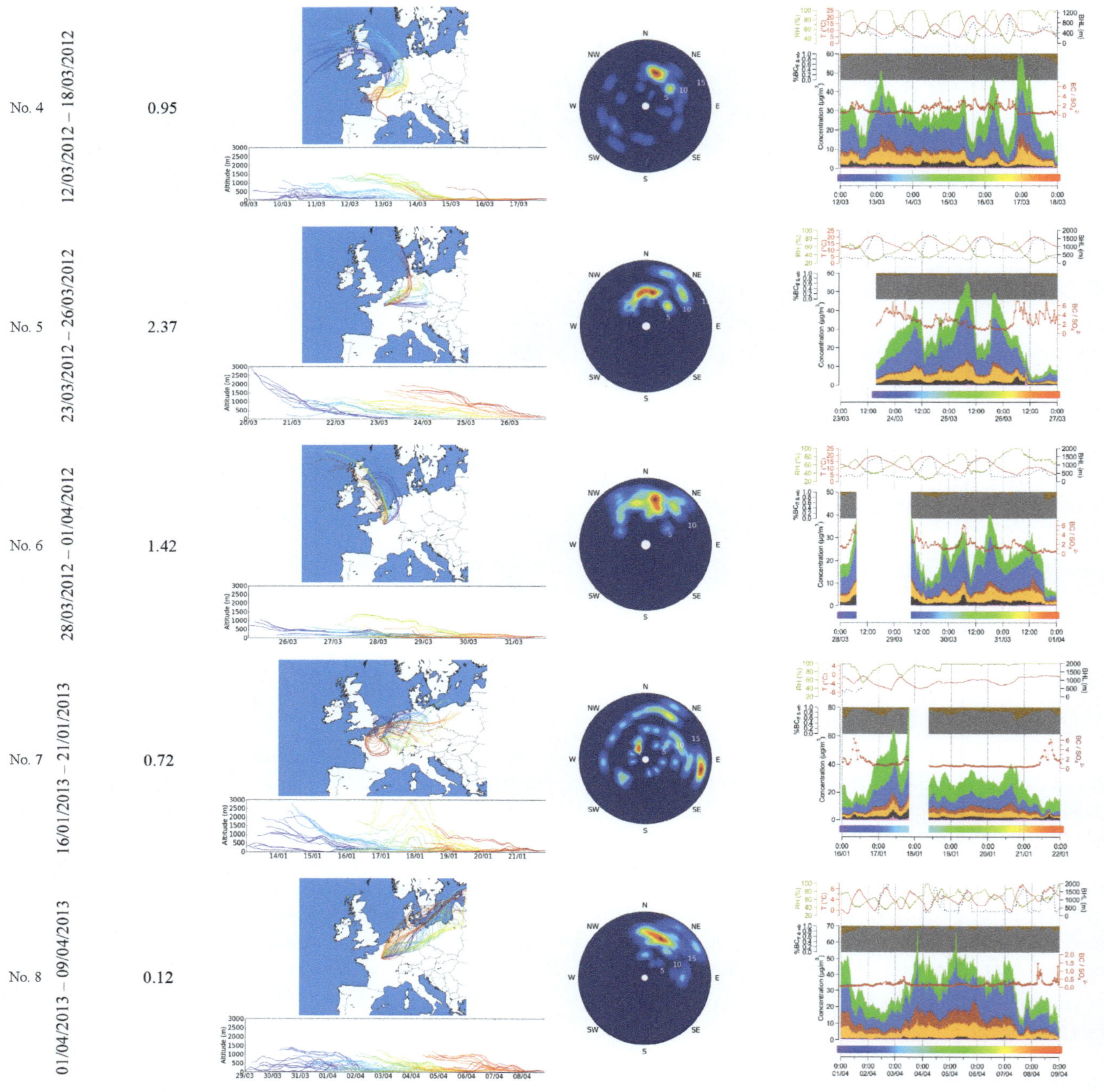

Figure 13. Illustration of meteorological conditions and chemical composition during the eight pollution episodes. Left graphs represent 72 h backtrajectories ending at SIRTA at 100 m a.g.l. every 3 h and their altitude; middle graphs illustrate the wind rose (radial axis in $km\,h^{-1}$); right graphs represent the chemical composition, in $\mu g\,m^{-3}$ of submicron particles (organic, nitrate, sulfate, ammonium, chloride and black carbon in green, blue, red, orange, pink and dark grey, respectively), the contribution of traffic and wood burning to BC, the BC / SO_4^{2-} ratio, and temperature, RH and BLH.

gins, making it difficult to draw general rules for these episodes. The BC / SO_4 ratio was used here to better separate local, regional (BC dominated) and advected (SO_4 dominated) contributions, and showed that, with very few exceptions, most of these persistent episodes were dominated by medium- to long-range transported pollution. However, it is interesting to note that the majority of the highest (time-limited) PM_1 concentrations (30 min ACSM data points with $PM_1 > 50\,\mu g\,m^{-3}$) fell within these persistent pollution episodes and were characterized by a significant local/regional contribution (high BC / SO_4 ratios). This result, obtained with real-time measurements, may offer new perspectives in the definition and the evaluation of the effectiveness of local mitigation policies such as emergency measures (traffic or wood-burning restrictions, for instance) taken to improve air quality during pollution events. In parallel, the long-term characterization of the organic fraction would surely lead to a better assessment of aerosol sources and some (trans)formation processes of secondary pollution in the Ile-de-France area.

In conclusion, these first 2-year quality-controlled measurements of ACSM clearly demonstrate their great potential to monitor on a long-term basis aerosol sources and their geographical origin, and provide strategic information in near real time during pollution episodes. They also support the capacity of the ACSM to be proposed as a robust and credible alternative to filter-based sampling techniques for long-term monitoring strategies. The networking of such instrumentation (ACSM and BC) throughout Europe – as is currently being built up within the European ACTRIS program – will certainly offer tremendous opportunities for modeling studies in order to improve prevision models, as well as large-scale spatially and temporally resolved source apportionment studies of organic aerosols using the high potential of ACSM organic fragments.

Acknowledgements. The research leading to these results has received funding from INERIS, CNRS, CEA, the French SOERE-ORAURE network, the European Union Seventh Framework Program (FP7/2007–2013) project ACTRIS under grant agreement no. 262254, the DIM-R2DS program for the funding of the ACSM equipment, and the PRIMEQUAL-PREQUALIF and ADEME-REBECCA programs for the long-term observations of black carbon at SIRTA.

Edited by: J. Allan

References

Airparif: Bilan des Émissions de Polluants Atmosphériques et de Gaz à Effet de Serre en Île- de-France pour l'Année 2010 et Historique 2000/05: Méthodologie et résultats, available at: http://www.airparif.asso.fr/_pdf/publications/Emissions_2010_CG75.pdf (last access: February 2015), 2013.

Appel, B. R., Tokiwa, Y., Haik, M., and Kothny, E. L.: Artifact particulate sulphate and nitrate formation on filter media, Atmos. Environ., 18, 409–416, 1984.

Arguez, A. and Vose, R. S.: The Definition of the Standard WMO Climate Normal: The Key to Deriving Alternative Climate Normals, B. Am. Meteorol. Soc., 92, 699–704, doi:10.1175/2010BAMS2955.1, 2011.

Bari, A., Ferraro, V., Wilson, L. R., Luttinger, D., and Husain, L.: Measurements of gaseous HONO, HNO_3, SO_2, HCl, NH_3, particulate sulphate and $PM_{2.5}$ in New York, NY, Atmos. Environ., 37, 2825–2835, doi:10.1016/S1352-2310(03)00199-7, 2003.

Barmpadimos, I., Keller, J., Oderbolz, D., Hueglin, C., and Prévôt, A. S. H.: One decade of parallel fine ($PM_{2.5}$) and coarse ($PM_{10} - PM_{2.5}$) particulate matter measurements in Europe: trends and variability, Atmos. Chem. Phys., 12, 3189–3203, doi:10.5194/acp-12-3189-2012, 2012.

Baum, M. M., Kiyomiya, E. S., Kumar, S., Lappas, A. M., Kapinus, V. A., and Lord, H. C.: Multicomponent remote sensing of vehicle exhaust by dispersive absorption spectroscopy. 2. Direct on-road ammonia measurements, Environ. Sci. Technol., 35, 3735–3741, doi:10.1021/es002046y, 2001.

Bessagnet, B., Hodzic, A., Blanchard, O., Lattuati, M., Lebihan, O., Marfaing, H. and Rouil, L.: Origin of particulate matter pollution episodes in wintertime over the Paris Basin, Atmos. Environ., 39, 6159–6174, doi:10.1016/j.atmosenv.2005.06.053, 2005.

Bond, T. C and Bergstrom, R. W.: Light Absorption by Carbonaceous Particles: An Investigative Review, Aerosol Sci. Tech., 40, 27–67, doi:10.1080/02786820500421521, 2006.

Bougiatioti, A., Stavroulas, I., Kostenidou, E., Zarmpas, P., Theodosi, C., Kouvarakis, G., Canonaco, F., Prévôt, A. S. H., Nenes, A., Pandis, S. N., and Mihalopoulos, N.: Processing of biomass-burning aerosol in the eastern Mediterranean during summertime, Atmos. Chem. Phys., 14, 4793–4807, doi:10.5194/acp-14-4793-2014, 2014.

Bressi, M., Sciare, J., Ghersi, V., Bonnaire, N., Nicolas, J. B., Petit, J.-E., Moukhtar, S., Rosso, A., Mihalopoulos, N., and Féron, A.: A one-year comprehensive chemical characterisation of fine aerosol ($PM_{2.5}$) at urban, suburban and rural background sites in the region of Paris (France), Atmos. Chem. Phys., 13, 7825–7844, doi:10.5194/acp-13-7825-2013, 2013.

Bressi, M., Sciare, J., Ghersi, V., Mihalopoulos, N., Petit, J.-E., Nicolas, J. B., Moukhtar, S., Rosso, A., Féron, A., Bonnaire, N., Poulakis, E., and Theodosi, C.: Sources and geographical origins of fine aerosols in Paris (France), Atmos. Chem. Phys., 14, 8813–8839, doi:10.5194/acp-14-8813-2014, 2014.

Budisulisticrini, S. H., Canagaratna, M. R., Croteau, P. L., Baumann, K., Edgerton, E. S., Kollman, M. S., Ng, N. L., Verma, V., Shaw, S. L., Knipping, E. M., Worsnop, D. R., Jayne, J. T., Weber, R.J., and Surratt, J. D.: Intercomparison of an Aerosol Chemical Speciation Monitor (ACSM) with ambient fine aerosol measurements in downtown Atlanta, Georgia, Atmos. Meas. Tech., 7, 1929–1941, doi:10.5194/amt-7-1929-2014, 2014.

Bussink, D. W., Harper, L. A., and Corré, W. J.: Ammonia Transport in a Temperate Grassland: II. Diurnal Fluctuations in Response to Weather and Management Conditions, Agron. J., 88, 621–626, 1996.

Cappa, C. D., Onasch, T. B., Massoli, P., Worsnop, D. R., Bates, T. S., Cross, E. S., Davidovits., P., Hakala, J., Hayden, K. L., Jobson, B. T., Kolesar, K. R., Lack, D. A., Lerner, B. M., Li, S.-M., Mellon, D., Nuaaman, I., Olfert, J. S., Petaja, T., Quinn, P. K., Song, C., Subramanian, R., Williams, E. J., and Zaveri, R. A.: Radiative absorption enhancements due to the mixing state of atmospheric black carbon, Science, 337, 1078–1081, 2012.

Carbone, S., Saarikoski, S., Frey, A., Reyes, F., Reyes, P., Castillo, M., Gramsch, E., Oyola, P., Jayne, J. T., Worsnop, D. R., and Hillamo, R.: Chemical Characterization of Submicron Aerosol Particles in Santiago de Chile, Aerosol Air Qual. Res., 13, 462–473, doi:10.4209/aaqr.2012.10.0261, 2013.

Carlton, A. G., Wiedinmyer, C., and Kroll, J. H.: A review of Secondary Organic Aerosol (SOA) formation from isoprene, Atmos. Chem. Phys., 9, 4987–5005, doi:10.5194/acp-9-4987-2009, 2009.

Cavalli, F., Viana, M., Yttri, K. E., Genberg, J., and Putaud, J.-P.: Toward a standardised thermal-optical protocol for measuring atmospheric organic and elemental carbon: the EUSAAR protocol, Atmos. Meas. Tech., 3, 79–89, doi:10.5194/amt-3-79-2010, 2010.

Cheng, Y., He, K. B., Duan, F. K., Zheng, M., Ma, Y. L., and Tan, J. H.: Positive sampling artifact of carbonaceous aerosols and its influence on the thermal-optical split of OC / EC, Atmos. Chem. Phys., 9, 7243–7256, doi:10.5194/acp-9-7243-2009, 2009.

Chow, J. C., Watson, J. G., Mauderly, J. L., Costa, D. L., Wyzga, R. E., Vedal, S., Hidy, G. M., Altshuler, S. L., Marrack, D., Heuss, J. M., Wolff, G. T., Pope, C. A., and Dockery, D. W.: Health Effects of Fine Particulate Air Pollution: Lines that Connect, J. Air Waste Manage., 56, 1368–1380, 2006.

Collaud Coen, M., Weingartner, E., Apituley, A., Ceburnis, D., Fierz-Schmidhauser, R., Flentje, H., Henzing, J. S., Jennings, S. G., Moerman, M., Petzold, A., Schmid, O., and Baltensperger, U.: Minimizing light absorption measurement artifacts of the Aethalometer: evaluation of five correction algorithms, Atmos. Meas. Tech., 3, 457–474, doi:10.5194/amt-3-457-2010, 2010.

Cowen, K., Sumner, A. L., Dinhal, A., Riggs, K., and Willenberg, Z.: Environmental Technology Verification Report, Mechatronics Instruments BV AiRRmonia Ammonia Analyzer, 2004.

Crenn, V., Sciare, J., Croteau, P. L., Favez, O., Verlhac, S., Belis, C. A., Fröhlich, R., Aas, W., Äijälä, M., Alastuey, A., Artiñano, B., Baisnée, D., Baltensperger, U., Bonnaire, N., Bressi, M., Canagaratna, M., Canonaco, F., Carbone, C., Cavalli, F., Coz, E., Cubison, M. J., Gietl, J. K., Green, D. C., Heikkinen, L., Lunder, C., Minguillón, M. C., Močnik, G., O'Dowd, C. D., Ovadnevaite, J., Petit, J.-E., Petralia, E., Poulain, L., Prévôt, A. S. H., Priestman, M., Riffault, V., Ripoll, A., Sarda-Estève, R., Slowik, J., Setyan, A., and Jayne, J. T.: ACTRIS ACSM intercomparison: Part 1 – Intercomparison of concentration and fragment results from 13 individual co-located aerosol chemical speciation monitors (ACSM), Atmos. Meas. Tech. Discuss., submitted, 2015.

Crippa, M., Canonaco, F., Slowik, J. G., El Haddad, I., DeCarlo, P. F., Mohr, C., Heringa, M. F., Chirico, R., Marchand, N., Temime-Roussel, B., Abidi, E., Poulain, L., Wiedensohler, A., Baltensperger, U., and Prévôt, A. S. H.: Primary and secondary organic aerosol origin by combined gas-particle phase source apportionment, Atmos. Chem. Phys., 13, 8411–8426, doi:10.5194/acp-13-8411-2013, 2013a.

Crippa, M., DeCarlo, P. F., Slowik, J. G., Mohr, C., Heringa, M. F., Chirico, R., Poulain, L., Freutel, F., Sciare, J., Cozic, J., Di Marco, C. F., Elsasser, M., Nicolas, J. B., Marchand, N., Abidi, E., Wiedensohler, A., Drewnick, F., Schneider, J., Borrmann, S., Nemitz, E., Zimmermann, R., Jaffrezo, J.-L., Prévôt, A. S. H., and Baltensperger, U.: Wintertime aerosol chemical composition and source apportionment of the organic fraction in the metropolitan area of Paris, Atmos. Chem. Phys., 13, 961–981, doi:10.5194/acp-13-961-2013, 2013b.

Crippa, M., El Haddad, I., Slowik, J. G., DeCarlo, P. F., Mohr, C., Heringa, M. F., Chirico, R., Marchand, N., Sciare, J., Baltensperger, U. and Prevot, A. S. H.: Identification of marine and continental aerosol sources in Paris using high resolution aerosol mass spectrometry, J. Geophys. Res.-Atmos., 118, 1950–1963, doi:10.1002/jgrd.50151, 2013c.

Draxler, R.: Hysplit4 User's Guide, available at: http://www.arl.noaa.gov/documents/reports/hysplit_user_guide.pdf, (last access: 14 May 2014), 1999.

Drinovec, L., Močnik, G., Zotter, P., Prévôt, A. S. H., Ruckstuhl, C., Coz, E., Rupakheti, M., Sciare, J., Müller, T., Wiedensohler, A., and Hansen, A. D. A.: The "dual-spot" Aethalometer: an improved measurement of aerosol black carbon with real-time loading compensation, Atmos. Meas. Tech. Discuss., 7, 10179–10220, doi:10.5194/amtd-7-10179-2014, 2014.

Edgerton, E. S., Saylor, R. D., Hartsell, B. E., Jansen, J. J., and Alan Hansen, D.: Ammonia and ammonium measurements from the southeastern United States, Atmos. Environ., 41, 3339–3351, doi:10.1016/j.atmosenv.2006.12.034, 2007.

El Haddad, I., D'Anna, B., Temime-Roussel, B., Nicolas, M., Boreave, A., Favez, O., Voisin, D., Sciare, J., George, C., Jaffrezo, J.-L., Wortham, H., and Marchand, N.: Towards a better understanding of the origins, chemical composition and aging of oxygenated organic aerosols: case study of a Mediterranean industrialized environment, Marseille, Atmos. Chem. Phys., 13, 7875–7894, doi:10.5194/acp-13-7875-2013, 2013.

Favez, O., Cachier, H., Sciare, J., Sarda-estève, R., and Martinon, L.: Evidence for a significant contribution of wood burning aerosols to $PM_{2.5}$ during the winter season in Paris, France, Atmos. Environ., 43, 3640–3644, 2009.

Favez, O., El Haddad, I., Piot, C., Boréave, A., Abidi, E., Marchand, N., Jaffrezo, J.-L., Besombes, J.-L., Personnaz, M.-B., Sciare, J., Wortham, H., George, C., and D'Anna, B.: Inter-comparison of source apportionment models for the estimation of wood burning aerosols during wintertime in an Alpine city (Grenoble, France), Atmos. Chem. Phys., 10, 5295–5314, doi:10.5194/acp-10-5295-2010, 2010.

Freutel, F., Schneider, J., Drewnick, F., von der Weiden-Reinmüller, S.-L., Crippa, M., Prévôt, A. S. H., Baltensperger, U., Poulain, L., Wiedensohler, A., Sciare, J., Sarda-Estève, R., Burkhart, J. F., Eckhardt, S., Stohl, A., Gros, V., Colomb, A., Michoud, V., Doussin, J. F., Borbon, A., Haeffelin, M., Morille, Y., Beekmann, M., and Borrmann, S.: Aerosol particle measurements at three stationary sites in the megacity of Paris during summer 2009: meteorology and air mass origin dominate aerosol particle composition and size distribution, Atmos. Chem. Phys., 13, 933–959, doi:10.5194/acp-13-933-2013, 2013.

Fröhlich, R., Crenn, V., Setyan, A., Belis, C. A., Canonaco, F., Favez, O., Riffault, V., Slowik, J. G., Aas, W., Aijälä, M., Alastuey, A., Artiñano, B., Bonnaire, N., Bozzetti, C., Bressi, M., Carbone, C., Coz, E., Croteau, P. L., Cubison, M. J., Esser-Gietl, J. K., Green, D. C., Gros, V., Heikkinen, L., Herrmann, H., Jayne, J. T., Lunder, C. R., Minguillón, M. C., Močnik, G., O'Dowd, C. D., Ovadnevaite, J., Petralia, E., Poulain, L., Priestman, M., Ripoll, A., Sarda-Estève, R., Wiedensohler, A., Baltensperger, U., Sciare, J., and Prévôt, A. S. H.: ACTRIS ACSM intercomparison – Part 2: Intercomparison of ME-2 organic source apportionment results from 15 individual, co-located aerosol mass spectrometers, Atmos. Meas. Tech. Discuss., 8, 1559–1613, doi:10.5194/amtd-8-1559-2015, 2015.

Ghersi, V., Rosso, A., Moukhtar, S., Léger, K., Sciare, J., Bressi, M., Nicolas, J., Féron, A., and Bonnaire, N.: Sources of fine aerosol($PM_{2.5}$) in the region of Paris, Pollution Atmosphérique, Numéro Spécial, Novembre, 2012.

Grover, B. D.: Measurement of total PM $_{2.5}$ mass (nonvolatile plus semivolatile) with the Filter Dynamic Measurement System tapered element oscillating microbalance monitor, J. Geophys. Res., 110, D07S03, doi:10.1029/2004JD004995, 2005.

Haeffelin, M., Barthès, L., Bock, O., Boitel, C., Bony, S., Bouniol, D., Chepfer, H., Chiriaco, M., Cuesta, J., Delanoë, J., Drobinski, P., Dufresne, J.-L., Flamant, C., Grall, M., Hodzic, A., Hourdin, F., Lapouge, F., Lemaître, Y., Mathieu, A., Morille, Y., Naud, C., Noël, V., O'Hirok, W., Pelon, J., Pietras, C., Protat, A., Romand, B., Scialom, G., and Vautard, R.: SIRTA, a ground-based atmospheric observatory for cloud and aerosol research, Ann. Geophys., 23, 253–275, doi:10.5194/angeo-23-253-2005, 2005.

Healy, R. M., Sciare, J., Poulain, L., Kamili, K., Merkel, M., Müller, T., Wiedensohler, A., Eckhardt, S., Stohl, A., Sarda-Estève, R., McGillicuddy, E., O'Connor, I. P., Sodeau, J. R., and Wenger, J. C.: Sources and mixing state of size-resolved elemental carbon particles in a European megacity: Paris, Atmos. Chem. Phys., 12, 1681–1700, doi:10.5194/acp-12-1681-2012, 2012.

Healy, R. M., Sciare, J., Poulain, L., Crippa, M., Wiedensohler, A., Prévôt, A. S. H., Baltensperger, U., Sarda-Estève, R., McGuire, M. L., Jeong, C.-H., McGillicuddy, E., O'Connor, I. P., Sodeau, J. R., Evans, G. J., and Wenger, J. C.: Quantitative determination of carbonaceous particle mixing state in Paris using single-particle mass spectrometer and aerosol mass spectrometer measurements, Atmos. Chem. Phys., 13, 9479–9496, doi:10.5194/acp-13-9479-2013, 2013.

Herich, H., Hueglin, C., and Buchmann, B.: A 2.5 year's source apportionment study of black carbon from wood burning and fossil fuel combustion at urban and rural sites in Switzerland, Atmos. Meas. Tech., 4, 1409–1420, doi:10.5194/amt-4-1409-2011, 2011.

Henry, R., Norris, G. A., Vedantham, R., and Turner, J. R.: Source Region Identification Using Kernel Smoothing, Environ. Sci. Technol., 43, 4090–4097, doi:10.1021/es8011723, 2009.

IARC: Outdoor air pollution a leading environmental cause of cancer deaths, press release no. 221, 2013.

Janssen, N. A., Hoek, G., Simic-Lawson, M., Fischer, P., van Bree, L., ten Brink, H., Keuken, M., Atkinson, R. W., Anderson, H. R., Brunekreef, B., and Cassee, F. R.: Black carbon as an additional indicator of the adverse health effects of airborne particles compared with PM_{10} and $PM_{2.5}$., Environ. Health Persp., 119, 1691–1699, 2011.

Jayne, J. T., Leard, D. C., Zhang, X., Davidovits, P., Smith, K. S., Kolb, C. E., and Worsnop, D. R.: Development of an Aerosol Mass Spectrometer for Size and Composition Analysis of Submicron Particles, Aerosol Sci. Tech., 33, 49–70, 2000.

Jimenez, J. L., Canagaratna, M. R., Donahue, N. M., Prevot, A. S. H., Zhang, Q., Kroll, J. H., DeCarlo, P. F., Allan, J. D., Coe, H., Ng, N. L., Aiken, A. C., Docherty, K. S., Ulbrich, I. M., Grieshop, A. P., Robinson, A. L., Duplissy, J., Smith, J. D., Wilson, K. R., Lanz, V. A., Hueglin, C., Sun, Y. L., Tian, J., Laaksonen, A., Raatikainen, T., Rautiainen, J., Vaattovaara, P., Ehn, M., Kulmala, M., Tomlinson, J. M., Collins, D. R., Cubison, M. J., Dunlea, E. J., Huffman, J. A., Onasch, T. B., Alfarra, M. R., Williams, P. I., Bower, K., Kondo, Y., Schneider, J., Drewnick, F., Borrmann, S., Weimer, S., Demerjian, K., Salcedo, D., Cottrell, L., Griffin, R., Takami, A., Miyoshi, T., Hatakeyama, S., Shimono, A., Sun, J. Y., Zhang, Y. M., Dzepina, K., Kimmel, J. R., Sueper, D., Jayne, J. T., Herndon, S. C., Trimborn, A. M., Williams, L. R., Wood, E. C., Middlebrook, A. M., Kolb, C. E., Baltensperger, U., and Worsnop, D. R.: Evolution of Organic Aerosols in the Atmosphere, Science, 326, 1525–1529, doi:10.1126/science.1180353, 2009.

Kai, Z., Yuesi, W., Tianxue, W., Yousef, M., and Frank, M.: Properties of nitrate, sulphate and ammonium in typical polluted atmospheric aerosols (PM_{10}) in Beijing, Atmos. Res., 84, 67–77, doi:10.1016/j.atmosres.2006.05.004, 2007.

Laborde, M., Crippa, M., Tritscher, T., Jurányi, Z., Decarlo, P. F., Temime-Roussel, B., Marchand, N., Eckhardt, S., Stohl, A., Baltensperger, U., Prévôt, A. S. H., Weingartner, E., and Gysel, M.: Black carbon physical properties and mixing state in the European megacity Paris, Atmos. Chem. Phys., 13, 5831–5856, doi:10.5194/acp-13-5831-2013, 2013.

Lack, D. A., Cappa, C. D., Covert, D. S., Baynard, T., Massoli, P., Sierau, B., Bates, T. S., Quinn, P. K., Lovejoy, E. R., and Ravishankara, A. R.: Bias in Filter-Based Aerosol Light Absorption Measurements Due to Organic Aerosol Loading: Evidence from Ambient Measurements, Aerosol Sci. Tech., 42, 1033–1041, doi:10.1080/02786820802389277, 2008.

Lanz, V. A., Alfarra, M. R., Baltensperger, U., Buchmann, B., Hueglin, C., and Prévôt, A. S. H.: Source apportionment of submicron organic aerosols at an urban site by factor analytical modelling of aerosol mass spectra, Atmos. Chem. Phys., 7, 1503–1522, doi:10.5194/acp-7-1503-2007, 2007.

Lin, Y., Cheng, M., Ting, W., and Yeh, C.: Characteristics of gaseous HNO_2, HNO_3, NH_3 and particulate ammonium nitrate in an urban city of Central Taiwan, Atmos. Environ., 40, 4725–4733, doi:10.1016/j.atmosenv.2006.04.037, 2006.

Liousse, C., Cachier, H., and Jennings, S. G.: Optical and thermal measurements of black carbon aerosol content in different environments: variation of the specific attenuation cross-section, sigma, Atmos. Environ., 27A, 1203–1211, 1993.

Middlebrook, A. M., Bahreini, R., Jimenez, J. L., and Canagaratna, M. R.: Evaluation of Composition-Dependent Collection Efficiencies for the Aerodyne Aerosol Mass Spectrometer using Field Data, Aerosol Sci. Tech., 46, 258–271, doi:10.1080/02786826.2011.620041, 2012.

Ng, N. L., Herndon, S. C., Trimborn, A., Canagaratna, M. R., Croteau, P. L., Onasch, T. B., Sueper, D., Worsnop, D. R., Zhang, Q., and Sun, Y. L.: An aerosol chemical speciation monitor (ACSM) for routine monitoring of the composition and mass concentrations of ambient aerosol, Aerosol Sci. Tech., 45, 780–794, 2011.

Nordin, E. Z., Eriksson, A. C., Roldin, P., Nilsson, P. T., Carlsson, J. E., Kajos, M. K., Hellén, H., Wittbom, C., Rissler, J., Löndahl, J., Swietlicki, E., Svenningsson, B., Bohgard, M., Kulmala, M., Hallquist, M., and Pagels, J. H.: Secondary organic aerosol formation from idling gasoline passenger vehicle emissions investigated in a smog chamber, Atmos. Chem. Phys., 13, 6101–6116, doi:10.5194/acp-13-6101-2013, 2013.

Olson, D. A., Vedantham, R., Norris, G. A., Brown, S. G., and Roberts, P.: Determining source impacts near roadways using wind regression and organic source markers, Atmos. Environ., 47, 261–268, doi:10.1016/j.atmosenv.2011.11.003, 2012.

Pal, S., Haeffelin, M., and Batchvarova, E.: Exploring a geophysical process-based attribution technique for the determination of the atmospheric boundary layer depth using aerosol lidar and near-surface meteorological measurements: NEW ATTRI-

BUTION LIDAR-DERIVED ABL DEPTH, J. Geophys. Res.-Atmos., 118, 9277–9295, doi:10.1002/jgrd.50710, 2013.

Pancras, J. P., Vedantham, R., Landis, M. S., Norris, G. A., and Ondov, J. M.: Application of EPA Unmix and Nonparametric Wind Regression on High Time Resolution Trace Elements and Speciated Mercury in Tampa, Florida Aerosol, Environ. Sci. Technol., 45, 3511–3518, doi:10.1021/es103400h, 2011.

Pandolfi, M., Amato, F., Reche, C., Alastuey, A., Otjes, R. P., Blom, M. J., and Querol, X.: Summer ammonia measurements in a densely populated Mediterranean city, Atmos. Chem. Phys., 12, 7557–7575, doi:10.5194/acp-12-7557-2012, 2012.

Pay, M. T., Jiménez-Guerrero, P., and Baldasano, J. M.: Assessing sensitivity regimes of secondary inorganic aerosol formation in Europe with the CALIOPE-EU modeling system, Atmos. Environ., 51, 146–164, doi:10.1016/j.atmosenv.2012.01.027, 2012.

Pathak, R., Yao, X., and Chan, C.: Sampling artifacts of acidity and ionic species in $PM_{2.5}$, Environ. Sci. Technol., 38, 254–259, doi:10.1021/es0342244, 2004.

Petetin, H., Beekmann, M., Sciare, J., Bressi, M., Rosso, A., Sanchez, O., and Ghersi, V.: A novel model evaluation approach focusing on local and advected contributions to urban $PM_{2.5}$ levels – application to Paris, France, Geosci. Model Dev., 7, 1483–1505, doi:10.5194/gmd-7-1483-2014, 2014.

Petit, J.-E., Favez, O., Sciare, J., Canonaco, F., Croteau, P., Močnik, G., Jayne, J., Worsnop, D., and Leoz-Garziandia, E.: Submicron aerosol source apportionment of wintertime pollution in Paris, France by double positive matrix factorization (PMF^2) using an aerosol chemical speciation monitor (ACSM) and a multi-wavelength Aethalometer, Atmos. Chem. Phys., 14, 13773–13787, doi:10.5194/acp-14-13773-2014, 2014.

Platt, S. M., El Haddad, I., Zardini, A. A., Clairotte, M., Astorga, C., Wolf, R., Slowik, J. G., Temime-Roussel, B., Marchand, N., Ježek, I., Drinovec, L., Močnik, G., Möhler, O., Richter, R., Barmet, P., Bianchi, F., Baltensperger, U., and Prévôt, A. S. H.: Secondary organic aerosol formation from gasoline vehicle emissions in a new mobile environmental reaction chamber, Atmos. Chem. Phys., 13, 9141–9158, doi:10.5194/acp-13-9141-2013, 2013.

Pope, C. A. and Dockery, D. W.: Health Effects of Fine Particulate Air Pollution: Lines that Connect, J. Air Waste Manage., 56, 709–742, doi:10.1080/10473289.2006.10464485, 2006.

Putaud, J.-P., Raes, F., Van Dingenen, R., Brüggemann, E., Facchini, M.-C., Decesari, S., Fuzzi, S., Gehrig, R., Hüglin, C., Laj, P., Lorbeer, G., Maenhaut, W., Mihalopoulos, N., Müller, K., Querol, X., Rodriguez, S., Schneider, J., Spindler, G., Brink, H. ten, Tørseth, K., and Wiedensohler, A.: A European aerosol phenomenology – 2: chemical characteristics of particulate matter at kerbside, urban, rural and background sites in Europe, Atmos. Environ., 38, 2579–2595, doi:10.1016/j.atmosenv.2004.01.041, 2004.

Putaud, J.-P., Dingenen, R. V., Alastuey, A., Bauer, H., Birmili, W., Cyrys, J., Flentje, H., Fuzzi, S., Gehrig, R., Hansson, H., Harrison, R., Herrmann, H., Hitzenberger, R., Hüglin, C., Jones, A., Kasper-Giebl, A., Kiss, G., Kousa, A., Kuhlbusch, T., Löschau, G., Maenhaut, W., Molnar, A., Moreno, T., Pekkanen, J., Perrino, C., Pitz, M., Puxbaum, H., Querol, X., Rodriguez, S., Salma, I., Schwarz, J., Smolik, J., Schneider, J., Spindler, G., ten Brink, H., Tursic, J., Viana, M., Wiedensohler, A., and Raes, F.: A European aerosol phenomenology – 3: Physical and chemical characteristics of particulate matter from 60 rural, urban, and kerbside sites across Europe, Atmos. Environ., 44, 1308–1320, doi:10.1016/j.atmosenv.2009.12.011, 2010.

Ramgolam, K., Favez, O., Cachier, H., Gaudichet, A., Marano, F., Martinon, L., and Baeza-Squiban, A.: Size-partitioning of an urban aerosol to identify particle determinants involved in the proinflammatory response induced in airway epithelial cells, Part. Fibre Toxicol., 6, doi:10.1186/1743-8977-6-10, 2009.

Rengarajan, R., Sudheer, A. K., and Sarin, M. M.: Wintertime $PM_{2.5}$ and PM_{10} carbonaceous and inorganic constituents from urban site in western India, Atmos. Res., 102, 420–431, doi:10.1016/j.atmosres.2011.09.005, 2011.

Sandradewi, J., Prévôt, A. S. H., Szidat, S., Perron, N., Alfarra, M. R., Lanz, V. A., Weingartner, E., and Baltensperger, U.: Using Aerosol Light Absorption Measurements for the Quantitative Determination of Wood Burning and Traffic Emission Contributions to Particulate Matter, Environ. Sci. Technol., 42, 3316–3323, doi:10.1021/es702253m, 2008.

Saylor, R. D., Edgerton, E. S., Hartsell, B. E., Baumann, K., and Hansen, D. A.: Continuous gaseous and total ammonia measurements from the southeastern aerosol research and characterization (SEARCH) study, Atmos. Environ., 44, 4994–5004, doi:10.1016/j.atmosenv.2010.07.055, 2010.

Sciare, J., Sarda-estève, R., Favez, O., Cachier, H., Aymoz, G., and Laj, P.: Nighttime residential wood burning evidenced from an indirect method for estimating real-time concentration of particulate organic matter (POM), Atmos. Environ., 42, 2158–2172, 2008.

Sciare, J., d'Argouges, O., Zhang, Q. J., Sarda-Estève, R., Gaimoz, C., Gros, V., Beekmann, M., and Sanchez, O.: Comparison between simulated and observed chemical composition of fine aerosols in Paris (France) during springtime: contribution of regional versus continental emissions, Atmos. Chem. Phys., 10, 11987–12004, doi:10.5194/acp-10-11987-2010, 2010.

Sciare, J., d' Argouges, O., Sarda-Estève, R., Gaimoz, C., Dolgorouky, C., Bonnaire, N., Favez, O., Bonsang, B., and Gros, V.: Large contribution of water-insoluble secondary organic aerosols in the region of Paris (France) during wintertime, J. Geophys. Res.-Atmos., 116, D22203, doi:10.1029/2011JD015756, 2011.

Takegawa, N., Miyazaki, Y., Kondo, Y., Komazaki, Y., Miyakawa, T., Jimenez, J. L., Jayne, J. T., Worsnop, D. R., Allan, J. D., and Weber, R. J.: Characterization of an Aerodyne Aerosol Mass Spectrometer (AMS): Intercomparison with Other Aerosol Instruments, Aerosol Sci. Tech., 39, 760–770, doi:10.1080/02786820500243404, 2005.

Turpin, B. J., Huntzicker, J. J., and Hering, S. V.: Investigation of organic aerosol sampling artifacts in the Los Angeles Basin, Atmos. Environ., 28, 3061–3071, 1994.

Waked, A., Favez, O., Alleman, L. Y., Piot, C., Petit, J.-E., Delaunay, T., Verlinden, E., Golly, B., Besombes, J.-L., Jaffrezo, J.-L., and Leoz-Garziandia, E.: Source apportionment of PM_{10} in a north-western Europe regional urban background site (Lens, France) using positive matrix factorization and including primary biogenic emissions, Atmos. Chem. Phys., 14, 3325–3346, doi:10.5194/acp-14-3325-2014, 2014.

Weingartner, E., Saathoff, H., Schnaiter, M., Streit, N., Bitnar, B., and Baltensperger, U.: Absorption of light by soot particles: determination of the absorption coefficient by means of aethalometers, J. Aerosol Sci., 34, 1445–1463, 2003.

Yiou, P. and Cattiaux, J.: Contribution of atmospheric circulation to wet north European summer precipitation of 2012, Bull. Amer. Meteorol. Soc., 94, 9, 2013.

Yu, K., Cheung, Y., Cheung, T., and Henry, R. C.: Identifying the impact of large urban airports on local air quality by nonparametric regression, Atmos. Environ., 38, 4501–4507, doi:10.1016/j.atmosenv.2004.05.034, 2004.

Zechmeister-Boltenstern, S.: Training on NH_3 measurement by wet chemistry techniques, ACTRIS TNA Activity Report, 2010.

Zhang, Q., Stanier, C. O., Canagaratna, M. R., Jayne, J. T., Worsnop, D. R., Pandis, S. N., and Jimenez, J. L.: Insights into the Chemistry of New Particle Formation and Growth Events in Pittsburgh Based on Aerosol Mass Spectrometry, Environ. Sci. Technol., 38, 4797–4809, doi:10.1021/es035417u, 2004.

Zhang, Q., Jimenez, J. L., Canagaratna, M. R., Allan, J. D., Coe, H., Ulbrich, I., Alfarra, M. R., Takami, A., Middlebrook, A. M., Sun, Y. L., Dzepina, K., Dunlea, E., Docherty, K., DeCarlo, P. F., Salcedo, D., Onasch, T., Jayne, J. T., Miyoshi, T., Shimono, A., Hatakeyama, S., Takegawa, N., Kondo, Y., Schneider, J., Drewnick, F., Borrmann, S., Weimer, S., Demerjian, K., Williams, P., Bower, K., Bahreini, R., Cottrell, L., Griffin, R. J., Rautiainen, J., Sun, J. Y., Zhang, Y. M., and Worsnop, D. R.: Ubiquity and dominance of oxygenated species in organic aerosols in anthropogenically-influenced Northern Hemisphere midlatitudes, Geophys. Res. Lett., 34, L13801, doi:10.1029/2007GL029979, 2007.

Zhang, Q. J., Beekmann, M., Drewnick, F., Freutel, F., Schneider, J., Crippa, M., Prevot, A. S. H., Baltensperger, U., Poulain, L., Wiedensohler, A., Sciare, J., Gros, V., Borbon, A., Colomb, A., Michoud, V., Doussin, J.-F., Denier van der Gon, H. A. C., Haeffelin, M., Dupont, J.-C., Siour, G., Petetin, H., Bessagnet, B., Pandis, S. N., Hodzic, A., Sanchez, O., Honoré, C., and Perrussel, O.: Formation of organic aerosol in the Paris region during the MEGAPOLI summer campaign: evaluation of the volatility-basis-set approach within the CHIMERE model, Atmos. Chem. Phys., 13, 5767–5790, doi:10.5194/acp-13-5767-2013, 2013.

7

Gaseous products and secondary organic aerosol formation during long term oxidation of isoprene and methacrolein

L. Brégonzio-Rozier[1], F. Siekmann[2], C. Giorio[3,4], E. Pangui[1], S. B. Morales[1], B. Temime-Roussel[2], A. Gratien[1], V. Michoud[1], S. Ravier[2], M. Cazaunau[1], A. Tapparo[4], A. Monod[2], and J.-F. Doussin[1]

[1]Laboratoire Interuniversitaire des Systèmes Atmosphériques (LISA), UMR7583, CNRS, Université Paris-Est-Créteil (UPEC) et Université Paris Diderot (UPD), Institut Pierre Simon Laplace (IPSL), Créteil, France
[2]Aix-Marseille Université, CNRS, LCE FRE 3416, 13331, Marseille, France
[3]Department of Chemistry, University of Cambridge, Cambridge CB2 1EW, UK
[4]Dipartimento di Scienze Chimiche, Università degli Studi di Padova, Padova, 35131, Italy

Correspondence to: L. Brégonzio-Rozier (lola.bregonzio@lisa.u-pec.fr)

Abstract. First- and higher order-generation products formed from the oxidation of isoprene and methacrolein with OH radicals in the presence of NO_x have been studied in a simulation chamber. Significant oxidation rates have been maintained for up to 7 h, allowing the study of highly oxidized products. Gas-phase product distribution and yields were obtained, and show good agreement with previous studies. Secondary organic aerosol (SOA) formation has also been investigated. SOA mass yields from previous studies show large discrepancies. The mass yields obtained here were consistent with the lowest values found in the literature, and more specifically in agreement with studies carried out with natural light or artificial lamps with emission similar to the solar spectrum. Differences in light source are therefore proposed to explain partially the discrepancies observed between different studies in the literature for both isoprene- and methacrolein-SOA mass yields. There is a high degree of similarity between the SOA mass spectra from isoprene and methacrolein photooxidation, thus strengthening the importance of the role of methacrolein in SOA formation from isoprene photooxidation under our experimental conditions (i.e., presence of NO_x and long term oxidation). According to our results, SOA mass yields from both isoprene and methacrolein in the atmosphere could be lower than suggested by most of the current chamber studies.

1 Introduction

Isoprene (2-methyl-1,3-butadiene) is a biogenic volatile organic compound (VOC) emitted by vegetation. It is one of the most abundant non-methane hydrocarbons emitted into the troposphere, with annual global emissions of 440 to 660 TgC (Guenther et al., 2006). As a diene, isoprene is highly reactive in the atmosphere, resulting in low atmospheric lifetimes due to its reaction with atmospheric oxidants, especially the hydroxyl radical (OH), with a lifetime of 1.7 h (Karl et al., 2006). Because of its large emission rates and high reactivity, isoprene can have a strong influence on tropospheric photochemistry on the local, regional and global scales. The OH-initiated oxidation of isoprene leads to the production of first-generation oxidation compounds, i.e., first stable products which result from the initial OH attack on isoprene and do not involve additional attack by atmospheric oxidants (OH, O_3 or NO_3). The major primary products, in the presence of nitrogen oxides ($NO_x = NO + NO_2$), are methyl vinyl ketone (MVK), methacrolein (MACR), and formaldehyde (HCHO) (Miyoshi et al., 1994; Paulson and Seinfeld, 1992; Sprengnether et al., 2002; Tuazon and Atkinson, 1990a). Isoprene photooxidation in the presence of sufficient NO_x is also known to result in the production of significant quantities of ozone on regional scales, in rural as well as in urban areas during summer (Biesenthal et al., 1997; Starn et al., 1998; Wiedinmyer et al., 2001).

For years, it was considered that, because of the high volatility of first-generation products, secondary organic aerosol (SOA) formation from isoprene photooxidation in the presence of NO_x was insignificant in the troposphere (Pandis et al., 1991). However, in the early 2000s, detailed analysis of natural aerosols from the Amazonian rain forest (Claeys et al., 2004) showed significant amounts of two diastereoisomeric 2-methyltetrols (2-methylerythritol and 2-methylthreitol). These compounds bear the isoprene skeleton, and their biogenic sources are not primary. Following this discovery, SOA formation from isoprene has been reconsidered: field observations (Edney et al., 2005; Ion et al., 2005; Kourtchev et al., 2005) and laboratory chamber studies (Boge et al., 2006; Edney et al., 2005; Kroll et al., 2005) confirmed the ability of isoprene (or its oxidation products) to contribute significantly to atmospheric SOA. In fact, even if isoprene leads to small SOA yields (few percent or less), the global contribution of isoprene to the total particulate organic matter could be important considering its large emissions on the global scale. Because organic matter (mostly with secondary origin) accounts for a large, and often dominant, fraction (between 20 and 90 %) of fine particulate mass in the atmosphere (Kanakidou et al., 2005; Zhang et al., 2007), it is important to fully understand the SOA formation processes from isoprene oxidation, and especially the impact of the experimental conditions on SOA yields.

Laboratory chamber studies investigated the dependence of isoprene-SOA yields on NO_x levels (Kroll et al., 2006; Surratt et al., 2006). They showed that SOA yields are higher in the absence of NO_x, thus suggesting an important role of peroxy radical chemistry (RO_2). When RO_2 chemistry is dominated by the RO_2+ NO reaction, small alkoxy radicals (RO, which tend to fragment), and organic nitrates are formed and are likely sufficiently volatile to remain in the gas phase. On the contrary, in the absence of NO_x (< 1 ppb), RO_2 radicals react preferentially with HO_2 radicals to form hydroxy hydroperoxides and peroxy acids with lower volatility, leading to higher SOA yields. Experiments performed in the presence of NO_x also showed that SOA yields are higher for high NO_2 / NO ratios (3 to 8) (Chan et al., 2010). This result is due to the dominating RO_2+ NO_2 reaction, which leads to PAN-like compounds. In particular, MPAN (peroxy methacryloyl nitrate) was identified as an important intermediate to SOA formation from isoprene and MACR in the presence of NO_x (Surratt et al., 2010). MPAN formation is thus suppressed or delayed in the presence of high initial concentrations of NO, leading to lower SOA yields at low (< 1) initial VOC / NO ratios (Zhang et al., 2012).

Due to the identified link between MPAN and SOA formation from MACR, and the high degree of similarity of SOA mass spectra from isoprene and MACR photooxidation, MACR was recognized as the major contributor to SOA formation from isoprene in the presence of NO_x (Kroll et al., 2006; Surratt et al., 2006). Its gas-phase primary oxidation products in the presence of NO_x are CO, CO_2, HCHO, hydroxyacetone, methylglyoxal and MPAN (Galloway et al., 2011; Orlando et al., 1999; Tuazon and Atkinson, 1990b). SOA yields from MACR are globally higher than SOA yields from isoprene and are also influenced by NO_2 / NO ratios (Chan et al., 2010).

Although the influence of NO_x levels on SOA yields from isoprene and MACR photooxidation is recognized, it cannot fully explain the high degree of variability observed among studies from the literature. As pointed out by Carlton et al. (2009), this variability can be explained by differences in OH concentrations (that are related to the employed radical precursor) which have an important impact on the extent of the reactions and the rate of formation of semi-volatile compounds. Other experimental parameters, far less studied, could also contribute to the observed variability in SOA yields, including the effects of different light sources as well as the role of the chamber walls. It was demonstrated by Zhang et al. (2014) that wall losses of semi-VOCs during photooxidation experiments can lead to lower SOA yields. Furthermore, Warren et al. (2008) used black lights and an argon arc lamp (which exhibits an emission spectrum more similar to the solar spectrum) on the *m*-xylene/NO_x photooxidation system and they observed an effect of the irradiation wavelength spectrum on the SOA yields. It was suggested that black lights may be missing photolysis reactions which are important in SOA formation, leading to an increase in SOA yields when the argon arc lamp was used.

In this work, we investigate the formation of gas-phase first- and higher-generation products and SOA during isoprene and MACR + OH reactions in the presence of NO_x. The experiments have been carried out in a stainless steel chamber with very realistic irradiation to study the possible effect of the light source used and the state of cleanliness of the walls on SOA yields.

2 Experimental section

Experiments were performed in the CESAM chamber (French acronym for experimental multiphasic atmospheric simulation chamber), described in detail elsewhere (Wang et al., 2011). This 4.2 m^3 cylindrical stainless steel chamber is equipped with three high-pressure xenon arc lamps and Pyrex® filters of 6.5 mm thickness that provide, inside the chamber, an irradiation with a spectrum that is very close to the solar spectrum at the ground level (Fig. S1 in the Supplement). For these experiments, NO_2 photolysis frequency was $2.8 \times 10^{-3}\ s^{-1}$. In order to avoid an increase of temperature during experiments due to irradiation, a cooling system was used. The system is based on the circulation of a liquid coolant (70 % water and 30 % ethylene glycol) in the chamber double wall. The liquid's circulation and temperature were controlled by a thermostat (LAUDA, Integral T10000 W). Temperature and relative humidity (RH) inside the chamber were continuously monitored by a Vaisala HU-

Table 1. Experimental conditions and results.

Experiment[a,b]	$[VOC]_0$[h] ppb	OH source	$[NO]_0$ ppb	$[NO_2]_0$ ppb	$[HONO]_0$ ppb	$[O_3]_{max}$ ppb	T °C	$V_{AS,0}$[e] $\mu m^3\,cm^{-3}$	ΔM_0[f,i] $\mu g\,m^{-3}$	SOA mass yield
					Isoprene					
I150211	473	NO_x	119	32	–	347	18.1	–	8.4	0.006 ($\pm 5.10^{-4}$)
I160211	500	NO_x	14	79	–	546	16.4	–	4.7	0.003 ($\pm 3.10^{-4}$)
I170211	485	NO_x	22	55	–	397	16.6	–	1.6	0.001 ($\pm 1.10^{-4}$)
I050411	465	NO_x	110	4	–	495	20	–	12.4	0.010 ($\pm 7.10^{-4}$)
I060411	458	NO_x	135	5	–	300	21.1	–	7.3	0.006 ($\pm 5.10^{-4}$)
I080411	465	NO_x	109	3	–	286	20.8	16.2	5.5	0.004 ($\pm 3.10^{-4}$)
I110411	462	NO_x	127	5	–	359	21.9	12.9	6.2	0.005 ($\pm 4.10^{-4}$)
I150512	452	NO_x	101	<1	–	174	21.4	–	7.8	0.006 ($\pm 5.10^{-4}$)
I160512	445	NO_x	117	<1	–	175	20.6	–	4.4	0.004 ($\pm 3.10^{-4}$)
I210512[g]	442	NO_x	110	<1	–	183	22.2	–	<0.1	0
I220512	444	NO_x	111	<1	–	113	21	–	0.3	0
I230512	439	NO_x	21	76	–	131	24.3	–	0.1	0
I160113[g]	846	HONO	143	27[d]	15	122	21.5	–	<0.1	0
I280113[g]	833	HONO	88	45[d]	125	201	18.3	–	2.8	0.001 ($\pm 9.10^{-5}$)
I130313[g]	840	HONO	66	<1[d]	45	54	17.5	–	2.4	0.001 ($\pm 8.10^{-5}$)
I250313[g]	802	HONO	137	48[d]	121	81	19.7	–	0.15	0
					Methacrolein					
M120411	474	NO_x	117	4	–	145	19	14.8	17.4	0.013 ($\pm 9.10^{-4}$)
M130411	480	NO_x	123	4	–	130	20.8	13.5	13.9	0.010 ($\pm 7.10^{-4}$)
M240512	457	NO_x	19	84	–	97	24.2	–	9.5	0.008 ($\pm 6.10^{-4}$)
M250512	405	NO_x	26	100	–	46	24	–	5.0	0.005 ($\pm 5.10^{-4}$)
M280512	403	NO_x	n.m.[c]	80	–	59	23.8	–	9.4	0.009 ($\pm 8.10^{-4}$)
M180113[g]	735	HONO	88	25[d]	124	94	19.8	–	58.8	0.03 ($\pm 1.10^{-3}$)
M210113[g]	927	HONO	118	81[d]	150	123	19.4	–	65.8	0.032 ($\pm 1.10^{-3}$)
M230113[g]	396	HONO	67	5[d]	125	51	19.6	–	27.3	0.028 ($\pm 3.10^{-3}$)
M250113[g]	445	HONO	39	8[d]	60	31	18.8	–	7.8	0.007 ($\pm 6.10^{-4}$)
M110313[g]	400	HONO	107	38[d]	91	17	21.8	–	44.8	0.042 ($\pm 3.10^{-3}$)

[a] All experiments were carried out at RH < 5%. [b] Experimental IDs starting with "I" indicate isoprene photooxidation experiments and experimental IDs starting with "M" indicate methacrolein photooxidation experiments. [c] Not measured. [d] Corrected from HONO interference. [e] Volume concentration of ammonium sulfate seed. [f] SOA mass concentration using an effective density of 1.4 $g\,cm^{-3}$ (see text). [g] Experiment with manual cleaning the day before. [h] Measurement uncertainty is ± 15 ppb. [i] Measurement uncertainty is $\pm 0.1\,\mu g\,m^{-3}$.

MICAP HMP234 probe. Due to the very low level of charges on the walls (conductive and grounded), aerosols exhibit a long lifetime (between 10 h and 4 days, depending on the particle size distribution) in the chamber (McMurry and Rader, 1985; Wang et al., 2011).

2.1 Chamber conditioning

Prior to each experiment, the chamber was cleaned by overnight pumping at a secondary vacuum in the range of 6×10^{-4} mbar. This procedure has shown very satisfactory results for most chemical systems (Wang et al., 2011). In the case of isoprene photooxidation, due to the expected low aerosol yield, several experiments were preceded by an additional manual cleaning (Table 1). This manual cleaning was performed using ultrapure water (18.2 MΩ, ELGA Maxima) and lint free wipes (Spec-Wipe® 3), then the chamber walls were heated at 40 °C prior to overnight pumping. This procedure leads to experiments with very high initial level of cleanliness of the chamber walls, leading to low OH formation due to low nitrous acid (HONO) formation from NO_x wall reaction at the beginning of the experiment ($<5 \times 10^5$ molec cm^{-3}, see Fig. S2), thus requiring additional HONO introduction (Table 1).

After overnight pumping, the chamber was filled with synthetic air produced by mixing approximately 800 mbar of N_2 produced from the evaporation of a pressurized liquid nitrogen tank, and around 200 mbar of O_2 (Linde, 5.0). A known pressure of isoprene (Sigma Aldrich, 99 %), or methacrolein (Sigma Aldrich, 95 %), prepared in a known volume glass bulb was then introduced into the chamber by flushing with a low flow of O_2. Two different OH precursors were used: NO_x wall reaction (Wang et al., 2011) and HONO. NO_2 injection (Air Liquide, Alphagaz 99.9 % purity) in the chamber was made using a gas syringe and a septum valve. NO was injected from a gas cylinder (Air Liquide, 8 ppm in N_2) using a mass flow controller. HONO was prepared by a dropwise addition of sulfuric acid (10^{-2} M) into a solution of $NaNO_2$ (0.1 M) and carried into the chamber with a flow of pure N_2. During this synthesis, NO_x were also formed and introduced in the chamber. For some experiments, inorganic seed particles were generated from a 0.0012 M aqueous ammonium sulfate solution using a constant output atomizer (TSI, model 3075) and were injected into the chamber through a diffusion dryer (TSI, model 3062). The irradiation started after these injections and lasted for 6 to 9 h. Unless specified, time 0 denotes the irradiation start.

In order to avoid a decrease in the OH production efficiency due to a fast consumption of NO in the first hours of the experiment (see Fig. S3), a low (0.3 L min^{-1}) flow of NO (Air Liquide, 8 ppm in N_2) was continuously introduced into the chamber. The NO flow was started only when NO mixing ratio in the system reached a concentration below 5 ppb and was manually adjusted to avoid an accumulation of NO in the system in order to maintain a NO mixing ratio between 2 and 5 ppb during the entire experiment. The pressure inside the chamber was maintained at a pressure slightly higher than the ambient by applying a flow of air (80 % N_2 and 20 % O_2) to offset the pressure loss due to the continuous sampling. The experimental initial conditions are summarized in Table 1.

2.2 Measurements

The gas-phase concentrations of isoprene, MACR, MVK, HCHO, PAN, methylglyoxal, MPAN, formic acid, carbon monoxide (CO) and NO_2 were monitored by Fourier transform infrared spectrometry (FTIR, Bruker®, TENSOR 37) interfaced with an in situ multiple reflection cell. To determine the contribution of the pure reference spectra to the mixture spectra, an automatic procedure based on matrix algebra was used and results were cross-checked by manual subtraction performed over selected spectra. Complementarily to FTIR measurements, a proton-transfer time of flight mass spectrometer (PTR-ToF-MS 8000, Ionicon Analytik®) was used for online gas-phase measurements in the m/z range 10–200 including isoprene, the sum of {methacrolein + methyl vinyl ketone}, formaldehyde, methylglyoxal, formic acid, 3-methylfuran (3-MF), acetaldehyde, the sum of {acetic acid + glycolaldehyde}, acetone, acrolein (using the $(C_3H_4O)H^+$ ion signal, contribution from fragmentation of higher-molecular weight products could not be excluded), hydroxyacetone, and a few other oxygenated VOCs. Pure standards tests were previously carried out to identify fragmentation patterns and sensitivities of the main oxidation products. The signal of the PTR-ToF-MS was calibrated using a certified gas standard mixture (EU Version TO-14A Aromatics 110 L, 100 ppbv each) before the set of experiments. Parameters of the PTR-ToF-MS during the series of measurements were as follows: drift tube voltage: 500 V, drift tube pressure ≈ 2.15 mbar, drift tube temperature: 353 K, resulting in a E/N of 120 125 Td (E being the electric field strength applied to the drift tube and N the density of the gas in the drift tube) (1 Td = 10^{-17} V cm^{-2}). Data analysis of the PTR-ToF-MS measurements was carried out using the ToFViewer® software. ToF-to-mass assignment was performed using hydronium ion isotope ($H_3^{18}O^+$ $m/z = 21.023$) and protonated acetone ($C_3H_7O^+$ $m/z = 59.049$). The mass resolution m/Δ m of 4000 (at full width at half maximum) was achieved with the PTR-ToF-MS during the series of experiments. This enabled separation and formula assignment for most of the ions comprising the mass spectra. Some VOCs were measured using both FTIR and PTR-ToF-MS (isoprene, formaldehyde, methyl vinyl ketone, methacrolein, ...), thus providing intercalibration of their PTR-ToF-MS signal with the FTIR derived concentrations. Ozone was measured by a commercial UV absorption monitor (Horiba®, APOA-370). A commercial chemiluminescence NO_x analyzer (Horiba®, APNA-370) was used to monitor NO. Interferences on the NO_2 signal from the NO_x monitor could occur due to the presence of NO_y during the experiments (Dunlea et al., 2007), NO_2 mixing ratio was therefore determined using FTIR data.

HONO was measured using an instrument constructed in-house (NitroMAC) based on a wet chemical derivatization technique and high performance liquid chromatography – visible (HPLC-VIS) detection (Zhou et al., 1999). Gaseous HONO was sampled by dissolution in a buffer phosphate solution followed by derivatization with an aqueous sulphanilamide/N-(1-naphthyl)-ethylenediamine solution (SA/NED).

A scanning mobility particle sizer (SMPS) was used to monitor aerosol total number and volume concentrations and size distributions from 10.9 to 478 nm. It consists of a differential mobility analyzer (DMA, TSI, model 3080) coupled with a condensation particle counter (CPC, TSI, model 3010). The non-refractory submicron particulate matter bulk chemical composition was measured using a high resolution time-of-flight aerosol mass spectrometer (AMS, Aerodyne) (Canagaratna et al., 2007; De Carlo et al., 2006). The instrument was used under standard conditions (vaporizer at 600 °C and electron ionization at 70 eV) and operated in MS mode (V and W, 30 s each) and PToF mode (60 s). Ammonium nitrate particles were used regularly to perform standard AMS calibration procedures (including brute force single particle (BFSP) ionization efficiency calibration and size calibration). The AMS data were analyzed using the standard fragmentation table with the corrected air fragment column for our carrier gas, the default values of relative ionization efficiency and a collection efficiency of 0.5 for the organics (Squirrel ToF-AMS Analysis 1.51H and Pika® ToF-AMS HR Analysis 1.10H packages for the software Igor® Pro 6.21).

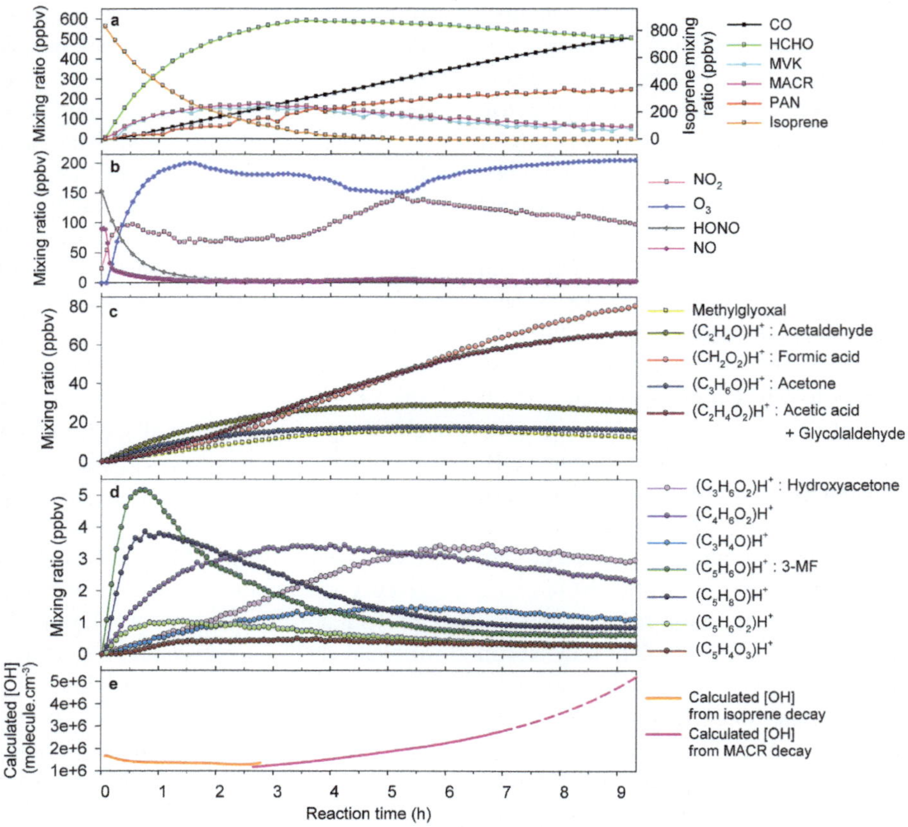

Figure 1. Time profiles of gas-phase measured compounds and calculated OH concentrations during isoprene photooxidation (experiment I280113) performed with no seeds and with HONO as OH source. PTR-ToF-MS measurements are represented by circles, and FTIR measurements, by squares. Calculated [OH] is represented by a dotted line after 7 h of irradiation due to low MACR mixing ratios which implies less precision in the calculation as the contribution from other VOCs is not negligible.

3 Results and discussion

3.1 Isoprene experiments

Figure 1 shows the time profiles of the gas phase reactants and reaction products during an isoprene photooxidation experiment performed without inorganic seed and using HONO as OH precursor. The irradiation induced a fast consumption of NO, leading to an increase of NO_2 concentrations via NO to NO_2 conversion (Fig. 1b). Isoprene concentration decay was also observed (Fig. 1a) and its lifetime due to reaction with OH ($\tau_{\text{isoprene-OH}}$) was determined considering the time needed to divide the initial isoprene concentration by a factor e. In all our experiments, $\tau_{\text{isoprene-OH}}$ was 1.7 ± 0.4 h (except for I160113 and I130313 in which isoprene lifetime was 1 and 2 h longer, respectively), thus close to the atmospheric isoprene lifetime (Karl et al., 2006). During isoprene photooxidation, NO reacts with RO_2 and HO_2 radicals to form NO_2 which photolyzes and leads to ozone production, thus explaining the observation of O_3 mixing ratios reaching up to several hundred ppbv (Fig. 1b). Despite these high O_3 mixing ratios, isoprene oxidation was dominated by OH radicals during all experiments, due to the higher reactivity of isoprene towards OH radicals than ozone (Karl et al., 2006).

3.1.1 Determination of OH concentrations

The OH concentrations and their time profiles were estimated from the observed decay rate of isoprene and methacrolein (using polynomial fit curves). The loss of VOC (i.e., either isoprene or methacrolein) was corrected from their reaction with ozone, photolysis and dilution, using Eq. (1):

$$[\text{OH}]_i = \frac{1}{k_{\text{OH+VOC}}}\left(-\frac{1}{[\text{VOC}]_i} \times \frac{\Delta[\text{VOC}]}{\Delta t} - k_{\text{O}_3\text{+VOC}}[\text{O}_3]_i - k_{\text{dil}} - J_{\text{VOC}}\right), \quad (1)$$

where $\frac{\Delta[\text{VOC}]}{\Delta t}$ is the time variation of the VOC mixing ratios, $k_{\text{OH+VOC}}$ and $k_{\text{O}_3\text{+VOC}}$ are the rate coefficients for reaction with, respectively, OH and O_3 (from Atkinson et al., 2006), k_{dil} is the dilution rate, and J_{VOC} is the photolysis rate of the VOC. J_{VOC} was determined for MACR (J_{MACR}) using MACR absorption cross-section and quantum yields (Atkinson et al., 2006), and xenon arc lamp irradiation spectrum with 6.5 mm Pyrex® filters (Fig. S1). The value for

k_{dil} was determined using the air flow rate used to offset the loss of pressure due to sampling and was found to be around $1.6 \times 10^{-5}\,s^{-1}$. The OH concentrations were calculated from the isoprene decay until its concentration became too low, and then the MACR decay was used. The resulting OH concentrations ranged between 1.5×10^6 and 6×10^6 molec cm^{-3}, thus showing that the protocol used (low flow of diluted NO continuously introduced) allowed to maintain an OH level in the chamber of the same order of magnitude as the one of the atmosphere during the entire experiment length (Fig. 1e).

3.1.2 Isoprene gas-phase reaction products

The major first generation products of isoprene OH-oxidation were MACR, HCHO and MVK (Fig. 1a, FTIR measurements), two minor first generation products (3-MF and C_5H_8O, see Fig. 1d) were also detected by the PTR-ToF-MS. Plotting the concentration (dilution corrected) of each first generation products versus the reacted isoprene concentration ($[\text{product}]_{\text{corr},t} = f([\text{isoprene}]_0 - [\text{isoprene}]_t)$) during the first hour of photooxidation (i.e., when photolysis and reaction with OH of the primary products were not significant), provided linear curves. The corresponding slope gave the yield for each first-generation product. Table 2 shows that, due to the indicated uncertainties, no significant differences between the obtained yields and those from the literature were observed. They were thus in good agreement.

Thanks to the continuous OH radical concentrations maintained throughout the experiments, the consumption of the primary products leading to the formation of compounds of higher generation was clearly seen (Fig. 1a, c, d). Among these compounds, PAN, methylglyoxal, glycolaldehyde and hydroxyacetone were previously identified as major primary oxidation products of MACR and MVK (Galloway et al., 2011; Orlando et al., 1999; Tuazon and Atkinson, 1990b, 1989). Other species generally observed in isoprene photooxidation experiments as acetaldehyde, formic acid and acetone (Nguyen et al., 2011b; Paulot et al., 2009) were also observed (Fig. 1c). Some other compounds were not clearly identified considering only their molecular formula given by PTR-ToF-MS measurements, but some assumptions can be made. We measured C_5H_8O (Fig. 1d), which seems to be a primary product (with a yield around 1 %), it may be attributed to 2-methylbut-3-enal. The latter was identified in the gas phase by Healy et al. (2008) in the same type of experiments. The yield for this compound was determined considering only the $(C_5H_8O)H^+$ ion signal and assuming no contribution from fragmentation of higher-molecular weight products. This unsaturated C_5 carbonyl compound can be formed, like the other primary products, from OH addition to one of the isoprene double bonds, followed by oxidation of the hydroxyalkyl radical produced. Compound $C_5H_6O_2$ (Fig. 1d) could correspond to methylbutandial, that was assumed to be formed by the δ-hydroxy channels including 3-MF reaction with OH (Paulot et al., 2009). It was also suggested by Paulot et al. (2009) that the δ-hydroxy channels lead to the formation of 3-oxobutanal, with a molecular formula corresponding to $C_4H_6O_2$, that we also observed (Fig. 1d). Species with this molecular formula could also be hydroxy methyl vinyl ketone (Galloway et al., 2011). MPAN concentrations were detected (but not quantified) by the PTR-ToF-MS at the $(C_4H_6O_3)H^+$ ion (m/z 103) (Hansel and Wisthaler, 2000). This compound was not detected by FTIR; thus, it was deduced that its mixing ratios were below the FTIR detection limit (i.e., 5 ppb).

Table 2. Yields of first-generation oxidation products during isoprene photooxidation compared with previous studies. Values in parentheses are 2-sigma uncertainties.

Compound	Yield	Reference
Formaldehyde	0.75 (±0.11)	This work
	0.63 (±0.10)	Tuazon and Atkinson (1990a)
	0.57 (±0.06)	Miyoshi et al. (1994)
	0.59 (±0.12)	Sprengnether et al. (2002)
Methacrolein	0.30 (±0.09)	This work
	0.22 (±0.05)	Tuazon and Atkinson (1990a)
	0.25 (±0.03)	Paulson et al. (1992)
	0.22 (±0.02)	Miyoshi et al. (1994)
	0.27 (±0.04)	Sprengnether et al. (2002)
	0.22 (±0.006)	Galloway et al. (2011)
	0.27 (±0.03)	Karl et al. (2006)
Methyl vinyl ketone	0.27 (±0.08)	This work
	0.32 (±0.07)	Tuazon and Atkinson (1990a)
	0.36 (±0.04)	Paulson et al. (1992)
	0.32 (±0.05)	Miyoshi et al. (1994)
	0.44 (±0.06)	Sprengnether et al. (2002)
	0.30 (±0.01)	Galloway et al. (2011)
	0.41 (±0.03)	Karl et al. (2006)
3-Methylfuran	0.033 (±0.014)	This work
	0.048 (±0.006)	Tuazon and Atkinson (1990a)
	0.04 (±0.02)	Paulson et al. (1992)
	<0.001	Sprengnether et al. (2002)

3.1.3 Isoprene-SOA yields

Figure 2 shows a typical time profile of SOA mass and number size distributions during isoprene OH-oxidation without seed particles. In all experiments, SOA formation started when a major part (>80 %) of isoprene was consumed, i.e.,

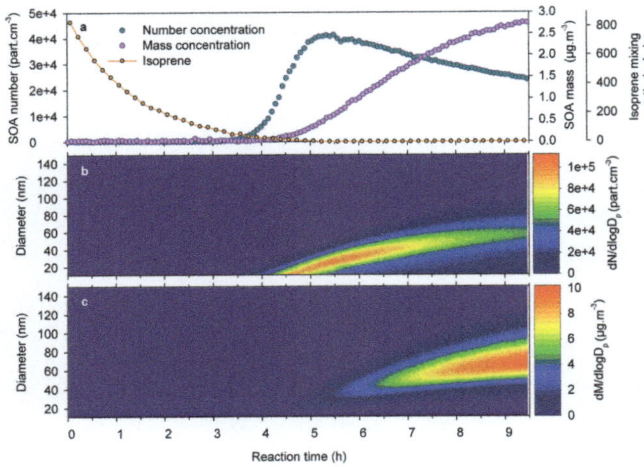

Figure 2. Time profiles of measured **(a)** SOA mass and number concentrations, **(b)** number size distribution and **(c)** mass size distribution during isoprene photooxidation (experiment I280113) performed with no seeds and with HONO as OH source. A particle density of $1.4\,g\,cm^{-3}$ was assumed (see text).

2 h (minimum) after the irradiation started. Particle number concentrations showed a sharp increase at the onset of SOA formation and then, a gradual decrease with a corresponding rise in average particle diameter due to coagulation (Fig. 2a, c). SOA growth continued even after isoprene complete consumption and the aerosol mass typically reached a maximum after approximately 7–9 h of irradiation (showing the importance of maintaining OH level during several hours). These observations are typical of an SOA formation induced from the oxidation of secondary products as observed by Ng et al. (2006). At the end of the experiment, particle mean mass diameter was around 85 nm. Table 1 shows that the maximum SOA mass observed during these experiments ranged from less than 0.1 to $12.4\,\mu g\,m^{-3}$. Even if sometimes this SOA formation was very low, it was considered significant due to observed differences with control experiments (irradiation of a synthetic air / NO_x / HONO mixture during several hours). Indeed, control experiments did not allow the detection of any significant aerosol formation with particle number remaining in the range of a few tens of particles per cubic centimeter when any experiment presented in Table 1 led to the formation of several thousands of particles per cubic centimeter (or more) even when the SOA mass remained very small.

Once SOA mass was stabilized, aerosol yields (Y) were calculated following Eq. (2):

$$Y = \frac{\Delta M_0}{\Delta\left[\text{isoprene}\right]}, \tag{2}$$

where ΔM_0 is the mass concentration of SOA formed and $\Delta\left[\text{isoprene}\right]$ is the mass concentration of isoprene reacted. All values were dilution corrected. Similarly to what has been shown by Wang et al. (2011), particles wall losses were not significant in our experiments (see Fig. S4). The comparison of the mobility diameter (obtained from the SMPS measurements) and the vacuum aerodynamic diameter (measured by the AMS) (as described by Bahreini et al., 2005) lead to the SOA effective density of $1.4\,g\,cm^{-3}$ in good agreement with previous studies carried out in the presence of NO_x (Dommen et al., 2006; Kroll et al., 2005). Aerosol volume concentrations were converted to mass concentrations using this value.

As shown in Table 1, the isoprene-SOA yields are low and range from 0.1 to 1 %. In our experiments, the initial NO_2 / NO or isoprene / NO ratios did not influence the SOA yields. Our initial NO_2 / NO ratios varied from 0.01 to 5.64, and isoprene / NO, from 3.4 to 35, it is thus possible that the lower values of these ratios were already too high in our experiments to observe any impact on SOA yields (Chan et al., 2010; Zhang et al., 2012). No direct effect of the average NO_2 / NO ratio (during isoprene decay), which ranged from 4 to 50, was detected. There was no obvious dependence of SOA yields on the presence of seed particles and on the OH radical precursor used. In order to explain our very low yields, especially those obtained after the manual cleaning (Table 1), we suspected an impact of the chamber walls' cleanliness on our SOA yields. Two hypotheses can be made: (1) when manual cleaning has been performed, high cleanliness could lead to an enhanced loss of semi-volatile reaction products that would affect the late and slow SOA growth observed for isoprene experiments; (2) in other experiments, lower cleanliness could contribute to particles formation and growth.

Concerning the first hypothesis, a higher degree of wall loss of semi-volatile species would be expected, leading to a shift in the gas and particle partitioning equilibrium, resulting in lower SOA yields. Significant loss of semi-volatile species on chamber walls was already observed in other studies carried out in Teflon film chambers (Loza et al., 2010; Matsunaga and Ziemann, 2010), and its influence on SOA yields was demonstrated by Zhang et al. (2014) in photooxidation experiments. In addition Saathoff et al. (2009) have shown that, in an aluminum chamber, considerable amounts of condensable material can be lost from the gas phase to the chamber wall in the course of an experiment and affect the SOA yield. These authors have modeled this effect and have shown that the gas mass lost to the wall could represent from 100 % (for low SOA concentration condition) to 25 % (at higher SOA concentration) of the measured airborne particle mass. Some pseudo-first order rates for loss processes of organic compounds in the CESAM chamber can also be found in Wang et al. (2011): although the data set is limited, these values are of the same order of magnitude as those obtained with other simulation chambers. Furthermore, an SOA yield study for a well-known system (α-pinene ozonolysis) is provided in this study without any significant difference with already published values.

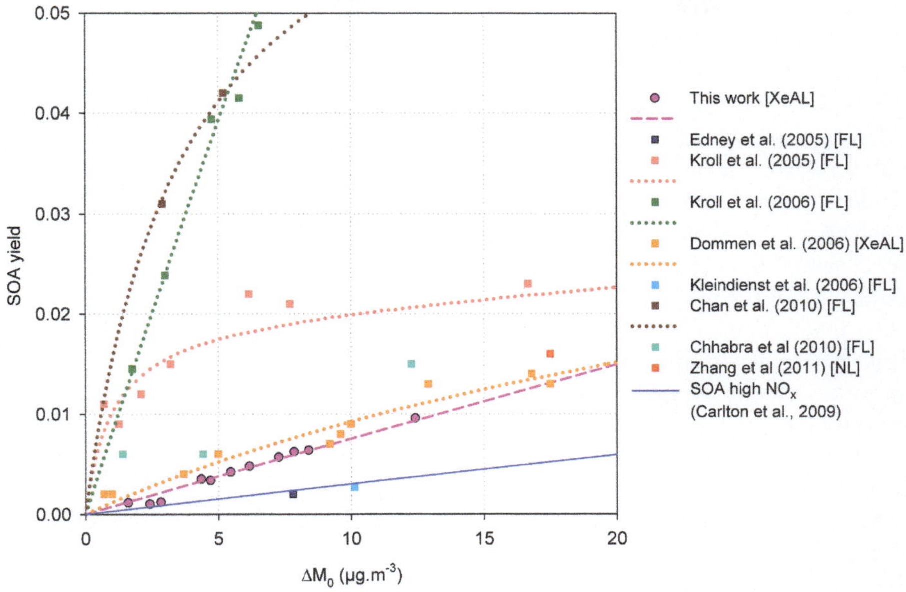

Figure 3. SOA mass yield curves from various isoprene photooxidation experiments in the presence of NO_x from the literature compared with the present study. An effective density of $1.4\,g\,cm^{-3}$ was used (see text) for conversion to mass of all our volume-based measurements. Light sources used are specified in square brackets (XeAL: xenon arc lamps; FL: fluorescent lamps; NL: natural light). The parameters of our two products yield curve (Odum et al., 1996) are as follows: $\alpha_1 = 0.508$; $K_{om,1} = 7.4 \times 10^{-4}$; $\alpha_2 = 0.509$; $K_{om,2} = 7.4 \times 10^{-4}$.

Nevertheless, if this first hypothesis would be verified, SOA yields obtained in our study would represent lower limits. On the contrary, if the second hypothesis is right, it strongly suggests that yields obtained in other studies for this very sensitive and low productive system could be overestimated since semi-volatile species adsorbed on the walls (even in small quantities) could re-partition into the reacting mixture and contribute to particle growth.

In order to rationalize our SOA yields and compare them to the literature, the aerosol yields were plotted as a function of the organic aerosol concentrations (Odum et al., 1996). Figure 3 shows a comparison between our SOA yields from isoprene photooxidation and those from previous studies (Chan et al., 2010; Chhabra et al., 2010; Dommen et al., 2006; Edney et al., 2005; Kleindienst et al., 2006; Kroll et al., 2006, 2005; Zhang et al., 2011). Also plotted on this graph are the two products yields curves for each data set determined using Eq. (3) (Odum et al., 1996).

$$Y = M_0 \left(\frac{\alpha_1 K_{om,1}}{1 + K_{om,1} M_0} + \frac{\alpha_2 K_{om,2}}{1 + K_{om,2} M_0} \right), \quad (3)$$

where α_i is a stoichiometric factor, and $K_{om,i}$ a gas-particle partitioning coefficient, defined according to semi-volatile partitioning theory (Pankow, 1994) for the species i. Despite the variability of SOA yields in this study, they were well reproduced by the two products model from Odum et al. (1996), showing that this variability was not due to a change in the chemical system but rather to a variability in its initiation or in equilibria between the walls, the gas and the particle phases. While the yields from previous studies exhibit some variation, our yields are consistent with the lowest values found in the literature. More specifically, they are very similar to those from Dommen et al. (2006) and Zhang et al. (2011). As strongly suggested by Carlton et al. (2009), the high sensitivity of the system to experimental and/or reaction conditions leads to a high degree of variability in yields measured in the different studies of isoprene photooxidation. These differences cannot be explained by the nature of the walls since studies from the literature all use Teflon chambers (Chan et al., 2010; Chhabra et al., 2010; Dommen et al., 2006; Kroll et al., 2006, 2005; Zhang et al., 2011), or stainless steel chambers with Teflon coating (Edney et al., 2005; Kleindienst et al., 2006).

Another parameter that might influence the SOA yields is the light intensity, determined as the NO_2 photolysis rate. Among the studies cited in Fig. 3 (including our study), J_{NO2} varied from 2×10^{-3} to $5.7 \times 10^{-3}\,s^{-1}$. Furthermore, it has been shown by Warren et al. (2008) that, for the *m*-xylene/NO_x photooxidation system, an increase in J_{NO2} of only $7 \times 10^{-4}\,s^{-1}$ induces an increase in SOA yields by a factor of 1.6. However, the comparison of isoprene-SOA yields obtained by Edney et al. (2005) with those by Kleindienst et al. (2006) who carried out experiments in the same simulation chamber under high NO_x conditions, but with different J_{NO2} ($5.7 \times 10^{-3}\,s^{-1}$ and $2.8 \times 10^{-3}\,s^{-1}$, respectively), shows that both yields follow the same yield curve as the one modeled by Carlton et al. (2009) for high NO_x conditions (Fig. 3). In contrast, it is possible that the type of light sources used in the different studies reported in Fig. 3 plays a role in the SOA yield variability: the only studies who used light sources with spectra representing the solar one are those by Zhang et al. (2011) (outdoor chamber), and by Dommen et al. (2006) who used xenon arc lamps like in our study. Although fluorescent lamps used as irradiation source in the other studies (Chan et al., 2010; Chhabra et al., 2010; Edney et al., 2005; Kleindienst et al., 2006; Kroll et al., 2006, 2005) deliver a light intensity equivalent to NO_2 photolysis rates which are close to natural light intensity, they exhibit emission spectra significantly different from the solar spectrum (with no emission in the longer wavelength regions, i.e above 400 nm). It is thus suggested that some oxidation products contributing to the aerosol formation and growth in studies using fluorescent lamps (under similar NO_x conditions), could be photolyzed in our experiments, leading to lower SOA yields. It can be noted that the photolysis of α-dicarbonyls, for example methylglyoxal and glyoxal, may occur outside the fluorescent lamp spectrum. Average photolysis wavelengths of methylglyoxal and glyoxal are at 417 and 383 nm, respectively (Carter et al., 1995). This hypothesis is thus opposite to the one from Warren et al. (2008) who observed higher SOA yields using an argon arc lamp (which presents a realistic irradiation spectrum) instead of black lights. However, atmospheric chemistry of aromatics is strongly different from that of alkenes, it is thus not surprising to observe a different behavior concerning relation between light source and SOA yields for isoprene / NO_x system.

3.1.4 Isoprene-SOA composition

The time profiles of elemental ratios (O / C, H / C and OM / OC) are shown in Fig. 4. The organic mass to organic carbon ratio (OM / OC) was calculated using the equation from Aiken et al. (2007). In our experiments performed without seed particles, the very small size of the formed SOA did not allow any reliable detection by the AMS before approximately 2 h of SOA formation (Fig. 4a). However, the use of ammonium sulfate seed particles for two experiments (I080411 and I110411) allowed an earlier detection (Fig. 4b). This figure shows that O / C and OM / OC ratios increased during the first hour of SOA formation while H / C decreased, thus exhibiting oxidation processing during the particle formation. After 2 h of SOA formation, all these ratios reached a plateau that remained stable until the end of the experiment. The comparison between different experiments performed under different conditions (seeds/no seeds, HONO/NO_x...) (Fig. 4a, b) reveals that the final O / C, H / C and OM / OC values were highly reproducible. Thus the observed variability of the SOA yields in our experiments was likely disconnected to the elemental ratios. This reproducibility also shows that all our experiments were performed in the same chemical system. Table 3 shows that the average elemental ratios (O / C, H / C and OM / OC) and their associated estimated uncertainties (as determined by Aiken et al., 2008) are in good agreement with previous studies who carried out isoprene-SOA formation under high-NO_x conditions (Aiken et al., 2008; Chhabra et al., 2010; Nguyen et al., 2011a). Furthermore, the stability of the evolution of our elemental ratios is in very good agreement with the observations by Chhabra et al. (2010). These authors related these observations to the findings by Surratt et al. (2006) who pointed out that an important pathway for isoprene-SOA formation under high NO_x conditions occurs via the reactivity of MACR and MPAN that were detected in the present study. Therefore, we hereafter investigate the MACR-SOA formation under identical conditions as those performed with isoprene, and we compare the two systems.

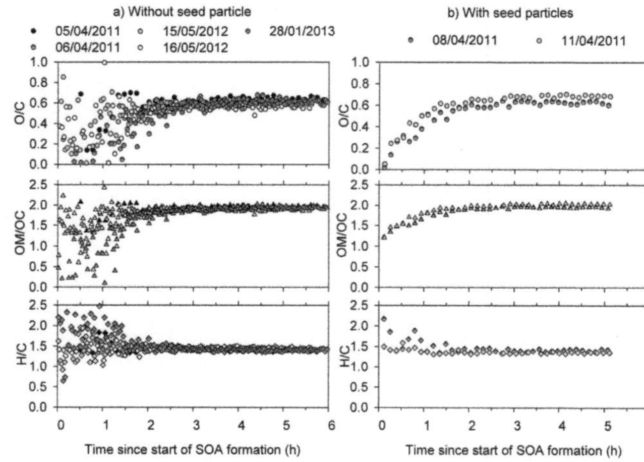

Figure 4. Time profiles of O / C, OM / OC and H / C ratios for seven different isoprene photooxidation experiments performed **(a)** without seed particle; **(b)** with ammonium sulfate seed particles.

3.2 Methacrolein experiments

Typical time profiles of gas-phase compounds in a MACR photooxidation experiment (M240512 in Table 1) without seeds and with NO_x as OH source is shown in Fig. 5. Compared to isoprene experiments, ozone production was slower and reached lower maximum concentrations (Table 1), and NO consumption was slower (compare Figs. 5b and 1b), thus denoting a slower oxidation process.

3.2.1 MACR gas-phase organic reaction products

Formaldehyde, hydroxyacetone, methylglyoxal, MPAN and CO (Fig. 5a, c, d) were observed in our study as the major primary MACR-oxidation products in the presence of NO_x. The variability in initial NO_x levels impacted primary yields, values obtained were thus different between experiments. Table 4 shows that their yields were in good agreement with previous studies, except for hydroxyacetone which showed yields 4 times lower in our study. Small hydroxyacetone yields could not be explained by wall loss in our chamber considering its low decrease in concentration after

Table 3. Average elemental ratios of SOA from isoprene and MACR photooxidation. Values in parentheses reflect the measurement uncertainty as determined by Aiken et al. (2008).

O / C	OM / OC	H / C	Reference
		Isoprene	
0.60 ($\pm$0.19)	1.92 ($\pm$0.12)	1.43 ($\pm$0.14)	This work without seeds
0.65 ($\pm$0.20)	1.99 ($\pm$0.12)	1.39 ($\pm$0.14)	This work with seeds
0.41 ($\pm$0.13)	1.75 ($\pm$0.10)	–	Aiken et al. (2008)
0.62 ($\pm$0.19)	2.00 ($\pm$0.12)	1.46 ($\pm$0.15)	Chhabra et al. (2010)
0.83	2.26	1.55	Nguyen et al. (2011a)
		Methacrolein	
0.61 ($\pm$0.19)	1.93 ($\pm$0.12)	1.43 ($\pm$0.14)	This work without seeds
0.72 ($\pm$0.22)	2.07 ($\pm$0.12)	1.32 ($\pm$0.13)	This work with seeds
0.54 ($\pm$0.17)	1.87 ($\pm$0.11)	1.53 ($\pm$0.15)	Chhabra et al. (2011)

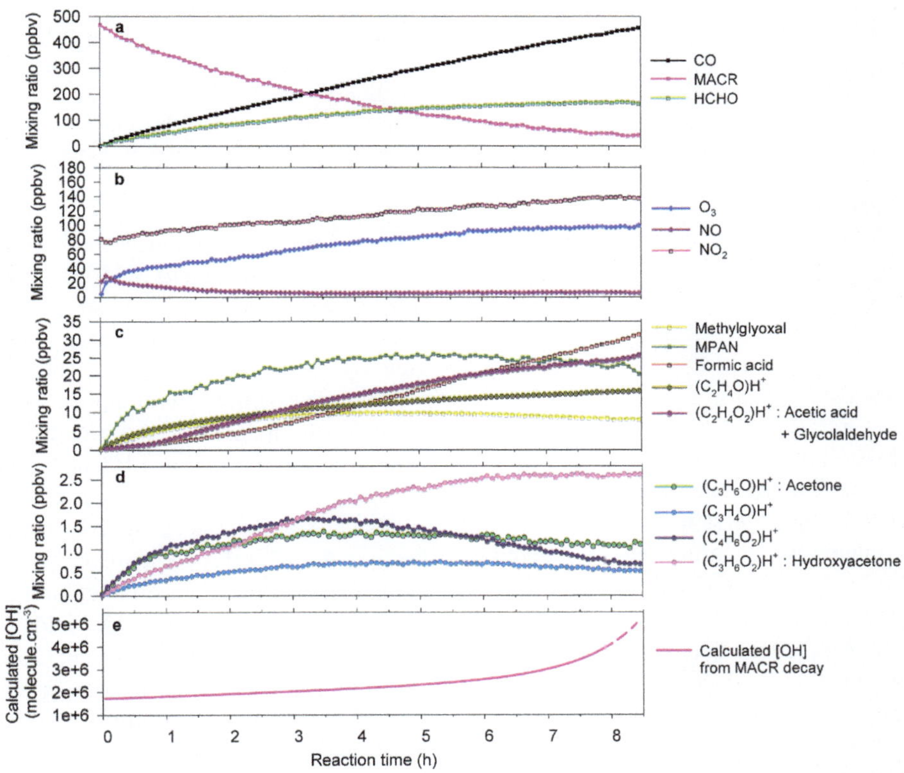

Figure 5. Time profiles of gas-phase measured compounds during MACR photooxidation (experiment M240512) performed without seed particles and with NO_X as OH source. PTR-ToF-MS measurements are represented by circles, and FTIR measurements, by squares. Calculated [OH] is represented by a dotted line after 8 h of irradiation due to low MACR mixing ratios which implies less precision in the calculation as the contribution from other VOCs is not negligible.

its production period (i.e., after 6 h of reaction in Fig. 5d). The $(C_2H_4O)H^+$ signal at m/z 45 measured by PTR-ToF-MS showed a primary production (Fig. 5c), it was attributed to acetaldehyde with a yield of 2–6 %. The origin of this primary behavior is difficult to explain since it implies an H transfer which is complicated in gas chemistry (Fig. S5). The presence of fragments of higher molecules contributing to this signal (like methylglyoxal; Müller et al., 2012) cannot be excluded. Unfortunately, the presence of acetaldehyde in the chamber could not be verified by FTIR measurements since the maximum concentrations observed throughout all experiments were below its detection limit (i.e., 20 ppb). Furthermore, a release from the walls is unlikely since acetaldehyde was not observed in control experiments (i.e., irradiation of a N_2 / O_2 mixture (80 / 20 %)).

Table 4. Yields of first-generation oxidation products during methacrolein photooxidation compared with previous studies.

Compound	Yield*	Reference
Formaldehyde	0.3–0.6	This work
	0.4–0.7	Orlando et al. (1999)
Methylglyoxal	0.02–0.06	This work
	0.08 (0.002)	Tuazon and Atkinson (1990b)
	<0.12	Orlando et al. (1999)
	0.08 (0.004)	Galloway et al. (2011)
Hydroxyacetone	0.01–0.1	This work
	0.41 (0.03)	Tuazon and Atkinson (1990b)
	0.47 (0.05)	Orlando et al. (1999)
	0.39 (0.017)	Galloway et al. (2011)
Carbon monoxide	0.45–0.85	This work
	0.51 (0.04)	Tuazon and Atkinson (1990b)
	0.6–0.8	Orlando et al. (1999)
MPAN	0.06–0.17	This work
	0.04–0.30	Orlando et al. (1999)

* The range of primary yield values from this work was partially caused by the variability in the initial NO_x levels.

3.2.2 MACR-SOA yields

SOA formation from MACR photooxidation was usually observed between 10 min and 1 h after the start of irradiation, depending on the OH level in the system. Since SOA production in these experiments began earlier than in isoprene experiments (less than 25 % of MACR was consumed before the start of the SOA formation), MACR can be considered as a more direct SOA precursor (compare Fig. 6a with 2a). In all experiments, SOA mass concentration reached a plateau between 5 and 8 h after the onset of irradiation. These mass concentrations were calculated using an effective density of 1.4 g cm^{-3} which was obtained from the comparison of the mobility diameter and the vacuum aerodynamic diameter (Bahreini et al., 2005). Table 1 shows that our SOA mass yields varied between 0.6 and 4 %, thus 4 to 6 times higher than in isoprene experiments. Except for experiment M250113, the use of HONO as OH precursor led to higher yields (at least twice higher) than in experiments using NO_x as OH source. This observation is directly linked to higher OH concentrations obtained at the beginning of the experiment when HONO was used (Fig. S6). Table 1 shows that the cleanliness of the walls did not affect the MACR-SOA mass yields, as opposed to isoprene experiments. It seems to suggest that the state of cleanliness of the walls would have a smaller impact on SOA yields when more direct SOA precursors are used: the extent of semi-volatile wall losses could be limited by the fast SOA production. The use of inorganic seed particles did not affect our SOA mass yields. Except for M240512 and M250512, initial NO_2 / NO ratios were below 1, so these ratios are probably too low to make any conclusion about the influence of this ratio on SOA mass yields (Chan et al., 2010). At the end of MACR experiments, we observed particles with higher size than in isoprene experiments (compare Fig. 6a, c with 2b, c) with mean mass diameters ranging between 100 nm and 180 nm.

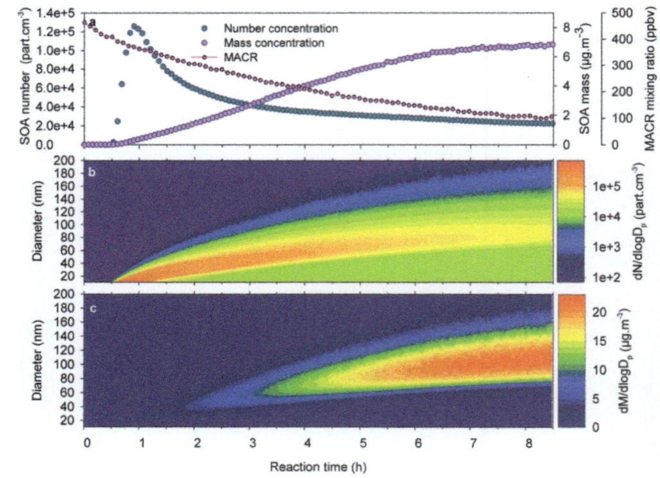

Figure 6. Time profiles of measured **(a)** SOA mass and number concentrations, **(b)** number size distribution and **(c)** mass size distribution during MACR photooxidation (experiment M240512) performed without seed particles and with NO_x as OH source. A particle density of 1.4 g cm^{-3} was assumed (see text).

Figure 7 shows a comparison of our MACR-SOA mass yields and the corresponding two products yield curves with the literature. For this comparison, fewer studies than for isoprene experiments are available, but it can be seen that, for MACR experiments too, SOA yields exhibit a fairly large variation. In their study, Chan et al. (2010) used two different OH sources (leading to a change in initial NO_2 / NO ratio) that affect SOA yields, resulting in two different yield curves. In our experiments, considering the differences observed between yields obtained with the two OH sources (which lead to similar initial NO_2 / NO ratio), two yield curves were also modeled. Like for isoprene experiments, our SOA yields are among the lowest values from the literature, i.e., comparable to those obtained by Zhang et al. (2012) in which an outdoor chamber was used (with NO_x as OH source), thus strengthening the hypothesis of an impact of light sources used on SOA yields.

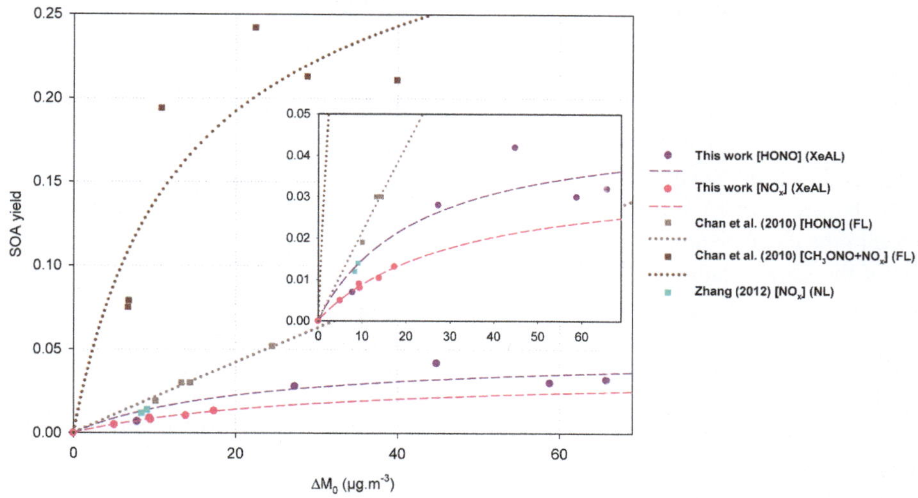

Figure 7. SOA mass yield curves from MACR photooxidation experiments in the presence of NO_X carried out by Chan et al. (2010) and Zhang et al. (2012) compared with the present study. An effective density of $1.4\,g\,cm^{-3}$ was used for conversion to mass of all our volume-based measurements. Molecules in square brackets are OH sources used. Light sources used are specified in parentheses (XeAL: xenon arc lamps; FL: fluorescent lamps; NL: natural light). The parameters determined for the two products model (Odum et al., 1996) in our study are, for experiments with NO_X as OH source: $\alpha_1 = 3.6 \times 10^{-2}$; $K_{om,1} = 3.2 \times 10^{-2}$; $\alpha_2 = 2.6 \times 10^{-11}$; $K_{om,2} = 1.63 \times 10^{-9}$. For experiments with HONO as OH source, these parameters are: $\alpha_1 = 4.83 \times 10^{-2}$; $K_{om,1} = 4.35 \times 10^{-2}$; $\alpha_2 = 6.2 \times 10^{-2}$; $K_{om,2} = 8.47 \times 10^{-10}$.

3.2.3 Composition of MACR-SOA

SOA mass spectrum obtained by AMS measurements (Fig. S7b) showed no significant variation over the experiment. Its comparison with the mass spectrum of SOA from isoprene photooxidation (Fig. S7a) exhibits a high degree of similarity which strongly suggests that methacrolein is a principal intermediate in SOA formation from isoprene photooxidation in our experimental conditions, as observed by previous studies performed in the presence of NO_X (Kroll et al., 2006; Surratt et al., 2006). Temporal variations of elemental ratios are presented in Figure 8. Experiments carried out with seeds showed that O / C and OM / OC ratios increased quickly during the first hour of SOA formation and then stabilized. Obtained ratios after stabilization were reproducible and no clear influence of OH source or of the presence of seeds was observed. Table 3 shows that these average values are in good agreement with elemental ratios from Chhabra et al. (2011), considering measurement uncertainties given by Aiken et al. (2008). Elemental ratios for MACR-SOA were very close to those measured for isoprene-SOA, confirming the role of MACR in SOA formation from isoprene photooxidation.

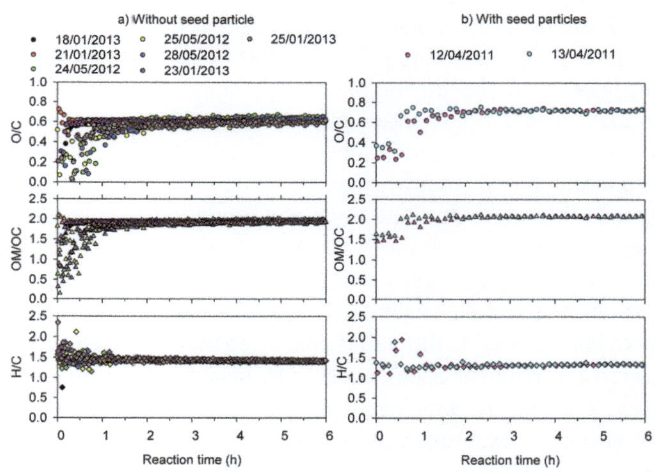

Figure 8. Time profiles of O / C, OM / OC and H / C ratios for nine different MACR photooxidation experiments performed **(a)** without seed particle; **(b)** with ammonium sulfate seed particles.

4 Atmospheric implications and conclusion

In the present study, gas-phase products and SOA formation from isoprene and MACR photooxidation were investigated in a stainless steel simulation chamber equipped with realistic artificial light sources (Fig. S1). A good agreement with the literature was observed for the gas-phase products and particularly for primary oxidation product yields.

Comparing the SOA mass formed during isoprene experiments performed before and after manual cleaning of the chamber walls suggested an impact of the state of cleanliness of the walls on the nucleation step. While this hypothesis has not been verified with other hydrocarbons or commercially available oxidized species, it is suggested that more oxidized species could be adsorbed on clean walls, preventing the initial nucleation step. It must also be noted that such an effect has not been observed with other chemical systems (such as α-pinene ozonolysis) in the same chamber (Wang et al., 2011). This may be due to the fact that the molecules produced during isoprene oxidation are much smaller and hence need to be much more oxidized to undergo partitioning; they are therefore more sensitive to wall effects. This delay in particle formation could lead to a deficit in the SOA mass because of the potentially higher proportion of isoprene oxidation products adsorbed to the clean chamber walls. Nevertheless, the partitioning of the semi-volatile compounds was reproducible within several experiments spanning over more than 3 years as shown by the high level of internal consistency of the measured data on the Odum plots (Fig. 3). Furthermore, it is striking to see that the SOA yield may vary significantly when measured in similar chambers with similar protocols (for example Kroll et al., 2005 and Dommen et al., 2006) and be in good agreement in two different chambers (Dommen et al., 2006, and this work). This somewhat contradicts the existence of a possible wall effect proposed above.

Our results for the particle phase show that SOA yields from isoprene and MACR photooxidation are in good agreement with the lowest values reported in the literature. They correspond to experiments carried out with natural irradiation or with artificial irradiation with an emission profile more similar to the solar spectrum (e.g., xenon arc lamps) than the fluorescent lamps used in other studies. It was thus suggested that the use of fluorescent lamps as irradiation sources would not activate photolysis reactions requiring longer wavelengths, such as those which break down oxidation products (e.g., methylglyoxal). These oxidation products would otherwise contribute to aerosol formation and growth, thus leading to a decrease in SOA mass yields when xenon arc lamp or natural light were used. The differences in light sources between environmental chambers may thus be able to explain partially the variability in SOA mass yield observed for identical compounds. SOA mass yields from isoprene and MACR in the atmosphere could therefore be overestimated by experiments carried out in simulation chambers equipped with fluorescent lamps. Further experiments and analysis are needed to determine the influence of the irradiation spectrum on SOA formation from isoprene and MACR photooxidation like experiments carried out by Warren et al. (2008) on the m-xylene/NO_x photooxidation system. Therefore, according to our results, isoprene and MACR SOA mass yields in the atmosphere could be lower than most current chamber studies suggest. Finally, this study highlights the need for additional work on the photochemical fate of SOA components.

Acknowledgements. The authors are grateful to Arnaud Allanic, Pascal Renard and Pascal Zapf for their helpful contributions with handling the simulation chamber experiments. The authors gratefully acknowledge the institutions that have provided financial support: French National Institute for Geophysical Research (CNRS-INSU) within the LEFE-CHAT program through the project "Impact de la chimie des nuages sur la formation d'aérosols organiques secondaires dans l'atmosphère" and the French National Agency for Research (ANR) project CUMULUS ANR-2010-BLAN-617-01. This work was also supported by the EC within the I3 project "Integrating of European Simulation Chambers for Investigating Atmospheric Processes" (EUROCHAMP-2, contract no. 228335). The authors gratefully acknowledge the MASSALYA instrumental platform (Aix Marseille Université, http://lce.univ-amu.fr) for the analysis and measurements used in this publication. We thank Clare Fitzgerald and Brendan Mahon (University of Cambridge, UK) for helpful revision of the manuscript.

Edited by: H. Saathoff

References

Aiken, A. C., DeCarlo, P. F., and Jimenez, J. L.: Elemental analysis of organic species with electron ionization high-resolution mass spectrometry, Anal. Chem., 79, 8350–8358, 2007.

Aiken, A. C., Decarlo, P. F., Kroll, J. H., Worsnop, D. R., Huffman, J. A., Docherty, K. S., Ulbrich, I. M., Mohr, C., Kimmel, J. R., Sueper, D., Sun, Y., Zhang, Q., Trimborn, A., Northway, M., Ziemann, P. J., Canagaratna, M. R., Onasch, T. B., Alfarra, M. R., Prevot, A. S. H., Dommen, J., Duplissy, J., Metzger, A., Baltensperger, U., and Jimenez, J. L.: O / C and OM / OC ratios of primary, secondary, and ambient organic aerosols with high-resolution time-of-flight aerosol mass spectrometry, Environ. Sci. Technol., 42, 4478–4485, 2008.

Atkinson, R., Baulch, D. L., Cox, R. A., Crowley, J. N., Hampson, R. F., Hynes, R. G., Jenkin, M. E., Rossi, M. J., Troe, J., and IUPAC Subcommittee: Evaluated kinetic and photochemical data for atmospheric chemistry: Volume II – gas phase reactions of organic species, Atmos. Chem. Phys., 6, 3625–4055, doi:10.5194/acp-6-3625-2006, 2006.

Bahreini, R., Keywood, M. D., Ng, N. L., Varutbangkul, V., Gao, S., Flagan, R. C., Seinfeld, J. H., Worsnop, D. R., and Jimenez, J. L.: Measurements of secondary organic aerosol from oxidation of cycloalkenes, terpenes, and m-xylene using an Aerodyne aerosol mass spectrometer, Environ. Sci. Technol., 39, 5674–5688, 2005.

Biesenthal, T. A., Wu, Q., Shepson, P. B., Wiebe, H. A., Anlauf, K. G., and Mackay, G. I.: A study of relationships between isoprene, its oxidation products, and ozone, in the Lower Fraser Valley, BC, Atmos. Environ., 31, 2049–2058, 1997.

Boge, O., Miao, Y., Plewka, A., and Herrmann, H.: Formation of secondary organic particle phase compounds from isoprene gas-phase oxidation products: An aerosol chamber and field study, Atmos. Environ., 40, 2501–2509, 2006.

Canagaratna, M. R., Jayne, J. T., Jimenez, J. L., Allan, J. D., Alfarra, M. R., Zhang, Q., Onasch, T. B., Drewnick, F., Coe, H., Middlebrook, A., Delia, A., Williams, L. R., Trimborn, A. M., Northway, M. J., DeCarlo, P. F., Kolb, C. E., Davidovits, P., and

Worsnop, D. R.: Chemical and microphysical characterization of ambient aerosols with the aerodyne aerosol mass spectrometer, Mass Spectrom. Rev., 26, 185–222, 2007.

Carlton, A. G., Wiedinmyer, C., and Kroll, J. H.: A review of Secondary Organic Aerosol (SOA) formation from isoprene, Atmos. Chem. Phys., 9, 4987–5005, doi:10.5194/acp-9-4987-2009, 2009.

Carter, W. P. L., Luo, D., Malkina, I. L., and Pierce, J. A.: Chamber Studies of Atmospheric Reactivities of Volatile Organic Compounds. Effects of Varying Chamber and Light Source, Final report to National Renewable Energy Laboratory, Coordinating Research Council, Inc., California Air Resources Board, South Coast Air Quality Management District, available at: http://www.cert.ucr.edu/~carter/pubs/explrept.pdf (last access: 20 July 2014), 1995.

Chan, A. W. H., Chan, M. N., Surratt, J. D., Chhabra, P. S., Loza, C. L., Crounse, J. D., Yee, L. D., Flagan, R. C., Wennberg, P. O., and Seinfeld, J. H.: Role of aldehyde chemistry and NO_X concentrations in secondary organic aerosol formation, Atmos. Chem. Phys., 10, 7169–7188, doi:10.5194/acp-10-7169-2010, 2010.

Chhabra, P. S., Flagan, R. C., and Seinfeld, J. H.: Elemental analysis of chamber organic aerosol using an aerodyne high-resolution aerosol mass spectrometer, Atmos. Chem. Phys., 10, 4111–4131, doi:10.5194/acp-10-4111-2010, 2010.

Chhabra, P. S., Ng, N. L., Canagaratna, M. R., Corrigan, A. L., Russell, L. M., Worsnop, D. R., Flagan, R. C., and Seinfeld, J. H.: Elemental composition and oxidation of chamber organic aerosol, Atmos. Chem. Phys., 11, 8827–8845, doi:10.5194/acp-11-8827-2011, 2011.

Claeys, M., Graham, B., Vas, G., Wang, W., Vermeylen, R., Pashynska, V., Cafmeyer, J., Guyon, P., Andreae, M. O., Artaxo, P., and Maenhaut, W.: Formation of secondary organic aerosols through photooxidation of isoprene, Science, 303, 1173–1176, 2004.

De Carlo, P. F., Kimmel, J. R., Trimborn, A., Northway, M. J., Jayne, J. T., Aiken, A. C., Gonin, M., Fuhrer, K., Horvath, T., Docherty, K. S., Worsnop, D. R., and Jimenez, J. L.: Field-Deployable, High-Resolution, Time-of-Flight Aerosol Mass Spectrometer, Anal. Chem., 78, 8281–8289, 2006.

Dommen, J., Metzger, A., Duplissy, J., Kalberer, M., Alfarra, M. R., Gascho, A., Weingartner, E., Prevot, A. S. H., Verheggen, B., and Baltensperger, U.: Laboratory observation of oligomers in the aerosol from isoprene/NO_X photooxidation, Geophys. Res. Lett., 33, L13805, doi:10.1029/2006GL026523, 2006.

Dunlea, E. J., Herndon, S. C., Nelson, D. D., Volkamer, R. M., San Martini, F., Sheehy, P. M., Zahniser, M. S., Shorter, J. H., Wormhoudt, J. C., Lamb, B. K., Allwine, E. J., Gaffney, J. S., Marley, N. A., Grutter, M., Marquez, C., Blanco, S., Cardenas, B., Retama, A., Ramos Villegas, C. R., Kolb, C. E., Molina, L. T., and Molina, M. J.: Evaluation of nitrogen dioxide chemiluminescence monitors in a polluted urban environment, Atmos. Chem. Phys., 7, 2691–2704, doi:10.5194/acp-7-2691-2007, 2007.

Edney, E. O., Kleindienst, T. E., Jaoui, M., Lewandowski, M., Offenberg, J. H., Wang, W., and Claeys, M.: Formation of 2-methyl tetrols and 2-methylglyceric acid in secondary organic aerosol from laboratory irradiated isoprene/NO_X/SO_2/air mixtures and their detection in ambient PM2.5 samples collected in the eastern United States, Atmos. Environ., 39, 5281–5289, 2005.

Galloway, M. M., Huisman, A. J., Yee, L. D., Chan, A. W. H., Loza, C. L., Seinfeld, J. H., and Keutsch, F. N.: Yields of oxidized volatile organic compounds during the OH radical initiated oxidation of isoprene, methyl vinyl ketone, and methacrolein under high-NO_X conditions, Atmos. Chem. Phys., 11, 10779–10790, doi:10.5194/acp-11-10779-2011, 2011.

Guenther, A., Karl, T., Harley, P., Wiedinmyer, C., Palmer, P. I., and Geron, C.: Estimates of global terrestrial isoprene emissions using MEGAN (Model of Emissions of Gases and Aerosols from Nature), Atmos. Chem. Phys., 6, 3181–3210, doi:10.5194/acp-6-3181-2006, 2006.

Hansel, A. and Wisthaler, A.: A method for real-time detection of PAN, PPN and MPAN in ambient air, Geophys. Res. Lett., 27, 895–898, 2000.

Healy, R. M., Wenger, J. C., Metzger, A., Duplissy, J., Kalberer, M., and Dommen, J.: Gas/particle partitioning of carbonyls in the photooxidation of isoprene and 1,3,5-trimethylbenzene, Atmos. Chem. Phys., 8, 3215–3230, doi:10.5194/acp-8-3215-2008, 2008.

Ion, A. C., Vermeylen, R., Kourtchev, I., Cafmeyer, J., Chi, X., Gelencsér, A., Maenhaut, W., and Claeys, M.: Polar organic compounds in rural $PM_{2.5}$ aerosols from K-puszta, Hungary, during a 2003 summer field campaign: Sources and diel variations, Atmos. Chem. Phys., 5, 1805–1814, doi:10.5194/acp-5-1805-2005, 2005.

Kanakidou, M., Seinfeld, J. H., Pandis, S. N., Barnes, I., Dentener, F. J., Facchini, M. C., Van Dingenen, R., Ervens, B., Nenes, A., Nielsen, C. J., Swietlicki, E., Putaud, J. P., Balkanski, Y., Fuzzi, S., Horth, J., Moortgat, G. K., Winterhalter, R., Myhre, C. E. L., Tsigaridis, K., Vignati, E., Stephanou, E. G., and Wilson, J.: Organic aerosol and global climate modelling: a review, Atmos. Chem. Phys., 5, 1053–1123, doi:10.5194/acp-5-1053-2005, 2005.

Karl, M., Dorn, H. P., Holland, F., Koppmann, R., Poppe, D., Rupp, L., Schaub, A., and Wahner, A.: Product study of the reaction of OH radicals with isoprene in the atmosphere simulation chamber SAPHIR, J. Atmos. Chem., 55, 167–187, 2006.

Kleindienst, T. E., Edney, E. O., Lewandowski, M., Offenberg, J. H., and Jaoui, M.: Secondary organic carbon and aerosol yields from the irradiations of isoprene and alpha-pinene in the presence of NO_X and SO_2, Environ. Sci. Technol., 40, 3807–3812, 2006.

Kourtchev, I., Ruuskanen, T., Maenhaut, W., Kulmala, M., and Claeys, M.: Observation of 2-methyltetrols and related photo-oxidation products of isoprene in boreal forest aerosols from Hyytiälä, Finland, Atmos. Chem. Phys., 5, 2761–2770, doi:10.5194/acp-5-2761-2005, 2005.

Kroll, J. H., Ng, N. L., Murphy, S. M., Flagan, R. C., and Seinfeld, J. H.: Secondary organic aerosol formation from isoprene photooxidation under high-NOx conditions, Geophys. Res. Lett., 32, L18808, doi:10.1029/2005GL023637, 2005.

Kroll, J. H., Ng, N. L., Murphy, S. M., Flagan, R. C., and Seinfeld, J. H.: Secondary organic aerosol formation from isoprene photooxidation, Environ. Sci. Technol., 40, 1869–1877, 2006.

Loza, C. L., Chan, A. W. H., Galloway, M. M., Keutsch, F. N., Flagan, R. C., and Seinfeld, J. H.: Characterization of Vapor Wall Loss in Laboratory Chambers, Environ. Sci.Technol., 44, 5074–5078, 2010.

Matsunaga, A. and Ziemann, P. J.: Gas-Wall Partitioning of Organic Compounds in a Teflon Film Chamber and Potential Effects on Reaction Product and Aerosol Yield Measurements, Aerosol Sci. Tech., 44, 881–892, 2010.

McMurry, P. H. and Rader, D. J.: Aerosol Wall Losses in Electrically Charged Chambers, Aerosol Sci. Tech., 4, 249–268, 1985.

Miyoshi, A., Hatakeyama, S., and Washida, N.: OH radical-initiated photooxidation of isoprene: an estimate of global CO production, J. Geophys. Res.-Atmos., 99, 18779–18787, 1994.

Müller, M., Graus, M., Wisthaler, A., Hansel, A., Metzger, A., Dommen, J., and Baltensperger, U.: Analysis of high mass resolution PTR-TOF mass spectra from 1,3,5-trimethylbenzene (TMB) environmental chamber experiments, Atmos. Chem. Phys., 12, 829–843, doi:10.5194/acp-12-829-2012, 2012.

Ng, N. L., Kroll, J. H., Keywood, M. D., Bahreini, R., Varutbangkul, V., Flagan, R. C., Seinfeld, J. H., Lee, A., and Goldstein, A. H.: Contribution of first- versus second-generation products to secondary organic aerosols formed in the oxidation of biogenic hydrocarbons, Environ. Sci. Technol., 40, 2283–2297, 2006.

Nguyen, T. B., Laskin, J., Laskin, A., and Nizkorodov, S. A.: Nitrogen-Containing Organic Compounds and Oligomers in Secondary Organic Aerosol Formed by Photooxidation of Isoprene, Environ. Sci. Technol., 45, 6908–6918, 2011a.

Nguyen, T. B., Roach, P. J., Laskin, J., Laskin, A., and Nizkorodov, S. A.: Effect of humidity on the composition of isoprene photooxidation secondary organic aerosol, Atmos. Chem. Phys., 11, 6931–6944, doi:10.5194/acp-11-6931-2011, 2011b.

Odum, J. R., Hoffmann, T., Bowman, F., Collins, D., Flagan, R. C., and Seinfeld, J. H.: Gas/Particle Partitioning and Secondary Organic Aerosol Yields, Environ. Sci. Technol., 30, 2580–2585, 1996.

Orlando, J. J., Tyndall, G. S., and Paulson, S. E.: Mechanism of the OH-initiated oxidation of methacrolein, Geophys. Res. Lett., 26, 2191–2194, 1999.

Pandis, S. N., Paulson, S. E., Seinfeld, J. H., and Flagan, R. C.: Aerosol formation in the photooxidation of isoprene and β-pinene, Atmos. Environ., Part A, General Topics, 25, 997–1008, 1991.

Pankow, J. F.: An absorption model of gas/particle partitioning of organic compounds in the atmosphere, Atmos. Environ., 28, 185–188, 1994.

Paulot, F., Crounse, J. D., Kjaergaard, H. G., Kroll, J. H., Seinfeld, J. H., and Wennberg, P. O.: Isoprene photooxidation: new insights into the production of acids and organic nitrates, Atmos. Chem. Phys., 9, 1479–1501, doi:10.5194/acp-9-1479-2009, 2009.

Paulson, S. E. and Seinfeld, J. H.: Development and evaluation of a photooxidation mechanism for isoprene, J. Geophys. Res.-Atmos., 97, 20703–20715, 1992.

Paulson, S. E., Flagan, R. C., and Seinfeld, J. H.: Atmospheric photooxidation of isoprene part I: The hydroxyl radical and ground state atomic oxygen reactions, Int. J. Chem. Kinet., 24, 79–101, 1992.

Saathoff, H., Naumann, K.-H., Möhler, O., Jonsson, Å. M., Hallquist, M., Kiendler-Scharr, A., Mentel, Th. F., Tillmann, R., and Schurath, U.: Temperature dependence of yields of secondary organic aerosols from the ozonolysis of α-pinene and limonene, Atmos. Chem. Phys., 9, 1551–1577, doi:10.5194/acp-9-1551-2009, 2009.

Sprengnether, M., Demerjian, K. L., Donahue, N. M., and Anderson, J. G.: Product analysis of the OH oxidation of isoprene and 1,3-butadiene in the presence of NO, J. Geophys. Res.-Atmos., 107, ACH 8-1–ACH 8-13, doi:10.1029/2001JD000716, 2002.

Starn, T. K., Shepson, P. B., Bertman, S. B., White, J. S., Splawn, B. G., Riemer, D. D., Zika, R. G., and Olszyna, K.: Observations of isoprene chemistry and its role in ozone production at a semirural site during the 1995 Southern Oxidants Study, J. Geophys. Res.-Atmos., 103, 22425–22435, 1998.

Surratt, J. D., Murphy, S. M., Kroll, J. H., Ng, N. L., Hildebrandt, L., Sorooshian, A., Szmigielski, R., Vermeylen, R., Maenhaut, W., Claeys, M., Flagan, R. C., and Seinfeld, J. H.: Chemical composition of secondary organic aerosol formed from the photooxidation of isoprene, J. Phys. Chem. A, 110, 9665–9690, 2006.

Surratt, J. D., Chan, A. W. H., Eddingsaas, N. C., Chan, M. N., Loza, C. L., Kwan, A. J., Hersey, S. P., Flagan, R. C., Wennberg, P. O., and Seinfeld, J. H.: Reactive intermediates revealed in secondary organic aerosol formation from isoprene, Proc. Natl. Acad. Sci. USA, 107, 6640–6645, 2010.

Tuazon, E. C. and Atkinson, R.: A product study of the gas-phase reaction of methyl vinyl ketone with the OH radical in the presence of NOx, Int. J. Chem. Kinet., 21, 1141–1152, 1989.

Tuazon, E. C. and Atkinson, R.: A product study of the gas-phase reaction of Isoprene with the OH radical in the presence of NOx, Int. J. Chem. Kinet., 22, 1221–1236, 1990a.

Tuazon, E. C. and Atkinson, R.: A product study of the gas-phase reaction of Methacrolein with the OH radical in the presence of NOx, Int. J. Chem. Kinet., 22, 591–602, 1990b.

Wang, J., Doussin, J. F., Perrier, S., Perraudin, E., Katrib, Y., Pangui, E., and Picquet-Varrault, B.: Design of a new multi-phase experimental simulation chamber for atmospheric photosmog, aerosol and cloud chemistry research, Atmos. Meas. Tech., 4, 2465–2494, doi:10.5194/amt-4-2465-2011, 2011.

Warren, B., Song, C., and Cocker III., D. R.,: Light intensity and light source influence on secondary organic aerosol formation for the m-xylene/NOx photooxidation system, Environ. Sci. Technol., 42, 5461–5466, 2008.

Wiedinmyer, C., Friedfeld, S., Baugh, W., Greenberg, J., Guenther, A., Fraser, M., and Allen, D.: Measurement and analysis of atmospheric concentrations of isoprene and its reaction products in central Texas, Atmos. Environ., 35, 1001–1013, 2001.

Zhang, H., Surratt, J. D., Lin, Y. H., Bapat, J., and Kamens, R. M.: Effect of relative humidity on SOA formation from isoprene/NO photooxidation: enhancement of 2-methylglyceric acid and its corresponding oligoesters under dry conditions, Atmos. Chem. Phys., 11, 6411–6424, doi:10.5194/acp-11-6411-2011, 2011.

Zhang, H. F., Lin, Y. H., Zhang, Z. F., Zhang, X. L., Shaw, S. L., Knipping, E. M., Weber, R. J., Gold, A., Kamens, R. M., and Surratt, J. D.: Secondary organic aerosol formation from methacrolein photooxidation: roles of NOx level, relative humidity and aerosol acidity, Environ. Chem., 9, 247–262, 2012.

Zhang, Q., Jimenez, J. L., Canagaratna, M. R., Allan, J. D., Coe, H., Ulbrich, I., Alfarra, M. R., Takami, A., Middlebrook, A. M., Sun, Y. L., Dzepina, K., Dunlea, E., Docherty, K., DeCarlo, P. F., Salcedo, D., Onasch, T., Jayne, J. T., Miyoshi, T., Shimono, A., Hatakeyama, S., Takegawa, N., Kondo, Y., Schneider, J., Drewnick, F., Borrmann, S., Weimer, S., Demerjian, K., Williams, P., Bower, K., Bahreini, R., Cottrell, L., Griffin, R. J., Rautiainen, J., Sun, J. Y., Zhang, Y. M., and Worsnop, D. R.: Ubiquity and dominance of oxygenated species in organic aerosols in anthropogenically-influenced Northern Hemisphere midlatitudes, Geophys. Res. Lett., 34, L13801, doi:10.1029/2007GL029979, 2007.

Zhang, X., Cappa, C. D., Jathar, S. H., McVay, R. C., Ensberg, J. J., Kleeman, M. J., and Seinfeld, J. H.: Influence of vapor wall loss in laboratory chambers on yields of secondary organic aerosol, Proc. Natl. Acad. Sci. USA, 111, 5802–5807, 2014.

Zhou, X. L., Qiao, H. C., Deng, G. H., and Civerolo, K.: A method for the measurement of atmospheric HONO based on DNPH derivatization and HPLC analysis, Environ. Sci. Technol., 33, 3672–3679, 1999.

8

Long-term real-time chemical characterization of submicron aerosols at Montsec (southern Pyrenees, 1570 m a.s.l.)

A. Ripoll[1,2], M. C. Minguillón[1], J. Pey[3], J. L. Jimenez[4], D. A. Day[4], Y. Sosedova[5], F. Canonaco[5], A. S. H. Prévôt[5], X. Querol[1], and A. Alastuey[1]

[1]Institute of Environmental Assessment and Water Research (IDAEA-CSIC), Jordi Girona 18–26, 08034, Barcelona, Spain
[2]Department of Astronomy and Meteorology, Faculty of Physics, University of Barcelona, Martí i Franquès 1, 08028, Barcelona, Spain
[3]Aix-Marseille Université, CNRS, LCE FRE 3416, Marseille, 13331, France
[4]Department of Chemistry and Biochemistry, and Cooperative Institute for Research in the Environmental Sciences (CIRES), University of Colorado at Boulder, CO 80309, USA
[5]Paul Scherrer Institute, Laboratory of Atmospheric Chemistry, 5232 Villigen PSI, Switzerland

Correspondence to: A. Ripoll (anna.ripoll@idaea.csic.es)

Abstract. Real-time measurements of inorganic (sulfate, nitrate, ammonium, chloride and black carbon (BC)) and organic submicron aerosols (particles with an aerodynamic diameter of less than 1 μm) from a continental background site (Montsec, MSC, 1570 m a.s.l.) in the western Mediterranean Basin (WMB) were conducted for 10 months (July 2011–April 2012). An aerosol chemical speciation monitor (ACSM) was co-located with other online and offline PM_1 measurements. Analyses of the hourly, diurnal, and seasonal variations are presented here, for the first time, for this region.

Seasonal trends in PM_1 components are attributed to variations in evolution of the planetary boundary layer (PBL) height, air mass origin, and meteorological conditions. In summer, the higher temperature and solar radiation increases convection, enhancing the growth of the PBL and the transport of anthropogenic pollutants towards high altitude sites. Furthermore, the regional recirculation of air masses over the WMB creates a continuous increase in the background concentrations of PM_1 components and causes the formation of reservoir layers at relatively high altitudes. The combination of all these atmospheric processes results in a high variability of PM_1 components, with poorly defined daily patterns, except for the organic aerosols (OA). OA was mostly composed (up to 90 %) of oxygenated organic aerosol (OOA), split in two types: semivolatile (SV-OOA) and low-volatility (LV-OOA), the rest being hydrocarbon-like OA (HOA). The marked diurnal cycles of OA components regardless of the air mass origin indicates that they are not only associated with anthropogenic and long-range-transported secondary OA (SOA) but also with recently produced biogenic SOA.

Very different conditions drive the aerosol phenomenology in winter at MSC. The thermal inversions and the lower vertical development of the PBL leave MSC in the free troposphere most of the day, being affected by PBL air masses only after midday, when the mountain breezes transport emissions from the adjacent valleys and plains to the top of the mountain. This results in clear diurnal patterns of both organic and inorganic concentrations. OA was also mainly composed (71 %) of OOA, with contributions from HOA (5 %) and biomass burning OA (BBOA; 24 %). Moreover, in winter sporadic long-range transport from mainland Europe is observed.

The results obtained in the present study highlight the importance of SOA formation processes at a remote site such as MSC, especially in summer. Additional research is needed to characterize the sources and processes of SOA formation at remote sites.

1 Introduction

Earth's climate system is modulated by atmospheric aerosols. Submicron particles (< 1 μm in aerodynamic diameter) play a dominant role in both cloud formation and scattering or absorbing solar radiation (IPCC, 2013). The complexity of aerosol sources and processes results in an uncertainty in the radiative forcing of climate. Aerosol optical properties are connected to direct and indirect climate forcing effects, and they are dependent on particle composition. Moreover, aerosol composition may provide valuable information on aerosol sources and processes. Consequently, long-term measurements of PM_1 chemical composition are needed to better understand aerosol sources, to quantify their lifetime in the atmosphere and to constrain the uncertainties of their climatic influence.

Long-term PM_1 chemical composition measurements are relatively scarce both offline and online. In the last decade, online PM_1 chemical composition measurements have been performed using aerosol mass spectrometers (AMS) at a number of locations. Measurements of online chemical composition are useful to study hourly variations and daily patterns. Most of these studies, however, correspond to short-term measurement campaigns (typically a month) (e.g., Crippa et al., 2014; Jimenez et al., 2009; Lanz et al., 2010; Ng et al., 2010; Zhang et al., 2007) given the intensive instrument maintenance required and the need of highly qualified personnel for a good quality data set.

In contrast to the use of the AMS in relatively short campaigns, the more recently developed Aerodyne aerosol chemical speciation monitor (ACSM) is becoming a widely used online instrument for long-term measurements of PM_1 chemical composition (Budisulistiorini et al., 2013; Canonaco et al., 2013; Carbone et al., 2013; Petit et al., 2014; Tiitta et al., 2014). The ACSM is built upon the same technology as the AMS, in which an aerodynamic particle focusing lens is combined with high vacuum thermal particle vaporization, electron impact ionization, and mass spectrometry. Modifications in the ACSM design (e.g., lack of particle sizing chamber and components, and use of a simple and compact RGA mass spectrometer detector), however, allow it to be smaller, less costly, and simpler to operate than the AMS (Ng et al., 2011c). The ACTRIS (Aerosols, Clouds, and Trace gases Research InfraStructure) European network is evaluating the use of the ACSM as a reliable instrument, which will provide the opportunity to study long-term data sets of PM_1 chemical composition across the continent.

Recent publications have investigated most of the existing worldwide AMS databases (e.g., Crippa et al., 2014; Jimenez et al., 2009; Lanz et al., 2010; Ng et al., 2010; Zhang et al., 2007) and reflected a prevalence of organic aerosols (20–90 %) in the submicron fraction, largely independent of the region and type of environment. However, our knowledge on organic aerosol (OA) formation, sources, and atmospheric processing is still very incomplete, especially for secondary organic aerosol (SOA) formed from chemical reactions of gas-phase species (e.g., Donahue et al., 2014; Hallquist et al., 2009; Kroll and Seinfeld, 2008; Robinson et al., 2007; Volkamer et al., 2006). Recent progress has been made in identifying primary organic aerosol (POA) sources (e.g., Elbert et al., 2007; Zhang et al., 2005), but significant gaps still remain in our understanding on the atmospheric evolution of POA after emission (de Gouw and Jimenez, 2009). For these reasons, OA measurements and analysis are required to better understand its chemical evolution in the atmosphere.

The lack of long-term online PM_1 chemical composition measurements is especially evident in the western Mediterranean Basin (WMB), which is characterized by particular atmospheric dynamics strongly influenced by its topography (Jorba et al., 2013; Millan et al., 1997). Over this region, arrival of natural and anthropogenic aerosols as a result of long-range transport from Africa and Europe is frequent (e.g., Pey et al., 2013; Querol et al., 2009; Ripoll et al., 2014; Rodríguez et al., 2001) and accumulation and recirculation processes are frequently observed (Rodriguez et al., 2002). The sources and meteorological controls of PM in the regional background of the WMB have been recently investigated during the DAURE (Determination of the sources of atmospheric Aerosols in Urban and Rural Environments in the western Mediterranean) study (Pandolfi et al., 2014a) using an AMS and ^{14}C analyses (Crippa et al., 2014; Minguillón et al., 2011). Furthermore, Ripoll et al. (2015) studied the PM_1 and PM_{10} chemical composition with daily time resolution in the continental and regional background environments in the WMB. In that study, a higher mineral contribution was identified in the continental background due to the preferential transport of African dust at high altitude layers and to the increased regional dust resuspension enhanced by the drier surface and higher convection. Nevertheless, aerosol chemical characterization with higher time resolution is needed to study the origin of specific PM components and the local and/or regional processes, in particular to exploit the information contained in diurnal cycles that is typically not accessible with offline measurements.

In this study we deployed an ACSM at a high altitude site (Montsec, 1570 m a.s.l.) in the NE of the Iberian Peninsula (42°03′ N, 0°44′ E), representative of the continental background conditions of the WMB (Ripoll et al., 2014). This environment is under free tropospheric (FT) influence most of the time, although it is exposed to regional pollutants during the summer time and/or under the influence of mountain breezes, and it is affected by transboundary incursions of natural and anthropogenic aerosols from Europe and North Africa (Ripoll et al., 2014, 2015). Co-located online and offline PM_1 measurements were also carried out to complement the ACSM data set. Hence, the work presented here interprets the real-time variation of inorganic and organic submicron components during 10 months (July 2011–April 2012), and the types of OA are also studied. Special emphasis is placed on the analysis of the diurnal pattern and seasonal variations of chemical components and the main factors influencing these variations.

2 Methodology

2.1 Sampling site

The Montsec site (MSC) is located on the highest part of the Montsec d'Ares mountain, at an altitude of 1570 m a.s.l., in a plain near the edge of a 1000 m cliff to the south, with no wind obstructions present around. It is located in the NE of the Iberian Peninsula (42°03′ N, 0°43′ E), 50 km S of the Pyrenees and 140 km NW of Barcelona (Fig. S1 in the Supplement.). A detailed description of this site can be found in Ripoll et al. (2014). This station is a member of the GAW (global atmospheric watch) network as a regional station named MSA.

The daily classification of atmospheric episodes affecting MSC was made following the procedure described by Ripoll et al. (2014) using the HYSPLIT (Hybrid Single Particle Lagrangian Integrated Trajectory Model) model from the NOAA Air Resources Laboratory (ARL). Air masses reaching MSC are mainly from the Atlantic (62 % of the days) throughout the year. From March to October, North African (NAF) episodes are more frequent (17 % of the days) and are very often alternated with the summer regional (SREG) scenarios (12 % of the days). The winter regional (WREG) scenarios are detected from October to March (11 % of the days), as well as the European (EU) episodes (11 % of the days). Conversely, the Mediterranean (MED) episodes are detected sporadically (4 % of the days).

The boundary layer height was calculated using the Global Data Assimilation System (GDAS1) model from the NOAA Air Resources Laboratory (http://www.ready.noaa.gov/READYamet.php) (Fig. 1).

2.2 ACSM sampling and data analysis

The ACSM (Aerodyne Research Inc.) was measuring continuously from July 2011 to April 2012. The ACSM provides real-time mass concentration of submicron particulate organics, nitrate, sulfate, ammonium and chloride via thermal vaporization and electron impact ionization, with detection by a quadrupole mass spectrometer (Ng et al., 2011c). The mass concentration of a given species is determined from the sum of the ion signals at each of its mass spectral fragments and its ionization efficiency (IE) (Canagaratna et al., 2007). Since calibration of IEs for all species is not feasible, the relative ionization efficiency (RIE) (compared to that of nitrate) is used (Jimenez et al., 2003). The ammonium nitrate calibration described by Ng et al. (2011b) was performed using an atomizer (TSI, Constant Output Atomizer model 3076) for primary aerosol generation, followed by a silica gel diffusion dryer, a differential mobility analyzer (DMA) model TSI 3081, and a condensation particle counter (CPC, TSI 3772). Monodisperse 300 nm ammonium nitrate aerosol particles were used, covering a range of nitrate concentrations from 2 to 15 $\mu g\, m^{-3}$. Several calibrations were conducted throughout the sampling period, and average values of 2.2×10^{-11} for nitrate IE and 5.4 for RIE for ammonium were used for the whole data set. The RIE values used in this study for the rest of the species were those usually applied in AMS ambient concentrations: 1.4 for OA and 1.1, 1.2, and 1.3 for nitrate, sulfate, and chloride, respectively (Canagaratna et al., 2007). RIE for sulfate was experimentally determined 1 year later and was found to be 1.26, although the default value was used for the current data set. A time resolution of 62 min was used as a result of 12 scans (one open and one filtered) per data point with a scan speed of 1 s amu^{-1}.

The ACSM data were analyzed with the standard ACSM data analysis software version 1.5.3.2 (Aerodyne Research Inc.) written in Igor Pro 6 (WaveMetrics, Inc., Lake Oswego, OR, USA). As the ACSM was measuring continuously for a long time, the standard correction for instrument sensitivity drifts was applied to the data set based on the inlet pressure and N_2 signal. Finally, mass concentrations were corrected using a collection efficiency (CE) to account for the particle bounce of aerosols on the vaporizer. The composition-dependent CE was calculated as described by Middlebrook et al. (2012) and was close to 0.45 for most of the time. Since for most ambient studies a 0.5 CE value is found to be representative with data uncertainties generally within $\pm 20\,\%$ (Canagaratna et al., 2007) and since our ACSM concentrations using CE = 0.5 were in good agreement with concentrations from other co-located instruments, a CE of 0.5 was used.

The organic components were further investigated by applying the Multilinear Engine (ME-2) (Paatero, 1999) to the organic mass spectra. With the ME-2, the user can introduce a priori information about sources with the so-called *a* value approach. Hence, the user inputs one or more factor profiles and a constraint defined by the *a* value, which determines the extent to which the output profile can differ from the profile fed to the model. The source apportionment of OA was performed applying ME-2 using the custom software tool of Source Finder (SoFi) version 4.8 developed by Canonaco et al. (2013). The ME-2 analysis was carried out separately for the summer period (14 July 2011–24 September 2011) and the winter period (10 January 2012–7 March 2012). Only $m/z \leq 100$ were used for source apportionment of OA because (a) the signals of $m/z > 100$ account for a minor fraction of the total organic mass (on average, 2 %), (b) the $m/z > 100$ have larger uncertainties, and (c) the large interference of naphthalene signals (used for m/z calibration of the ACSM) at these m/z (e.g., m/z 127, 128, and 129) (Sun et al., 2012).

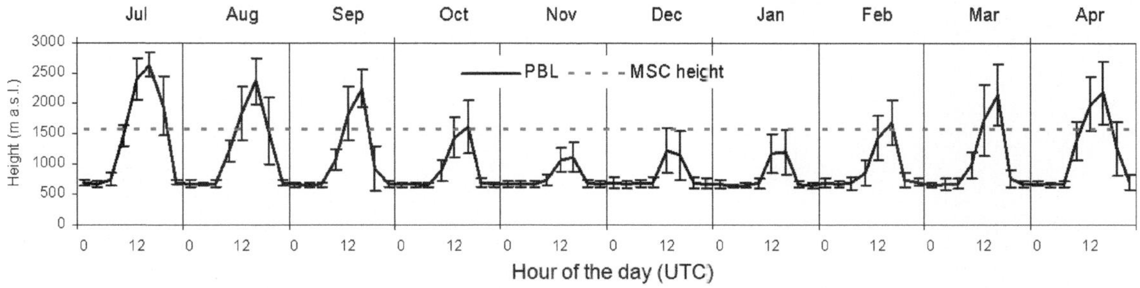

Figure 1. Diurnal variation of the boundary layer height (computed with HYSPLIT model) averaged for each month during the study period at Montsec. Variation bars indicate ± standard deviation.

2.3 Co-located measurements used in this study

MSC has been permanently equipped with aerosol monitoring instrumentation since January 2010 and some of these measurements were used in this study. The 24 h PM_1 samples were collected every 4 days on 150 mm quartz microfiber filters (Pallflex QAT) using high volume samplers ($30\,m^3\,h^{-1}$, MCV CAV-A/MSb) equipped with MCV PM_1 cutoff inlets. Daily PM_1 mass concentrations were determined by offline gravimetric procedures according to the EN 12341 standard (CEN, 1999), i.e., at 20 °C temperature and 50 % relative humidity. Furthermore, PM_1 chemical composition was obtained as described by Ripoll et al. (2015) using inductively coupled plasma atomic emission spectroscopy (ICP-AES) and mass spectrometry (ICP-MS) for major and trace elements, respectively, ion high performance liquid chromatography (HPLC) and selective electrode for ions concentrations, and the thermal–optical method (using the EUSSAR 2 protocol) for elemental carbon (EC) and organic carbon (OC) concentrations. Real-time PM_1 mass concentrations were continuously measured by an optical particle counter (OPC, model GRIMM 1.107 calibrated with a different latex PSL refraction index 1.59). PM_1 30 min data were averaged daily and subsequently corrected by comparison with 24 h standard gravimetric mass measurements. The absorption coefficient was measured continuously at 637 nm using a multi-angle absorption photometer (MAAP, model 5012, Thermo). Equivalent black carbon (BC) mass concentrations (Petzold et al., 2013) were calculated by the MAAP instrument software by dividing the measured absorption coefficient σ_{ap} (λ) by $6.6\,m^2\,g^{-1}$, which is the instrument default mass absorption cross section (MAC) at 637 nm (Müller et al., 2011; Petzold and Schönlinner, 2004). Particle scattering (σ_{sp}; 0–360°) and hemispheric backscattering (σ_{bsp}; 90–270°) coefficients at three wavelengths (450, 525, 635 nm) were measured with a LED-based integrating nephelometer (model Aurora 3000, ECOTECH Pty, Ltd, Knoxfield, Australia).

Finally, all meteorological data were measured by the Catalonian Meteorological Service from the Montsec d'Ares station. Gaseous pollutants (O_3, NO, NO_2, CO and SO_2) were measured using real-time monitors belonging to the Network of Control and Surveillance of Air Quality of the Autonomous Government of Catalonia. NO and NO_2 concentrations were measured using a Thermo Scientific instrument model 42i-TL, CO using a Teledyne 300 EU Gas filter correlation analyzer, O_3 using a MCV 48AV UV photometry analyzer, and SO_2 using a Teledyne 100 EU UV fluorescence analyzer.

In addition to these routine measurements, two intensive campaigns were performed in July–August 2011 and January–February 2012. During these intensive campaigns PM_1 filters were collected daily and a scanning mobility particle sizer (SMPS) was installed to measure particle number size distribution of mobility diameters of 11–350 nm in the summer campaign and of 8–450 nm in the winter campaign. The SMPS system comprises a classifier unit (model TSI 3080) and a differential mobility analyzer (DMA, model TSI 3081) connected to a condensation particle counter (CPC, model TSI 3772). The SMPS data for the winter campaign were also used to estimate the mass concentration to compare with the ACSM data. To this end, the volume size distributions were calculated from the measured particle number distributions assuming sphericity. The total volume concentrations were computed by integrating over the measured particle range and converted to mass concentration using the estimated composition-dependent density, calculated using the chemical composition given by the ACSM and the equation of Salcedo et al. (2006). Average concentrations shown in the whole paper are arithmetic averages unless otherwise specified.

3 Results and discussion

3.1 Submicron aerosol mass concentrations

In order to establish the consistency of the different measurements during this study, the sum of the ACSM species (i.e., sulfate + nitrate + ammonium + OA + chloride) and the BC mass concentrations was compared with the co-located PM_1 and light scattering measurements (Fig. 2). The scatter plots of ACSM plus BC concentrations vs. PM_1 con-

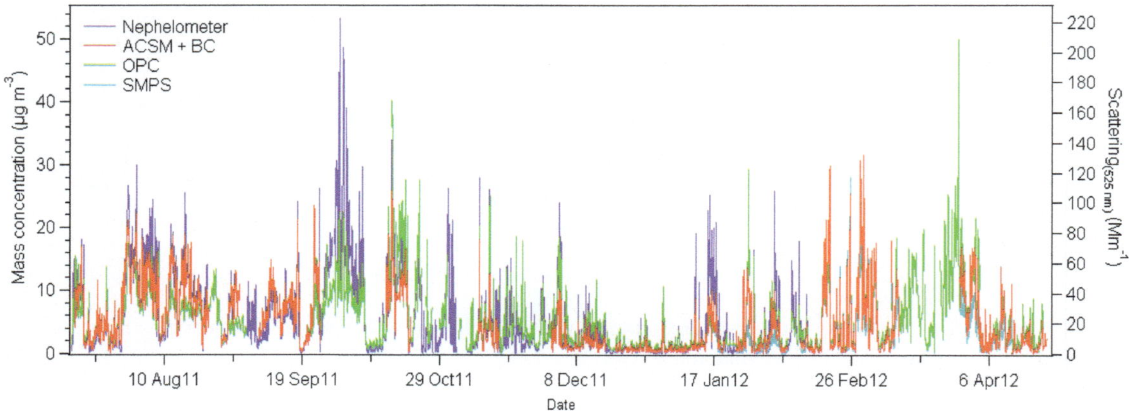

Figure 2. Time series of PM_1 total mass from co-located measurements and light scattering at 525 nm.

centrations from the OPC and SMPS showed strong correlations ($R^2 = 0.72$ and $R^2 = 0.87$, respectively) and slopes close to unity (slope = 0.94 and intercept = 0.09 for the ACSM + BC vs. OPC, and slope = 1.21 and intercept = 0.86 for the ACSM + BC vs. SMPS) (Fig. 3). The differences in the particle size range measured by the different instruments need to be considered when assessing these comparisons. Moreover, ACSM plus BC concentrations were also highly correlated with light scattering at 525 nm determined by the nephelometer ($R^2 = 0.85$, Fig. 3). The high degree of agreement is also apparent in the time series plots of PM_1 shown in Fig. 2.

The average concentration (25th, 50th, 75th percentiles) of the ACSM + BC mass during this study (July 2011–April 2012) was 4.9 $\mu g\, m^{-3}$ (0.9, 2.8, 7.9 $\mu g\, m^{-3}$) (Table 1), which is similar to the 2010–2012 average reported by Ripoll et al. (2014) from OPC measurements (5.0 $\mu g\, m^{-3}$). For the sake of brevity only summer (14 July 2011–24 September 2011) and winter (10 January 2012–7 March 2012) hourly variations will be discussed in the following sections, given that hourly variation in spring was similar to that in summer and that in fall was similar to that in winter. The seasonal average PM_1 concentrations were higher in summer (7.5 $\mu g\, m^{-3}$ (3.4, 7.1, 10.5 $\mu g\, m^{-3}$), Table 1) than in winter (4.1 $\mu g\, m^{-3}$ (0.8, 1.7, 5.6 $\mu g\, m^{-3}$), Table 1). A similar seasonal pattern has been described at other high altitude sites in Europe (e.g., Carbone et al., 2014; Cozic et al., 2008; Freney et al., 2011; Tositti et al., 2013), which are associated with differences in the air mass origin from summer to winter, and also to variations in the planetary boundary layer (PBL) height. The seasonal variation at MSC has been described in detail in recent works (Pandolfi et al., 2014b; Ripoll et al., 2014, 2015), and it has been principally attributed to the seasonal variation of the PBL (Fig. 1). In summer, the stable anticyclonic conditions over this continental area enhance convection, increasing the development of the PBL and favoring the transport of anthropogenic pollutants towards high altitude sites such as MSC. The situation in winter is notably different, as the lower vertical development of the PBL over this area leaves high altitude sites in the FT, isolating MSC from polluted air masses. The seasonal variation of PM concentrations at MSC has been also connected to mesoscale and synoptic processes. At MSC, southern flows and regional recirculation episodes are more frequent in summer, whereas clean Atlantic advections and northeastern winds from mainland Europe are more common in winter (Fig. S2, Table S1) (Pandolfi et al., 2014b; Ripoll et al., 2014, 2015). Moreover, the summer maximum has been ascribed to the more vigorous photochemistry in the atmosphere that enhances the formation of secondary inorganic and organic aerosols (Querol et al., 1999).

3.2 Submicron aerosol chemical composition

Concentrations of ACSM species were averaged daily and compared with offline measurements from 24 h PM_1 samples, and all species showed strong correlations (R^2 between 0.77 and 0.96; Fig. S3). Different slopes (ACSM vs. offline measurements) were found for each of the species: 1.12 for sulfate, 1.31 for ammonium and 1.35 for nitrate. The relatively higher slope for nitrate, with respect to sulfate, could be attributed to a sampling negative artifact due to the volatilization of nitrate on the offline samples (Schaap et al., 2004). Ammonium is present as a counterion for sulfate and nitrate and, thus, its slope is in between those of the two species.

The OM-to-OC ratio cannot be determined for ACSMs based on f_{44} following Aiken et al. (2008) as concluded from an intercomparison of 13 ACSMs (Fröhlich et al., 2015). Therefore, the OM-to-OC in the present study was estimated from the slope of OA, measured by ACSM, vs. OC, obtained from offline measurements. This slope was found to be 3.39; however, values of 2.2 are more common for aged aerosol (e.g., Aiken et al., 2008; Minguillón et al., 2011; Takahama et al., 2011). This disagreement of a factor of 1.54 may be attributed to different reasons. A negative volatilization ar-

Table 1. Diurnal variation of the boundary layer height (computed with HYSPLIT model) averaged for each month during the study period at Montsec. Variation bars indicate ± standard deviation.

	Whole period (μg m^{-3})	Summer (μg m^{-3})	Winter (μg m^{-3})
Period	14 July 2011–23 April 2012	14 July 2011–24 September 2011	10 January 2012–7 March 2012
Organics	2.4 (0.5, 1.4, 3.8)	4.0 (2.1, 4.0, 5.6)	1.9 (0.4, 1.0, 2.6)
Sulfate	1.0 (0.2, 0.5, 1.4)	1.9 (0.7, 1.6, 2.7)	0.5 (0.1, 0.3, 0.7)
Nitrate	0.7 (0.05, 0.2, 0.7)	0.5 (0.1, 0.3, 0.6)	1.0 (0.05, 0.2, 1.0)
Ammonium	0.6 (0.09, 0.3, 0.9)	0.9 (0.3, 0.8, 1.2)	0.5 (0.06, 0.2, 0.7)
Chloride	0.05 (0.01, 0.02, 0.05)	0.04 (0.01, 0.02, 0.05)	0.1 (0.004, 0.02, 0.1)
BC	0.2 (0.04, 0.1, 0.3)	0.2 (0.1, 0.2, 0.3)	0.2 (0.03, 0.1, 0.2)
Total	4.9 (0.9, 2.8, 7.9)	7.5 (3.4, 7.1, 10.5)	4.1 (0.8, 1.7, 5.6)

tifact may occur in the filters, hence resulting in an underestimation of OC. Alternatively it is possible that the RIE for OA in the ACSM is larger than the value of 1.4 determined for the AMS, a topic currently being investigated by the ACSM manufacturer, which would result in an overestimation of OA. Similar series of intercomparisons with a similar discrepancy for OA have been found for a 1-year data set (June 2012–July 2013) with the same instrument at the Montseny site (Minguillón et al., 2015) and for a recently reported study in Atlanta, USA, by Budisulistiorini et al. (2014). Assuming that the disagreement was due to the overestimation by the ACSM, the OA concentrations were corrected dividing by the disagreement factor (1.54) to compare the results with co-located measurements (Fig. 3). The resulting slopes were very similar and, hence, OA concentrations reported in the present paper were not corrected since further research is needed to better estimate the RIE for OA in the ACSM.

The average of PM_1 chemical composition at MSC during this study (July 2011–April 2012) is given in Fig. 4. On average, OA was the largest PM_1 constituent (50 %), followed by sulfate (20 %), nitrate (14 %), ammonium (12 %), BC (4 %) and chloride (1 %). As was the case of PM_1 concentrations, all chemical components increased in summer and decreased in winter, with the exception of nitrate (Fig. 4, Table 1). The higher nitrate concentrations in winter than in summer were also observed in other studies in the Mediterranean region (e.g., Pey et al., 2009; Querol et al., 2009; Ripoll et al., 2015) and this variation was attributed to the high volatility of ammonium nitrate at low humidity and high temperature (Zhuang et al., 1999). At MSC, the summer maximum of the rest of PM components has been mainly ascribed to the higher temperature and solar radiation in summer (Table S1), which enhances atmospheric photochemistry, promoting the formation of secondary inorganic and organic aerosols. All these seasonal characteristics are described in detail in Ripoll et al. (2015).

3.2.1 Summer trends

Time series of PM_1 components during summertime (14 July–24 September 2011) are shown in Fig. 5. Wind direction and wind speed, temperature, precipitation and concentrations of nitrogen oxides, sulfur dioxide, and ozone are also depicted. The daily classification of atmospheric episodes affecting MSC is also illustrated in different background colors. On average, during summer the lowest concentrations of all PM_1 components and gases were recorded under the Atlantic advection conditions since these air masses are associated with precipitation, decreased temperature and solar radiation, and strong winds, leading to cleaner atmospheric conditions. Conversely, summer regional episodes lasted for 6–11 consecutive days and led to sustained increases of the background concentrations of sulfate, OA and BC at MSC. Despite the limited ACSM data availability during North African episodes, relatively high concentrations of PM_1 components were observed under this type of episodes, especially BC (Fig. 5).

The diurnal cycles of PM_1 components, gaseous pollutants and meteorological variables are shown in Fig. 6. The summer concentrations of PM_1 components and gases showed no clear diurnal patterns, except for ozone and OA. This lack of defined daily patterns is similar to the findings obtained at the high altitude Puy-de-Dôme station in central France (Freney et al., 2011, the only similar study found at a remote site). For a high altitude site as MSC, the lack of diurnal cycles can be explained by a combination of factors. In summer, the recirculation of air masses over the WMB induced by an abrupt orography (Fig. S1) causes the formation of reservoir layers at any time at relatively high altitudes (Millan et al., 1997; Rodríguez et al., 2003). Moreover, long-range transport from North Africa, which can be more intense at high altitude layers (Ripoll et al., 2015), could also blur the daily patterns since the occurrence of this transport does not depend on the time of the day. These factors result in a lack of well-defined daily patterns but in a high variability of diurnal cycles even within the same type of episode, which is

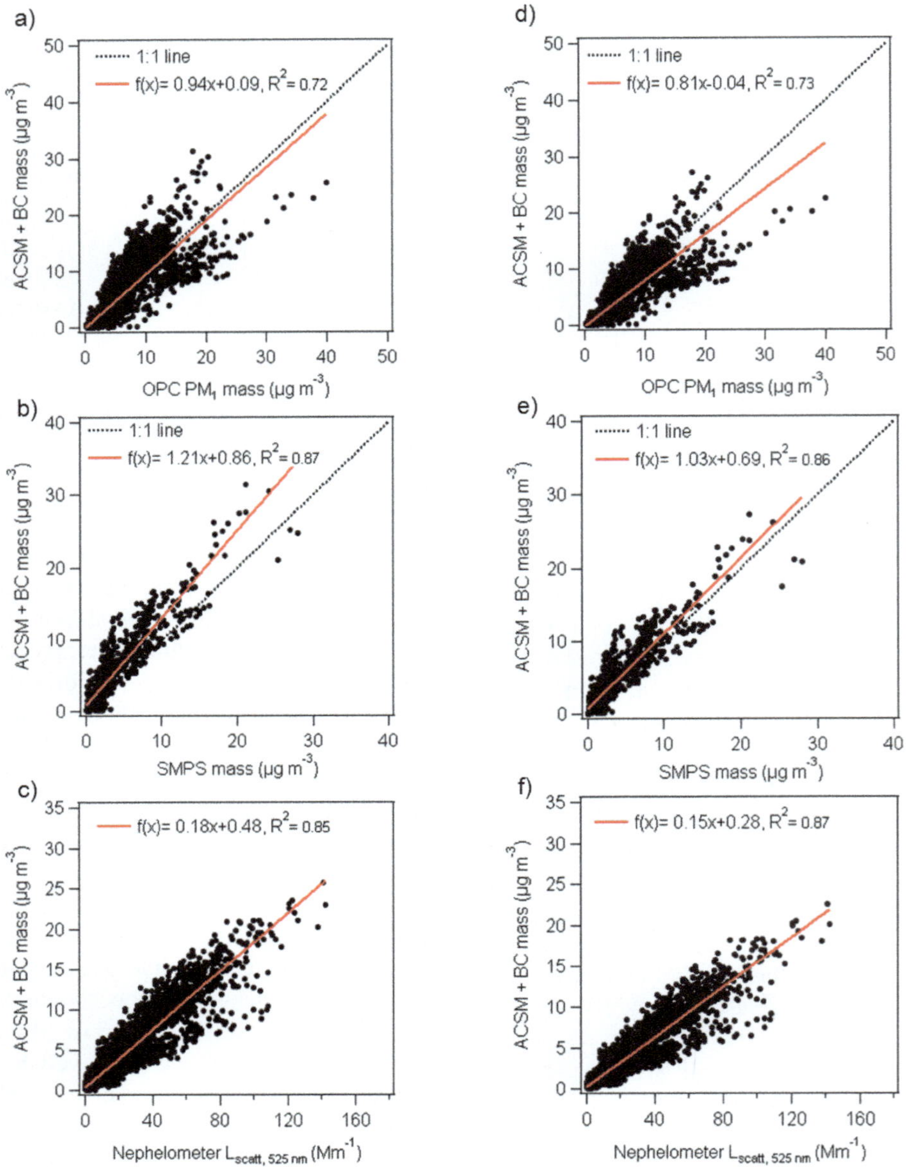

Figure 3. Scatter plots of ACSM plus BC mass vs. PM_1 from **(a)** OPC and **(b)** SMPS (winter campaign), and vs. light scattering at 525 nm from **(c)** nephelometer. Scatter plots of ACSM plus BC mass after dividing OA by 1.54 vs. PM_1 from **(d)** OPC and **(e)** SMPS (winter campaign), and vs. light scattering at 525 nm from **(f)** nephelometer. Data points correspond to hourly values. Equations and red lines correspond to linear regression fits.

reflected in the similar average daily evolutions and the high standard deviations calculated for the average daily patterns when separated by air mass origin (Fig. S4).

The ozone and OA concentrations had a marked diurnal cycle regardless of the air mass origin (Fig. 6, Fig. S4) in summer. These different daily patterns with respect to the rest of the gases and chemical components points to the fact that ozone and OA variations are strongly influenced by local/regional processes and not just dominated by long-range transport. Minimum ozone concentrations were recorded between 08:00 and 09:00 UTC (coordinated universal time, which is local time minus 1 h and local summer time minus 2 h), whereas maximum concentrations were measured between 16:00 and 17:00 UTC. In contrast, the highest OA concentrations were observed around 12:00 UTC and the lowest during the night and in the early morning. The ozone variations may influence those of OA, although the complete understanding of the ozone diurnal evolution is outside the scope of this study. The average increase in OA during the day is likely due to the photooxidation of volatile organic compounds (VOCs). Given that MSC is a remote site, in summer VOCs are most likely dominated by local bio-

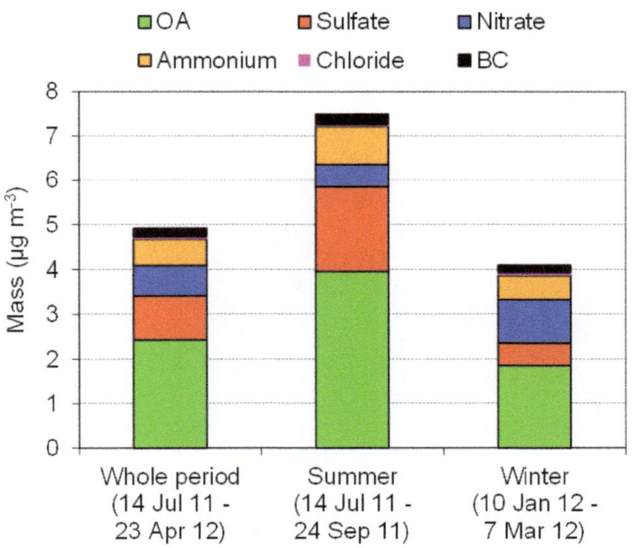

Figure 4. Average concentrations of PM_1 chemical species measured at Montsec during the whole study, in summer and in winter.

genic emissions (BVOCs), as it was found in the Mediterranean forested area of Montseny (Seco et al., 2011). Hence, the midday increase is likely due to recently produced biogenic SOA and, to a lesser degree, photooxidation of anthropogenic VOCs.

Despite the marked diurnal cycle of OA regardless of air mass origin, the average increase during the day with respect to average concentrations during the night was higher under summer regional ($2.6\,\mu g\,m^{-3}$) and North African ($3.0\,\mu g\,m^{-3}$) episodes than during Atlantic advections ($1.3\,\mu g\,m^{-3}$) (Fig. S4). This difference could be caused by the higher SOA formation. This is due to the increase in BVOC emissions and atmospheric photooxidation caused by the higher temperature and solar radiation (Paasonen et al., 2013; Seco et al., 2011) under summer regional and North African episodes. Furthermore, under these episodes higher concentrations of ozone were measured, which also favor the formation of SOA (via direct oxidation and also by leading to higher OH concentrations). The SOA formation registered at MSC is relatively high when compared to other high altitude sites such as Puy-de-Dôme (Freney et al., 2011). This is in agreement with the modeled SOA emissions over Europe, which identified higher SOA concentrations in Mediterranean environments (Bessagnet et al., 2008). This higher SOA formation is probably due to the higher emissions of BVOCs in the Mediterranean forested areas (up to 3 times higher than Boreal forested areas) (Bessagnet et al., 2008; Lang-Yona et al., 2010; Steinbrecher et al., 2009) and the comparable concentrations of tropospheric ozone with other high altitude European sites (Chevalier et al., 2007). On the other hand, the extra formation of SOA under summer regional and North African episodes might also have a contribution from the photooxidation of anthropogenic VOCs since Atlantic advections are associated with cleaner atmospheric conditions

3.2.2 Winter trends

As in summer, the lowest concentrations of all PM_1 components and gases in winter (10 January 2012–7 March 2012) were recorded under the Atlantic advections, whereas the highest were measured when MSC was affected by air masses from mainland Europe and sporadically under regional conditions (Fig. 7). Mediterranean air masses were detected very infrequently and therefore conclusions on their characteristics will not be drawn in the present paper. The relative contribution of different components was similar, with OA representing a somewhat smaller fraction than in summer.

In contrast to what was found in summer, in winter concentrations of most PM_1 components and gaseous pollutants showed much clearer diurnal patterns, with a minimum around 07:00 UTC and a maximum between 14:00 and 15:00 UTC (Fig. 6). Similar patterns have been observed at the Puy-de-Dôme station during a winter campaign (Freney et al., 2011). These daily cycles are probably caused by the fact that MSC is located most of the day within the FT in winter, whereas PBL air masses are only injected upwards after midday (Fig. 1) Moreover, thermal inversions are very frequent from 20:00 to 07:00 UTC. These situations prevent the transport of pollutants from the lower populated areas towards higher altitudes, especially at night. During the morning, the thermal inversions dissipate due to the radiative warming of the ground and mountain upslope winds develop (e.g., Henne et al., 2004). These mountain winds transport anthropogenic emissions from the adjacent valleys and plains to the top of the mountain, with a maximum upslope transport in the afternoon. Moreover, the biogenic emissions influence cannot be ruled out as average winter temperatures are high enough for them to occur (Seco et al., 2011; Steinbrecher et al., 2009). Thus, mountain breezes play an important role in determining the diurnal variation of PM_1 components in winter (Fig. 6), especially under regional conditions. A clear example of the PM_1 components' diurnal pattern under winter regional episodes was observed from 22 to 25 February 2012 with PM_1 concentrations (and NO_x) increasing several fold during the afternoon (Fig. 7).

The study of the daily cycles as a function of air mass origin (Fig. S5) showed clear diurnal patterns under winter regional episodes, as mentioned above, and less marked daily patterns when MSC is affected by Atlantic advections and long-range transport from mainland Europe. Under Atlantic episodes, the concentrations of PM_1 components were very low and the standard deviations with respect to the average pattern were quite high, resulting in unclear diurnal patterns compared to those under winter regional conditions. During European episodes, which can be more intense at high altitude layers (Ripoll et al., 2015; Sicard et al., 2011), back-

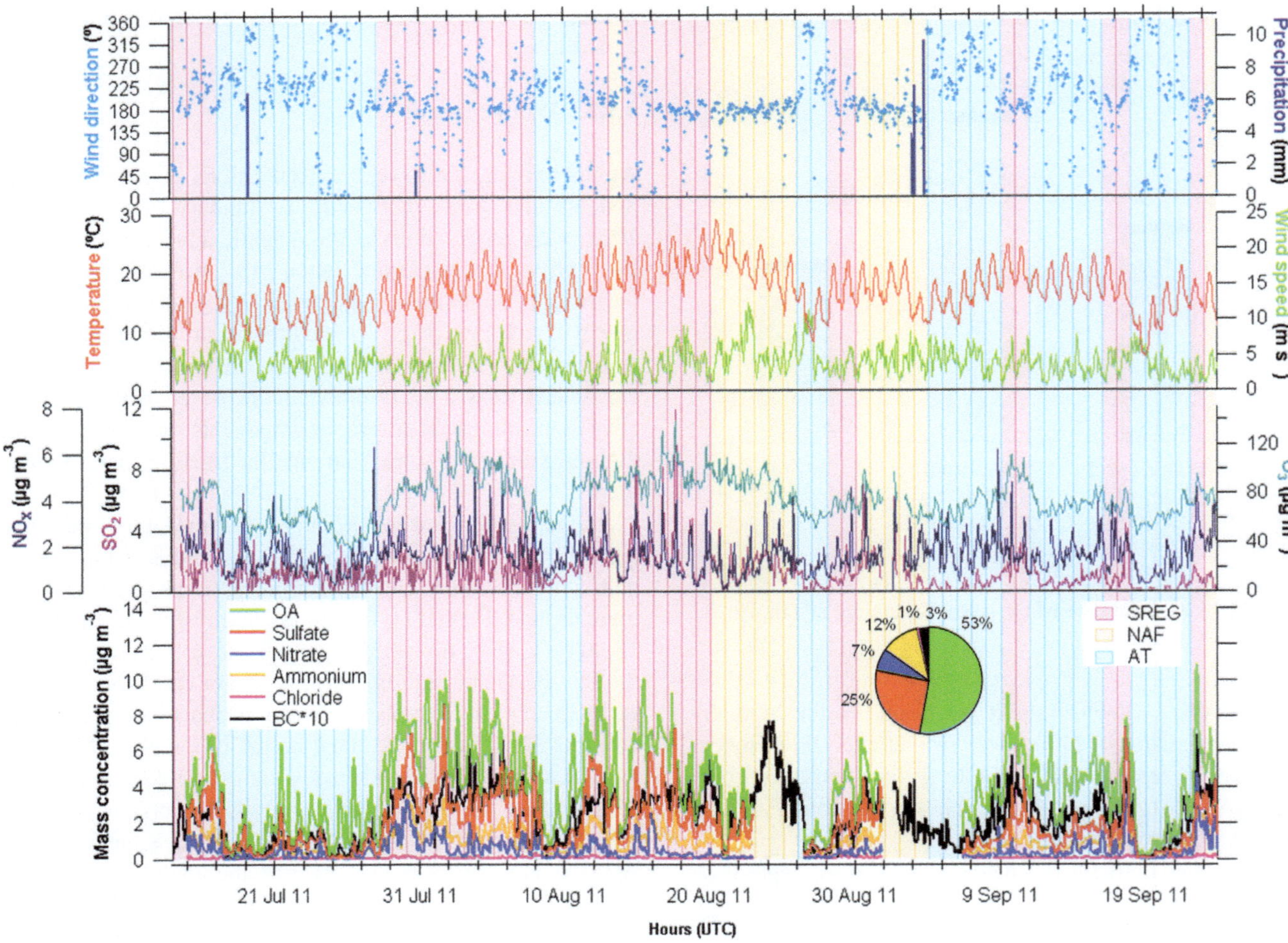

Figure 5. Time series of wind direction, precipitation, temperature, wind speed, concentrations of nitrogen oxides (NO_X), sulfur dioxide (SO_2), ozone (O_3), and PM_1 chemical species (OA, sulfate, nitrate, ammonium, chloride and black carbon (BC)) in summer (14 July 2011–24 September 2011). Background colors correspond to daily classification of atmospheric episodes (summer regional (SREG), North African (NAF), and Atlantic (AT)) and the pie chart corresponds to the average chemical composition for the summer period.

ground concentrations of PM_1 components were higher and the midday increment was lower compared to those under winter regional conditions, resulting in less marked daily patterns. These less-marked diurnal cycles are probably due to the fact that the increase of PM_1 components occurs during these episodes regardless of the time of the day. A good example of this less-marked diurnal variation during European episodes was observed from 17 to 19 February 2012 (Fig. 7).

3.3 Characterization of OA components

In order to better characterize the profiles of OA components, the ME-2 analysis was performed separately for the summer period (14 July–24 September 2011) and the winter period (10 January 2012–7 March 2012), since OA components are expected to vary throughout the year. The solution of the ME-2 analysis selected for each season was based on several tests, with a different number of factors and different a values, taking into account the correlations with external tracers (including nitrate, sulfate, BC and ozone), the daily patterns of each factor, and the residuals. As a result, a solution of three factors was selected for each season. In summer, a hydrocarbon-like OA (HOA), a semivolatile oxygenated OA (SV-OOA) and a low-volatility oxygenated OA (LV-OOA) (Fig. S6) were resolved. The HOA factor was constrained using an average HOA factor from different data sets (Ng et al., 2011b), with an a value of 0.1. The SV-OOA was characterized by a high 43 : 44 ratio, and the LV-OOA was defined by having a dominant peak at m/z 44. In winter the 3 factors identified were HOA, biomass burning OA (BBOA) and OOA (Fig. S6). The HOA and the BBOA factors were constrained based on the profiles from different data sets (Ng et al., 2011b), with an a value of 0.1 in both factors, and including BC as an additional variable to the ME-2 analysis. The OOA was characterized by a 43 : 44 ratio between those found for the LV-OOA and the SV-OOA in summer, and by having a dominant peak at m/z 44. This high signal

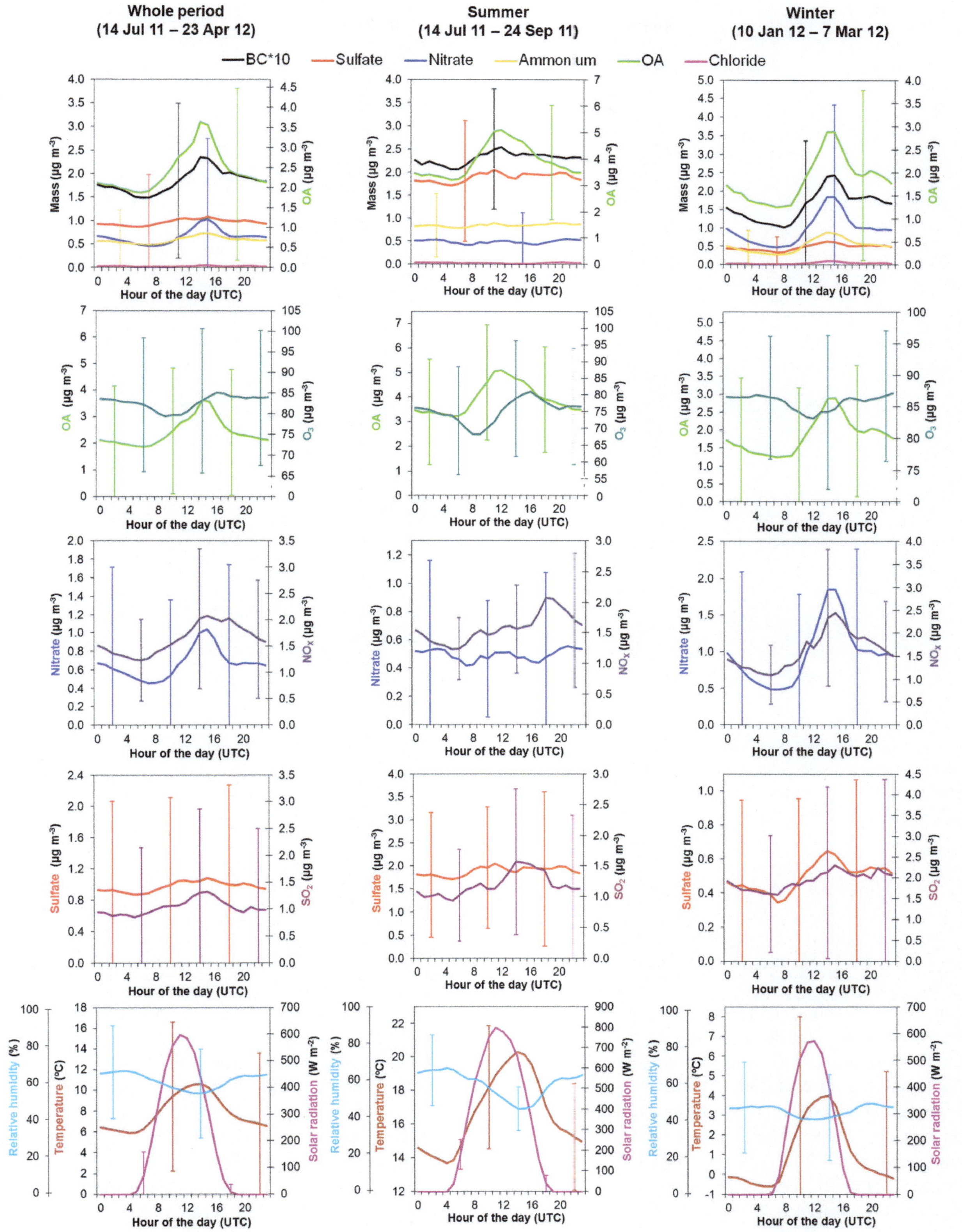

Figure 6. Diurnal cycles of PM_1 chemical species (BC, sulfate, nitrate, ammonium, chloride and OA), gaseous pollutants (O_3, NO_X, and SO_2), and meteorological parameters (relative humidity, temperature and solar radiation) averaged for the whole period, summer and winter. Variation bars indicate ± standard deviation.

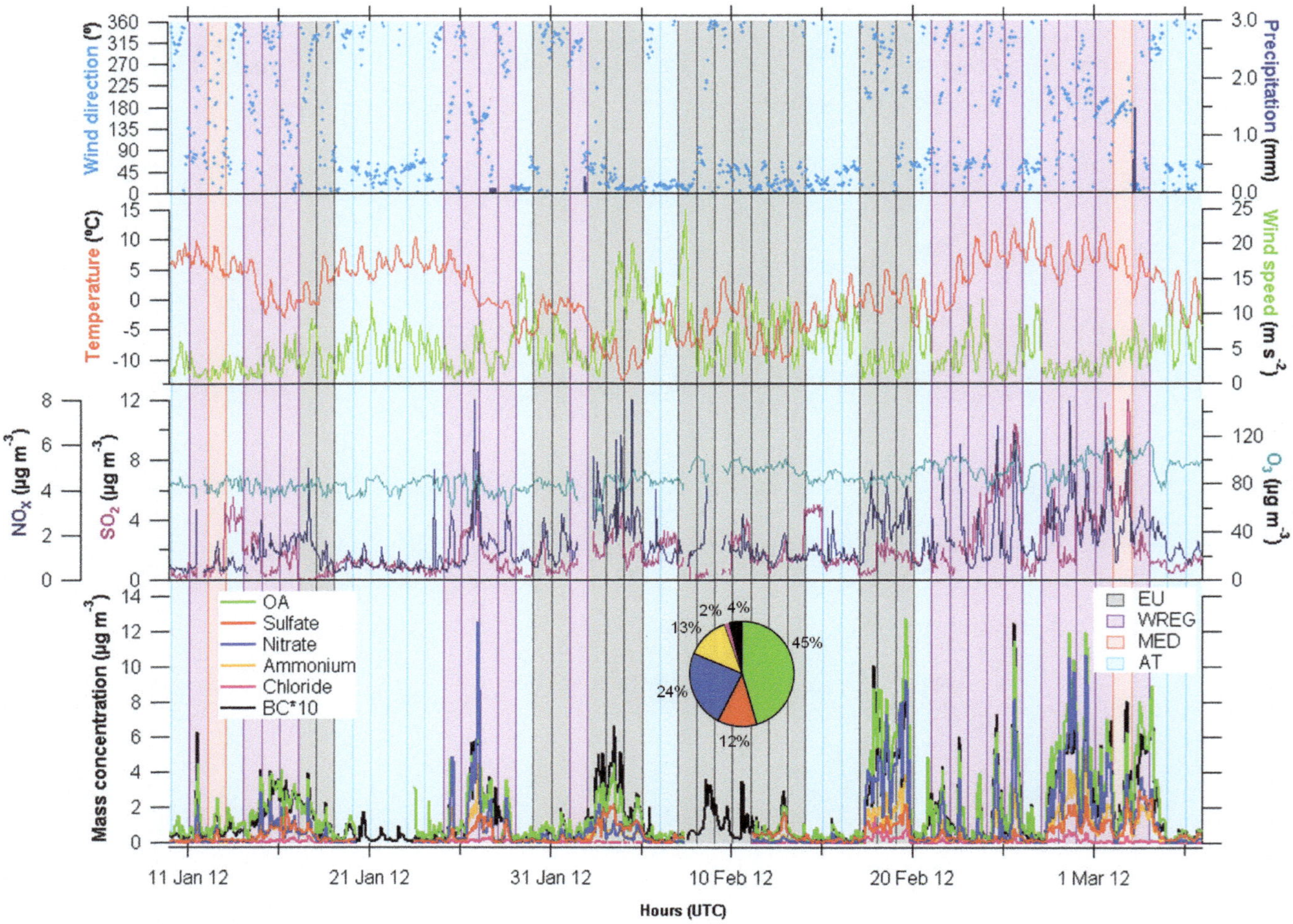

Figure 7. Time series of wind direction (WD) and speed (WS), temperature (T), precipitation (PP), concentrations of NO_x, SO_2, O_3, and PM_1 chemical species (organics, sulfate, nitrate, ammonium, chloride and black carbon (BC)) in winter (10 January 2012–7 March 2012). Background colors correspond to daily classification of atmospheric episodes (European (EU), winter regional (WREG), Mediterranean (MED) and Atlantic (AT)) and the pie chart corresponds to the average chemical composition for the winter period.

of m/z 44 in the winter OOA indicates a high degree of oxidation and, therefore, a dominant aged character during winter. A solution with four factors in winter was investigated in order to split the OOA into SV-OOA and LV-OOA, but the resulting profiles did not represent two different OOA types, and the time series showed that one of the factors was mainly representing noise. The HOA profiles obtained for the summer and winter periods were similar and showed similar deviations from the Ng et al. (2011b) spectrum (Fig. S7), since they were constrained with the same anchor HOA profile.

On average, LV-OOA dominated the OA fraction in summer, contributing 64 %, followed by SV-OOA (26 %) and the HOA (10 %) (Fig. 8), whereas in winter OOA accounted for 71 %, BBOA contributed 24 %, and HOA contribution decreased to 5 % (Fig. 9). The high contribution of OOA components confirms the initial hypothesis of MSC organic components being mostly secondary in their origin, and it is in agreement with what was found in other remote sites (Freney et al., 2011; Raatikainen et al., 2010). Furthermore, the origin of OA has been recently investigated in the Mediterranean forested area of Montseny and it has been found that SOA accounted for 91 % (Minguillón et al., 2011). The low contribution of primary organic components at MSC is in agreement with the location, since the primary organic emissions are mixed and oxidized during their transport from industrial and urban areas to the remote site of MSC.

The diurnal cycles of OA components were studied as a function of air mass origin (Fig. 8, 9). A clear daily pattern of OA components was found regardless of the air mass origin, except for Atlantic advections in winter. In summer, the maximum concentrations of LV-OOA and HOA were measured between 12:00 and 13:00 UTC, whereas those of SV-OOA were observed between 11:00 and 12:00 UTC. The LV-OOA has been generally associated with highly oxidized, aged, and long-range-transported aerosol particles (Lanz et al., 2010). Conversely, the SV-OOA has been described as the less oxy-

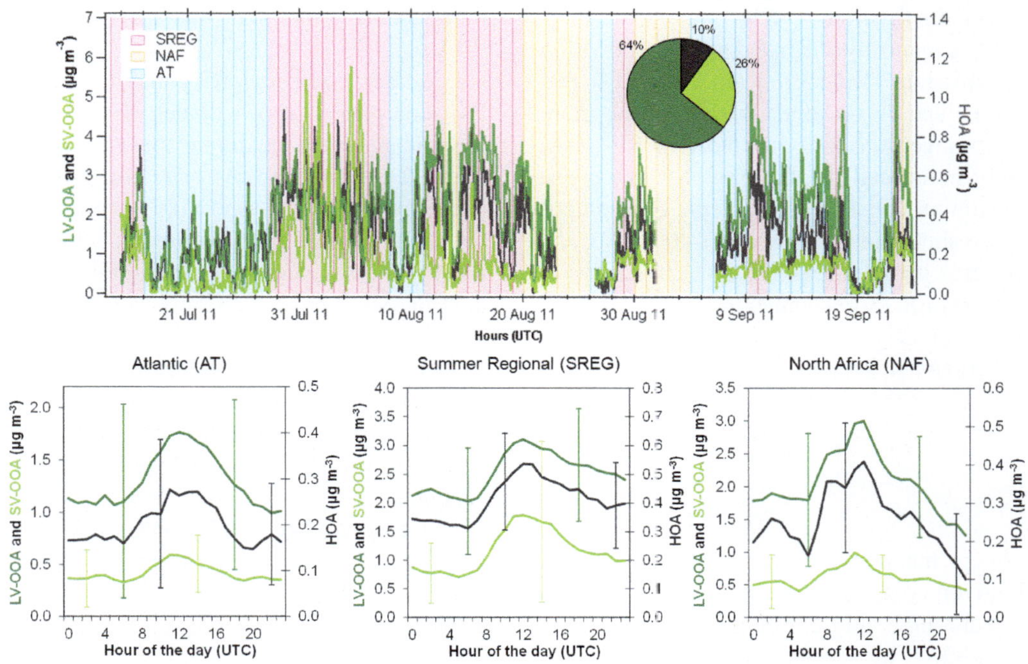

Figure 8. Top: time series of organic species (hydrocarbon-like organic aerosol (HOA), semivolatile oxygenated organic aerosol (SV-OOA) and low-volatility oxygenated organic aerosol (LV-OOA)) concentrations in summer (14 July 2011–24 September 2011). Background colors correspond to daily classification of atmospheric episodes and the pie chart correspond to the average organic species composition for the summer period. Bottom: diurnal cycles of organic species concentrations averaged as a function of atmospheric episode for the summer period. Variation bars indicate ± standard deviation.

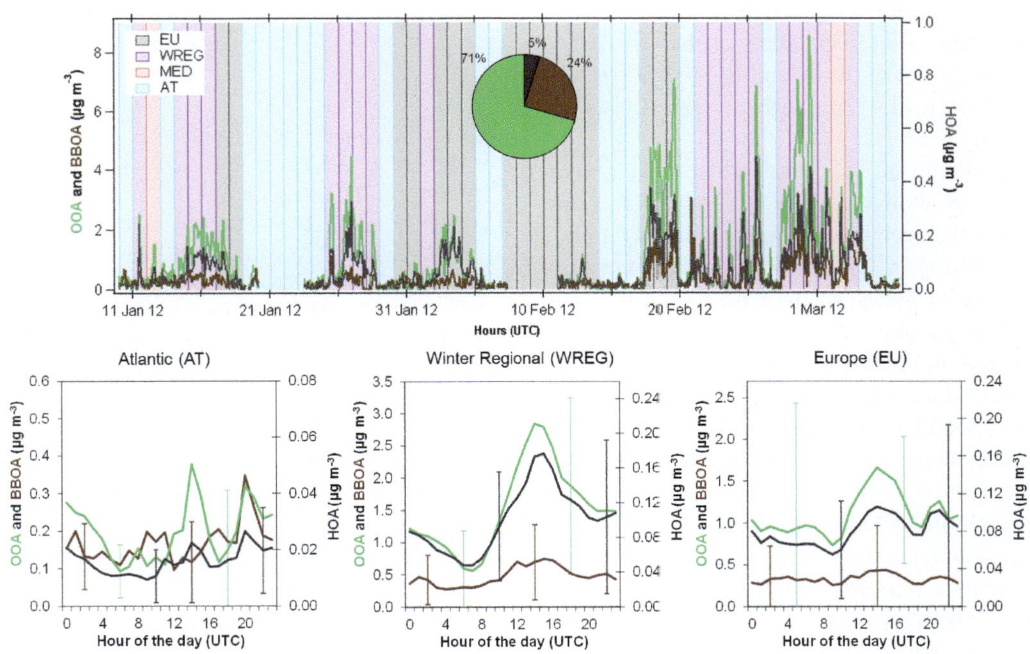

Figure 9. Top: time series of organic species (hydrocarbon-like organic aerosol (HOA), biomass burning organic aerosol (BBOA) and oxygenated organic aerosol (OOA)) concentrations in winter (10 January 2012–7 March 2012). Background colors correspond to daily classification of atmospheric episodes and the pie chart correspond to the average organic species composition for the winter period. Bottom: diurnal cycles of organic species concentrations averaged as a function of atmospheric episode for the winter period. Variation bars indicate ± standard deviation.

genated and semivolatile fraction of OOA (Ng et al., 2011a) and therefore it has been mostly attributed to SOA formation from more local emissions (Jimenez et al., 2009). For this reason, the LV-OOA and HOA hourly variations are more influenced by long-range transport than those of SV-OOA, which are strongly influenced by local/regional processes. In winter, the maximum concentrations of the OA components were observed simultaneously around 14:00 UTC. The different daily patterns between seasons can be attributed to the higher production of SOA in summer as opposed to winter, when the maximum daily concentrations are reached and later driven by the mountain breezes.

4 Conclusions

This work interprets the real-time variation of inorganic and organic submicron components during 10 months (July 2011–April 2012) at a high altitude site in southern Europe (Montsec, 1570 m a.s.l.). The aerosol chemical composition was obtained with an ACSM, and co-located online and offline PM_1 measurements were also carried out. The average concentration of the ACSM plus BC mass during this study was 4.9 $\mu g\, m^{-3}$ and, on average, OA was the foremost PM_1 constituent (50 %), followed by sulfate (20 %), nitrate (14 %), ammonium (12 %), BC (4 %) and chloride (1 %). Discrepancies of OA determined by ACSM with co-located measurements pointed to an overestimation by the ACSM probably caused by the use of the default RIE for OA, which could be lower than the actual one. Further research is needed to better address this issue.

The seasonal variation of PM_1 mass and chemical component concentrations showed similar patterns, with an increase in summer and a decrease in winter, except for nitrate which has high volatility in summer. The seasonal variation was attributed to the evolution of the PBL height throughout the year and to synoptic circulation and meteorological factors. At MSC the higher temperature and solar radiation in summer enhances the convection processes, incrementing the development of the PBL, and augments atmospheric photochemistry, promoting the formation of secondary inorganic and organic aerosols.

The diurnal variation of PM_1 components had no clear diurnal patterns in summer, except for organics. This lack of defined daily patterns was ascribed to the recirculation of air masses that causes the formation of reservoir layers at relatively high altitudes and to the long-range transport from North Africa. These factors result in a high variability of diurnal cycles even within the same type of episode. Nevertheless, organic concentrations had a marked diurnal cycle regardless of the air mass origin, with maximum concentrations around 12:00 UTC. The OA was dominated by LV-OOA (64 %), followed by SV-OOA (26 %), and HOA (10 %). Hence, the midday increase with respect to average concentrations during the night was attributed to the formation of SOA.

In winter, under regional conditions, concentrations of all PM_1 components showed much clearer diurnal patterns than in summer, with a maximum between 14:00 and 15:00 UTC. These daily cycles were connected to the fact that MSC is located most of the day within the FT, whereas PBL air masses are only injected upwards after midday. However, when MSC was affected by long-range transport from mainland Europe, less marked daily patterns of PM_1 components were observed.

The OA in winter was also mainly secondary (71 %), with contributions from BBOA (24 %), and HOA (5 %). The hourly variation of these factors showed a clear diurnal pattern regardless of the air mass origin, except for Atlantic advections.

To the authors' knowledge, this is one of the first times that real-time submicron aerosol chemical composition is characterized and its variation interpreted during a period of almost a year in a continental background environment. The results obtained in the present study highlight the importance of the SOA formation processes at a remote site such as MSC, which could be the objective of further investigations.

Acknowledgements. This study was supported by the Ministry of Economy and Competitiveness and FEDER funds under the PRISMA (CGL2012-39623-C02-1) and CARIATI (CGL2008-06294/CLI) projects, and by the Generalitat de Catalunya (AGAUR 2009 SGR8 and the DGQA). The research received funding from the European Union Seventh Framework Programme (FP7/2007-2013) ACTRIS under grant agreement no. 262254. We would like to extend their gratitude to the personnel from the COU and the OAdM. We would also like to express our gratitude to the NOAA Air Resources Laboratory (ARL) for the provision of the HYSPLIT transport and dispersion model, and boundary layer height calculation, used in this publication. D. A. Day and J. L. Jimenez thank the DOE (BER/ASR) DE-SC0011105 and NOAA NA13OAR4310063.

Edited by: R. McLaren

References

Aiken, A. C., Decarlo, P. F., Kroll, J. H., Worsnop, D. R., Huffman, J. A., Docherty, K. S., Ulbrich, I. M., Mohr, C., Kimmel, J. R., Sueper, D., Sun, Y., Zhang, Q., Trimborn, A., Northway, M., Ziemann, P. J., Canagaratna, M. R., Onasch, T. B., Alfarra, M. R., Prevot, A. S. H., Dommen, J., Duplissy, J., Metzger, A., Baltensperger, U., and Jimenez, J. L.: O/C and OM/OC ratios of primary, secondary, and ambient organic aerosols with high-resolution time-of-flight aerosol mass spectrometry, Environ. Sci. Technol., 42, 4478–4485, doi:10.1021/es703009q, 2008.

Bessagnet, B., Menut, L., Curci, G., Hodzic, A., Guillaume, B., Liousse, C., Moukhtar, S., Pun, B., Seigneur, C., and Schulz, M.: Regional modeling of carbonaceous aerosols over Europe-focus on secondary organic aerosols, J. Atmos. Chem., 61, 175–202, doi:10.1007/s10874-009-9129-2, 2008.

Budisulistiorini, S. H., Canagaratna, M. R., Croteau, P. L., Baumann, K., Edgerton, E. S., Kollman, M. S., Ng, N. L., Verma, V., Shaw, S. L., Knipping, E. M., Worsnop, D. R., Jayne, J. T., Weber, R.J., and Surratt, J. D.: Intercomparison of an Aerosol Chemical Speciation Monitor (ACSM) with ambient fine aerosol measurements in downtown Atlanta, Georgia, Atmos. Meas. Tech., 7, 1929–1941, doi:10.5194/amt-7-1929-2014, 2014.

Budisulistiorini, S. H., Canagaratna, M. R., Croteau, P. L., Marth, W. J., Baumann, K., Edgerton, E. S., Shaw, S. L., Knipping, E. M., Worsnop, D. R., Jayne, J. T., Gold, A. and Surratt, J. D.: Real-time continuous characterization of secondary organic aerosol derived from isoprene epoxydiols in downtown Atlanta, Georgia, using the Aerodyne Aerosol Chemical Speciation Monitor., Environ. Sci. Technol., 47, 5686–5694, doi:10.1021/es400023n, 2013.

Canagaratna, M. R., Onasch, T. B., Wood, E. C., Herndon, S. C., Jayne, J. T., Cross, E. S., Miake-Lye, R. C., Kolb, C. E., and Worsnop, D. R.: Chemical and microphysical characterization of ambient aerosols with the aerodyne aerosol mass spectometer, Wiley Intersci., 26, 185–222, doi:10.1002/mas.20115, 2007.

Canonaco, F., Crippa, M., Slowik, J. G., Baltensperger, U., and Prévôt, A. S. H.: SoFi, an IGOR-based interface for the efficient use of the generalized multilinear engine (ME-2) for the source apportionment: ME-2 application to aerosol mass spectrometer data, Atmos. Meas. Tech., 6, 3649–3661, doi:10.5194/amt-6-3649-2013, 2013.

Carbone, S., Saarikoski, S., Frey, A., Reyes, F., Reyes, P., Castillo, M., Gramsch, E., Jayne, J., Worsnop, D. R., and Hillamo, R.: Chemical Characterization of Submicron Aerosol Particles in Santiago de Chile, Aerosol Air Qual. Res., 13, 462–473, doi:10.4209/aaqr.2012.10.0261, 2013.

Carbone, C., Decesari, S., Paglione, M., Giulianelli, L., Rinaldi, M., Marinoni, A., Cristofanelli, P., Didiodato, A., Bonasoni, P., Fuzzi, S., and Facchini, M. C.: 3-year chemical composition of free tropospheric PM1 at the Mt. Cimone GAW global station – South Europe – 2165 m a.s.l., Atmos. Environ., 87, 218–227, doi:10.1016/j.atmosenv.2014.01.048, 2014.

CEN: Air quality – Determination of the PM_{10} fraction of suspended particulate matter. Reference method and field test procedure to demonstrate reference equivalence of measurement methods, EN12341:1999, 1999.

Chevalier, A., Gheusi, F., Delmas, R., Ordóñez, C., Sarrat, C., Zbinden, R., Thouret, V., Athier, G., and Cousin, J.-M.: Influence of altitude on ozone levels and variability in the lower troposphere: a ground-based study for western Europe over the period 2001–2004, Atmos. Chem. Phys., 7, 4311–4326, doi:10.5194/acp-7-4311-2007, 2007.

Cozic, J., Verheggen, B., Weingartner, E., Crosier, J., Bower, K. N., Flynn, M., Coe, H., Henning, S., Steinbacher, M., Henne, S., Collaud Coen, M., Petzold, A., and Baltensperger, U.: Chemical composition of free tropospheric aerosol for PM1 and coarse mode at the high alpine site Jungfraujoch, Atmos. Chem. Phys., 8, 407–423, doi:10.5194/acp-8-407-2008, 2008.

Crippa, M., Canonaco, F., Lanz, V. A., Äijälä, M., Allan, J. D., Carbone, S., Capes, G., Ceburnis, D., Dall'Osto, M., Day, D. A., DeCarlo, P. F., Ehn, M., Eriksson, A., Freney, E., Hildebrandt Ruiz, L., Hillamo, R., Jimenez, J. L., Junninen, H., Kiendler-Scharr, A., Kortelainen, A.-M., Kulmala, M., Laaksonen, A., Mensah, A. A., Mohr, C., Nemitz, E., O'Dowd, C., Ovadnevaite, J., Pandis, S. N., Petäjä, T., Poulain, L., Saarikoski, S., Sellegri, K., Swietlicki, E., Tiitta, P., Worsnop, D. R., Baltensperger, U., and Prévôt, A. S. H.: Organic aerosol components derived from 25 AMS data sets across Europe using a consistent ME-2 based source apportionment approach, Atmos. Chem. Phys., 14, 6159–6176, doi:10.5194/acp-14-6159-2014, 2014.

De Gouw, J. and Jimenez, J. L.: Organic aerosols in the Earth's atmosphere., Environ. Sci. Technol., 43, 7614–7618, doi:10.1021/es9006004, 2009.

Donahue, N. M., Robinson, A. L., Trump, E. R., Riipinen, I., and Kroll, J. H.: Volatility and Aging of Atmospheric Organic Aerosol, Top. Curr. Chem., 339, 97–144, doi:10.1007/128_2012_355, 2014.

Elbert, W., Taylor, P. E., Andreae, M. O., and Pöschl, U.: Contribution of fungi to primary biogenic aerosols in the atmosphere: wet and dry discharged spores, carbohydrates, and inorganic ions, Atmos. Chem. Phys., 7, 4569–4588, doi:10.5194/acp-7-4569-2007, 2007.

Freney, E. J., Sellegri, K., Canonaco, F., Boulon, J., Hervo, M., Weigel, R., Pichon, J. M., Colomb, A., Prévôt, A. S. H., and Laj, P.: Seasonal variations in aerosol particle composition at the puy-de-Dôme research station in France, Atmos. Chem. Phys., 11, 13047–13059, doi:10.5194/acp-11-13047-2011, 2011.

Fröhlich, R., Crenn, V., Setyan, A., Belis, C. A., Canonaco, F., Favez, O., Riffault, V., Slowik, J. G., Aas, W., Aijälä, M., Alastuey, A., Artiñano, B., Bonnaire, N., Bozzetti, C., Bressi, M., Carbone, C., Coz, E., Croteau, P. L., Cubison, M. J., Esser-Gietl, J. K., Green, D. C., Gros, V., Heikkinen, L., Herrmann, H., Jayne, J. T., Lunder, C. R., Minguillón, M. C., Močnik, G., O'Dowd, C. D., Ovadnevaite, J., Petralia, E., Poulain, L., Priestman, M., Ripoll, A., Sarda-Estève, R., Wiedensohler, A., Baltensperger, U., Sciare, J., and Prévôt, A. S. H.: ACTRIS ACSM intercomparison – Part 2: Intercomparison of ME-2 organic source apportionment results from 15 individual, co-located aerosol mass spectrometers, Atmos. Meas. Tech. Discuss., 8, 1559–1613, doi:10.5194/amtd-8-1559-2015, 2015.

Hallquist, M., Wenger, J. C., Baltensperger, U., Rudich, Y., Simpson, D., Claeys, M., Dommen, J., Donahue, N. M., George, C., Goldstein, A. H., Hamilton, J. F., Herrmann, H., Hoffmann, T., Iinuma, Y., Jang, M., Jenkin, M. E., Jimenez, J. L., Kiendler-Scharr, A., Maenhaut, W., McFiggans, G., Mentel, Th. F., Monod, A., Prévôt, A. S. H., Seinfeld, J. H., Surratt, J. D., Szmigielski, R., and Wildt, J.: The formation, properties and impact of secondary organic aerosol: current and emerging issues, Atmos. Chem. Phys., 9, 5155–5236, doi:10.5194/acp-9-5155-2009, 2009.

Henne, S., Furger, M., Nyeki, S., Steinbacher, M., Neininger, B., de Wekker, S. F. J., Dommen, J., Spichtinger, N., Stohl, A., and Prévôt, A. S. H.: Quantification of topographic venting of boundary layer air to the free troposphere, Atmos. Chem. Phys., 4, 497–509, doi:10.5194/acp-4-497-2004, 2004.

IPCC 2013 in: Climate Change 2013: The Physical Science Basis (Contribution of Working Group I to the Fifth Assessment Report of the Intergovernmental Panel on Climate Change), edited by: Stocker, T. F., Qin, D., Plattner, G. K., Tignor, M. M. B., Allen, S. K., Boschung, J., Nauels, A., Xia, Y., Bex, V., and Midgley, P. M., Cambridge Univ. Press, UK and New York, USA, 2013.

Jimenez, J. L., Jayne, J. T., Shi, Q., Kolb, C. E., Worsnop, D. R., Yourshaw, I., Seinfeld, J. H., Flagan, R. C., Zhang, X., Smith, K. A., Morris, J. W., and Davidovits, P.: Ambient aerosol sampling using the Aerodyne Aerosol Mass Spectrometer, J. Geophys. Res., 108, 8425, doi:10.1029/2001JD001213, 2003.

Jimenez, J. L., Canagaratna, M. R., Donahue, N. M., Prevot, A. S. H., Zhang, Q., Kroll, J. H., DeCarlo, P. F., Allan, J. D., Coe, H., Ng, N. L., Aiken, A. C., Docherty, K. S., Ulbrich, I. M., Grieshop, A. P., Robinson, A. L., Duplissy, J., Smith, J. D., Wilson, K. R., Lanz, V. A., Hueglin, C., Sun, Y. L., Tian, J., Laaksonen, A., Raatikainen, T., Rautiainen, J., Vaattovaara, P., Ehn, M., Kulmala, M., Tomlinson, J. M., Collins, D. R., Cubison, M. J., Dunlea, E. J., Huffman, J. A., Onasch, T. B., Alfarra, M. R., Williams, P. I., Bower, K., Kondo, Y., Schneider, J., Drewnick, F., Borrmann, S., Weimer, S., Demerjian, K., Salcedo, D., Cottrell, L., Griffin, R., Takami, A., Miyoshi, T., Hatakeyama, S., Shimono, A., Sun, J. Y., Zhang, Y. M., Dzepina, K., Kimmel, J. R., Sueper, D., Jayne, J. T., Herndon, S. C., Trimborn, A. M., Williams, L. R., Wood, E. C., Middlebrook, A. M., Kolb, C. E., Baltensperger, U., and Worsnop, D. R.: Evolution of organic aerosols in the atmosphere., Science, 326, 1525–1529, doi:10.1126/science.1180353, 2009.

Jorba, O., Pandolfi, M., Spada, M., Baldasano, J. M., Pey, J., Alastuey, A., Arnold, D., Sicard, M., Artiñano, B., Revuelta, M. A., and Querol, X.: Overview of the meteorology and transport patterns during the DAURE field campaign and their impact to PM observations, Atmos. Environ., 77, 607–620, doi:10.1016/j.atmosenv.2013.05.040, 2013.

Kroll, J. H. and Seinfeld, J. H.: Chemistry of secondary organic aerosol: Formation and evolution of low-volatility organics in the atmosphere, Atmos. Environ., 42, 3593–3624, doi:10.1016/j.atmosenv.2008.01.003, 2008.

Lang-Yona, N., Rudich, Y., Mentel, Th. F., Bohne, A., Buchholz, A., Kiendler-Scharr, A., Kleist, E., Spindler, C., Tillmann, R., and Wildt, J.: The chemical and microphysical properties of secondary organic aerosols from Holm Oak emissions, Atmos. Chem. Phys., 10, 7253–7265, doi:10.5194/acp-10-7253-2010, 2010.

Lanz, V. A., Prévôt, A. S. H., Alfarra, M. R., Weimer, S., Mohr, C., DeCarlo, P. F., Gianini, M. F. D., Hueglin, C., Schneider, J., Favez, O., D'Anna, B., George, C., and Baltensperger, U.: Characterization of aerosol chemical composition with aerosol mass spectrometry in Central Europe: an overview, Atmos. Chem. Phys., 10, 10453–10471, doi:10.5194/acp-10-10453-2010, 2010.

Middlebrook, A. M., Bahreini, R., Jimenez, J. L., and Canagaratna, M. R.: Evaluation of Composition-Dependent Collection Efficiencies for the Aerodyne Aerosol Mass Spectrometer using Field Data, Aerosol Sci. Technol., 46, 258–271, doi:10.1080/02786826.2011.620041, 2012.

Millan, M. M., Salvador, R., Mantilla, E., and Kallos, G.: Photooxidant dynamics in the Mediterranean basin in summer: Results from European research projects, J. Geophys. Res., 102, 8811–8823, 1997.

Minguillón, M. C., Perron, N., Querol, X., Szidat, S., Fahrni, S. M., Alastuey, A., Jimenez, J. L., Mohr, C., Ortega, A. M., Day, D. A., Lanz, V. A., Wacker, L., Reche, C., Cusack, M., Amato, F., Kiss, G., Hoffer, A., Decesari, S., Moretti, F., Hillamo, R., Teinilä, K., Seco, R., Peñuelas, J., Metzger, A., Schallhart, S., Müller, M., Hansel, A., Burkhart, J. F., Baltensperger, U., and Prévôt, A. S. H.: Fossil versus contemporary sources of fine elemental and organic carbonaceous particulate matter during the DAURE campaign in Northeast Spain, Atmos. Chem. Phys., 11, 12067–12084, doi:10.5194/acp-11-12067-2011, 2011.

Minguillón, M. C., Ripoll, A., Pérez, N., Prévôt, A. S. H., Canonaco, F., Querol, X., and Alastuey, A.: Chemical characterization of submicron regional background aerosols in the Western Mediterranean using an Aerosol Chemical Speciation Monitor, Atmos. Chem. Phys. Discuss., 15, 965–1000, doi:10.5194/acpd-15-965-2015, 2015.

Müller, T., Henzing, J. S., de Leeuw, G., Wiedensohler, A., Alastuey, A., Angelov, H., Bizjak, M., Collaud Coen, M., Engström, J. E., Gruening, C., Hillamo, R., Hoffer, A., Imre, K., Ivanow, P., Jennings, G., Sun, J. Y., Kalivitis, N., Karlsson, H., Komppula, M., Laj, P., Li, S.-M., Lunder, C., Marinoni, A., Martins dos Santos, S., Moerman, M., Nowak, A., Ogren, J. A., Petzold, A., Pichon, J. M., Rodriquez, S., Sharma, S., Sheridan, P. J., Teinilä, K., Tuch, T., Viana, M., Virkkula, A., Weingartner, E., Wilhelm, R., and Wang, Y. Q.: Characterization and intercomparison of aerosol absorption photometers: result of two intercomparison workshops, Atmos. Meas. Tech., 4, 245–268, doi:10.5194/amt-4-245-2011, 2011.

Ng, N. L., Canagaratna, M. R., Zhang, Q., Jimenez, J. L., Tian, J., Ulbrich, I. M., Kroll, J. H., Docherty, K. S., Chhabra, P. S., Bahreini, R., Murphy, S. M., Seinfeld, J. H., Hildebrandt, L., Donahue, N. M., DeCarlo, P. F., Lanz, V. A., Prévôt, A. S. H., Dinar, E., Rudich, Y., and Worsnop, D. R.: Organic aerosol components observed in Northern Hemispheric datasets from Aerosol Mass Spectrometry, Atmos. Chem. Phys., 10, 4625–4641, doi:10.5194/acp-10-4625-2010, 2010.

Ng, N. L., Canagaratna, M. R., Jimenez, J. L., Chhabra, P. S., Seinfeld, J. H., and Worsnop, D. R.: Changes in organic aerosol composition with aging inferred from aerosol mass spectra, Atmos. Chem. Phys., 11, 6465–6474, doi:10.5194/acp-11-6465-2011, 2011a.

Ng, N. L., Canagaratna, M. R., Jimenez, J. L., Zhang, Q., Ulbrich, I. M., and Worsnop, D. R.: Real-Time Methods for Estimating Organic Component Mass Concentrations from Aerosol Mass Spectrometer Data, Environ. Sci. Technol., 45, 910–916, doi:10.1021/es102951k, 2011b.

Ng, N. L., Herndon, S. C., Trimborn, A., Canagaratna, M. R., Croteau, P. L., Onasch, T. B., Sueper, D., Worsnop, D. R., Zhang, Q., Sun, Y. L., and Jayne, J. T.: An Aerosol Chemical Speciation Monitor (ACSM) for routine monitoring of the composition and mass concentrations of ambient aerosol, Aerosol Sci. Technol., 45, 780–794, doi:10.1080/02786826.2011.560211, 2011c.

Paasonen, P., Asmi, A., Petäjä, T., Kajos, M. K., Äijälä, M., Junninen, H., Holst, T., Abbatt, J. P. D., Arneth, A., Birmili, W., van der Gon, H. D., Hamed, A., Hoffer, A., Laakso, L., Laaksonen, A., Richard Leaitch, W., Plass-Dülmer, C., Pryor, S. C., Räisänen, P., Swietlicki, E., Wiedensohler, A., Worsnop, D. R., Kerminen, V.-M., and Kulmala, M.: Warming-induced increase in aerosol number concentration likely to moderate climate change, Nat. Geosci., 6, 438–442, doi:10.1038/ngeo1800, 2013.

Paatero, P.: The multilinear engine – a table-driven, least squares program for solving multilinear problems, including the n-way parallel factor analysis model, J. Comput. Graph. Stat., 8, 854–888, doi:10.2307/1390831, 1999.

Pandolfi, M., Querol, X., Alastuey, A., Jimenez, J. L., Jorba, O., Day, D., Ortega, A., Cubison, M. J., Comerón, A., Sicard, M., Mohr, C., Prévôt, A. S. H., Minguillón, M. C., Pey, J., Baldasano, J. M., Burkhart, J. F., Seco, R., Peñuelas, J., van Drooge, B. L., Artiñano, B., Di Marco, C., Nemitz, E., Schallhart, S., Metzger, A., Hansel, A., Lorente, J., Ng, S., Jayne, J., and Szidat, S.: Effects of sources and meteorology on particulate matter in the Western Mediterranean Basin: An overview of the DAURE campaign, J. Geophys. Res. Atmos., 119, 4978–5010, doi:10.1002/2013JD021079, 2014a.

Pandolfi, M., Ripoll, A., Querol, X., and Alastuey, A.: Climatology of aerosol optical properties and black carbon mass absorption cross section at a remote high-altitude site in the western Mediterranean Basin, Atmos. Chem. Phys., 14, 6443–6460, doi:10.5194/acp-14-6443-2014, 2014b.

Petit, J.-E., Favez, O., Sciare, J., Crenn, V., Sarda-Estève, R., Bonnaire, N., Moc˘(nik, G., Dupont, J.-C., Haeffelin, M., and Leoz-Garziandia, E.: Two years of near real-time chemical composition of submicron aerosols in the region of Paris using an Aerosol Chemical Speciation Monitor (ACSm) and a multi-wavelength Aethalometer, Atmos. Chem. Phys. Discuss., 14, 24221–24271, doi:10.5194/acpd-14-24221-2014, 2014.

Petzold, A. and Schönlinner, M.: Multi-angle absorption photometry – a new method for the measurement of aerosol light absorption and atmospheric black carbon, J. Aerosol Sci., 35, 421–441, doi:10.1016/j.jaerosci.2003.09.005, 2004.

Petzold, A., Ogren, J. A., Fiebig, M., Laj, P., Li, S.-M., Baltensperger, U., Holzer-Popp, T., Kinne, S., Pappalardo, G., Sugimoto, N., Wehrli, C., Wiedensohler, A., and Zhang, X.-Y.: Recommendations for reporting"black carbon" measurements, Atmos. Chem. Phys., 13, 8365–8379, doi:10.5194/acp-13-8365-2013, 2013.

Pey, J., Pérez, N., Castillo, S., Viana, M., Moreno, T., Pandolfi, M., López-Sebastián, J. M., Alastuey, A., and Querol, X.: Geochemistry of regional background aerosols in the Western Mediterranean, Atmos. Res., 94, 422–435, doi:10.1016/j.atmosres.2009.07.001, 2009.

Pey, J., Querol, X., Alastuey, A., Forastiere, F., and Stafoggia, M.: African dust outbreaks over the Mediterranean Basin during 2001–2011: PM_{10} concentrations, phenomenology and trends, and its relation with synoptic and mesoscale meteorology, Atmos. Chem. Phys., 13, 1395–1410, doi:10.5194/acp-13-1395-2013, 2013.

Querol, X., Alastuey, A., Lopez-soler, A., Plana, F., and Puicercus, J. A.: Daily evolution of sulphate aerosols in a rural area, northeastern Spain – elucidation of an atmospheric reservoir effect, Environ. Pollut., 105, 397–407, doi:10.1016/S0269-7491(99)00037-8, 1999.

Querol, X., Alastuey, A., Pey, J., Cusack, M., Pérez, N., Mihalopoulos, N., Theodosi, C., Gerasopoulos, E., Kubilay, N., and Koçak, M.: Variability in regional background aerosols within the Mediterranean, Atmos. Chem. Phys., 9, 4575–4591, doi:10.5194/acp-9-4575-2009, 2009.

Raatikainen, T., Vaattovaara, P., Tiitta, P., Miettinen, P., Rautiainen, J., Ehn, M., Kulmala, M., Laaksonen, A., and Worsnop, D. R.: Physicochemical properties and origin of organic groups detected in boreal forest using an aerosol mass spectrometer, Atmos. Chem. Phys., 10, 2063–2077, doi:10.5194/acp-10-2063-2010, 2010.

Ripoll, A., Pey, J., Minguillón, M. C., Pérez, N., Pandolfi, M., Querol, X., and Alastuey, A.: Three years of aerosol mass, black carbon and particle number concentrations at Montsec (southern Pyrenees, 1570 m a.s.l.), Atmos. Chem. Phys., 14, 4279–4295, doi:10.5194/acp-14-4279-2014, 2014.

Ripoll, A., Minguillón, M. C., Pey, J., Pérez, N., Querol, X., and Alastuey, A.: Joint analysis of continental and regional background environments in the western Mediterranean: PM_1 and PM_{10} concentrations and composition, Atmos. Chem. Phys., 15, 1129–1145, doi:10.5194/acp-15-1129-2015, 2015.

Robinson, A. L., Donahue, N. M., Shrivastava, M. K., Weitkamp, E. A., Sage, A. M., Grieshop, A. P., Lane, T. E., Pierce, J. R., and Pandis, S. N.: Rethinking organic aerosols: semivolatile emissions and photochemical aging., Science, 315, 1259–1262, doi:10.1126/science.1133061, 2007.

Rodríguez, S., Querol, X., Alastuey, A., Kallos, G., and Kakaliagou, O.: Saharan dust contributions to PM_{10} and TSP levels in Southern and Eastern Spain, Atmos. Environ., 35, 2433–2447, doi:10.1016/S1352-2310(00)00496-9, 2001.

Rodriguez, S., Querol, X., Alastuey, A., and Mantilla, E.: Origin of high summer PM10 and TSP concentrations at rural sites in Eastern Spain, Atmos. Environ., 36, 3101–3112, doi:10.1016/S1352-2310(02)00256-X, 2002.

Rodríguez, S., Querol, X., Alastuey, A., Viana, M.-M., and Mantilla, E.: Events affecting levels and seasonal evolution of airborne particulate matter concentrations in the Western Mediterranean., Environ. Sci. Technol., 37, 216–222, doi:10.1021/es020106p, 2003.

Salcedo, D., Onasch, T. B., Dzepina, K., Canagaratna, M. R., Zhang, Q., Huffman, J. A., DeCarlo, P. F., Jayne, J. T., Mortimer, P., Worsnop, D. R., Kolb, C. E., Johnson, K. S., Zuberi, B., Marr, L. C., Volkamer, R., Molina, L. T., Molina, M. J., Cardenas, B., Bernabé, R. M., Márquez, C., Gaffney, J. S., Marley, N. A., Laskin, A., Shutthanandan, V., Xie, Y., Brune, W., Lesher, R., Shirley, T., and Jimenez, J. L.: Characterization of ambient aerosols in Mexico City during the MCMA-2003 campaign with Aerosol Mass Spectrometry: results from the CENICA Supersite, Atmos. Chem. Phys., 6, 925–946, doi:10.5194/acp-6-925-2006, 2006.

Schaap, M., Spindler, G., Schulz, M., Acker, K., Maenhaut, W., Berner, A., Wieprecht, W., Streit, N., Müller, K., Brüggemann, E., Chi, X., Putaud, J.-P., Hitzenberger, R., Puxbaum, H., Baltensperger, U., and ten Brink, H.: Artefacts in the sampling of nitrate studied in the "INTERCOMP" campaigns of EUROTRAC-AEROSOL, Atmos. Environ., 38, 6487–6496, doi:10.1016/j.atmosenv.2004.08.026, 2004.

Seco, R., Peñuelas, J., Filella, I., Llusià, J., Molowny-Horas, R., Schallhart, S., Metzger, A., Müller, M., and Hansel, A.: Contrasting winter and summer VOC mixing ratios at a forest site in the Western Mediterranean Basin: the effect of local biogenic emissions, Atmos. Chem. Phys., 11, 13161–13179, doi:10.5194/acp-11-13161-2011, 2011.

Sicard, M., Rocadenbosch, F., Reba, M. N. M., Comerón, A., Tomás, S., García-Vízcaino, D., Batet, O., Barrios, R., Kumar, D., and Baldasano, J. M.: Seasonal variability of aerosol optical properties observed by means of a Raman lidar at an EARLINET site over Northeastern Spain, Atmos. Chem. Phys., 11, 175–190, doi:10.5194/acp-11-175-2011, 2011.

Steinbrecher, R., Smiatek, G., Köble, R., Seufert, G., Theloke, J., Hauff, K., Ciccioli, P., Vautard, R., and Curci, G.: Intra- and inter-annual variability of VOC emissions from natural and semi-natural vegetation in Europe and neighbouring countries, Atmos. Environ., 43, 1380–1391, doi:10.1016/j.atmosenv.2008.09.072, 2009.

Sun, Y., Wang, Z., Dong, H., Yang, T., Li, J., Pan, X., Chen, P., and Jayne, J. T.: Characterization of summer organic and inorganic aerosols in Beijing, China with an Aerosol Chemical Speciation Monitor, Atmos. Environ., 51, 250–259, doi:10.1016/j.atmosenv.2012.01.013, 2012.

Takahama, S., Schwartz, R. E., Russell, L. M., Macdonald, A. M., Sharma, S., and Leaitch, W. R.: Organic functional groups in aerosol particles from burning and non-burning forest emissions at a high-elevation mountain site, Atmos. Chem. Phys., 11, 6367–6386, doi:10.5194/acp-11-6367-2011, 2011.

Tiitta, P., Vakkari, V., Croteau, P., Beukes, J. P., van Zyl, P. G., Josipovic, M., Venter, A. D., Jaars, K., Pienaar, J. J., Ng, N. L., Canagaratna, M. R., Jayne, J. T., Kerminen, V.-M., Kokkola, H., Kulmala, M., Laaksonen, A., Worsnop, D. R., and Laakso, L.: Chemical composition, main sources and temporal variability of PM_1 aerosols in southern African grassland, Atmos. Chem. Phys., 14, 1909–1927, doi:10.5194/acp-14-1909-2014, 2014.

Tositti, L., Riccio, A., Sandrini, S., Brattich, E., Baldacci, D., Parmeggiani, S., Cristofanelli, P., and Bonasoni, P.: Short-term climatology of PM10 at a high altitude background station in southern Europe, Atmos. Environ., 65, 142–152, doi:10.1016/j.atmosenv.2012.10.051, 2013.

Volkamer, R., Jimenez, J. L., San Martini, F., Dzepina, K., Zhang, Q., Salcedo, D., Molina, L. T., Worsnop, D. R., and Molina, M. J.: Secondary organic aerosol formation from anthropogenic air pollution: Rapid and higher than expected, Geophys. Res. Lett., 33, L17811, doi:10.1029/2006GL026899, 2006.

Zhang, Q., Worsnop, D. R., Canagaratna, M. R., and Jimenez, J. L.: Hydrocarbon-like and oxygenated organic aerosols in Pittsburgh: insights into sources and processes of organic aerosols, Atmos. Chem. Phys., 5, 3289–3311, doi:10.5194/acp-5-3289-2005, 2005.

Zhang, Q., Jimenez, J. L., Canagaratna, M. R., Allan, J. D., Coe, H., Ulbrich, I., Alfarra, M. R., Takami, A., Middlebrook, a. M., Sun, Y. L., Dzepina, K., Dunlea, E., Docherty, K., DeCarlo, P. F., Salcedo, D., Onasch, T., Jayne, J. T., Miyoshi, T., Shimono, A., Hatakeyama, S., Takegawa, N., Kondo, Y., Schneider, J., Drewnick, F., Borrmann, S., Weimer, S., Demerjian, K., Williams, P., Bower, K., Bahreini, R., Cottrell, L., Griffin, R. J., Rautiainen, J., Sun, J. Y., Zhang, Y. M., and Worsnop, D. R.: Ubiquity and dominance of oxygenated species in organic aerosols in anthropogenically-influenced Northern Hemisphere midlatitudes, Geophys. Res. Lett., 34, L13801, doi:10.1029/2007GL029979, 2007.

Zhuang, H., Chan, C. K., Fang, M., and Wexler, A. S.: Size distributions of particulate sulfate , nitrate, and ammonium at a coastal site in Hong Kong, Atmos. Environ., 33, 843–853, doi:10.1016/S1352-2310(98)00305-7, 1999.

9

Properties and evolution of biomass burning organic aerosol from Canadian boreal forest fires

M. D. Jolleys[1], H. Coe[1], G. McFiggans[1], J. W. Taylor[1], S. J. O'Shea[1], M. Le Breton[1], S. J.-B. Bauguitte[2], S. Moller[3], P. Di Carlo[4,5], E. Aruffo[4,5], P. I. Palmer[6], J. D. Lee[3], C. J. Percival[1], and M. W. Gallagher[1]

[1]Centre for Atmospheric Science, School of Earth, Atmospheric and Environmental Science, University of Manchester, Manchester, UK
[2]Facility for Airborne Atmospheric Measurements, Bedford, UK
[3]National Centre for Atmospheric Science (NCAS), Department of Chemistry, University of York, York, UK
[4]Centre of Excellence CETEMPS, University of L'Aquila, L'Aquila, Italy
[5]Department of Physical and Chemical Sciences, University of L'Aquila, L'Aquila, Italy
[6]School of Geosciences, University of Edinburgh, Edinburgh, UK

Correspondence to: M. D. Jolleys (matthew.jolleys@manchester.ac.uk)

Abstract. Airborne measurements of biomass burning organic aerosol (BBOA) from boreal forest fires reveal highly contrasting properties for plumes of different ages. These measurements, performed using an Aerodyne Research Inc. compact time-of-flight aerosol mass spectrometer (C-ToF-AMS) during the BORTAS (quantifying the impact of BOReal forest fires on Tropospheric oxidants over the Atlantic using Aircraft and Satellites) experiment in the summer of 2011, have been used to derive normalised excess organic aerosol (OA) mass concentrations (ΔOA / ΔCO), with higher average ratios observed closer to source (0.190 ± 0.010) than in the far-field (0.097 ± 0.002). The difference in ΔOA / ΔCO between fresh and aged plumes is influenced by a change in dominant combustion conditions throughout the campaign. Measurements at source comprised 3 plume interceptions during a single research flight and sampled largely smouldering fires. Twenty-three interceptions were made across four flights in the far-field, with plumes originating from fires occurring earlier in the campaign when fire activity had been more intense, creating an underlying contrast in emissions prior to any transformations associated with aging. Changing combustion conditions also affect the vertical distribution of biomass burning emissions, as aged plumes from more flaming-dominated fires are injected to higher altitudes of up to 6000 m. Proportional contributions of the mass-to-charge ratio (m/z) 60 and 44 peaks in the AMS mass spectra to the total OA mass (denoted f_{60} and f_{44}) are used as tracers for primary and oxidised BBOA, respectively. f_{44} is lower on average in near-field plumes than those sampled in the far-field, in accordance with longer aging times as plumes are transported a greater distance from source. However, high levels of ΔO_3 / ΔCO and $-\log(NO_x / NO_y)$ close to source indicate that emissions can be subject to very rapid oxidation over short timescales. Conversely, the lofting of plumes into the upper troposphere can lead to the retention of source profiles after transportation over extensive temporal and spatial scales, with f_{60} also higher on average in aged plumes. Evolution of OA composition with aging is comparable to observations of BB tracers in previous studies, revealing a consistent progression from f_{60} to f_{44}. The elevated levels of oxygenation in aged plumes, and their association with lower average ΔOA / ΔCO, are consistent with OA loss through evaporation during aging due to a combination of dilution and chemical processing, while differences in combustion conditions throughout the campaign also have a significant influence on BBOA production and composition.

1 Introduction

The BORTAS (quantifying the impact of BOReal forest fires on Tropospheric oxidants over the Atlantic using Aircraft and Satellites) campaign was a major international research effort to improve understanding of the properties and evolution of biomass burning (BB) plumes. BB emissions form a major source of atmospheric particulate matter on a global scale, contributing around 90 % of the total primary organic aerosol (OA) (Bond et al., 2004). The radiative effects of atmospheric aerosols represent one of the major sources of uncertainty with regard to influences on climate change (Textor et al., 2006; Forster et al., 2007). Given the prominence of OA in global aerosol budgets (Zhang et al., 2007; Jimenez et al., 2009), limited understanding of BB emissions, and more specifically biomass burning organic aerosol (BBOA) emissions, forms an important component of this uncertainty. Improved projection of climate change impacts through global climate model simulation is dependent on more robust parameterisation of the constituent drivers, constrained by direct measurements. Several fundamental aspects of the BBOA lifecycle remain poorly characterised (Hallquist et al., 2009), including the conditions and processes controlling formation and the effects of transformations occurring during aging, such as gas-particle partitioning of low volatility organic compounds following photo-oxidation, heterogeneous reactions with existing OA and losses through dilution-based evaporation or volatilisation (Reid et al., 2005; Grieshop et al., 2009; DeCarlo et al., 2010; Hennigan et al., 2011). Variability at source has been shown to be extensive, in response to changes in both fuel properties and combustion conditions (McMeeking et al., 2009; Jolleys et al., 2012, 2014). The importance of secondary organic aerosol (SOA) in aging plumes is also particularly unclear. Substantial SOA formation as a result of photochemical processing has been demonstrated in laboratory experiments, increasing OA concentrations by up to a factor of 4 over several hours (Grieshop et al., 2009; Hennigan et al., 2011; Heringa et al., 2011). However, under ambient conditions the importance of SOA addition relative to primary (POA) emissions is more disputable. Despite widespread evidence for the increasing oxygenation of BBOA with aging (Capes et al., 2008; DeCarlo et al., 2008; Cubison et al., 2011; Jolleys et al., 2012), net mass enhancements are not observed consistently. The underlying causes of this variable SOA contribution, including the implications of initial OA composition, also remain ambiguous and require further refinement.

The BORTAS campaign is described in detail by Palmer et al. (2013), with an overview of measurements used within this analysis given here. BORTAS took place across several regions of Canada between the 12 July and 3 August in both 2010 and 2011, although activity during the 2010 deployment (BORTAS-A) was limited to ground-based measurements at a main ground station located at Dalhousie University in Halifax, Nova Scotia, along with ozonesonde launches from a network of seven sites across central and eastern Canada and supporting satellite observations (Parrington et al., 2012). Airborne measurements were carried out during BORTAS-B in 2011, providing all data contributing towards this study. The UK Facility for Airborne Atmospheric Measurements (FAAM) BAe-146 Atmospheric Research Aircraft (ARA) performed a total of 15 flights, including 11 dedicated science flights between the 15 and 31 July. Research flights primarily originated from Halifax and largely involved surveying areas adjacent to the Gulf of St. Lawrence and the North Atlantic. A predominant source region in northwestern Ontario (approximately 52.5° N, 93.5° W) has been identified for the majority of plumes sampled throughout BORTAS, although more disperse fires were also active in northern Alberta and the Northwest Territories (Palmer et al., 2013; Parrington et al., 2013). As the majority of plumes from fires in this region were encountered at a distance of several thousand kilometres downwind, emissions would have undergone substantial processing prior to sampling, with estimated photochemical ages between 1–11 days. A single flight to the Ontario source region also sampled active fires directly at source, providing a valuable inventory of fresh plume measurements and enabling comparison of emissions in the near and far-field. Tracks of all flights included within this analysis are shown in Fig. 1. Back trajectories for air masses encountered throughout the durations of flights B621-624 are presented by O'Shea et al. (2013a), detailing the transport pathways of plumes and agreement with active fire locations, while Taylor et al. (2014) provide air mass histories for individual plumes sampled during flight B626. Comparison of emissions of different ages is subject to potential contrasts in fire behaviour, given that each set of measurements were obtained at different stages of the campaign. Fire activity within the region peaked between the 17 and 19 July (Fig. 2), with emissions from these fires intercepted far downwind. Plumes from active fires within this region were also sampled at source on the 26 July (flight B626), representing the only measurements of fresh plumes from BORTAS. However, by this time fire activity had significantly abated, bringing about a change in combustion conditions to yield smaller, less intense fires more typically dominated by smouldering combustion (O'Shea et al., 2013a). The more intense period of fires earlier in the campaign is expected to have involved larger events with a more prominent flaming combustion phase, as indicated by the detection of pronounced smoke plumes at altitudes of up to approximately 7000 m over the North Atlantic (Palmer et al., 2013). As a result, any comparison of fresh and aged plumes during BORTAS must also account for this disparity in source conditions. While such a scenario would reduce the potential to evaluate the continuous evolution of smoke plumes from source into the ambient atmosphere, and prevent direct comparison of near and far-field plumes derived from similar

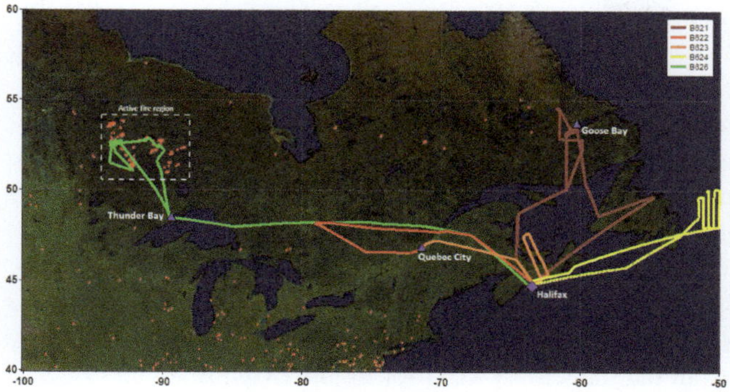

Figure 1. Flight tracks for flights B621-624 and B626 during BORTAS, overlain on a MODIS accumulated 10 day fire map for eastern Canada during the period 20–29 July 2011. Images courtesy of MODIS Rapid Response Project at NASA/GSFC.

combustion conditions, it also provides a baseline for conditions at source.

2 Background

2.1 Instrumentation and measurements

A wide array of instrumentation performing particulate and gas phase measurements were deployed throughout BORTAS. This study focuses primarily on the analysis of OA mass and composition data obtained from an Aerodyne Research Inc. compact time-of-flight aerosol mass spectrometer (C-ToF-AMS; Drewnick et al., 2005; Canagaratna et al., 2007). The AMS provides highly time-resolved mass concentrations of sub-micron, non-refractory aerosol, and a broad chemical characterisation across a complete range of constituent ion mass-to-charge ratios (m/z). Operation of the AMS, including calibration and necessary correction factors, during aircraft deployment (Bahreini et al., 2003) and specifically onboard the BAe-146 (Crosier et al., 2007; Morgan et al., 2009; Taylor et al., 2014) have been described in detail. Refractory black carbon (BC) was measured using a Droplet Measurement Technologies single particle soot photometer (SP-2; Schwarz et al., 2006; Taylor et al., 2014). Although analysis of the chemical and optical properties of single BC particles was not performed as part of this study, mass concentrations in smoke plumes, particularly in relation to OA concentrations, were used as a means of evaluating the proportional contributions of different combustion phases. A range of gas phase measurements were undertaken on the BAe-146, including species used as tracers for both primary emissions and photochemical processing. CO mixing ratios were measured with an Aerolaser AL5002 UV fluorescence analyser and O_3 by a Thermo Scientific TEi49C UV photometric analyser as part of the standard complement of instrumentation for BAe-146 science flights. Additional instrumentation included a chemical ionisation mass spectrometer

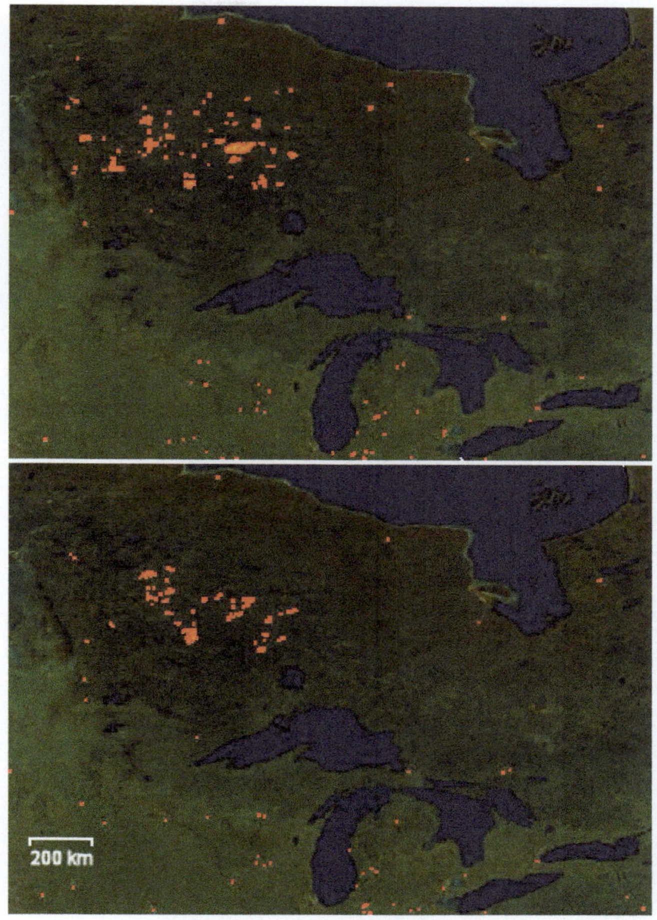

Figure 2. MODIS fire maps showing the reduction in fire activity in north-western Ontario between the 10 day periods of 10–19 July 2011 (top) and 20–29 July 2011 (bottom).

(CIMS; Nowak et al., 2007; Le Breton et al., 2012) providing real-time measurements of HCN, which is widely used as a tracer for BB emissions given that vegetation fires constitute its primary global source (Li et al., 2000; Sinha et al., 2003; Yokelson et al., 2007). NO_x ($NO + NO_2$) and NO_y (NO_x oxidation products, including HNO_3 and N_2O_5) act as important tracers for oxidation in aging plumes, and were measured respectively by an Air Quality Design Inc. chemiluminescence NO_x analyser and by thermal dissociation-laser induced fluorescence (TD-LIF; Di Carlo et al., 2013). The assembly of gas phase measurements used within this analysis was completed by CO_2 mixing ratios from a Los Gatos Research Inc. cavity enhanced absorption spectrometer-based fast greenhouse gas analyser (FGGA; O'Shea et al., 2013b). Aerosol size distributions in the range 20–350 nm were obtained from a scanning mobility particle sizer (SMPS), with integrated distributions over this size range used as an approximation of particle number concentration.

2.2 Data selection

Measurements from 5 BORTAS flights (B621-B624 and B626) were included in this analysis. Flight B626 provided the only measurements of fresh BB plumes throughout the campaign, with all other flights sampling air masses downwind of the source region at ages of several days. Data were screened in order to isolate emissions with a biomass burning influence, resulting in a total number of 26 valid plume interceptions (3 fresh and 23 aged) across the 5 flights. Screening was performed using the guidelines proposed by Capes et al. (2008) and Jolleys et al. (2012) based upon minimum ΔCO (the excess CO concentration above background levels) and absolute number concentrations. Respective thresholds of 20 ppb and 2000 cm^{-3} were applied for ΔCO and number concentration. Background concentrations for CO and other species were calculated for each flight according to minimum observed concentrations, which were applied to all measurements throughout the full vertical extent of sampling, given the limited variation in background concentrations with altitude. CO_2 was the only exception, with high variability both in and out of plume making it difficult to define an appropriate background concentration. As a result, only absolute concentrations are reported for CO_2, as opposed to excess values. A threshold of 0.003 was used for f_{60}, representing the ratio of levoglucosan-like species, which correspond to the m/z 60 peak in the AMS mass spectra (Schneider et al., 2006; Alfarra et al., 2007), to the total OA mass. This threshold is based upon observed background levels of f_{60} in OA emissions from urban and biogenic sources where BB influences are absent (Cubison et al., 2011; Aiken et al., 2009; DeCarlo et al., 2008). Levoglucosan and other anhydrous sugars such as mannosan and galactosan have been shown to be strongly associated with primary BB emissions (Simoneit et al., 1999; Iinuma et al., 2007; Sullivan et al., 2008; Lee et al., 2010). All data were also averaged to the temporal resolution of the AMS (~8 second time step on average) to enable direct comparison of different species.

Alternative screening procedures for BB influences have been applied throughout separate analyses of BORTAS data (Palmer et al., 2013). Concentrations of trace gases primarily produced by fire sources, including HCN and CH_3CN, are commonly used as indicators for BB plumes (Li et al., 2000; Yokelson et al., 2007; Crounse et al., 2009; Yokelson et al., 2009; Akagi et al., 2011). A scheme using a HCN concentration threshold of six times the standard deviation (6σ) has been used during BORTAS in an analysis of high sensitivity 1 Hz CIMS measurements and their consistency with CO and CH_3CN concentrations (Le Breton et al., 2013). However, as many previous data sets do not include HCN measurements, a screening procedure using only ΔOA, ΔCO and number concentration data has been applied here, so that the approach can be used consistently across a broader range of data. This approach performs well when compared to the Le Breton et al. method, producing similarly strong correlations between ΔHCN and ΔCO for flights B621, B622, B624 and B626 ($R^2 = 0.64, 0.52, 0.84$ and 0.93) as the 6σ technique ($R^2 = 0.83, 0.46, 0.82$ and 0.81). HCN was not measured during B623, preventing comparison of classification schemes for this flight. Several flights carried out later in the campaign (B628-B630) also measured highly aged plumes with a photochemical age of up to 11 days (Palmer et al., 2013). However, correlations between ΔOA and ΔCO throughout these flights were exceptionally weak, yielding R^2 values consistently well below 0.1, contrasting with values in the range 0.39–0.74 for flights B621-B624 and B626. These weak correlations from later flights suggest that sampled air masses lack a common emission source and instead represent extensive mixtures of different plumes following dispersion, or that emissions have been differentially processed to the extent that representative properties can no longer be distinguished. As a result, data from these flights were omitted from this analysis.

Throughout this study, extensive use is made of normalised measurements as a means of assessing the relative abundances of different species. Normalising to a co-emitted, non-reactive tracer such as ΔCO provides an emission ratio (ER) when calculated at source. Normalised excess mixing ratios (NEMR) are used to represent these values for any other point in a plume away from source along a Lagrangian trajectory, and account for the effects of dispersion as concentrations in plumes decrease through dilution. These ratios can also be used as a marker for potential SOA formation, as the longer atmospheric lifetime of CO (~1 month) relative to that of OA (on the order of several weeks) makes it likely that any enhancement of the ratio between the two species will be a result of the addition of OA, rather than increased removal of CO in isolation.

3 Results and discussion

3.1 Spatial and temporal variability in BB emissions properties

Measurements of OA in BB plumes during the BORTAS flights included within this analysis encompassed a wide range of ages, from at source to up to 5 days after emission. The extent of this diversity in age contributed to a high level of variability in plume properties, both across separate research flights and between individual plumes encountered in different periods of the same flight. Excess OA concentrations measured in-plume ranged from close to zero to around 180 $\mu g\,m^{-3}$, with maximum ΔCO concentrations approaching 1000 ppb. Vertical profiles of both species are shown in Fig. 3, revealing an overall increase in concentrations throughout the boundary layer to a peak at around 2000 m, before declining to background levels through the free troposphere. Significant elevations in both ΔOA and ΔCO occurred close to ground level, most likely as a result of influ-

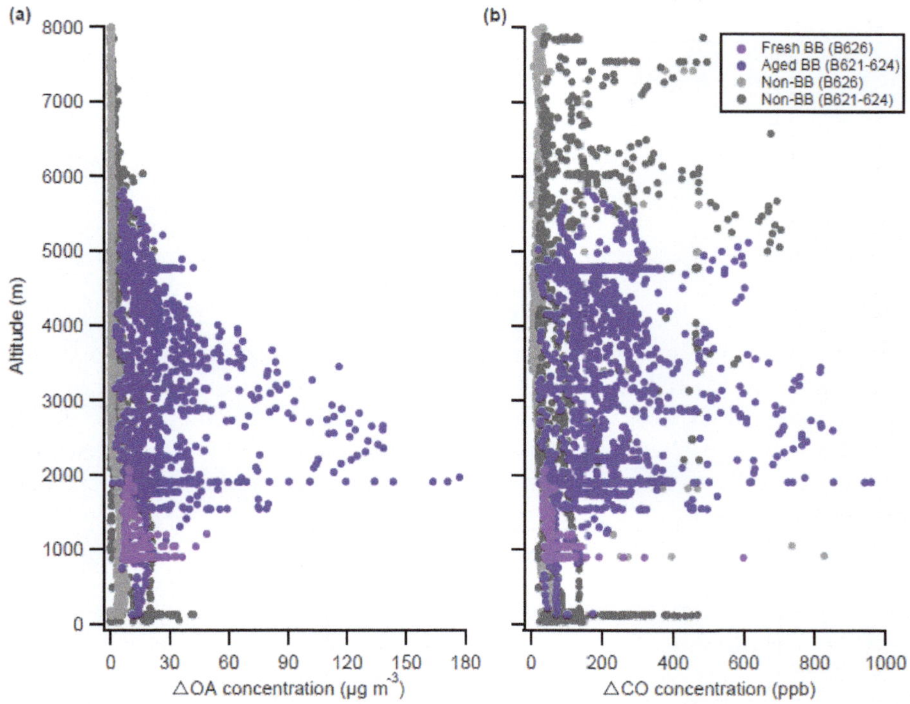

Figure 3. Vertical profiles of **(a)** ΔOA and **(b)** ΔCO in fresh and aged plumes, together with concentrations in air masses free from the influence of biomass burning.

ence from local sources. The observed decrease in concentration with altitude is more marked for ΔOA, which returns to background levels by 6000 m. Variability in ΔCO is much greater than ΔOA at higher altitudes. ΔCO concentrations of up to 800 ppb were observed between 5000–8000 m, whereas ΔOA did not exceed $30\,\mu g\,m^{-3}$. This disparity is attributed to the removal of OA from plumes encountered during flight B622 (20 July) by precipitation prior to sampling following advection through clouds, as corroborated by meteorological observations and back trajectory models (Griffin et al., 2013; Taylor et al., 2014). Wet deposition of aerosol reduced ΔOA to background levels, while ΔCO concentrations remained elevated to similar levels as observed at lower altitudes. Despite their biomass burning origin, the absence of OA and BC from these plumes resulted in their omission from this analysis.

The change in combustion conditions between different periods of BORTAS is reflected in the contrast between loadings of particulate and gas-phase species. Concentrations of the majority of sampled species in aged plumes during flights B621-B624, including ΔOA, CO, CO_2 and BC, consistently exceeded those at source from B626, irrespective of the effect of dilution as plumes dispersed into the ambient atmosphere. During flight B626, ΔOA peaked at around $50\,\mu g\,m^{-3}$, with concentrations in more aged plumes exceeding this level by a factor of 3.6. ΔCO concentrations were also significantly elevated in aged plumes relative to fresh emissions. The contrast in properties between plumes of different ages is likely to be primarily affected by a change in the size and intensity of fires, rather than combustion phase alone, given the stronger association of both OA and CO production with predominantly smouldering combustion in the latter stages of fire evolution (Reid et al., 2005).

While the higher concentrations identified in aged plumes may be influenced to some extent by contributions from SOA, initial indications from calculated ΔOA / ΔCO ratios suggest this contribution did not provide any net increase in OA loadings. Figure 4 shows ΔOA / ΔCO for all five analysed BORTAS flights, with average values determined from the gradient of linear least squares regressions. Using this approach reveals that the average ΔOA / ΔCO close to source (0.190 ± 0.010, where uncertainty represents the standard deviation in the fit) exceeds that for aged plumes (0.097 ± 0.002) by around 50 %, with an overall campaign average of 0.104 ± 0.003. Average ratios for individual flights sampling aged emissions range from 0.056 ± 0.003 (B624) to 0.114 ± 0.003 (B622), giving an overall range of 0.058. The level of average ΔOA / ΔCO for fresh emissions from boreal forest fires during BORTAS falls between the upper extent derived from the eucalypt forests of northern Australia during the Aerosol and Chemical Transport in Tropical Convection (ACTIVE) campaign (0.329), and lower ratios from several other campaigns where OA enhancements were comparatively reduced (0.019–0.065; Jolleys et al., 2012). Average ΔOA / ΔCO from aged plumes during BORTAS was again within the range identified from

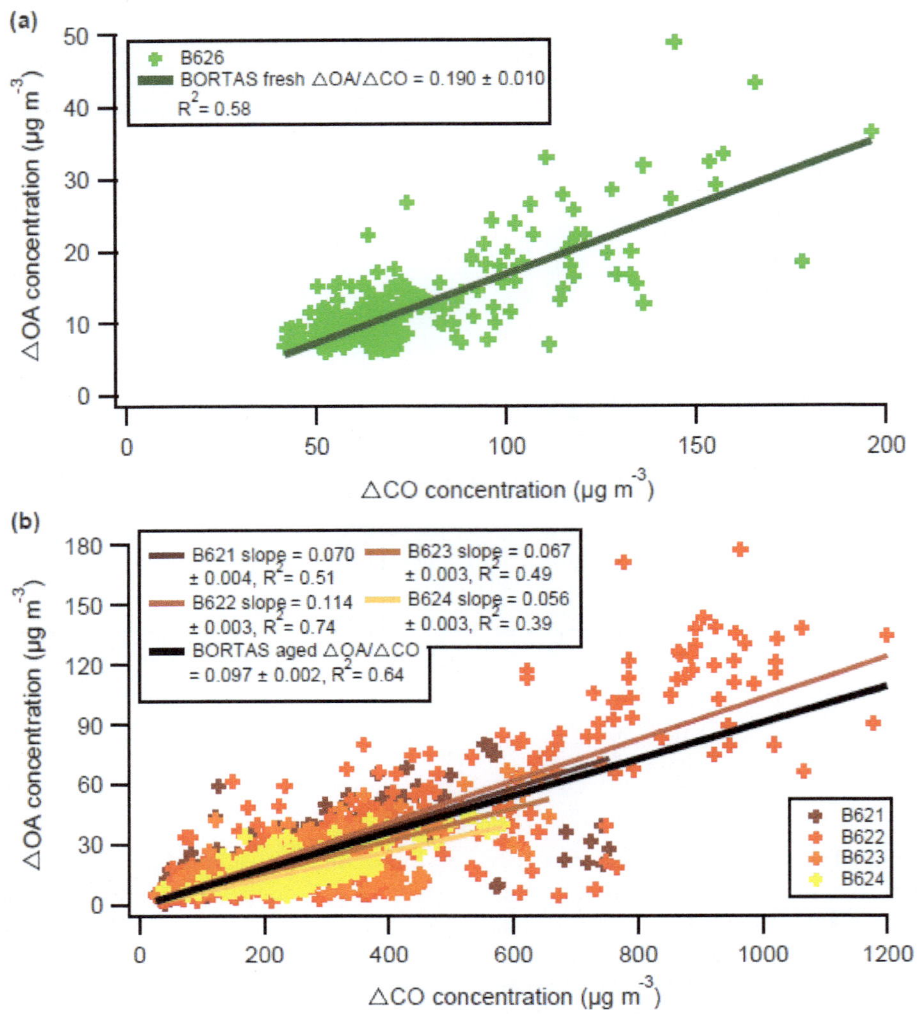

Figure 4. ΔOA vs. ΔCO for **(a)** fresh and **(b)** aged plumes. Coefficients are for linear regressions, from which average ΔOA / ΔCO ratios are derived, with uncertainties of $\pm 1\sigma$.

previous field observations, although with closer proximity to ratios from the lower extent of the observed range, including aged boreal forest fire plumes sampled during the Arctic Research of the Composition of the Troposphere from Aircraft and Satellites (ARCTAS) campaign (Hecobian et al., 2011). The extent of variability amongst aged emissions during ACTIVE also exceeded that observed during BORTAS, with flights throughout the former campaign sampling plumes from fires in a number of different source regions. However, analysis of ERs from vegetation fires performed under laboratory conditions during the second Fire Lab At Missoula Experiment (FLAME II) also revealed extensive variability in ΔOA / ΔCO directly at source, even amongst single plant species (Jolleys et al., 2014). The single source region from which BORTAS plumes originated could therefore still be expected to give rise to significantly contrasting ΔOA / ΔCO, while the effects of atmospheric processing during transportation provide further perturbation of initial ERs. Flight-average ΔOA / ΔCO decreased progressively for aged emissions as the distance from source at which plumes were intercepted increased, with B622 performing a transit between Halifax and Quebec City, and B624 primarily sampling plumes over the North Atlantic off the eastern coast of Nova Scotia and Newfoundland, suggestive of OA losses during aging in these plumes from predominantly flaming sources.

3.2 Tracers for combustion conditions

While the evolution of ΔOA / ΔCO in aging plumes would appear to be strongly influenced by the effects of atmospheric processing, source conditions remain a critical factor in controlling OA production. Contrasts in ΔOA / ΔCO between fresh and aged OA are accompanied by varying properties with respect to the location and composition of plumes. Proportional contributions of OA mass fragment marker species differ between near and far-field measurements. f_{60} represents the prevalence of primary combustion products such as levoglucosan and is used as an indicator for fresh BB emissions (Schneider et al., 2006; Alfarra et al., 2007). Conversely, f_{44} is associated with the CO_2^+ ion derived from more aged OA as hydrocarbon fragments are oxidised to form organic acids (Zhang et al., 2005; Aiken et al., 2008), although m/z 44 is also a constituent of fresh smoke and has been shown to be significantly elevated at source, dependent on combustion conditions (Weimer et al., 2008; Jolleys et al., 2014). While strongly associated with saturated hydrocarbon fragments, m/z 43 can also originate from oxidised compounds such as aldehydes and ketones (Alfarra et al., 2004). Large contributions of m/z 43 have been observed within the mass spectra of OA during the laboratory combustion of a range of biomass fuels, typically accounting for a greater proportion of total OA mass than any other detected fragments (Schneider et al., 2006). This dominance of m/z 43 above m/z 44 amongst even compounds with high oxygen contents suggests the former can be produced preferentially during the fragmentation of oxygenated molecules, and as such f_{43} may prove to be an appropriate indicator of OA oxygenation at source.

Variations in the average proportions of m/z 43, 44 and 60 in OA are widely observed between fresh and aged plumes throughout BORTAS, emphasising the contrasting properties of aerosol of different ages. Mean f_{44} for B626, which comprised the only measurements of fresh OA during BORTAS, was lower than all other flights at 0.086 ± 0.014, with mean values for B621-B624 ranging from 0.104 to 0.139. This trend between the near and far-field is consistent with observations of boreal forest fire plumes during ARCTAS, where f_{44} was shown to increase as a function of plume transport time (Cubision et al., 2011; Hecobian et al., 2011). f_{60} was also shown to decrease concurrently with increasing f_{44} during ARCTAS, as a result of the oxidation of primary levoglucosan-type species with aging. However, mean f_{60} for BORTAS flight B626 was also amongst the lowest throughout the campaign at 0.007 ± 0.004. Averages were higher for B621-B623 (0.010–0.017), although B624 provided the lowest f_{60} with a mean of 0.005 ± 0.001.

While the higher mean f_{44} observed in the far-field is likely to primarily result from more extensive oxidation of OA after longer periods of aging, the transition to more smouldering-dominated combustion prior to sampling of near-field plumes could also have influenced observed changes in composition. Elevated levels of f_{60} in aged plumes are indicative of such an effect, as m/z 60 would be expected to constitute a greater proportion of fresh OA, given its typical progressive depletion through oxidation (Cubison et al., 2011). However, the relationships between f_{44}, f_{60} and combustion phase are known to be complex and subject to considerable uncertainty. Weimer et al. (2008) showed f_{60} to be more strongly associated with the initial flaming phase of combustion in wood burners used for domestic heating, while f_{44} was higher during the later stages of the burning process when smouldering combustion dominated. These trends are attributed to changes in combustion behaviour and the consumption of different fuel components at each stage of the fire. In contrast, Gao et al. (2003) reported significantly elevated levoglucosan concentrations from smouldering fires in southern Africa, and severe depletion in emissions from flaming fires. Furthermore, Lee et al. (2010) reported overall similarity in f_{60} across flaming and smouldering phases for open biomass fires carried out in a laboratory setting as part of FLAME II, while the ratio of levoglucosan to total organic carbon in filter samples from the same experiment shows a dependence on the fuel component burned (Sullivan et al., 2008). Although both f_{44} and f_{60} were more frequently at a maximum during flaming combustion in FLAME II burns (Jolleys et al., 2014), differences between phases were more pronounced for f_{44}, with less variation amongst f_{60}. This behaviour is expected to result from greater fire intensity during flaming combustion, although the specific effects of increased intensity on OA composition through changing oxygen availability remain unclear.

Further indication of a shift in combustion phase is provided by the differences in f_{43} between fresh and aged plumes, for which respective mean values were 0.123 ± 0.013 and 0.088 ± 0.012. The low f_{44} and f_{60} for fresh OA suggest a dominance of smouldering fires, in agreement with the trends identified by Jolleys et al. (2014). Additional variations in plume properties appear to substantiate an association between f_{43} and smouldering combustion, including the correlation between periods of high f_{43} (>0.1) and low $\Delta BC / \Delta OA$ (<0.02) in both fresh and aged plumes, with production of BC expected to be at a maximum during flaming combustion (Reid and Hobbs, 1998). Absence of a prominent flaming phase close to source is also corroborated by very low BC mass loadings, and reduced $\Delta BC / \Delta OA$ relative to aged emissions, while the elevated $\Delta OA / \Delta CO$ from these fires is consistent with the enhanced OA production typical of smouldering combustion (Yokelson et al., 1997). The lower $\Delta OA / \Delta CO$ and f_{43}, but higher f_{44} and f_{60}, for aged OA would therefore be expected to derive from more intense, flaming-dominated combustion, which would also account for the significantly higher concentrations observed for ΔOA, ΔBC, ΔCO and ΔHCN despite plumes being progressively diluted over several days.

3.3 Effects of combustion conditions on vertical distributions

Altitudinal variations in plume composition further emphasise the importance of combustion conditions as a control on BB emissions and their propagation within the atmosphere. Profiles of aged plumes during BORTAS shown in Fig. 5 highlight the shift in properties between the upper and lower troposphere, with higher altitude plumes more typical of flaming combustion. The $\Delta BC / \Delta OA$ ratio is used as an indicator for the comparative contributions from flaming and

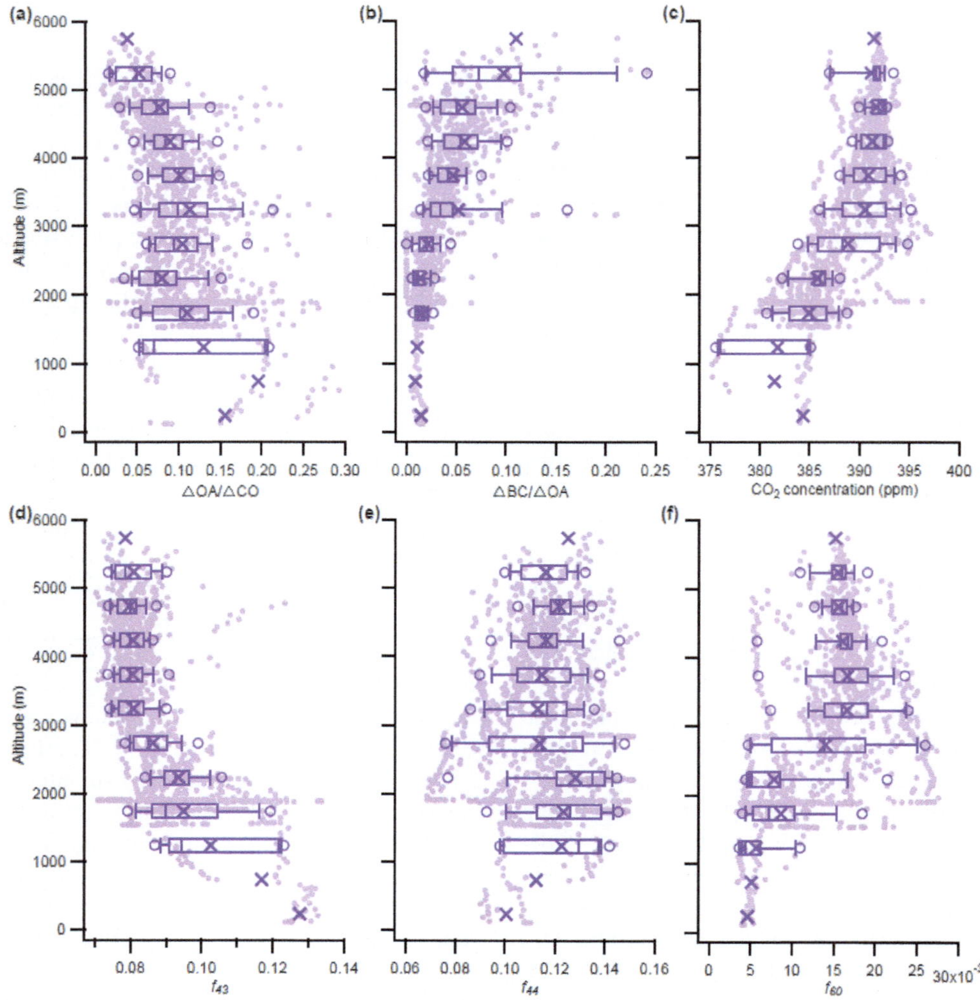

Figure 5. Vertical profiles of **(a)** $\Delta OA / \Delta CO$, **(b)** $\Delta BC / \Delta OA$, **(c)** CO_2, **(d)** f_{43}, **(e)** f_{44} and **(f)** f_{60} in aged plumes. Circles represent the 5th and 95th percentiles, vertical lines the 10th, 25th, 50th, 75th and 90th percentile, with crosses denoting mean values in each 500 m altitudinal bin.

smouldering combustion phases to the overall BB particulate loading (e.g. Grieshop et al., 2009), and is shown to increase through successive 500 m bins from 0.015 ± 0.003 at 500 m to 0.110 ± 0.055 at 6000 m, with the interquartile range increasing from 0.006 (2000 m) to 0.068 (5500 m). In contrast, $\Delta OA / \Delta CO$ decreases over the same range, revealing the stark contrasts in plume composition at different altitudes and the apparent influence of fire properties at source. Mean $\Delta OA / \Delta CO$ stands at 0.155 ± 0.061 and 0.196 ± 0.103 for the two bins closest to the surface, declining to 0.038 ± 0.015 at 6000 m. The interquartile range also decreases from 0.147 to 0.044 between 1500 and 5500 m, reflecting an overall reduction in variability with altitude.

With the exception of a few isolated points, $\Delta BC / \Delta OA$ only rises above 0.02 in higher plumes where f_{43} is below 0.09, and remains consistently low when f_{43} is above this level (Fig. 6). This trend is in part due to the greater production of BC under flaming conditions, as reflected by corresponding distributions of high ΔBC mass concentration, f_{60} and CO_2 (Fig. 6a–c). Conversely, plumes sampled at lower altitudes exhibit characteristics more strongly associated with smouldering combustion (Fig. 6d), and are comparable to fresh plumes with regards to low levels of $\Delta BC / \Delta OA$ and f_{60}, and high f_{43}. Weaker convection resulting from smouldering fires limits vertical transportation, retaining plumes within the boundary layer (Andreae et al., 1996; Warneke et al., 2006; Burling et al., 2011). The presence of flaming-derived emissions at higher altitudes alludes to an elevated injection height resulting from increased buoyancy and pyroconvection (Fromm et al., 2005; Damoah et al., 2006) driven by more intense fires earlier in the BORTAS campaign period (Fig. 2). A similar dependence on combustion phase has previously been observed for the altitudinal distribution of different combustion products from boreal forest fires during ARCTAS (Kondo et al., 2011).

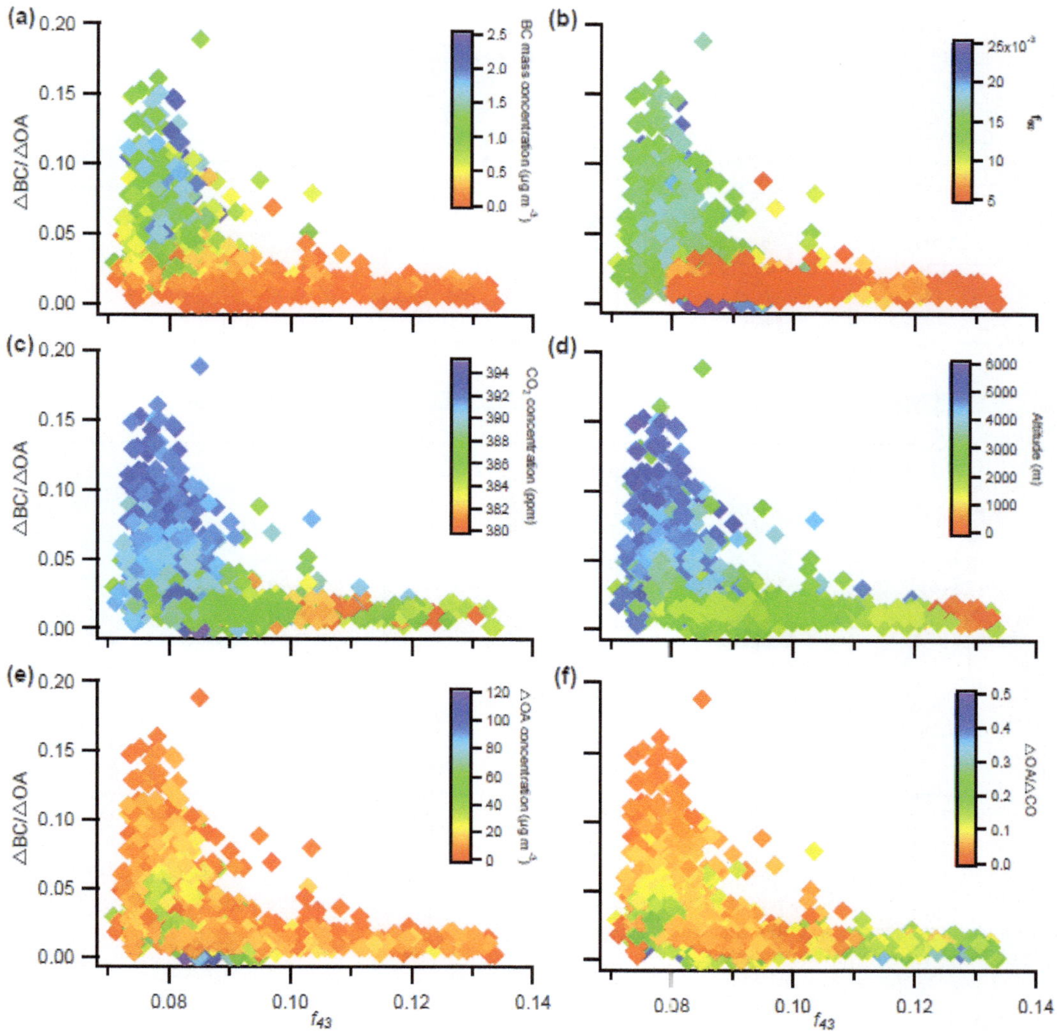

Figure 6. $\Delta BC / \Delta OA$ vs. f_{43} for aged emissions. Data points are coloured to show relationships with **(a)** ΔBC, **(b)** f_{60}, **(c)** CO_2, **(d)** altitude, **(e)** ΔOA and **(f)** $\Delta OA / \Delta CO$.

The altitudinal trends identified for $\Delta OA / \Delta CO$ and $\Delta BC / \Delta OA$ also show broad agreement with those of f_{43} and f_{60} respectively, with mean values for the former decreasing from 0.128 ± 0.006 to 0.078 ± 0.003 and latter increasing from 0.005 ± 0.001 to 0.015 ± 0.002. The directly opposing profiles of f_{43} and CO_2 (Fig. 5c–d), along with the correlation of increased f_{60} with CO_2 and ΔBC mass at high altitudes (Fig. 7h–i), further underline the importance of initial combustion conditions for aged emissions. Minimum CO_2 concentrations within aged plumes were around 375 ppm, representing a minimal elevation above typical background levels for boreal Canadian forest environments (Vay et al., 2011, Higuchi et al., 2003). Although the distribution of CO_2 clearly reflects the influence of the biosphere closer to the surface through uptake in photosynthesis, expected source profiles also appear to be largely conserved, further corroborated by the sustained correlation between periods of high f_{43} and low CO_2, and vice versa, relative to levels throughout the rest of the aged plumes.

3.4 Aging as a driver for plume variability

Despite the apparent influence of combustion conditions on the vertical distribution and composition of aged emissions, the effects of transformations associated with atmospheric processing cannot be entirely discounted. Certain contrasting properties between emissions of different ages could also be less dependent on source conditions and more strongly influenced by processing throughout plume evolution. Differences between fresh and aged plumes in the respective relationships of total ΔOA loadings, and those normalised to ΔCO, with a number of tracers highlight the combined effects of source conditions and processing, and their changing influence with aging. Both ΔOA and ΔCO concentrations show a negative correlation with f_{44} (e.g. Fig. 7a–b) and

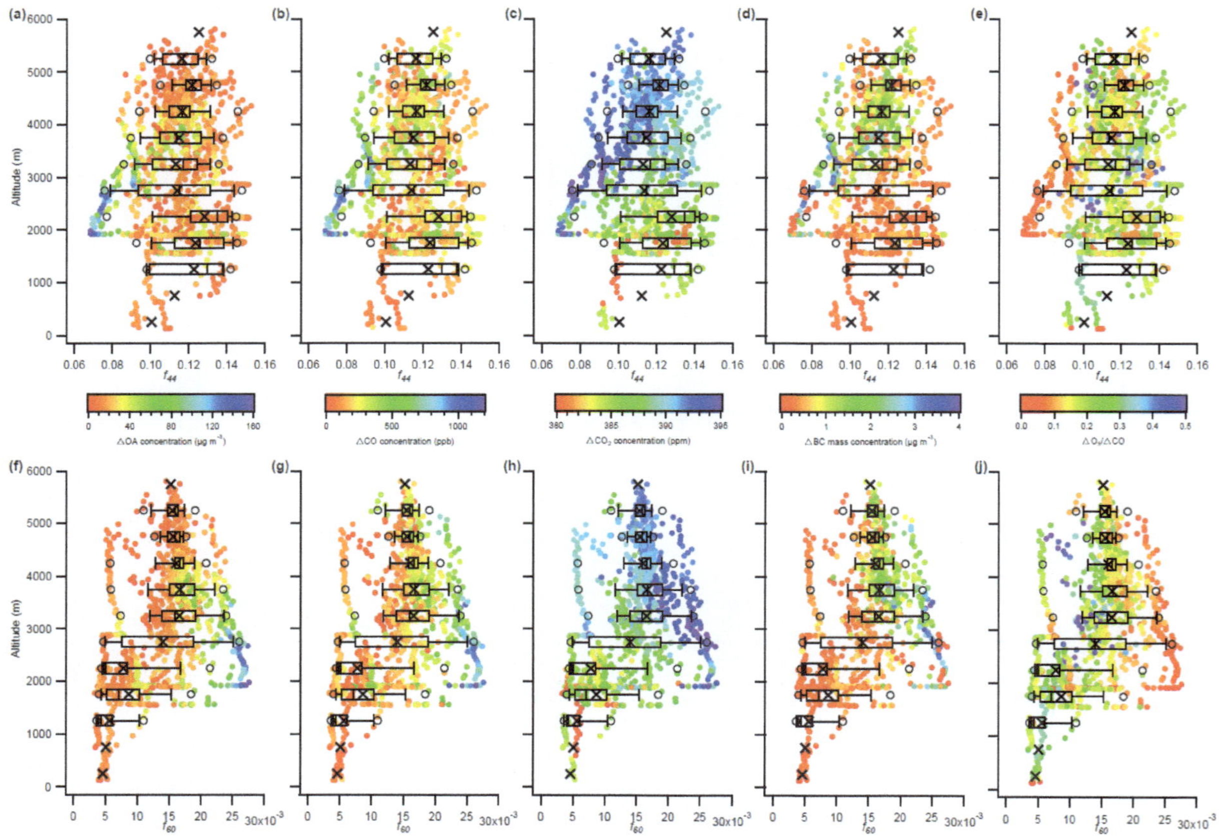

Figure 7. Vertical profiles of **(a–e)** f_{44} and **(f–j)** f_{60} in aged plumes. Data points are coloured by **(a + f)** ΔOA, **(b + g)** ΔCO, **(c + h)** CO_2, **(d + i)** ΔBC and **(e + j)** ΔO_3 / ΔCO. Circles represent the 5th and 95th percentiles, vertical lines the 10th, 25th, 50th, 75th and 90th percentile, with crosses denoting mean values in each 500 m altitudinal bin.

positive correlation with f_{60} (e.g. Fig. 7f–g). Furthermore, when binning concentrations by f_{44}, maximum binned ΔOA and ΔCO both coincide with minimum f_{44} and vice versa. These overriding trends remain consistent for emissions of all ages, although the nature of the relationship changes in each case. Linear relationships appear consistently for ΔOA ($R^2 = 0.51$ and 0.80 with f_{44} and f_{60} respectively) and ΔCO ($R^2 = 0.23$ and 0.49) in fresh emissions. The same relationships also persist to an extent for ΔOA / ΔCO in fresh plumes ($R^2 = 0.42$ and 0.47). However, at a greater distance from source, correlations for ΔOA and ΔCO are consistently below 0.3, while there is no relationship between ΔOA / ΔCO and either f_{44} or f_{60}.

In addition to providing a tracer for source profiles in aged BB emissions, ΔBC / ΔOA can also be used as an indicator for OA processing. Observations of increasing ΔBC / ΔOA with aging have previously been attributed to the loss of OA mass through evaporation (Liousse et al., 1995). Similar behaviour has also been proposed as a possible cause of the overall reduction in ΔOA / ΔCO between BB plumes in the near and far-field throughout several campaigns across different global regions (Jolleys et al., 2012). Any decrease in ΔBC / ΔOA could therefore be considered a product, to a certain degree, of the addition of secondary organic mass from either the processing of BBOA or external sources. Measurements performed at lower altitudes (< 2000 m) during flights B621 and B622 provide possible evidence to support such an effect. Both ΔBC mass and f_{60} remain low during these periods, at less than 0.5 μg m^{-3} and 0.01 respectively, consistent with wider observations of smouldering fire emissions at low altitude during BORTAS. However, ΔCO concentrations are also diminished and consistently below 100 ppb, relative to an average of 200 ppb for aged plumes, while ΔOA concentrations are comparable to levels in higher altitude, flaming-type plumes ($\sim 20\,\mu$g m^{-3}). These trends, which diverge from the expected characteristics for emissions of this origin, are further compounded by high ΔO_3 / ΔCO (> 0.2), indicative of an elevated level of oxygenation and photochemical activity (Mason et al., 2001; Parrington et al., 2013). Formation of SOA from biogenic precursors has previously been observed in the forests of Ontario (Slowik et al., 2010). These SOA events were also characterised by ΔOA / ΔCO levels far in excess of those derived for BB emissions during the same study. In accordance with this trend, ΔOA / ΔCO within the low altitude plumes in B621 and B622 were consistently above the average for aged emissions (0.097), reaching as high as ~0.4. There is subsequently considerable evidence to support biogenic SOA as a potential contributor to the OA burden during BORTAS, which could provide further enhancement of ΔOA / ΔCO as demonstrated by the Slowik et al. (2010) Ontario study. Whilst the further properties of aged plumes discussed here would suggest this effect is isolated and limited in its overall impact, it presents a further source of uncertainty for any attempts to develop parameterisations for the contribution of forest fires to regional and global OA budgets.

Although f_{60} displays a level of consistency with flaming combustion products in upper troposphere plumes during BORTAS, and increases on average with increasing altitude, further trends oppose the expected relationships for different combustion phases. Maximum concentrations of ΔOA, ΔBC, ΔCO and CO_2 all coincide with high f_{60} (0.025–0.300) and show a reduction as f_{60} decreases (Fig. 7f–i). Overall correlations between each species and f_{60} are all positive, albeit with varying fit coefficients. R^2 values were highest for CO_2 and ΔBC (0.52 and 0.47), reflecting their stronger associations with flaming combustion (Crutzen and Andreae, 1990; Reid et al., 2005). Correlations with ΔOA and ΔCO were weaker ($R^2 = 0.28$ and 0.23), as would be expected given production of each is greatest during the smouldering phase (Ferek et al., 1998; Andreae and Merlet, 2001; Gao et al., 2003). While trends with ΔO_3 / ΔCO show f_{60} to decrease with aging (Fig. 7j), the underlying relationships identified with all other species suggest f_{60} may prove to be a more resilient tracer for overall plume intensity rather than combustion conditions at long aging times. However, Petzold et al. (2007) demonstrated export efficiencies of up to 90 % for BC following intercontinental transport of boreal forest fire plumes. In the absence of significant removal through wet deposition, ΔBC / ΔCO in plumes encountered at altitudes above 4 km remained consistent with typical source values, indicating that mixing of emissions can be suppressed where fire intensity is sufficient to generate elevated injection heights. Conversely, the weaker convection associated with smouldering combustion may lead to emissions being retained within the boundary layer, contributing to the typically higher levels of ΔOA / ΔCO and f_{43} at low altitudes.

3.5 Tracer evolution during BORTAS

The progression from f_{60} to f_{44} can provide a useful metric to assess the evolution of OA composition with aging. Figure 8 shows the nature of this progression for both fresh and aged OA, together with further trends with several additional parameters. A strong linear relationship ($R^2 = 0.72$) is identified for emissions close to source. However, these observations comprised measurements of three separate periods of flight B626, and reveal a clear discrepancy for one of these periods. Measurements performed further to the east of the source region, on a transect from approximately 52.3° N, 90.0° W to 52.8° N, 91.3° W (a "downwind" plume) yielded higher f_{44} than any other fresh plumes, with f_{60} not exceeding 0.045. The two remaining sets of plumes ("source" plumes) were both encountered within a region where active fires were present at around 52.4–52.8° N, 93.0–93.7° W. Despite sampling taking place roughly two hours apart, and over a slightly different geographical extent, the f_{44}/f_{60} relationship remains highly consistent across all "source" plumes, with an R^2 of 0.82. The higher levels of f_{44} and absence of a trend with f_{60} in the "downwind" plume indicate OA is more heavily oxidised than in the fresher "source" plumes. This contrast in oxygenation is linked to other changes between plumes, including apparent photochemical age, which in this instance is represented by the $-\log(NO_x / NO_y)$ ratio (Kleinman et al., 2008; DeCarlo et al., 2008). Levels of the ratio are significantly elevated in the "downwind" plume from B626 (Fig. 8f, left panel), with an average of 1.45 ± 0.43, exceeding the mean value of 1.09 ± 0.29 for highly aged plumes sampled during flights B621-B624. CO_2 concentrations during this period are also higher than for the remainder of B626 (Fig. 8b, left panel), with an average of 378.6 ± 0.6 ppm compared to 375.0 ± 1.3 ppm closer to source. The $-\log(NO_x / NO_y)$ photochemical clock is also shown to increase throughout "source" plumes, with f_{44} and f_{60} changing in a manner consistent with the increasing oxidation of OA, and is further corroborated by a trend of increasing ΔO_3 / ΔCO (Fig. 8e, left panel). However, these changes also coincide with a trend of decreasing ΔOA / ΔCO (Fig. 8c, left panel), belying the expected addition of OA mass resulting from increasing oxygenation as semivolatile products condense to the particle phase. Average ΔOA / ΔCO is similar for the two "source" plumes (0.165 ± 0.042 and 0.180 ± 0.045), but is lower in the more photochemically aged "downwind" plume (0.114 ± 0.015). It is difficult to speculate on the significance of any link between a higher rate of oxidation and an overall reduction in ΔOA / ΔCO given the continuing uncertainty regarding the processes affecting OA in aging BB plumes. Yokelson et al. (2009) reported that ΔOA / ΔCO increased by a factor of 2.3 over a period of 1.4 h for plumes from fires in the Yucatán region of Mexico, coinciding with a comparably high f_{44}/f_{60} gradient to that in the "downwind" section of B626. Conversely, an increase in f_{44} has also been shown in conjunction with stable or even decreasing levels of ΔOA / ΔCO (Capes et al., 2008; Cubison et al., 2011; Akagi et al., 2012; Jolleys et al., 2012), suggesting OA loss through evaporation has an equally important effect throughout plume evolution.

Linear relationships between f_{44} and f_{60} are weaker for more aged plumes sampled at a greater distance downwind, with an overall R^2 value for all plumes of 0.44 and individual flights ranging from 0.01 (B624) to 0.32 (B622). The overall decline of f_{60} again appears to be strongly influenced by distance from the source region and the physical age of plumes, decreasing from a maximum of ~0.027 in B622 to a minimum of ~0.004 in B624. An effect of dilution is evident, given the concurrent reduction in both ΔCO

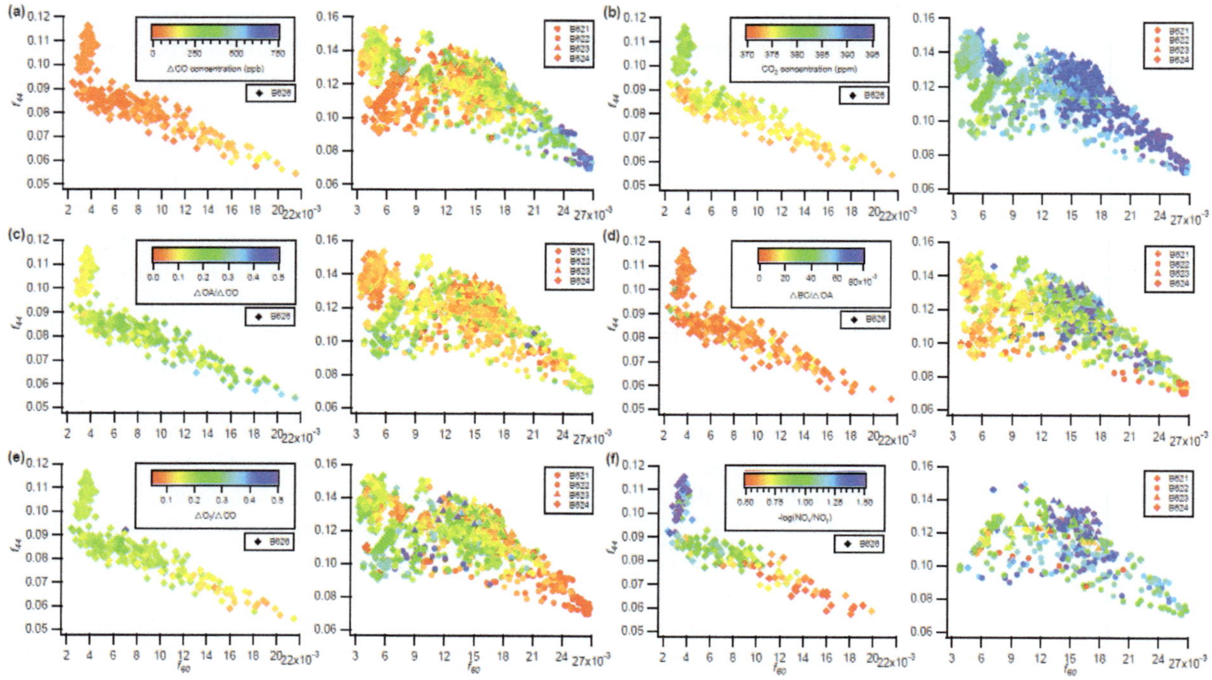

Figure 8. f_{44} vs. f_{60} with data points coloured by **(a)** ΔCO, **(b)** CO_2, **(c)** ΔOA / ΔCO, **(d)** ΔBC / ΔOA, **(e)** ΔO_3 / ΔCO and **(f)** $-\log(NO_x / NO_y)$. Data from fresh and aged plumes are shown on the left and right hand side of each panel, respectively.

and CO_2 with decreasing f_{60} and increasing f_{44} (Fig. 8a–b, right panels). ΔO_3 / ΔCO shows the reverse progression, increasing with the oxygenation of OA (Fig. 8e, right panel), while $-\log(NO_x / NO_y)$ does not exhibit the strong trend observed for near-field measurements but is typically higher and reflects longer aging times (Fig. 8f). However, $-\log(NO_x / NO_y)$ ratios were highest on average for the "downwind" plume in B626, which did not provide any significant indication of net addition of OA mass. While the highest average ΔOA / ΔCO ratio for an individual plume throughout the entirety of BORTAS was derived for one of the "source" plumes during B626, two aged plumes from B622 exhibited average ΔOA / ΔCO of a similar magnitude (0.120 ± 0.080 and 0.119 ± 0.042), with each plume representing a different region of f_{44}/ f_{60} space.

The contrasting behaviours of various tracers throughout fresh and aged plumes highlights the different ways in which these properties can be used to evaluate influences on BBOA evolution. With regards to f_{44}, the consistently higher values observed in aged plumes, and the strong trends identified with indicators of photochemical aging such as $-\log(NO_x / NO_y)$ and ΔO_3 / ΔCO close to source (Fig. 8e–f, left panels), substantiate its use as a tracer for OA aging. Although f_{60} exhibits the same clear relationship with $-\log(NO_x / NO_y)$ and ΔO_3 / ΔCO in fresh plumes, albeit reversed and decreasing with aging, values are higher on average amongst aged plumes. Given the overall trend of increasing f_{44} with decreasing f_{60} remains for aged OA, the longer periods of aging to which these plumes have been exposed would be expected to bring about a more extensive reduction in the latter tracer. The elevation in f_{60} relative to fresh plumes would therefore seem to stem from the contrasting dominant combustion phases associated with plumes of different ages, and the persistence of high levels in flaming-derived OA at greater altitudes. In contrast, f_{43} shows an overall reduction with aging, with mean values of 0.123 ± 0.013 and 0.088 ± 0.012 for near and far-field plumes respectively, consistent with the oxidation of primary OA components over time. However, overall trends with $-\log(NO_x / NO_y)$ and ΔO_3 / ΔCO in fresh plumes are generally positive, albeit with fairly weak correlation coefficients ($R^2 = 0.12$ and 0.34), resulting in f_{43} peaking at greater photochemical ages. This relationship contradicts that which would be expected in aging OA (Ng et al., 2010; Morgan et al., 2010), and suggests additional factors may be contributing to the observed variability in f_{43}. However, fragmentation of oxygenated aldehyde and ketone molecules has been shown to produce elevated levels of f_{43} relative to f_{44} in BB emissions (Schneider et al., 2006), suggesting possible contributions from secondary formation. Values of f_{43} in fresh OA are almost entirely greater than 0.1, while this threshold is most frequently exceeded amongst aged OA in plumes below around 3000 m (Fig. 5d). These lower altitude plumes exhibit the same low ΔBC / ΔOA levels as identified close to source (< 0.02), in contrast to the greater range in ΔBC / ΔOA (up to 0.15) coinciding with lower f_{43} (Fig. 6). Differing distributions of f_{43} in aged plumes, and the prescribed similarities with near-field observations, may reflect an influence of changing combustion conditions, with f_{43} seemingly more prominent in OA from smouldering fires. As a result, f_{43} may prove to be a more suitable tracer for source conditions rather than the effects of aging, although comparison between different combustion phases at source would be required in order to fully constrain any such relationship.

3.6 Campaign intercomparison and evaluation of f_{44} and f_{60} tracers

The progression of f_{44} and f_{60} throughout BORTAS shows a number of similarities with observations from other field campaigns and laboratory experiments. Distributions for fresh and aged emissions from BORTAS and montane forest fires during the Megacities Initiative: Local and Global Research Observations (MILAGRO) campaign are presented in Fig. 9, along with data from numerous plumes measured during ARCTAS-B. Data are also shown for the combustion of boreal forest plant species under laboratory conditions as part of FLAME II. Similar trends in f_{44}/f_{60} for fresh and aged emissions are identified for BORTAS and MILAGRO, with average f_{44} increasing with aging in both cases. A significant contrast is also evident in the distributions of f_{60}, which is higher on average for fresh plumes in MILAGRO and aged plumes in BORTAS, possibly as a result of the reduced intensity of fires sampled close to source. Average $\Delta OA / \Delta CO$ is again lower for the aged fraction in MILAGRO, decreasing from 0.051 ± 0.001 in fresh plumes to 0.041 ± 0.001 (Jolleys et al., 2012), consistent with a loss of OA. The lower magnitude of these ratios is likely to be a consequence of different fuel properties and resulting combustion conditions, as strongly flaming grass fires are expected to have made a significant contribution to smoke plumes sampled at the Paso de Cortes measurement site (Jolleys et al., 2014). Figure 9 also emphasises the differences in emissions from boreal forest fires during ARCTAS-B and BORTAS. Plumes encountered close to source in each campaign exhibit contrasting levels of f_{60}, reflecting the dominance of different combustion phases in each set of measurements. Unlike the heavily smouldering fires sampled in flight B626, the plume from a fire at Lake McKay in north-western Saskatchewan was produced by highly intense, flaming fires (Cubison et al., 2011). The Lake McKay fires subsequently yielded higher f_{60} than was observed for any BORTAS plumes, peaking at around 0.05. As the Lake McKay plume was tracked downwind, f_{44} increased to ~ 0.12, comparable to the upper extent for fresh plumes in BORTAS. Although f_{60} decreased to ~ 0.015, this level remained above the majority of the distribution from BORTAS. Similarly high levels of f_{60} were observed for black spruce fires during FLAME II. However, f_{44} from these burns was generally exceptionally low, as would be expected given the direct measurement at source and lack of aging. Higher f_{44} comparable to the range identified in ambient emissions did occur in chamber fires for plant species representing environments other than boreal forests, with average

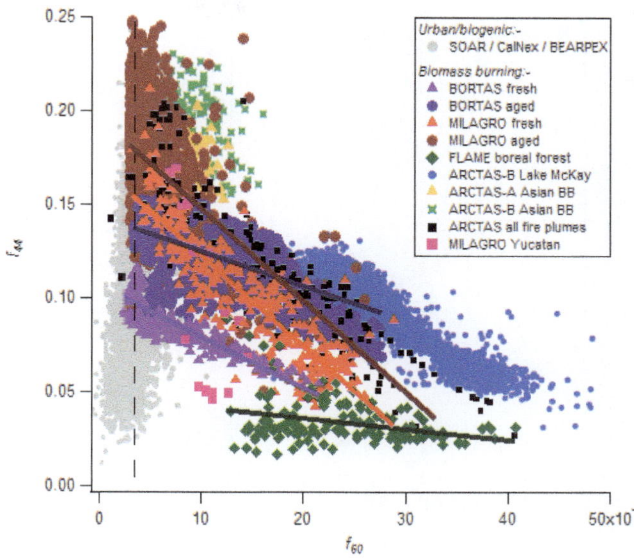

Figure 9. Synthesis of f_{44} vs. f_{60} from a range of ambient and laboratory measurements of BBOA, along with data from non-BB sources. The specified background f_{60} value of 0.003 used to identify BB influences is shown as the dashed vertical line. Coloured lines denote linear regressions for corresponding data sets. Adapted from Cubison et al. (2011).

values particularly high for chaparral fuels. Montane forest fuels, which like the boreal equivalent comprised samples of coniferous species, also yielded f_{44} up to ~ 0.15, although such fires largely involved drier, woody plant material leading to more flaming-dominated combustion (Jolleys et al., 2014).

The different f_{44} and f_{60} regimes in ambient and chamber fires, and their conflicting relationships with combustion phases, suggest their use as tracers for processing of BBOA is highly dependent on both fire properties and experimental conditions. Throughout FLAME II, f_{44} was shown to be more strongly associated with flaming combustion, as increased intensity and turbulent mixing enhanced the supply of oxygen to fires. In contrast, the rapid increase of f_{44} in fresh OA from smouldering fires during BORTAS, to levels comparable to more extensively aged plumes, indicate that f_{44} is strongly influenced by post-emission processing under ambient conditions. Relationships with f_{60} are more consistent, being higher on average more frequently for flaming-dominated fires under laboratory conditions, and showing a stronger association with seemingly flaming-derived aged emissions during BORTAS. Probability density functions (PDFs) for f_{44} and f_{60} in fresh and aged emissions from BORTAS, along with source emissions from fires involving boreal and montane forest fuels during FLAME II, are shown in Fig. 10. The clear separation in f_{44} distributions between chamber and ambient measurements reflects the role of aging in determining the level of oxidation in BBOA, as further evidenced by the enhancement in plumes in the far-

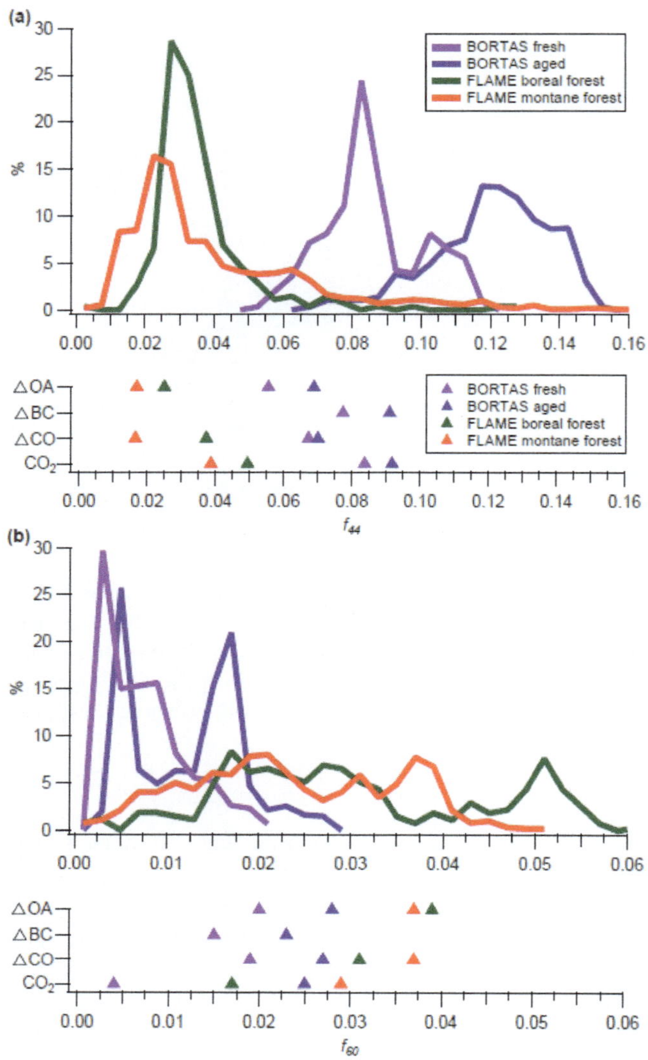

Figure 10. Probability density functions for **(a)** f_{44} and **(b)** f_{60} from a range of ambient and laboratory BB measurements. Also shown in the lower sections of each panel are the f_{44} and f_{60} values corresponding to maximum concentrations of ΔOA, ΔBC, ΔCO and CO_2 (ΔCO_2 during FLAME).

field above those at source. However, the trend of increasing f_{44} in fresh plumes suggests that this processing can occur over very short timescales under certain atmospheric conditions. Rapid oxidation of BB smoke plumes has previously been inferred from the addition of secondary OA mass within ~1 h of emission (Gao et al., 2003; Yokelson et al., 2009), corroborating the BORTAS trend. Values of f_{44} coinciding with peak concentrations for a number of combustion products are also shown in Fig. 10. These peak concentrations show a good agreement with prescribed combustion phase relationships for FLAME II data, where ΔCO_2 reaches a maximum when f_{44} is higher, and hence combustion more flaming-dominated, while ΔOA and ΔCO peak at a lower f_{44}. The same trends are also observed throughout BORTAS, with peak concentrations for CO_2 and ΔBC coinciding with higher levels of f_{44} than those of ΔCO or ΔOA. PDFs for f_{60} exhibit the same trend amongst ambient plumes, shifting to higher values with aging. Distributions are also broadened for emissions from chamber burns, for which levoglucosan-type species constitute a larger proportion of the total OA mass. The very low peak for near-field BORTAS plumes could be influenced by both the absence of a significant flaming phase and subsequent oxidation of primary OA (Cubison et al., 2011), contributing to the increase in f_{44}. The variable gradients for f_{44}/f_{60} regressions (Fig. 9) indicate a slower rate of decay for levoglucosan-type OA in aged BORTAS plumes compared to their equivalents from MILAGRO. Furthermore, mean f_{60} in aged MILAGRO plumes (0.006 ± 0.003) was lower than in fresh plumes (0.018 ± 0.006), while the opposite was true for BORTAS plumes (0.012 ± 0.005 and 0.007 ± 0.004 respectively). As such, the slower decline of f_{60} and potential influences from more strongly flaming combustion may contribute towards the observed enhancement in aged BORTAS plumes, while a faster rate of oxidation and largely smouldering fires reduce levels closer to source.

4 Conclusions

Smoke plumes from Canadian boreal forest fires have been shown to exhibit highly variable properties over a range of ages and combustion phases. Average ΔOA / ΔCO in 3 plumes sampled close to source (0.190 ± 0.010) exceed ratios in the far-field from 23 interceptions (0.056 ± 0.003 to 0.114 ± 0.003), reaffirming an absence of significant net SOA formation for aging BB emissions, at least to an extent that provides an elevation above initial OA production at source. While contrasting aging behaviours and significant SOA formation have been identified in some studies, an absence of increasing ΔOA / ΔCO has been observed in several previous BB assessments. The trend of decreasing ΔOA / ΔCO with increasing distance from source in BORTAS further emphasises the importance of source conditions for aging plumes. High levels of typical flaming combustion products were identified in highly aged plumes following transportation over a period of several days. Enhancements in ΔBC / ΔOA and f_{60} were most prominent within the free troposphere, typically displaying an overall increase with altitude, while aged OA sampled within the boundary layer showed stronger evidence for production by smouldering combustion.

Aging of BBOA during BORTAS has been extensively evaluated using the key tracers f_{44} and f_{60} from the AMS mass spectrum. An enhancement in f_{44} was determined for far-field plumes, where the mean value of 0.121 ± 0.016 significantly exceeded that in the near-field (0.086 ± 0.014). Similarly, f_{60} remained higher in aged plumes (0.012 ± 0.005) than those close to source

(0.007 ± 0.004), in spite of the concurrent increase in oxygenation and expected processing of primary OA components. These trends highlight the importance of both source conditions and processing for OA composition in BB plumes. While the influence of combustion phase on f_{44} remains highly uncertain given contrasting relationships with smouldering and flaming combustion reported in different studies, increases observed close to source suggest oxidation can occur over very short timescales after emission. This rapid processing is further corroborated by concurrent increases in photochemical tracers such as $\Delta O_3 / \Delta CO$ and $-\log(NO_x / NO_y)$ ratios in plumes sampled near to source. The increasing oxygenation of BBOA is not accompanied by an increase in $\Delta OA / \Delta CO$, which shows no significant change with $\Delta O_3 / \Delta CO$ and decreases on average with $-\log(NO_x / NO_y)$ over short aging times. A lack of $\Delta OA / \Delta CO$ enhancement irrespective of evidence for wider transformations therefore further substantiates the impact of OA losses in these aging BB plumes.

Presenting the changing composition of BBOA in f_{44}/f_{60} space reveals a consistent progression from high f_{60} to high f_{44} as primary levoglucosan-like species are lost through oxidation. Similar transitions occur across multiple data sets encompassing smoke plumes of varying origins and ages, although the gradients and extents of distributions show some variability between campaigns. Levels of f_{44} are also comparatively depleted in chamber burns of boreal forest fuels. The absence of aging and a strong association with flaming combustion, and hence oxygen supply through entrainment, in these experiments denote alternative tracer functions under laboratory and ambient conditions. While f_{44} can act as an indicator for oxygenation through combustion processes in chamber experiments, the influence of aging is likely to limit such application for ambient emissions. However, f_{60} has been shown to act as a long-lived tracer for BB emissions, despite evidence for an overall reduction with increasing f_{44}.

Analysis of measurements performed during the BORTAS campaign has provided further insight to the variability associated with BB emissions and the processes affecting changes in BBOA loadings and composition over time. However, there remains considerable uncertainty regarding the main drivers of OA processing. While data from BORTAS provide evidence for the influence of a range of source and aging processes, the extents of any effects on aging BBOA are unclear, particularly with regard to their consistency across different environments and fire types. Key trends identified in this analysis, such as the comparatively lower levels of f_{60} close to source, contradict previous findings and highlight the lack of consistency prevalent amongst many aspects of investigations focusing on BB emissions. Further research specifically targeting these areas of uncertainty is therefore essential in order to understand the cause of these disparities and provide more reliable parameterisations of BB contributions to the atmospheric aerosol burden.

Acknowledgements. The authors would like to acknowledge financial support from the United Kingdom Natural Environment Research Council (NERC) under grant number NE/F017391/1, and thank all those involved in the BORTAS project. Airborne data was obtained using the FAAM BAe-146 Atmospheric Research Aircraft (ARA) operated by Directflight Ltd (DFL) and managed by the Facility for Airborne Atmospheric Measurements (FAAM), which is a joint entity of NERC and the UK Meteorological Office. M. D. Jolleys was supported by a NERC studentship NE/H525162/1. P. I. Palmer also acknowledges support from the Leverhulme Trust.

Edited by: S. Matthiesen

References

Aiken, A. C., DeCarlo, P. F., Kroll, J. H., Worsnop, D. R., Huffman, J. A., Docherty, K. S., Ulbrich, I. M., Mohr, C., Kimmel, J. R., Sueper, D., Sun, Y., Zhang, Q., Trimborn, A., Northway, M., Ziemann, P. J., Canagaratna, M. R., Onasch, T. B., Alfarra, M. R., Prevot, A. S. H., Dommen, J., Duplissy, J., Metzger, A., Baltensperger, U., and Jimenez, J. L.: O/C and OM/OC ratios of primary, secondary, and ambient organic aerosols with high-resolution time-of-flight aerosol mass spectrometry, Environ. Sci. Tech., 42, 44780-4485, doi:10.1021/es703009q, 2008.

Aiken, A. C., Salcedo, D., Cubison, M. J., Huffman, J. A., DeCarlo, P. F., Ulbrich, I. M., Docherty, K. S., Sueper, D., Kimmel, J. R., Worsnop, D. R., Trimborn, A., Northway, M., Stone, E. A., Schauer, J. J., Volkamer, R. M., Fortner, E., de Foy, B., Wang, J., Laskin, A., Shutthanandan, V., Zheng, J., Zhang, R., Gaffney, J., Marley, N. A., Paredes-Miranda, G., Arnott, W. P., Molina, L. T., Sosa, G., and Jimenez, J. L.: Mexico City aerosol analysis during MILAGRO using high resolution aerosol mass spectrometry at the urban supersite (T0) – Part 1: Fine particle composition and organic source apportionment, Atmos. Chem. Phys., 9, 6633–6653, doi:10.5194/acp-9-6633-2009, 2009.

Akagi, S. K., Yokelson, R. J., Wiedinmyer, C., Alvarado, M. J., Reid, J. S., Karl, T., Crounse, J. D., and Wennberg, P. O.: Emission factors for open and domestic biomass burning for use in atmospheric models, Atmos. Chem. Phys., 11, 4039–4072, doi:10.5194/acp-11-4039-2011, 2011.

Akagi, S. K., Craven, J. S., Taylor, J. W., McMeeking, G. R., Yokelson, R. J., Burling, I. R., Urbanski, S. P., Wold, C. E., Seinfeld, J. H., Coe, H., Alvarado, M. J., and Weise, D. R.: Evolution of trace gases and particles emitted by a chaparral fire in California, Atmos. Chem. Phys., 12, 1397–1421, doi:10.5194/acp-12-1397-2012, 2012.

Alfarra, M. R., Coe, H., Allan, J. D., Bower, K. N., Boudries, H., Canagaratna, M. R., Jimenez, J. L., Jayne, J. T., Garforth, A. A., Li, S. M., and Worsnop, D. R.: Characterization of urban and rural organic particulate in the Lower Fraser Valley using two Aerodyne Aerosol Mass Spectrometers, Atmos. Environ.,38, 5745–5758, doi:10.1016/j.atmosenv.2004.01.054, 2004.

Alfarra, M. R., Prévôt, A. S. H., Szidat, S., Sandradewi, J., Weimer, S., Lanz, V. A., Schreiber, D., Mohr, M., and Baltensperger, U.: Identification of the mass spectral signature of organic aerosols from wood burning emissions, Environ. Sci. Tech., 41, 5770–5777, doi:10.1021/es062289b, 2007.

Andreae, M. O., Atlas, E. Cachier, H., Cofer III, W. R., Harris, G. W., Helas, G., Koppmann, R., Lacaux, J.-P., and Ward, D. E.: Trace gas and aerosol emissions from savanna fires, in: Biomass Burning and Global Change, edited by: Levine, S., 278–295, MIT Press, Cambridge, Mass., 1996.

Andreae, M. O. and Merlet, P.: Emission of trace gases and aerosols from biomass burning, Global Biogeochem. Cy., 15, 955–966, 2001.

Bahreini, R., Jimenez, J. L., Wang, J., Flagan, R. C., Seinfeld, J. H., Jayne, J. T., and Worsnop, D. R.: Aircraft-based aerosol size and composition measurements during ACE-Asia using an Aerodyne aerosol mass spectrometer, J. Geophys. Res.-Atmos., 108, 8645, doi:10.1029/2002jd003226, 2003.

Bond, T. C., Streets, D. G., Yarber, K. F., Nelson, S. M., Woo, J. H., and Klimont, Z.: A technology-based global inventory of black and organic carbon emissions from combustion, J. Geophys. Res.-Atmos., 109, D14203, doi:10.1029/2003jd003697, 2004.

Burling, I. R., Yokelson, R. J., Akagi, S. K., Urbanski, S. P., Wold, C. E., Griffith, D. W. T., Johnson, T. J., Reardon, J., and Weise, D. R.: Airborne and ground-based measurements of the trace gases and particles emitted by prescribed fires in the United States, Atmos. Chem. Phys., 11, 12197–12216, doi:10.5194/acp-11-12197-2011, 2011.

Canagaratna, M. R., Jayne, J. T., Jimenez, J. L., Allan, J. D., Alfarra, M. R., Zhang, Q., Onasch, T. B., Drewnick, F., Coe, H., Middlebrook, A., Delia, A., Williams, L. R., Trimborn, A. M., Northway, M. J., DeCarlo, P. F., Kolb, C. E., Davidovits, P., and Worsnop, D. R.: Chemical and microphysical characterization of ambient aerosols with the aerodyne aerosol mass spectrometer, Mass Spectrom. Rev., 26, 185–222, doi:10.1002/mas.20115, 2007.

Capes, G., B. Johnson, G. McFiggans, P. I. Williams, J. Haywood, and H. Coe: Aging of biomass burning aerosols over West Africa: Aircraft measurements of chemical composition, microphysical properties, and emission ratios, J. Geophys. Res.-Atmos., 113, D00C15, doi:10.1029/2008jd009845, 2008.

Crosier, J., Allan, J. D., Coe, H., Bower, K. N., Formenti, P., and Williams, P. I.: Chemical composition of summertime aerosol in the Po Valley (Italy), northern Adriatic and Black Sea, Q. J. Roy. Meteor. Soc., 133, 61–75, doi:10.1002/qj.88, 2007.

Crounse, J. D., DeCarlo, P. F., Blake, D. R., Emmons, L. K., Campos, T. L., Apel, E. C., Clarke, A. D., Weinheimer, A. J., McCabe, D. C., Yokelson, R. J., Jimenez, J. L., and Wennberg, P. O.: Biomass burning and urban air pollution over the Central Mexican Plateau, Atmos. Chem. Phys., 9, 4929–4944, doi:10.5194/acp-9-4929-2009, 2009.

Crutzen, P. J. and Andreae, M. O.: Biomass Burning in the Tropics – Impact on Atmospheric Chemistry and Biogeochemical Cycles, Science, 250, 1669–1678, doi:10.1126/science.250.4988.1669, 1990.

Cubison, M. J., Ortega, A. M., Hayes, P. L., Farmer, D. K., Day, D., Lechner, M. J., Brune, W. H., Apel, E., Diskin, G. S., Fisher, J. A., Fuelberg, H. E., Hecobian, A., Knapp, D. J., Mikoviny, T., Riemer, D., Sachse, G. W., Sessions, W., Weber, R. J., Weinheimer, A. J., Wisthaler, A., and Jimenez, J. L.: Effects of aging on organic aerosol from open biomass burning smoke in aircraft and laboratory studies, Atmos. Chem. Phys., 11, 12049–12064, doi:10.5194/acp-11-12049-2011, 2011.

Damoah, R., Spichtinger, N., Servranckx, R., Fromm, M., Eloranta, E. W., Razenkov, I. A., James, P., Shulski, M., Forster, C., and Stohl, A.: A case study of pyro-convection using transport model and remote sensing data, Atmos. Chem. Phys., 6, 173–185, doi:10.5194/acp-6-173-2006, 2006.

DeCarlo, P. F., Dunlea, E. J., Kimmel, J. R., Aiken, A. C., Sueper, D., Crounse, J., Wennberg, P. O., Emmons, L., Shinozuka, Y., Clarke, A., Zhou, J., Tomlinson, J., Collins, D. R., Knapp, D., Weinheimer, A. J., Montzka, D. D., Campos, T., and Jimenez, J. L.: Fast airborne aerosol size and chemistry measurements above Mexico City and Central Mexico during the MILAGRO campaign, Atmos. Chem. Phys., 8, 4027–4048, doi:10.5194/acp-8-4027-2008, 2008.

DeCarlo, P. F., Ulbrich, I. M., Crounse, J., de Foy, B., Dunlea, E. J., Aiken, A. C., Knapp, D., Weinheimer, A. J., Campos, T., Wennberg, P. O., and Jimenez, J. L.: Investigation of the sources and processing of organic aerosol over the Central Mexican Plateau from aircraft measurements during MILAGRO, Atmos. Chem. Phys., 10, 5257–5280, doi:10.5194/acp-10-5257-2010, 2010.

Di Carlo, P., Aruffo, E., Busilacchio, M., Giammaria, F., Dari-Salisburgo, C., Biancofiore, F., Visconti, G., Lee, J., Moller, S., Reeves, C. E., Bauguitte, S., Forster, G., Jones, R. L., and Ouyang, B.: Aircraft based four-channel thermal dissociation laser induced fluorescence instrument for simultaneous measurements of NO_2, total peroxy nitrate, total alkyl nitrate, and HNO_3, Atmos. Meas. Tech., 6, 971–980, doi:10.5194/amt-6-971-2013, 2013.

Drewnick, F., Hings, S. S., DeCarlo, P. F., Jayne, J. T., Gonin, M., Fuhrer, K., Weimer, S., Jimenez, J. L., Demerjian, K. L., Borrmann, S., and Worsnop, D. R.: A new time-of-flight aerosol mass spectrometer (TOF-AMS) – Instrument description and first field deployment, Aerosol Sci. Tech., 39, 637–658, doi:10.1080/02786820500182040, 2005.

Ferek, R. J., Reid, J. S., Hobbs, P. V., Blake, D. R., and Liousse, C.: Emission factors of hydrocarbons, halocarbons, trace gases and particles from biomass burning in Brazil, J. Geophys. Res.-Atmos., 103, 32107–32118, 1998.

Forster, P., Artaxo, P., Berntse, T., Betts, R., Fahey, D. W., Haywood, J., Lean, J., Lowe, D. C., Myhre, G., Nganga, J., Prinn, R., Raga, G., Schulz, M., Van Dorland, R., Bodeker, G., Boucher, O., Collins, W. D., Conway, T. J., Dlugokencky, E., Elkins, J. W., Etheridge, D., Foukal, P., Fraser, P., Geller, M., Joos, F., Keeling, C. D., Keeling, R., Kinne, S., Lassey, K., Lohmann, U., Manning, A. C., Montzka, S., Oram, D., O'Shaughnessy, K., Piper, S., Plattner, G.-K., Ponater, M., Ramankutty, N., Reid, G., Rind, D., Rosenlof, K., Sausen, R., Schwarzkopf, D., Solanki, S. K., Stenchikov, G., Stuber, N., Takemura, T., Textor, C., Wang, R., Weiss, R., and Whorf, T.: Changes in atmospheric constituents and in radiative forcing, in: Climate Change 2007: The Physical Science Basis, Contribution of Working Group I to the Fourth Assessment Report of the Intergovernmental Panel on Climate Change, edited by: Solomon, S. D., Qin, M., Manning, Z., Chen, M., Marquis, K. B., Averyt, M. T., and Miller, H. L., Cambridge University Press, Cambridge, UK and New York, NY, USA, 129–234, 2007.

Fromm, M., R. Bevilacqua, R. Servranckx, J. Rosen, J. P. Thayer, J. Herman, and D. Larko: Pyro-cumulonimbus injection of smoke to the stratosphere: Observations and impact of a super blowup in northwestern Canada on 3–4 August 1998, J. Geophys. Res.-Atmos., 110, D08205, doi:10.1029/2004jd005350, 2005.

Gao, S., Hegg, D. A., Hobbs, P. V., Kirchstetter, T. W., Magi, B. I., and Sadilek, M.: Water-soluble organic components in aerosols associated with savanna fires in southern Africa: Identification, evolution, and distribution, J. Geophys. Res.-Atmos., 108, 8491, doi:10.1029/2002jd002324, 2003.

Grieshop, A. P., Logue, J. M., Donahue, N. M., and Robinson, A. L.: Laboratory investigation of photochemical oxidation of organic aerosol from wood fires 1: measurement and simulation of organic aerosol evolution, Atmos. Chem. Phys., 9, 1263–1277, doi:10.5194/acp-9-1263-2009, 2009.

Griffin, D., Walker, K. A., Franklin, J. E., Parrington, M., Whaley, C., Hopper, J., Drummond, J. R., Palmer, P. I., Strong, K., Duck, T. J., Abboud, I., Bernath, P. F., Clerbaux, C., Coheur, P.-F., Curry, K. R., Dan, L., Hyer, E., Kliever, J., Lesins, G., Maurice, M., Saha, A., Tereszchuk, K., and Weaver, D.: Investigation of CO, C_2H_6 and aerosols in a boreal fire plume over eastern Canada during BORTAS 2011 using ground- and satellite-based observations and model simulations, Atmos. Chem. Phys., 13, 10227–10241, doi:10.5194/acp-13-10227-2013, 2013.

Hallquist, M., Wenger, J. C., Baltensperger, U., Rudich, Y., Simpson, D., Claeys, M., Dommen, J., Donahue, N. M., George, C., Goldstein, A. H., Hamilton, J. F., Herrmann, H., Hoffmann, T., Iinuma, Y., Jang, M., Jenkin, M. E., Jimenez, J. L., Kiendler-Scharr, A., Maenhaut, W., McFiggans, G., Mentel, Th. F., Monod, A., Prévôt, A. S. H., Seinfeld, J. H., Surratt, J. D., Szmigielski, R., and Wildt, J.: The formation, properties and impact of secondary organic aerosol: current and emerging issues, Atmos. Chem. Phys., 9, 5155–5236, doi:10.5194/acp-9-5155-2009, 2009.

Hecobian, A., Liu, Z., Hennigan, C. J., Huey, L. G., Jimenez, J. L., Cubison, M. J., Vay, S., Diskin, G. S., Sachse, G. W., Wisthaler, A., Mikoviny, T., Weinheimer, A. J., Liao, J., Knapp, D. J., Wennberg, P. O., Kürten, A., Crounse, J. D., Clair, J. St., Wang, Y., and Weber, R. J.: Comparison of chemical characteristics of 495 biomass burning plumes intercepted by the NASA DC-8 aircraft during the ARCTAS/CARB-2008 field campaign, Atmos. Chem. Phys., 11, 13325–13337, doi:10.5194/acp-11-13325-2011, 2011.

Hennigan, C. J., Miracolo, M. A., Engelhart, G. J., May, A. A., Presto, A. A., Lee, T., Sullivan, A. P., McMeeking, G. R., Coe, H., Wold, C. E., Hao, W.-M., Gilman, J. B., Kuster, W. C., de Gouw, J., Schichtel, B. A., Collett Jr., J. L., Kreidenweis, S. M., and Robinson, A. L.: Chemical and physical transformations of organic aerosol from the photo-oxidation of open biomass burning emissions in an environmental chamber, Atmos. Chem. Phys., 11, 7669–7686, doi:10.5194/acp-11-7669-2011, 2011.

Heringa, M. F., DeCarlo, P. F., Chirico, R., Tritscher, T., Dommen, J., Weingartner, E., Richter, R., Wehrle, G., Prévôt, A. S. H., and Baltensperger, U.: Investigations of primary and secondary particulate matter of different wood combustion appliances with a high-resolution time-of-flight aerosol mass spectrometer, Atmos. Chem. Phys., 11, 5945–5957, doi:10.5194/acp-11-5945-2011, 2011.

Higuchi, K., Worthy, D., Chan, D., and Shashkov, A.: Regional source/sink impact on the diurnal, seasonal and inter-annual variations in atmospheric CO_2 at a boreal forest site in Canada, Tellus B, 55, 115–125, doi:10.1034/j.1600-0889.2003.00062.x, 2003.

Iinuma, Y., Bruggemann, E., Gnauk, T., Muller, K., Andreae, M. O., Helas, G., Parmar, R., and Herrmann, H.: Source characterization of biomass burning particles: The combustion of selected European conifers, African hardwood, savanna grass, and German and Indonesian peat, J. Geophys. Res.-Atmos., 112, D08209, doi:10.1029/2006jd007120, 2007.

Jimenez, J. L., Canagaratna, M. R., Donahue, N. M., Prévôt, A. S. H., Zhang, Q., Kroll, J. H., DeCarlo, P. F., Allan, J. D., Coe, H., Ng, N. L., Aiken, A. C., Docherty, K. S., Ulbrich, I. M., Grieshop, A. P., Robinson, A. L., Duplissy, J., Smith, J. D., Wilson, K. R., Lanz, V. A., Hueglin, C., Sun, Y. L., Tian, J., Laaksonen, A., Raatikainen, T., Rautiainen, J., Vaattovaara, P., Ehn, M., Kulmala, M., Tomlinson, J. M., Collins, D. R., Cubison, M. J., Dunlea, E. J., Huffman, J. A., Onasch, T. B., Alfarra, M. R., Williams, P. I., Bower, K., Kondo, Y., Schneider, J., Drewnick, F., Borrmann, S., Weimer, S., Demerjian, K., Salcedo, D., Cottrell, L., Griffin, R., Takami, A., Miyoshi, T., Hatakeyama, S., Shimono, A., Sun, J. Y., Zhang, Y. M., Dzepina, K., Kimmel, J. R., Sueper, D., Jayne, J. T., Herndon, S. C., Trimborn, A. M., Williams, L. R., Wood, E. C., Middlebrook, A. M., Kolb, C. E., Baltensperger, U., Worsnop, D. R.: Evolution of Organic Aerosols in the Atmosphere, Science, 326, 1525–1529, doi:10.1126/science.1180353, 2009.

Jolleys, M. D., Coe, H., McFiggans, G., Capes, G., Allan, J. D., Crosier, J., Williams, P. I., Allen, G., Bower, K. N., Jimenez, J. L., Russell, L. M., Grutter, M., and Baumgardner, D.: Characterizing the Aging of Biomass Burning Organic Aerosol by Use of Mixing Ratios: A Meta-analysis of Four Regions, Environ. Sci. Tech., 46, 13093–13102, doi:10.1021/es302386v, 2012.

Jolleys, M. D., Coe, H., McFiggans, G., McMeeking, G. R., Lee, T., Kreidenweis, S. M., Collett, J. L., and Sullivan, A. P.: Organic aerosol emission ratios from the laboratory combustion of biomass fuels, J. Geophys. Res. Atmos., 119, 22, 12850–12871, doi:10.1002/2014JD021589, 2014.

Kleinman, L. I., Springston, S. R., Daum, P. H., Lee, Y.-N., Nunnermacker, L. J., Senum, G. I., Wang, J., Weinstein-Lloyd, J., Alexander, M. L., Hubbe, J., Ortega, J., Canagaratna, M. R., and Jayne, J.: The time evolution of aerosol composition over the Mexico City plateau, Atmos. Chem. Phys., 8, 1559–1575, doi:10.5194/acp-8-1559-2008, 2008.

Kondo, Y., Matsui, H., Moteki, N., Sahu, L., Takegawa, N., Kajino, M., Zhao, Y., Cubison, M. J., Jimenez, J. L., Vay, S., Diskin, G. S., Anderson, B. , Wisthaler, A., Mikoviny, T., Fuelberg, H. E., Blake, D. R., Huey, G. , Weinheimer, A. J., Knapp, D. J., Brune, W. H.: Emissions of black carbon, organic, and inorganic aerosols from biomass burning in North America and Asia in 2008, J. Geophys. Res.-Atmos., 116, D08204, doi:10.1029/2010jd015152, 2011.

Le Breton, M., McGillen, M. R., Muller, J. B. A., Bacak, A., Shallcross, D. E., Xiao, P., Huey, L. G., Tanner, D., Coe, H., and Percival, C. J.: Airborne observations of formic acid using a chemical ionization mass spectrometer, Atmos. Meas. Tech., 5, 3029–3039, doi:10.5194/amt-5-3029-2012, 2012.

Le Breton, M., Bacak, A., Muller, J. B. A., O'Shea, S. J., Xiao, P., Ashfold, M. N. R., Cooke, M. C., Batt, R., Shallcross, D. E., Oram, D. E., Forster, G., Bauguitte, S. J.-B., Palmer, P. I., Parrington, M., Lewis, A. C., Lee, J. D., and Percival, C.

J.: Airborne hydrogen cyanide measurements using a chemical ionisation mass spectrometer for the plume identification of biomass burning forest fires, Atmos. Chem. Phys., 13, 9217–9232, doi:10.5194/acp-13-9217-2013, 2013.

Lee, T., Sullivan, A. P., Mack, L., Jimenez, J. L., Kreidenweis, S. M. , Onasch, T. B., Worsnop, D. R., Malm, W., Wold, C. E., Hao, W. M., Collett, J. L..: Chemical Smoke Marker Emissions During Flaming and Smoldering Phases of Laboratory Open Burning of Wildland Fuels, Aerosol Sci. Tech., 44, I–V, doi:10.1080/02786826.2010.499884, 2010.

Li, Q. B., Jacob, D. J., Bey, I., Yantosca, R. M., Zhao, Y. J., Kondo, Y., and Notholt, J.: Atmospheric hydrogen cyanide (HCN): Biomass burning source, ocean sink?, Geophys. Res. Lett., 27, 357–360, doi:10.1029/1999gl010935, 2000.

Liousse, C., Devaux, C., Dulac, F., and Cachier, H.: Aging of Savanna Biomass Burning Aerosols – Consequences on their Optical Properties, J. Atmos. Chem., 22, 1–17, doi:10.1007/bf00708178, 1995.

Mason, S. A., Field, R. J., Yokelson, R. J., Kochivar, M. A., Tinsley, M. R., Ward, D. E., and Hao, W. M.: Complex effects arising in smoke plume simulations due to inclusion of direct emissions of oxygenated organic species from biomass combustion, J. Geophys. Res.-Atmos., 106, 12527-12539, 2001.

McMeeking, G. R., Kreidenweis, S. M., Baker, S., Carrico, C. M., Chow, J. C., Collett, Jr., J. L., Hao, W. M. , Holden, A. S., Kirchstetter, T. W., Malm, W. C., Moosmueller, H., Sullivan, A. P., and Wold, C. E.: Emissions of trace gases and aerosols during the open combustion of biomass in the laboratory, J. Geophys. Res.-Atmos., 114, D19210, doi:10.1029/2009jd011836, 2009.

Morgan, W. T., Allan, J. D., Bower, K. N., Capes, G., Crosier, J., Williams, P. I., and Coe, H.: Vertical distribution of sub-micron aerosol chemical composition from North-Western Europe and the North-East Atlantic, Atmos. Chem. Phys., 9, 5389–5401, doi:10.5194/acp-9-5389-2009, 2009.

Morgan, W. T., Allan, J. D., Bower, K. N., Highwood, E. J., Liu, D., McMeeking, G. R., Northway, M. J., Williams, P. I., Krejci, R., and Coe, H.: Airborne measurements of the spatial distribution of aerosol chemical composition across Europe and evolution of the organic fraction, Atmos. Chem. Phys., 10, 4065–4083, doi:10.5194/acp-10-4065-2010, 2010.

Ng, N. L., Canagaratna, M. R., Zhang, Q., Jimenez, J. L., Tian, J., Ulbrich, I. M., Kroll, J. H., Docherty, K. S., Chhabra, P. S., Bahreini, R., Murphy, S. M., Seinfeld, J. H., Hildebrandt, L., Donahue, N. M., DeCarlo, P. F., Lanz, V. A., Prévôt, A. S. H., Dinar, E., Rudich, Y., and Worsnop, D. R.: Organic aerosol components observed in Northern Hemispheric datasets from Aerosol Mass Spectrometry, Atmos. Chem. Phys., 10, 4625–4641, doi:10.5194/acp-10-4625-2010, 2010.

Nowak, J. B., Neuman, J. A., Kozai, K., Huey, L. G., Tanner, D. J. , Holloway, J. S., Ryerson, T. B., Frost, G. J., McKeen, S. A., and Fehsenfeld, F. C.: A chemical ionization mass spectrometry technique for airborne measurements of ammonia, J. Geophys. Res.-Atmos., 112, D10S02, doi:10.1029/2006jd007589, 2007.

O'Shea, S. J., Allen, G., Gallagher, M. W., Bauguitte, S. J.-B., Illingworth, S. M., Le Breton, M., Muller, J. B. A., Percival, C. J., Archibald, A. T., Oram, D. E., Parrington, M., Palmer, P. I., and Lewis, A. C.: Airborne observations of trace gases over boreal Canada during BORTAS: campaign climatology, air mass analysis and enhancement ratios, Atmos. Chem. Phys., 13, 12451–12467, doi:10.5194/acp-13-12451-2013, 2013a.

O'Shea, S. J., Bauguitte, S. J.-B., Gallagher, M. W., Lowry, D., and Percival, C. J.: Development of a cavity-enhanced absorption spectrometer for airborne measurements of CH_4 and CO_2, Atmos. Meas. Tech., 6, 1095–1109, doi:10.5194/amt-6-1095-2013, 2013b.

Palmer, P. I., Parrington, M., Lee, J. D., Lewis, A. C., Rickard, A. R., Bernath, P. F., Duck, T. J., Waugh, D. L., Tarasick, D. W., Andrews, S., Aruffo, E., Bailey, L. J., Barrett, E., Bauguitte, S. J.-B., Curry, K. R., Di Carlo, P., Chisholm, L., Dan, L., Forster, G., Franklin, J. E., Gibson, M. D., Griffin, D., Helmig, D., Hopkins, J. R., Hopper, J. T., Jenkin, M. E., Kindred, D., Kliever, J., Le Breton, M., Matthiesen, S., Maurice, M., Moller, S., Moore, D. P., Oram, D. E., O'Shea, S. J., Owen, R. C., Pagniello, C. M. L. S., Pawson, S., Percival, C. J., Pierce, J. R., Punjabi, S., Purvis, R. M., Remedios, J. J., Rotermund, K. M., Sakamoto, K. M., da Silva, A. M., Strawbridge, K. B., Strong, K., Taylor, J., Trigwell, R., Tereszchuk, K. A., Walker, K. A., Weaver, D., Whaley, C., and Young, J. C.: Quantifying the impact of BOReal forest fires on Tropospheric oxidants over the Atlantic using Aircraft and Satellites (BORTAS) experiment: design, execution and science overview, Atmos. Chem. Phys., 13, 6239–6261, doi:10.5194/acp-13-6239-2013, 2013.

Parrington, M., Palmer, P. I., Henze, D. K., Tarasick, D. W., Hyer, E. J., Owen, R. C., Helmig, D., Clerbaux, C., Bowman, K. W., Deeter, M. N., Barratt, E. M., Coheur, P.-F., Hurtmans, D., Jiang, Z., George, M., and Worden, J. R.: The influence of boreal biomass burning emissions on the distribution of tropospheric ozone over North America and the North Atlantic during 2010, Atmos. Chem. Phys., 12, 2077–2098, doi:10.5194/acp-12-2077-2012, 2012.

Parrington, M., Palmer, P. I., Lewis, A. C., Lee, J. D., Rickard, A. R., Di Carlo, P., Taylor, J. W., Hopkins, J. R., Punjabi, S., Oram, D. E., Forster, G., Aruffo, E., Moller, S. J., Bauguitte, S. J.-B., Allan, J. D., Coe, H., and Leigh, R. J.: Ozone photochemistry in boreal biomass burning plumes, Atmos. Chem. Phys., 13, 7321–7341, doi:10.5194/acp-13-7321-2013, 2013.

Petzold, A., Weinzierl, B., Huntrieser, H., Stohl, A., Real, E., Cozic, J., Fiebig, M., Hendricks, J., Lauer, A., Law, K., Roiger, A., Schlager, H., and Weingartner, E.: Perturbation of the European free troposphere aerosol by North American forest fire plumes during the ICARTT-ITOP experiment in summer 2004, Atmos. Chem. Phys., 7, 5105–5127, doi:10.5194/acp-7-5105-2007, 2007.

Reid, J. S. and Hobbs, P. V.: Physical and optical properties of young smoke from individual biomass fires in Brazil, J. Geophys. Res.-Atmos., 103, 32013–32030, 1998.

Reid, J. S., Koppmann, R., Eck, T. F., and Eleuterio, D. P.: A review of biomass burning emissions part II: intensive physical properties of biomass burning particles, Atmos. Chem. Phys., 5, 799–825, doi:10.5194/acp-5-799-2005, 2005.

Schneider, J., Weimer, S., Drewnick, F., Borrmann, S., Helas, G., Gwaze, P., Schmid, O., Andreae, M. O., and Kirchner, U.: Mass spectrometric analysis and aerodynamic properties of various types of combustion-related aerosol particles, Int. J. Mass Spectrom., 258, 37–49, doi:10.1016/j.ijms.2006.07.008, 2006.

Schwarz, J. P., Gao, R. S., Fahey, D. W., Thomson, D. S., Watts, L. A., Wilson, J. C., Reeves, J. M., Darbeheshti, M., Baumgardner, D. G., Kok, G. L., Chung, S. H., Schulz, M., Hendricks, J., Lauer, A., Kaercher, B., Slowik, J. G., Rosenlof, K. H., Thompson, T. L., Langford, A. O., Loewenstein, M., and Aikin, K. C.: Single-particle measurements of midlatitude black carbon and light-scattering aerosols from the boundary layer to the lower stratosphere, J. Geophys. Res.-Atmos., 111, D16207, doi:10.1029/2006jd007076, 2006.

Simoneit, B. R. T., Schauer, J. J., Nolte, C. G., Oros, D. R., Elias, V. O., Fraser, M. P., Rogge, W. F., and Cass, G. R.: Levoglucosan, a tracer for cellulose in biomass burning and atmospheric particles, Atmos. Environ., 33, 173–182, doi:10.1016/s1352-2310(98)00145-9, 1999.

Sinha, P., Hobbs, P. V., Yokelson, R. J., Bertschi, I. T., Blake, D. R., Simpson, I. J., Gao, S., Kirchstetter, T. W., and Novakov, T.: Emissions of trace gases and particles from savanna fires in southern Africa, J. Geophys. Res.-Atmos., 108, 32, doi:10.1029/2002jd002325, 2003.

Slowik, J. G., Stroud, C., Bottenheim, J. W., Brickell, P. C., Chang, R. Y.-W., Liggio, J., Makar, P. A., Martin, R. V., Moran, M. D., Shantz, N. C., Sjostedt, S. J., van Donkelaar, A., Vlasenko, A., Wiebe, H. A., Xia, A. G., Zhang, J., Leaitch, W. R., and Abbatt, J. P. D.: Characterization of a large biogenic secondary organic aerosol event from eastern Canadian forests, Atmos. Chem. Phys., 10, 2825–2845, doi:10.5194/acp-10-2825-2010, 2010.

Sullivan, A. P., Holden, A. S., Patterson, L. A., McMeeking, G. R., Kreidenweis, S. M., Malm, W. C., Hao, W. M., Wold, C. E., and Collett, Jr., J. L.: A method for smoke marker measurements and its potential application for determining the contribution of biomass burning from wildfires and prescribed fires to ambient $PM_{2.5}$ organic carbon, J. Geophys. Res.-Atmos., 113, D22302, doi:10.1029/2008jd010216, 2008.

Taylor, J. W., Allan, J. D., Allen, G., Coe, H., Williams, P. I., Flynn, M. J., Le Breton, M., Muller, J. B. A., Percival, C. J., Oram, D., Forster, G., Lee, J. D., Rickard, A. R., Parrington, M., and Palmer, P. I.: Size-dependent wet removal of black carbon in Canadian biomass burning plumes, Atmos. Chem. Phys., 14, 13755–13771, doi:10.5194/acp-14-13755-2014, 2014.

Textor, C., Schulz, M., Guibert, S., Kinne, S., Balkanski, Y., Bauer, S., Berntsen, T., Berglen, T., Boucher, O., Chin, M., Dentener, F., Diehl, T., Easter, R., Feichter, H., Fillmore, D., Ghan, S., Ginoux, P., Gong, S., Grini, A., Hendricks, J., Horowitz, L., Huang, P., Isaksen, I., Iversen, I., Kloster, S., Koch, D., Kirkevåg, A., Kristjansson, J. E., Krol, M., Lauer, A., Lamarque, J. F., Liu, X., Montanaro, V., Myhre, G., Penner, J., Pitari, G., Reddy, S., Seland, Ø., Stier, P., Takemura, T., and Tie, X.: Analysis and quantification of the diversities of aerosol life cycles within AeroCom, Atmos. Chem. Phys., 6, 1777–1813, doi:10.5194/acp-6-1777-2006, 2006.

Vay, S. A., Choi, Y., Vadrevu, K. P., Blake, D. R., Tyler, S. C., Wisthaler, A., Hecobian, A., Kondo, Y., Diskin, G. S., Sachse, G. W., Woo, J. H., Weinheimer, A. J., Burkhart, J. F., Stohl, A., Wennberg, P. O.: Patterns of CO_2 and radiocarbon across high northern latitudes during International Polar Year 2008, J. Geophys. Res.-Atmos., 116, D14301, doi:10.1029/2011jd015643, 2011.

Warneke, C., de Gouw, J. A., Stohl, A., Cooper, O. R., Goldan, P. D., Kuster, W. C., Holloway, J. S., Williams, E. J., Lerner, B. M., McKeen, S. A., Trainer, M., Fehsenfeld, F. C., Atlas, E. L., Donnelly, S. G., Stroud, V., Lueb, A., and Kato, S.: Biomass burning and anthropogenic sources of CO over New England in the summer 2004, J. Geophys. Res.-Atmos., 111, D23S15, doi:10.1029/2005jd006878, 2006.

Weimer, S., Alfarra, M. R., Schreiber, D., Mohr, M., Prévôt, A. S. H., and Baltensperger, U.: Organic aerosol mass spectral signatures from wood-burning emissions: Influence of burning conditions and wood type, J. Geophys. Res.-Atmos., 113, D10304, doi:10.1029/2007jd009309, 2008.

Yokelson, R. J., Susott, R., Ward, D. E., Reardon, J., and Griffith, D. W. T.: Emissions from smoldering combustion of biomass measured by open-path Fourier transform infrared spectroscopy, J. Geophys. Res.-Atmos., 102, 18865–18877, doi:10.1029/97jd00852, 1997.

Yokelson, R. J., Urbanski, S. P., Atlas, E. L., Toohey, D. W., Alvarado, E. C., Crounse, J. D., Wennberg, P. O., Fisher, M. E., Wold, C. E., Campos, T. L., Adachi, K., Buseck, P. R., and Hao, W. M.: Emissions from forest fires near Mexico City, Atmos. Chem. Phys., 7, 5569–5584, doi:10.5194/acp-7-5569-2007, 2007.

Yokelson, R. J., Crounse, J. D., DeCarlo, P. F., Karl, T., Urbanski, S., Atlas, E., Campos, T., Shinozuka, Y., Kapustin, V., Clarke, A. D., Weinheimer, A., Knapp, D. J., Montzka, D. D., Holloway, J., Weibring, P., Flocke, F., Zheng, W., Toohey, D., Wennberg, P. O., Wiedinmyer, C., Mauldin, L., Fried, A., Richter, D., Walega, J., Jimenez, J. L., Adachi, K., Buseck, P. R., Hall, S. R., and Shetter, R.: Emissions from biomass burning in the Yucatan, Atmos. Chem. Phys., 9, 5785–5812, doi:10.5194/acp-9-5785-2009, 2009.

Zhang, Q., Worsnop, D. R., Canagaratna, M. R., and Jimenez, J. L.: Hydrocarbon-like and oxygenated organic aerosols in Pittsburgh: insights into sources and processes of organic aerosols, Atmos. Chem. Phys., 5, 3289–3311, doi:10.5194/acp-5-3289-2005, 2005.

Zhang, Q., Jimenez, J. L., Canagaratna, M. R., Allan, J. D., Coe, H., Ulbrich, I., Alfarra, M. R., Takami, A., Middlebrook, A. M., Sun, Y. L., Dzepina, K., Dunlea, E., Docherty, K., DeCarlo, P. F., Salcedo, D., Onasch, T., Jayne, J. T., Miyoshi, T., Shimono, A., Hatakeyama, S., Takegawa, N., Kondo, Y., Schneider, J., Drewnick, F., Borrmann, S., Weimer, S., Demerjian, K., Williams, P., Bower, K., Bahreini, R., Cottrell, L., Griffin, R. J., Rautiainen, J., Sun, J. Y., Zhang, Y. M., and Worsnop, D. R.: Ubiquity and dominance of oxygenated species in organic aerosols in anthropogenically-influenced Northern Hemisphere midlatitudes, Geophys. Res. Lett., 34, L13801, doi:10.1029/2007gl029979, 2007.

10

Intercomparison of vertically resolved merged satellite ozone data sets: interannual variability and long-term trends

F. Tummon[1], B. Hassler[2,3], N. R. P. Harris[4], J. Staehelin[1], W. Steinbrecht[5], J. Anderson[6], G. E. Bodeker[7], A. Bourassa[8], S. M. Davis[2,3], D. Degenstein[8], S. M. Frith[9], L. Froidevaux[10], E. Kyrölä[11], M. Laine[11], C. Long[12], A. A. Penckwitt[7], C. E. Sioris[8], K. H. Rosenlof[3], C. Roth[8], H.-J. Wang[13], and J. Wild[12,14]

[1]ETH Zurich, Zurich, Switzerland
[2]Cooperative Institute for Research in Environmental Sciences, University of Colorado, Boulder, Colorado, USA
[3]Chemical Sciences Division, NOAA Earth System Research Laboratory, Boulder, Colorado, USA
[4]University of Cambridge, Chemistry Department, Cambridge, UK
[5]Deutscher Wetterdienst, Hohenpeissenberg, Germany
[6]Department of Atmospheric and Planetary Science, Hampton University, Hampton, Virginia, USA
[7]Bodeker Scientific, Alexandra, New Zealand
[8]Institute of Space and Atmospheric Studies, University of Saskatchewan, Saskatoon, Saskatchewan, Canada
[9]Science Systems and Applications, Inc., Lanham, Maryland, USA
[10]Jet Propulsion Laboratory, California Institute of Technology, Pasadena, California, USA
[11]Finnish Meteorological Institute, Helsinki, Finland
[12]NOAA/NWS/NCEP/Climate Prediction Center, College Park, Maryland, USA
[13]Georgia Institute of Technology, Atlanta, Georgia, USA
[14]Innovim, Greenbelt, Maryland, USA

Correspondence to: F. Tummon (fiona.tummon@env.ethz.ch)

Abstract. In the framework of the SI2N (SPARC (Stratosphere-troposphere Processes And their Role in Climate)/IO_3C (International Ozone Commission)/IGACO-O3 (Integrated Global Atmospheric Chemistry Observations – Ozone)/NDACC (Network for the Detection of Atmospheric Composition Change)) initiative, several long-term vertically resolved merged ozone data sets produced from satellite measurements have been analysed and compared. This paper presents an overview of the methods, assumptions, and challenges involved in constructing such merged data sets, as well as the first thorough intercomparison of seven new long-term satellite data sets. The analysis focuses on the representation of the annual cycle, interannual variability, and long-term trends for the period 1984–2011, which is common to all data sets. Overall, the best agreement amongst data sets is seen in the mid-latitude lower and middle stratosphere, with larger differences in the equatorial lower stratosphere and the upper stratosphere globally. In most cases, differences in the choice of underlying instrument records that were merged produced larger differences between data sets than the use of different merging techniques. Long-term ozone trends were calculated for the period 1984–2011 using a piecewise linear regression with a change in trend prescribed at the end of 1997. For the 1984–1997 period, trends tend to be most similar between data sets (with largest negative trends ranging from -4 to $-8\,\%\,\text{decade}^{-1}$ in the mid-latitude upper stratosphere), in large part due to the fact that most data sets are predominantly (or only) based on the SAGE-II record. Trends in the middle and lower stratosphere are much smaller, and, particularly for the lower stratosphere, large uncertainties remain. For the later period (1998–2011), trends vary to a greater extent, ranging from approximately -1 to $+5\,\%\,\text{decade}^{-1}$ in the upper stratosphere. Again, middle and lower stratospheric trends are smaller and for most data sets not significantly different from zero. Overall, however, there is a clear shift from mostly negative to mostly positive trends between the two periods over much of the profile.

1 Introduction

The phase-out of ozone-depleting substances (ODSs) through the Montreal Protocol and its subsequent amendments and adjustments (World Meteorological Organization (WMO), 2015; 2011) resulted in peak stratospheric ODS concentrations in the late 1990s or early 2000s; the exact timing of the peak depends on which part of the stratosphere is considered. Since then, ODS concentrations have been declining and stratospheric ozone is expected to return to 1980 levels throughout most of the stratosphere at various times during the 21st century (WMO, 2015; 2011; Eyring et al., 2010a; Austin and Butchart, 2003). Detecting such a response and attributing it to decreasing halogen levels requires long-term, temporally homogeneous ozone profile measurements. This is particularly important if the attribution is being done in the context of a changing climate, with increasing greenhouse gas concentrations and concomitant changes in stratospheric temperature and mean meridional transport affecting ozone in addition to the decrease in stratospheric halogen loading (e.g. Waugh et al., 2009; Eyring et al., 2007; Austin and Wilson, 2006).

While satellite instruments have provided continuous near-global measurements of the ozone profile since late 1978, no single instrument has provided continuous and stable global coverage for the entire period (e.g. Hassler et al., 2014). To date, the longest single-instrument space-based records are those of SAGE-II (Stratospheric Aerosol and Gas Experiment II) and HALOE (Halogen Occultation Experiment); these provided quasi-independent data from 1984 to 2005 (Damadeo et al., 2013; Grooß and Russell, 2005; McCormick et al., 1989). Over the past decade, a number of new satellite-based instruments have made ozone profile measurements (Hassler et al., 2014; Tegtmeier et al., 2013), but few continuous data sets extending the satellite record beyond 2005 using these new observations were available for consideration in WMO (2011). Therefore, the Stratosphere-troposphere Processes And their Role in Climate (SPARC), the International Ozone Commission (IO_3C), the ozone focus area of the Integrated Global Atmospheric Chemistry Observations (IGACO-O3), and the Network for the Detection of Atmospheric Composition Change (NDACC) supported the SPARC/IO_3C/IGACO-O3/NDACC (SI2N) initiative, with a major aim of developing and promoting long-term vertically resolved data sets for updated knowledge of long-term changes in the vertical distribution of ozone.

Included as part of the SI2N initiative are multiple data sets where SAGE-II has been merged with newer measurements (Adams et al., 2014; Davis et al., 2015; Froidevaux et al., 2015; Penckwitt et al., 2015; Kyrölä et al., 2013), as well as data sets based on the consistently processed SBUV and SBUV/2 measurements (e.g. Frith et al., 2014; Wild and Long, 2015; Kramarova et al., 2013; McPeters et al., 2013). Constructing these merged data sets requires careful consideration of several methodological choices so as to avoid introducing artefacts that may contaminate derived trends. Such factors include sampling biases, the use of different native vertical coordinate systems, a combination of different vertical and horizontal resolutions, measurements made on different segments of the diurnal cycle in ozone, and relative calibration – all of which must be considered when combining measurements from two or more satellite records.

Issues related to merging data from several records

Sampling by an individual satellite instrument can vary in space and/or time during its lifetime (e.g. McPeters et al., 2013; Toohey et al., 2013) and can thereby introduce biases in analyses of single-instrument records if not addressed properly (Damadeo et al., 2014). When merging measurements from more than one instrument, not only changing sampling patterns but also differences in sampling between the various instruments need to be taken into account. This is especially important in regions where there are strong ozone gradients such as across mixing barriers in the stratosphere or in the upper troposphere/lower stratosphere, where small differences in sampling may result in large differences in observed ozone (Lambert et al., 2015; Toohey et al., 2013). Ensuring that differences in temporal sampling are adequately taken into account is also important in the upper stratosphere, where the diurnal cycle in ozone is significant (Parrish et al., 2014; Schanz et al., 2014; Sakazaki et al., 2013; Studer et al., 2014) and has the potential to introduce biases when merging instruments that measure at different times of the day.

Ideally, biases between individual instruments would be traceable to fundamental differences in the instruments and/or retrieval algorithms, which could then be corrected for during the merging process. However, in practice, this remains challenging. Alternatively, individual data sets can be adjusted using a completely independent set of measurements, for example, with ground-based total column observations (e.g. Bodeker et al., 2005) or using a model as a transfer function (e.g. Hegglin et al., 2014). As yet, it has not been possible to a priori eliminate all systematic biases in any vertically resolved ozone data set, and therefore one data set is typically chosen as a reference, with others bias-corrected with respect to that reference. This requires a sufficiently long overlap between the instrument records to derive a statistically significant estimate of the bias, which can then be propagated to regions (in space and time) outside of the overlap. This is particularly important if the adjustments have a seasonal or spatial dependence. Such corrections are not always possible; linking the BUV record to the multi-instrument SBUV record or the SAGE-I record to the SAGE-II record is not possible as a result of lack of overlap.

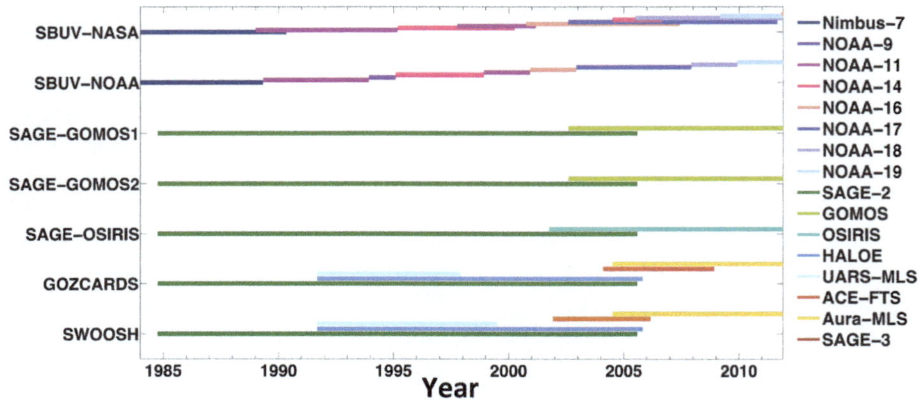

Figure 1. Time coverage of the individual instruments used in each of the seven data sets considered in this study for the 1984–2011 period.

Assuming that the temporal and spatial sampling bias has been corrected, the data sets to be merged may need to be transformed to a common coordinate system if they do no have the same intrinsic vertical coordinates and ozone units. For example, the native vertical coordinate for the solar occultation technique used by the SAGE-II instrument is geometric altitude with ozone being retrieved in units of number density, while the thermal emission measurements of the MLS (Microwave Limb Sounder) instruments provide ozone amounts on pressure levels and in units of mixing ratio. To combine two such data sets, one of them needs to be converted to the vertical coordinate of the other, requiring knowledge of the vertical temperature profile, and also to a common concentration unit, requiring local temperatures. For long-term ozone trends, where changes are on the order of a few percent per decade, the uncertainties in long-term stratospheric temperature records can confound such conversions between vertical coordinate systems and concentration units (Davis and Rosenlof, 2012; Thompson et al., 2012; Wang et al., 2011; McLinden and Fioletov, 2011; Xu and Powell, 2011). Any artificial trend in stratospheric temperature structure, which affects the altitudes of pressure levels, can alias into ozone trends (WMO, 2015; McLinden and Fioletov, 2011; Rosenfield et al., 2005). The different vertical and horizontal resolutions of each data set also need to be taken into account; adding data with a relatively low resolution to a high-resolution data set involves either a degradation of the high-resolution measurement or the need for additional information to justify interpolating the low-resolution product to a higher resolution.

These methodological aspects are discussed further below in the context of an in-depth intercomparison of seven newly available merged satellite ozone profile data sets. A detailed assessment of these data sets is not only useful for analyses of long-term ozone changes; it is also useful because such data sets are commonly used to validate chemistry and transport in numerical models (e.g. Eyring et al., 2010a) and can be used to prescribe ozone boundary conditions for models that do not explicitly include interactive stratospheric chemistry. The merged data sets and methods used in this paper are described in Sect. 2, while the analyses of the annual cycles and interannual variability of each data set are covered in Sect. 3. A comparison of the long-term ozone changes estimated from each data set is then presented in Sect. 4 and, finally, conclusions are presented in Sect. 5.

2 Data and methods

2.1 Merged data sets used in this study

In the absence of a single perfectly calibrated satellite instrument with complete and continuous global coverage and decadal stability, measurements from multiple instruments need to be combined. A critical factor when combining multiple data sets is the relative calibration of the measurements from the different instruments. In principle, data sets can be merged by combining data from a series of instruments of the same type, or by merging data from instruments of different types. Seven merged data sets are considered here: two are based on the series of SBUV instruments, while the other five use measurements from a number of different recent instruments which all have the long SAGE-II record (1984–2005) as their backbone. The seven data sets are briefly described below, while further details can be found in the corresponding references provided. Their characteristics are summarised in Tables 1–7, the temporal coverage of the underlying satellite instruments used in each data set are shown in Fig. 1, and the spatial coverage over time as well as the number of observations used in each data set is shown in Fig. 2.

2.1.1 BUV, SBUV, and SBUV/2

A series of BUV (Backscatter Ultraviolet Radiometer) and SBUV (Solar Backscatter Ultraviolet Radiometer) instruments have flown onboard NASA and NOAA satellites since 1970 and provided a continuous ozone record since 1978.

Table 1. Summary of the latitudinal coverage and resolution as well as the temporal coverage and resolution of each of the seven data sets.

Data set	Latitude coverage	Latitudinal resolution	Temporal coverage	Temporal resolution
v8.6 SBUV MOD	80° N–80° S	5°	1970/01–2013/12	Monthly means and daily overpass data at selected stations. Daily zonal means available on request.
SBUV Merged Cohesive	80° N–80° S	5°	1978/11–2013/12	Monthly and daily means
SAGE-GOMOS1	60° N–60° S	10°	1984/10–2011/12	Monthly means
SAGE-GOMOS2	90° N–90° S	5°	1984/10–2012/03	Monthly means
SAGE-OSIRIS	65° N–65° S	5°	1984/10–2011/12	Monthly means
GOZCARDS	90° N–90° S	10°	1979/01–2012/12	Monthly means
SWOOSH	90° N–90° S	10° (also 2.5 °)	1984/10-2013/12	Monthly means

Table 2. Summary of the vertical range, units, and number of levels, as well as the ozone units of each of the seven data sets.

Data set	Vertical range and unit	No. vertical levels (vertical resolution)	Ozone units
v8.6 SBUV MOD	50–0.5 hPa or 9 partial columns in DU	15 (~10 km near 0.75 hPa, ~6–7 km near 3 hPa, ~15 km near 50hPa)	ppmv or DU
SBUV Merged Cohesive	50–0.5 hPa or 13 Dobson layers	15 (~10 km near 0.75 hPa, ~6–7 km near 3 ppmv, ~15 km near 50hPa)	ppmv or DU
SAGE-GOMOS1	10–70 km	61 (1 km)	molecules cm^{-3}
SAGE-GOMOS2	10–69 km	60 (1 km)	10^{18}molecules m^{-3}
SAGE-OSIRIS	18.5–49.5 km	32 (1 km)	molecules cm^{-3}
GOZCARDS	215–0.2 hPa	25 (~2.7 km)	ppmv
SWOOSH	316.23–1 hPa	31 (~1.3 km, using a 7 km scale height)	ppmv

Measurements are made with the same instrument type and generally there has been a good temporal overlap between individual instruments (Bhartia et al., 2013; McPeters et al., 2013, and references therein). A new version of the retrieval algorithm (v8.6) was developed in which a number of improvements were made, including new ozone absorption cross sections, a new a priori ozone climatology, and a cloud-height climatology derived from Aura OMI (Ozone Monitoring Instrument) measurements (Bhartia et al., 2013). Inter-instrument calibration at the radiance level (as opposed to ozone measurement calibration) is accomplished during periods of overlap between the SBUV instruments and with the SSBUV (Shuttle SBUV) flown periodically on several different space shuttles (DeLand et al., 2012).

Validation of the reprocessed ozone shows that total column ozone is consistent with measurements from the Brewer–Dobson network to 1 % (Labow et al., 2013). Comparisons with Aura MLS, SAGE, ozonesondes, and lidar observations show that ozone at individual levels in the stratosphere is generally consistent to within 5 % (Kramarova et al., 2013). Inter-instrument differences are generally less than the differences compared to independent data sets. However, despite the common instrument design and algorithm, differences in data quality exist between instruments. Kramarova et al. (2013) report larger biases and drifts for NOAA-9, as well as portions of NOAA-11 and NOAA-14 data, from the mid- to late 1990s, primarily as a result of a slow orbit drift of these instruments. The drifting orbits cause the geometric positions of the instruments with respect to the Sun to vary in time, leading to changing error characteristics as a function of altitude and latitude. Agreement with independent measurements is largely within 10 % during this period.

It is important to note that the primary source of error in the SBUV retrieval is the smoothing error due to the instrument's limited vertical resolution, particularly in the troposphere and lower stratosphere (Kramarova et al., 2013; Bhartia et al., 2013). The SBUV instrument has a resolution of 6–7 km near 3 hPa, degrading to 15 km in the troposphere (Bhartia et al., 2013). Thus SBUV reliably measures the partial column of ozone from the ground to the lower stratosphere, but must use a priori information to resolve the signal within that range. The v8.6 a priori derives from an ozone climatology constructed from Aura MLS and ozonesonde data which vary seasonally (monthly) but has no trend or interannual variability component (McPeters and Labow, 2012; Bhartia et al., 2013). The SBUV resolution is also somewhat reduced in the upper stratosphere, degrading to ~10 km above 1 hPa.

Two merged data sets, described below, have been constructed using the SBUV v8.6 reprocessed observations. We limit the vertical range of our analysis of the SBUV data sets

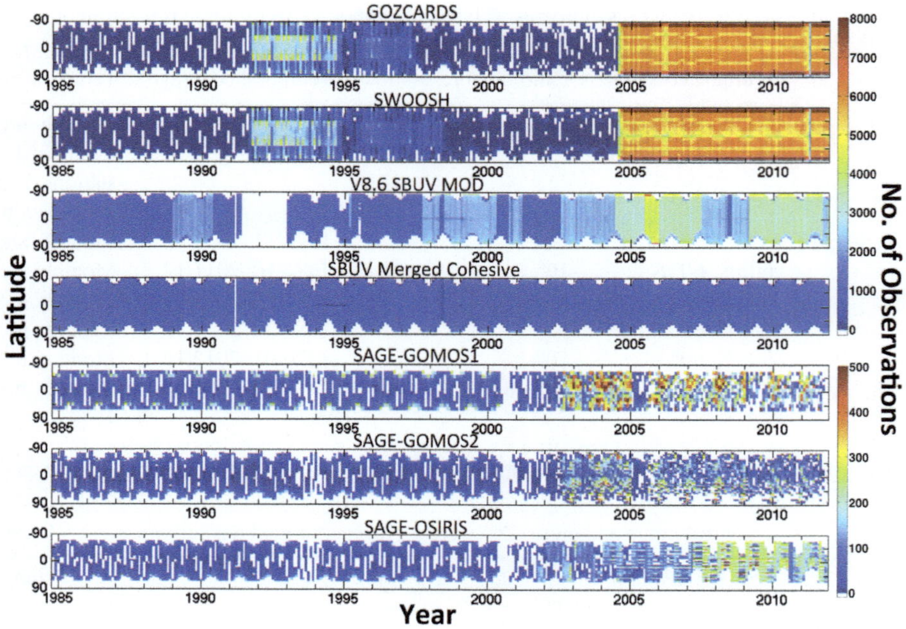

Figure 2. Latitudinal availability of data over the 1984–2011 period. The colours represent the sum of measurements used per latitudinal grid box over the entire profile for each merged data set (note that two separate colour scales are used).

Table 3. Summary of the uncertainties provided and individual satellite data sets merged for each of the seven data sets.

Data set	Uncertainties given with values	Data sources merged
v8.6 SBUV MOD	Smoothing error	BUV (1970–1976) – reduced coverage from 1972 onwards, SBUV (1978–1990), SBUV/2 (1985 onward) (NOAA 11, 14, 16, 17, 18, 19)
SBUV Merged Cohesive	Standard error	SBUV (1978–1990), SBUV/2 (1985 onward) (NOAA 9, 11, 14, 16, 17, 18, 19)
SAGE-GOMOS1	Standard error	SAGE-II v7 (1984–2005), GOMOS vIPF6 (2002–2011)
SAGE-GOMOS2	Standard error	SAGE-II v7 (1984–2005), GOMOS vIPF6 (2002–2011)
SAGE-OSIRIS	Standard error	SAGE-II v7 (1984–2005), OSIRIS v5.07 (2002–2011)
GOZCARDS	Standard deviations and standard error	SAGE-I v5.9rev (1979–1981), SAGE-II v6.2 (1984–2005), HALOE v19 (1991–2005), UARS MLS v5 (1991-1997), ACE-FTS v2.2 update (2004–2009), Aura MLS v2.2 (2004 onward)
SWOOSH	Standard deviation and mean profile uncertainty	SAGE-II v7.0 (1984–2005), SAGE-III v4.0 (2002–2005), HALOE v19 (1991–2005), UARS MLS v6 (1991–2005), Aura MLS v3.3 (2004 onward)

to pressures < 20 hPa in the tropics (20° N–20° S) and to pressures < 30 hPa outside the tropics to exclude regions where the low vertical resolution may affect derived trends.

V8.6 SBUV MOD

This data set, produced by researchers at NASA, is based on measurements from the BUV instrument on Nimbus 4 covering the period 1970–1976 (with reduced coverage after mid-1972), and the SBUV on Nimbus 7 and SBUV/2 instruments on NOAA-11, -14, -16, -17, -18, and -19 covering the period 1979–2013 (McPeters et al., 2013). Measurements from the SBUV/2 instrument onboard NOAA-9 are not included (see Fig. 2). Since the v8.6 algorithms include the aforementioned upgraded inter-instrument calibration, the approach taken is to average all data that meet quality standards set by calibration analysis and comparisons with external instruments rather than attempt to apply further external offsets (Frith et al., 2014). The largest measurement uncertainties occur in the 1990s, when instrument orbital drift led to less reliable measurements overall (Kramarova et al., 2013; DeLand et al., 2012). The Nimbus 4 BUV is included in the data set but cannot be intercalibrated as a result of the lack of temporal

Table 4. Summary of the merging approaches used for each of the seven data sets.

Data set	Merging approach
v8.6 SBUV MOD	Based on the reprocessed V8.6 SBUV(/2) data. No external calibration is carried out. Intercalibration between instruments is achieved at the radiance level, while remaining offsets are not adjusted for.
SBUV Merged Cohesive	Using the reprocessed V8.6 SBUV(/2) data, a correlation graph of the overlap period is examined and a line fitted to the comparison. This approach models an overall bias and accounts for differences in representation of the annual cycle. A single satellite is chosen to represent a period based on the best latitudinal coverage.
SAGE-GOMOS1	SAGE-II sunset and sunrise observations adjusted separately to GOMOS data. Individual profiles shifted using an average bias profile (for the entire SAGE-II period).
SAGE-GOMOS2	SAGE-II used as reference. GOMOS adjusted to SAGE-II using statistically estimated offsets (latitude/altitude/seasonally varying) for the overlap period (04/2002–08/2005).
SAGE-OSIRIS	SAGE-II used as reference, OSIRIS (only AM measurements used) is adjusted to SAGE-II for the overlap period (2001–2005) using monthly mean anomalies in 9° latitude bands.
GOZCARDS	SAGE-II used as reference, most other data sets adjusted for offset versus SAGE-II and then averaged.
SWOOSH	Aura MLS used as reference, SAGE and UARS adjusted to Aura MLS (offset corrections varying with latitude and altitude). A weighted mean of all available data points for each latitude/altitude/time bin is then calculated, with weighting based on the number of observations from the given sensor.

Table 5. Summary of the data screening applied to each of the seven data sets.

Data set	Data screening applied
v8.6 SBUV MOD	SBUV(/2) data are screened as outlined in ftp://toms.gsfc.nasa.gov/pub/sbuv/zonal_means/README.SBUVL3MZM.pdf. Monthly means are discarded if the number of measurements is less than $2/3$ the nominal number, if the average latitude is greater than 1° from the centre of the zonal bin, if the average time within the month is more than 4 days from the centre of the month, or if the mean equator crossing time of the orbit is outside the 08:00–16:00UTC. range (with the exception of 1994–1995 to avoid a data gap).
SBUV Merged Cohesive	SBUV screened according to NESDIS readme document: http://www.star.nesdis.noaa.gov/smcd/spb/ozone/dvd_v8/DVDhtml/V8_Data_Documentation.html with profile error flag set to 0 or 100, maximum solar zenith angle of 84°, and maximum ResQC of 20. Monthly means are discarded if there are less than 20 measurements in a zone or if the mean latitude is greater than 1° from the zone centre.
SAGE-GOMOS1	SAGE-II screened according to the SAGE-II v7.00 release notes. GOMOS measurements screened according to the GOMOS level 2 product quality readme file.
SAGE-GOMOS2	SAGE-II screened according to Hassler et al. (2008). GOMOS screened according to the GOMOS level 2 product quality readme file.
SAGE-OSIRIS	SAGE-II screened according to the SAGE-II v7.00 release notes. Only OSIRIS morning (descending node) observations at solar zenith angles $< 89.5°$ are used.
GOZCARDS	SAGE-I screened for aerosol/cloud effects. SAGE-II screened for aerosol/cloud effects as per Wang et al. (2002) – also includes other details (e.g. for high beta angle issues). HALOE screened for aerosol/cloud effects following Bhatt et al. (1999) and Hervig and McHugh (1999). UARS MLS: screened as per Livesey et al. (2003). Aura MLS screened as per Froidevaux et al. (2008). ACE-FTS: generally screened as per ACE-FTS team guidelines; outliers screened as per Froidevaux et al. (2015). Full details can be found in Froidevaux et al. (2015).
SWOOSH	SAGE-II data screened according to Wang et al. (2002) and then with a 3σ filter. Only UARS MLS data between 100 and 0.22hPa used and screened following Livesey et al. (2003). "Trip angle" and "constant lockdown angle" events are removed from the HALOE data, as are any points with uncertainties $\geqslant 100\,\%$. Aura MLS measurements are filtered according to the v3.3 data quality document. See full screening description and further references in Davis et al. (2015).

Table 6. Summary of the treatments of the diurnal cycle and references for each of the seven data sets.

Data set	Diurnal cycle treated?	Description and validation references
v8.6 SBUV MOD	No.	McPeters et al. (2013); Frith et al. (2014); Kramarova et al. (2013)
SBUV Merged Cohesive	Not explicitly.	McPeters et al. (2013); Kramarova et al. (2013); Wild and Long (2015)
SAGE-GOMOS1	Differentiating between sunrise and sunset SAGE-II profiles.	Kyrölä et al. (2013)
SAGE-GOMOS2	Above 2 hPa data normalised to solar zenith angle to account for diurnal ozone cycle.	Penckwitt et al. (2015)
SAGE-OSIRIS	Above ~50 km coincidence criteria are chosen very narrow to avoid the diurnal cycle problem.	Adams et al. (2013, 2014); Sioris et al. (2014)
GOZCARDS	Not explicitly, but data sets are adjusted to the mean of SAGE II sunsets and sunrises and sampling is most often quite uniform in time for each instrument's data set (when used).	Froidevaux et al. (2015)
SWOOSH	Data set only extends up to 1 hPa, therefore diurnal cycle problems are assumed to be negligible.	Davis et al. (2015).

Table 7. URLs for each of the seven data sets.

Data set	URL
v8.6 SBUV MOD	http://acd-ext.gsfc.nasa.gov/Data_services/merged/
SBUV Merged Cohesive	ftp://ftp.cpc.ncep.noaa.gov/SBUV_CDR/
SAGE-GOMOS1	http://www.esa-spin.org/index.php/spin-data-sets or http://igaco-o3.fmi.fi/VDO/data.html
SAGE-GOMOS2	http://www.esa-spin.org/index.php/spin-data-sets
SAGE-OSIRIS	Available upon request.
GOZCARDS	https://gozcards.jpl.nasa.gov or http://mirador.gsfc.nasa.gov
SWOOSH	http://www.esrl.noaa.gov/csd/groups/csd8/swoosh/

overlap with other SBUV instruments. These data are useful for investigating pre-1980 ozone levels but should not be used in a formal trend analysis.

SBUV Merged Cohesive

Though the improved calibrations of v8.6 have reduced the inter-satellite differences, it has not removed them. A second SBUV-based data set was developed by researchers at NOAA with the aim of removing the remaining differences. This approach varies from that of v8.6 SBUV MOD in that (i) adjustments are made to individual instrument records based on periods of overlap in order to account for any variations in the observed annual cycle as well as an overall bias; (ii) rather than an average of all available observations, a single satellite is chosen for each period based on the best latitudinal coverage allowing the clean retention of satellite characteristics such as time of measurement, solar zenith angle, etc. to be identified with an ozone value; (iii) measurements from NOAA-9 are included in a short period to allow greater global coverage in the bridge from NOAA-11 to -14 (Fig. 2); and (iv) measurements from BUV are excluded since there is no overlap with the subsequent instruments (Wild and Long, 2015). The resulting differences between the two SBUV-based data sets are further described in Sects. 3 and 4.

2.1.2 SAGE-based data sets

The 21-year SAGE-II record is a natural candidate to form the basis of a merged data set using multiple instrument types, and it is used in five of the seven records described here (Tables 1–7 and Fig. 1). The conceptually simplest extension is to add a single data set to the SAGE-II record to cover the years following 2005 after which SAGE-II was turned off. The period from 2005 to the present has had many operational satellite instruments (Hassler et al., 2014; Tegtmeier et al., 2013). To date, three single-instrument extensions have been made to the SAGE-II record. These include one case extended with limb-scattered measurements from OSIRIS (Optical Spectrograph and Infrared Imager System) onboard the Odin satellite (2001–present) (Bourassa et al., 2014; Adams et al., 2014; Sioris et al., 2014) and two cases using the stellar occultation measurements from GOMOS (Global Ozone Monitoring by Occultation of Stars) onboard the ENVISAT satellite (2002–2012) (Penckwitt et al., 2015; Kyrölä et al.,

2013). Both OSIRIS and GOMOS have similar vertical resolution to SAGE-II as well as a multi-year period of overlap.

SAGE-OSIRIS

The SAGE-OSIRIS merged data set combines the SAGE-II v7.0 and OSIRIS v5.07 measurements into a continuous data series covering the period 1984–2013 (Bourassa et al., 2014; Sioris et al., 2014). Data from each instrument are individually deseasonalised and thereafter the differences (varying in latitude and altitude) between the two sets of anomalies are calculated for the overlap period (January 2002–August 2005). The OSIRIS data, of which only the morning (sunrise) measurements are used since they show smaller bias compared to SAGE-II (Adams et al., 2013), are then shifted by this difference to produce a consistent time series of SAGE-II and OSIRIS data covering October 1984 to December 2013; typically values were shifted by less than 3 % (Bourassa et al., 2014).

SAGE-GOMOS

Two different merged data sets were produced by combining the SAGE-II v7.0 (using both sunrise and sunset observations) and GOMOS IPF6.0 measurements. The data set described and analysed by Kyrölä et al. (2013) is constructed taking into account the difference between SAGE-II sunrise and sunset profiles and the GOMOS night-time stellar occultation measurements, i.e. the SAGE-II sunrise and sunset measurements are adjusted separately to GOMOS (used as reference) for each latitude and altitude bin. For the overlap period (April 2002–August 2005), the weighted mean of the medians from both instruments is used, with the weights being determined from the error estimates of each median (Kyrölä et al., 2013). This is hereafter referred to as the SAGE-GOMOS1 data set.

The second SAGE-GOMOS data set was constructed using SAGE-II as reference. The GOMOS data were adjusted to SAGE-II using latitude- and altitude-varying offsets, which were estimated statistically for the overlap period from April 2002 to August 2005. The offsets vary with season, but not with year. At pressures <2 hPa, where the diurnal cycle in ozone may create differences between SAGE-II and GOMOS because of differences in the solar zenith angle of the measurements rather than because of instrument/retrieval-related biases between instruments, the SAGE-II and GOMOS data were normalised to a solar zenith angle of 90° using scaling factors derived from a high-resolution chemistry–climate model simulation. The data were accumulated into 5° latitude bands and then averaged to monthly means after having been corrected for zonal and monthly mean representativeness (Penckwitt et al., 2015). This data set will be referred to hereafter as the SAGE-GOMOS2 data set.

GOZCARDS

The SAGE-II (v6.2) record is also the central plank in the GOZCARDS (Global OZone Chemistry And Related trace gas Data records for the Stratosphere) data set, which is combined with data from five other instruments (SAGE-I v5.9rev, HALOE v19, UARS MLS v5, Aura MLS v2.2, and ACE-FTS v2.2 update) (Froidevaux et al., 2015). GOZCARDS extends the record back to include the SAGE-I data (although the data from this period are not considered in this study). The record is extended beyond the end of SAGE-II using a combination of Aura MLS and ACE-FTS measurements (see Tables 1–7 for details). The SAGE-II version 6.2 ozone number density versus height profiles were converted to mixing ratios on pressure levels using interpolated NCEP (National Centers for Environmental Prediction) temperature profiles obtained and archived by the SAGE-II instrument team. The converted SAGE-II monthly means are used as the reference data set, and other data sets are essentially bias-corrected against this reference.

SWOOSH

Similar to the GOZCARDS data set, SWOOSH (Stratospheric Water and OzOne Satellite Homogenized) combines SAGE-II (v7.0) ozone measurements with data from several other satellite instruments (Davis et al., 2015). The choice of instruments and data versions used are, however, slightly different than GOZCARDS (see Tables 11–13). Data sets originally in number density and altitude coordinates were converted to mixing ratios on pressure levels using the MERRA (Modern Era Retrospective analysis for Research and Applications) reanalysis (Rienecker et al., 2011). Aura MLS v3.3 is chosen as the reference data set, and the other instruments are adjusted by offsets that vary with latitude and height (but not time). Offsets are calculated from coincident observations during overlap periods between instruments. The final combined product is a mean of all available measurements in each latitude/height/time bin, but with greater weight given to instruments that sample more frequently (e.g. Aura MLS). Filled and unfilled versions of the data set exist on both geographical and equivalent latitude coordinates. Here we use the unfilled version on geographical coordinates averaged into 10° latitude bins.

Each of the five SAGE-II-based records use a different data set to fill in the recent and pre-1984 data (if extending back that far) and/or a different merging approach; differences between them will therefore reflect the use of these different instrument records, some differences in data versions, and differences in merging techniques. The relative importance of an individual instrument is reduced as more measurement sets are added. We examine the impact of these differences in Sects. 3 and 4.

2.2 Vertical coordinates and a common grid

The natural vertical coordinate of the limb and solar/stellar occultation techniques is geometric altitude with ozone concentration calculated as number density, while the nadir-viewing backscatter technique used by the SBUV instruments and thermal emission measurements by the MLS instruments provide ozone amounts in mixing ratio on pressure levels. The three data sets produced on an altitude–number-density grid (SAGE-OSIRIS and the two SAGE-GOMOS data sets) were converted to pressure and mixing ratio coordinates using the MERRA reanalysis (Rienecker et al., 2011). Sensitivity tests using the JRA-55 reanalysis (Ebita et al., 2011) for conversion showed negligible differences between monthly mean profiles compared to conversion with MERRA (not shown). MERRA stratospheric temperatures have, furthermore, been shown to compare well with other recent reanalyses (e.g. Simmons et al., 2014; Bosilovich et al., 2011; Rienecker et al., 2011). Once converted to the same vertical coordinates and ozone concentration units, the seven data sets were interpolated in log pressure space to a common grid to facilitate comparison. No spatial interpolation in the horizontal was applied; instead, data were averaged from their native resolution (5 or 10°) into latitude bands of interest (see Sect. 3). Furthermore, no additional screening was applied to any of the seven data sets.

2.3 Multi-data-set mean

A multi-data-set mean (MDM) is constructed to provide a common point of reference to compare the data sets. The MDM is by no measure the best representation of ozone but provides a simple average of all available data, rather than favouring one particular merged data set. It is calculated by averaging all available data for each time step and latitude/pressure bin, with no weighting applied. A weakness of the MDM is that five of the seven data sets averaged are based on the SAGE-II record; therefore, despite some of the data sets either using another instrument as reference or including other observations during the SAGE-II period, the MDM is largely dominated by the SAGE-II signal from 1984 to 2005.

2.4 Multiple linear regression model

In this study we use the multiple linear regression model described by Hassler et al. (2013), which in turn is an update of the model used by Bodeker et al. (1998). To quantify ozone variability and trends we include basis functions representing: the QBO (quasi-biennial oscillation), specified as monthly mean 50 hPa Singapore zonal wind and a synthetic basis function orthogonal to this – to allow for a time lag at different latitudes and altitudes (Austin et al., 2008); ENSO (El Niño–Southern Oscillation), using the monthly mean Southern Oscillation index as proxy; the solar cycle, as represented by monthly mean F10.7 solar flux data from NOAA's National Geophysical Data Center; and a proxy for ozone perturbations forced by aerosols from the Mt Pinatubo volcanic eruption based on a synthetic time series representing the approximate temporal evolution of stratospheric aerosol concentrations following the eruption (see Bodeker et al., 1998, for further details). Fioletov (2008) provides further details and a general overview of how each of the processes described by these proxies affects ozone variability. Equation (1) presents the simplest form of how the model is applied:

$$\begin{aligned}\Omega_t =\ & A_{(\mathrm{NA}=4)}+\\ & B_{(\mathrm{NB}=2)} \times t+\\ & C_{(\mathrm{NC}=2)} \times t_{t=0 \text{ for } t<\text{inflectionpoint}}+\\ & D_{(\mathrm{ND}=2)} \times \mathrm{QBO}(t)+\\ & E_{(\mathrm{NE}=2)} \times \mathrm{QBO}_{\mathrm{orthog}}(t)+\\ & F \times F10.7(t)+\\ & G \times \mathrm{Pinatubo}(t)+\\ & H \times \mathrm{ENSO}(t)+\\ & R(t), \end{aligned} \tag{1}$$

where Ω_t is the ozone for a particular month t for a particular data set; A–E are the model coefficients corresponding to an offset term (to account for the annual average ozone amount), linear trend, and other basis functions used; and $R(t)$ represents the residuals (difference between the measured and statistically modelled ozone values). Furthermore, each coefficient has a constant term which represents the mean value (seasonally unvarying) of each predictor. The subscript on each term A–E indicates how many Fourier pairs the term was expanded into to account for the seasonal dependence of the ozone anomalies on the basis functions (Bodeker et al., 1998); for example, NB = 2 indicates two Fourier pairs (two sine, two cosine). An autoregressive model is applied to the residuals $R(t)$ following Eq. (2):

$$R(t) = \epsilon_1(t) \times R(t-1) + \epsilon_2(t) \times R(t-2) + e_t, \tag{2}$$

where ϵ_1 and ϵ_2 are the model coefficients and e_t represents the independent random errors with zero mean and variances that are allowed to change from month to month (see Reinsel et al., 1994).

Piecewise linear regression is chosen for the analysis because a central point of interest is whether there is any evidence for a change in the ozone trend after the peak in EESC (Jones et al., 2009; Steinbrecht et al., 2006; Newchurch et al., 2003). The break point was chosen at the end of 1997, as has been used in a number of other recent studies (e.g. Bourassa et al., 2014; Chehade et al., 2014; Kyrölä et al., 2013; Laine et al., 2013; Jones et al., 2009). The regression was applied to each level of data on the common pressure grid and trends were only estimated if more than 50 % of the data for a particular level were available (i.e. more than half of all months had data). We also calculate the uncertainty associated with

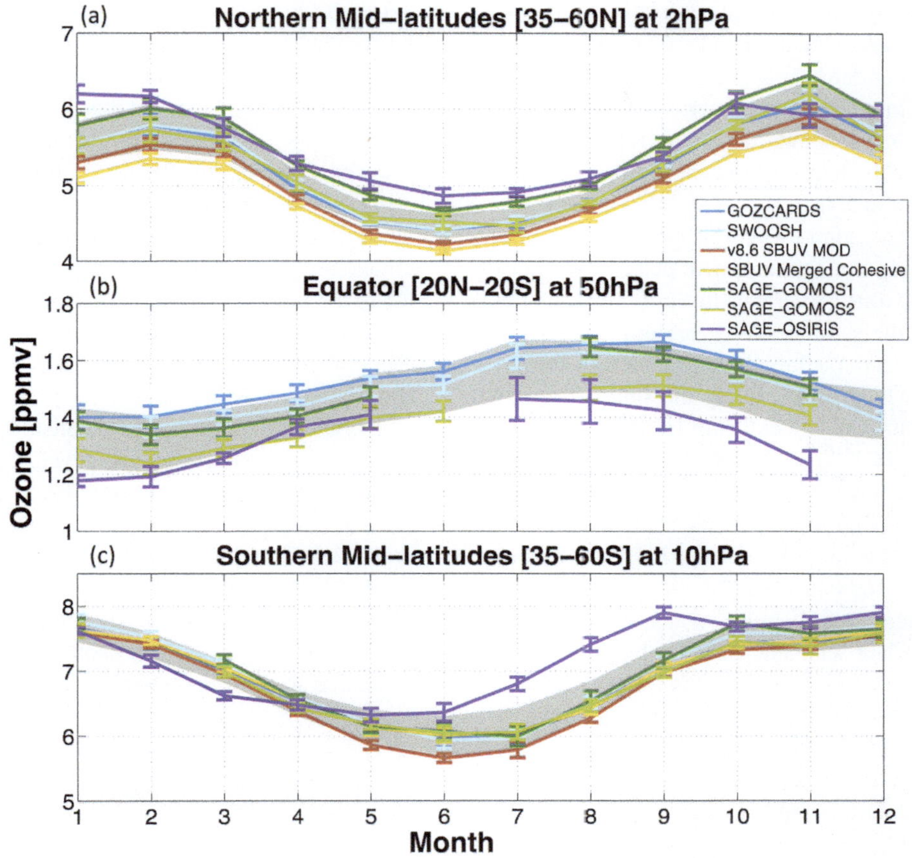

Figure 3. Annual cycle averaged over the 1984–2011 period for **(a)** the Northern Hemisphere mid-latitudes (35–60° N) at 2 hPa, **(b)** the tropics (20° N–20° S) at 50 hPa, and **(c)** the Southern Hemisphere mid-latitudes (35–60° S) at 10 hPa. Error bars indicate ±2 standard errors of the mean, while the grey range shows ±1 standard deviation (mean of all data sets). Values (monthly values, standard deviation, and standard error) are shown only if data are available for more than half (14) of the 28 years in the time series.

each trend estimate based on the variance in the residual time series and present the 2σ uncertainties on the trends throughout this paper.

3 Data set intercomparison

The core analysis presented in this section is a comparison of the seven merged data sets for the period common to all data sets: October 1984–December 2011. We highlight similar features and major differences between the data sets before examining the derived trends in Sect. 4. Results are shown for three latitude zones: northern mid-latitudes (35–60° N), tropics (20° N–20° S), and southern mid-latitudes (35–60° S) similar to those used in WMO (2015, 2011). To ensure representativeness, data were area-weighted by the cosine of latitude and averaged for each region only if more than two-thirds of the data for each latitude band were available.

3.1 Annual cycle

Figure 3 presents the annual cycle averaged over 1984–2011 for three selected levels in the three latitude regions (i.e. climatological average, not the annual cycle derived from the multiple regression model). These levels were chosen because they represent a spread of the variability between regions and are particular levels of interest. Ozone at 2 hPa (upper stratosphere) in the mid-latitudes has been most strongly affected by chemical depletion by ODSs. In the equatorial regions at 50 hPa (lower stratosphere), measurements are most uncertain because of the strong vertical ozone gradient (Tegtmeier et al., 2013). The 10 hPa level was chosen to give an indication of ozone changes in the middle stratosphere. The vertical error bars indicate ±2 standard error of the mean for each individual data set, while the grey-shaded region indicates the 1 standard deviation range (mean value of all data sets ± the mean of the standard deviations, which are calculated for each data set individually before averaging). The standard error of the mean provides a useful approximation of the uncertainty of the mean, although it may not represent the true uncertainty since individual samples of the population may exhibit autocorrelations (Toohey and von Clarmann, 2013). The standard deviation represents the ozone variability for each month, averaged across all systems. Averages, standard error, and standard deviations were only calculated for months that had data for more than 14 of the 28 years available for analysis.

Figure 3a shows the annual cycle of ozone in the northern mid-latitude upper stratosphere, with peak values during the winter months (from November through February) and minimum values in the summer (May through August). The annual cycle in this region is largely determined by catalytic ozone destruction, which peaks during the summer months, resulting in maximum values occurring in winter (Brasseur and Solomon, 2005 Perliski et al., 1989). All seven data sets show similar annual cycles, both in terms of phase and amplitude, and mostly lie within the ± 1 standard deviation range shown by the grey shading (the standard deviation across all data sets gives an indication of the spread between data sets). However, in terms of absolute values, there is quite a spread between the data sets. The SBUV Merged Cohesive data set shows consistently lower values for all months of the year, while SAGE-OSIRIS shows the opposite tendency, with mostly consistently higher values, although for only 5 of the 12 months the ± 2 standard error bars do not overlap with the standard deviation range of all data sets. SWOOSH, GOZCARDS, and SAGE-GOMOS2 show remarkably similar annual cycles, with mean values not significantly different from each other in all months.

In the tropical lower stratosphere, where the ozone seasonal cycle is essentially determined by vertical transport associated with tropical upwelling, the peak ozone values from July through October (Fig. 3b) correspond to the months when upward transport of ozone-poor air from the troposphere is at a minimum (Randel et al., 2007). As for the upper stratosphere, differences are large, but the variability is large in this region as well, as seen in the large $\pm 1\sigma$ range (grey shading), which ranges up to 0.2 ppmv ($\sim$15 %). SWOOSH and GOZCARDS are again most similar to each other for most of the year, and show consistently higher values than the other data sets. SAGE-GOMOS1 and SAGE-OSIRIS show lowest values, with the latter having mean values that fall outside of the $\pm 1\sigma$ range in nearly all months. The three data sets that extend the SAGE-II record with just one data set (GOMOS or OSIRIS) have considerably more missing data in this region of the stratosphere, with less than half of the data available for certain months of the year (where no data are shown; see caption Fig. 3). This is at least in part because of data being filtered out after the eruption of Mt Pinatubo. As mentioned above, the SBUV data sets are not shown at this pressure level.

In the southern mid-latitude middle stratosphere nearly all data sets agree remarkably well with each other throughout the year (Fig. 3c). In this region of the atmosphere, the annual cycle is opposite in phase to that in the upper stratosphere (Fioletov, 2008; Perliski et al., 1989), with peak ozone values in the summer (October through February) resulting from photochemical ozone production during this season (Perliski et al., 1989). The only data set that shows significantly different ozone values for much of the year is SAGE-OSIRIS, which has values up to 1 ppmv ($\sim$15 %) higher in the austral winter season and slightly lower values in February and March. This feature is also evident in the northern mid-latitudes during winter (not shown), although to a lesser extent, and is likely due to the reduced sampling of the OSIRIS instrument in the winter hemisphere (see Fig. 2), which seems to quite strongly affect the mid-latitude zonal mean values in the merged data set.

Overall, in the three regions considered, the seven merged data sets show similar annual cycles, particularly in terms of phase and amplitude. Furthermore, with the exception of the SAGE-OSIRIS data set, the biases between data sets are largely consistent throughout the year. Agreement is best in the mid-latitude middle stratosphere, while there are larger differences between data sets in the tropical lower stratosphere and in the upper stratosphere globally – results which are similar to those reported by Tegtmeier et al. (2013) on an instrument-by-instrument basis.

3.2 Interannual variability

Figure 4 shows the time series of monthly mean ozone values for the same three regions and pressure levels as shown in Fig. 3. As for the mean annual cycles, the consistency between data sets varies somewhat from location to location, but, in general, agreement is best in the mid-latitude middle stratosphere, while differences between data sets are somewhat larger in the lower tropical stratosphere and in the upper stratosphere. In the northern mid-latitude upper stratosphere (Fig. 4a), differences between data sets range up to approximately 1 ppmv ($\sim$15 %) and are larger prior to about 1995. Differences are of similar magnitude (up to 0.25 ppmv or $\sim$15 %) in the equatorial lower stratosphere and tend to be more consistent over time (Fig. 4b). What is also particularly noticeable in this region is the large number of missing data in the SAGE-GOMOS and SAGE-OSIRIS data sets, which is in part due to the 1991 Pinatubo eruption, but also because the solar occultation technique used by SAGE-II does not allow measurements in all months of the year in the tropics (Damadeo et al., 2013). SWOOSH and GOZCARDS ameliorate the relatively sparse SAGE-II sampling in the tropics, particularly in the post-Pinatubo period, through the inclusion of HALOE and UARS-MLS observations after 1991. These two data sets show a considerably reduced peak in maximum ozone values in 1992 compared to other years (Fig. 4b) – a direct result of the Pinatubo eruption. The sparser sampling of the GOMOS instrument is also evident in the SAGE-GOMOS data sets after the end of the SAGE-II record in 2005, as is the summer-only sampling of

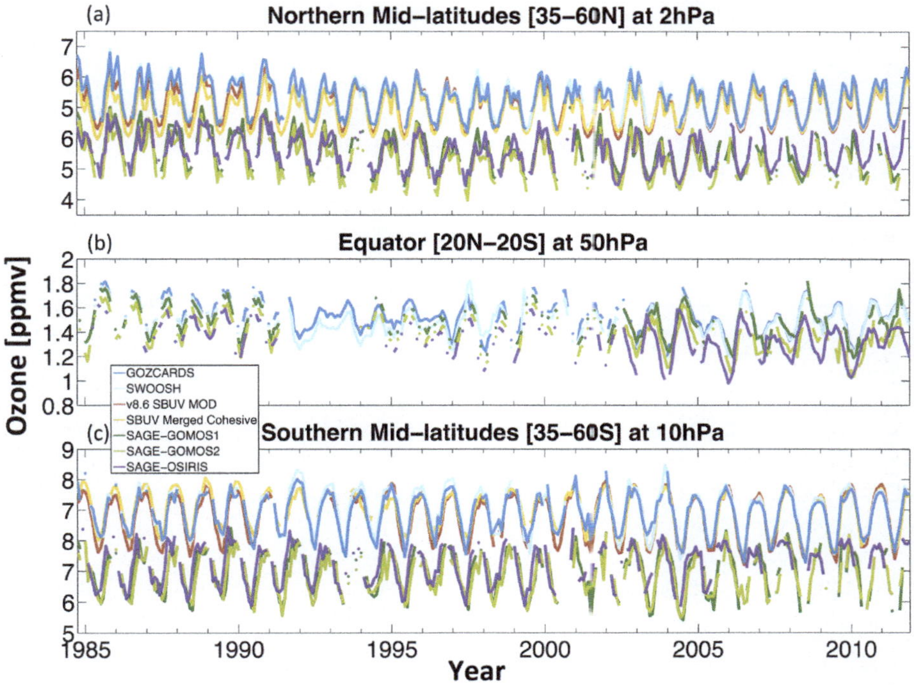

Figure 4. Monthly mean ozone (ppmv) in **(a)** the Northern Hemisphere mid-latitudes (35–60° N) at 2 hPa, **(b)** the tropics (20° N–20° S) at 50 hPa, and **(c)** the Southern Hemisphere mid-latitudes (35–60° S) at 10 hPa. Note that the data sets are separated in the two mid-latitude plots **(a)** and **(c)** for clarity, and thus the *y*-axis values are shifted.

the OSIRIS instrument clearly visible in the mid-latitudes (see especially Fig. 4c).

Figure 5 shows the monthly mean percentage differences from the MDM (multi-data-set mean; see Sect. 2.3) for the same three regions and pressure levels. Relative differences are estimated by dividing the difference between a particular data set and the MDM by the MDM (i.e. (data set X-MDM)/MDM). A 13-month running mean is applied, with values only being shown if data from more than 7 of the 13 months are available. In the upper stratosphere (Fig. 5a), agreement between data sets is, for most of the period, within ±8 % of the MDM for most data sets and well within ±5 % for SWOOSH, GOZCARDS, SAGE-GOMOS2, and v8.6 SBUV MOD. The SBUV Merged Cohesive data set shows a trend towards smaller and smaller differences from the MDM – a feature which is also to some extent visible in the monthly mean values in Fig. 4a but that is made more evident in terms of percent difference from the MDM. SAGE-GOMOS1 shows a slight reduction in differences from about 2001 onwards, which coincides with the introduction of the GOMOS data. Interestingly, the small improvement in agreement with the MDM also continues past the end of the SAGE-II record, so it is not simply the better spatial sampling when combining two instruments (see Fig. 2) that causes this. SAGE-OSIRIS shows a somewhat opposite tendency, with initially better agreement with the MDM during the overlap period between SAGE-II and OSIRIS (2001–2005) but then an increased difference from the MDM after 2005, when the SAGE-II record comes to an end. This may, as mentioned above, be related to the sparser sampling of the OSIRIS instrument during the winter months. Again, as discussed previously, despite SAGE-GOMOS1 using GOMOS as reference and SAGE-OSIRIS using SAGE-II as reference, the two data sets are remarkably similar in the upper stratosphere for the entire SAGE-II record from 1984 to 2005 (Fig. 5a). The SAGE-GOMOS2 data set, despite being referenced to SAGE-II, similar to SAGE-OSIRIS, shows lower values for the whole 1984–2011 period, perhaps because of the different treatment of the diurnal cycle of ozone in the upper stratosphere.

In the equatorial lower stratosphere (Fig. 5b), where the SBUV data sets are not used to construct the MDM, the data sets differ relatively consistently over the entire 1984–2011 period. SWOOSH and GOZCARDS show mostly positive differences from the MDM, which increase slightly over time. The main difference between these two data sets is in the post-Pinatubo period (1992–1995), when the MDM is based only on the mean of these two data sets because no other data are available, and GOZCARDS shows higher values than SWOOSH. SAGE-GOMOS2 and SAGE-OSIRIS show mainly negative differences from the MDM, indicating they have consistently lower values compared to the other data sets. The SAGE-GOMOS1 data set is closest to the mean, remaining within about ±2 % of the MDM until the introduction of the GOMOS data in 2002, when the difference from the MDM becomes larger and positive (up to +5 %

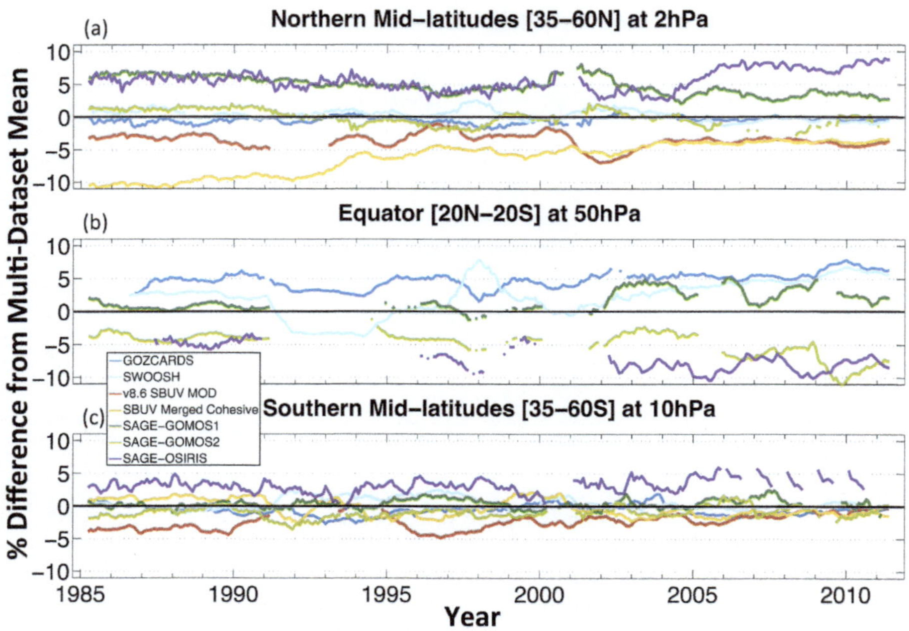

Figure 5. Monthly mean percentage differences from the multi-data-set mean (MDM; see text for details) for **(a)** the Northern Hemisphere mid-latitudes (35–60° N) at 2 hPa, **(b)** the tropics (20° N–20° S) at 50 hPa, and **(c)** the Southern Hemisphere mid-latitudes (35–60° S) at 10 hPa.

higher than the MDM). As already mentioned, agreement between data sets is best in the mid-latitude middle stratosphere. This can clearly be seen in Fig. 5c, where the seven data sets agree to within ±5 % and even to within ±3 % from about 2000 onwards (excluding SAGE-OSIRIS).

For the most part, the seven data sets agree to within ±10 % (or better) of each other throughout much of the stratosphere. In certain regions, some data sets show changes in time compared to the MDM. These tendencies are also evident in the anomalies shown in Sect. 3.3 and will be further discussed there, in particular in terms of the implications these might have on derived trends.

3.3 Interannual anomalies

The time series of ozone anomalies (all proxies used in the multiple linear regression, except the linear trend, are removed over the entire period 1984–2011) are presented to identify how differences between data sets may lead to differences in the calculated trends. For comparison, a mean of all data sets, the MDM, is again created from averaging the anomalies from all available data sets for each month/region/pressure level. Plots of the southern mid-latitude upper stratosphere are also included because this region is of particular interest in terms of long-term trends since it has been the most strongly affected by ODS-related ozone depletion (WMO, 2011).

In the northern mid-latitude upper stratosphere, all data sets, with the exception of SBUV Merged Cohesive, show a clear downward trend in anomalies until about 2001 in both the monthly and annual mean values (Figs. 6a and 7a, respectively). Thereafter, the anomalies remain mostly negative but show a tendency towards less negative and even some positive values. This is perhaps more evident in the annual mean values and in particular for SAGE-OSIRIS and GOZCARDS, which both show a return to positive anomalies for most months of the year by the end of the record. As discussed in Sect. 3.2, in this region the SBUV Merged Cohesive data set shows very different behaviour to any of the other data sets, with anomalies starting off negative and then becoming more positive over time, i.e. no downward and upward trend as seen in the other data sets. Similar but more pronounced features are also shown in the southern mid-latitude upper stratosphere (Figs. 6d and 7d). In contrast, the six other data sets show similar tendencies to the northern mid-latitudes, with decreasing anomalies through until about 2000, and then a tendency towards more positive values thereafter.

In the equatorial lower stratosphere the anomalies of all five SAGE-II-based data sets become more negative with time. During the first 5 years of the record nearly all data sets show almost exclusively positive anomalies, while during the last 6 years of the record most data sets show negative anomalies for most months of the year (Fig. 6b). This trend is even clearer in the annual means (Fig. 7b). The two SAGE-GOMOS data sets follow very similar trajectories, except for the last 3 years of the record, where differences between the two data sets are somewhat larger, with the SAGE-GOMOS2 data set showing more negative anomalies than any of the other data sets. SWOOSH and GOZCARDS are also mostly

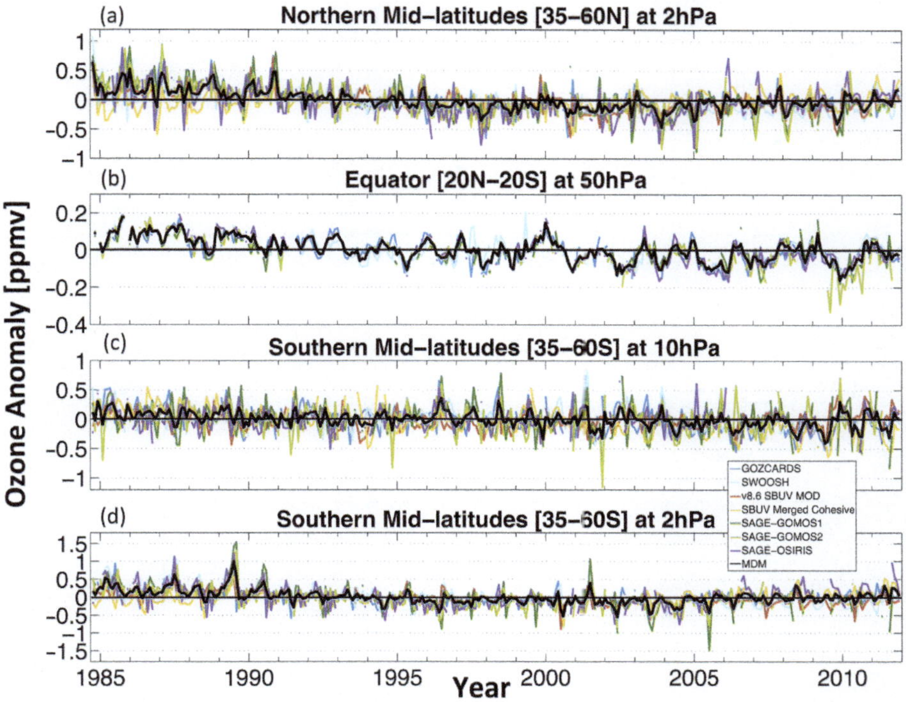

Figure 6. Monthly mean anomalies for **(a)** the Northern Hemisphere mid-latitudes (35–60° N) at 2 hPa, **(b)** the tropics (20° N–20° S) at 50 hPa, and **(c)** the Southern Hemisphere mid-latitudes (35–60° S) at 10 hPa. Note the different *y*-axis ranges for each plot **(a)–(d)**.

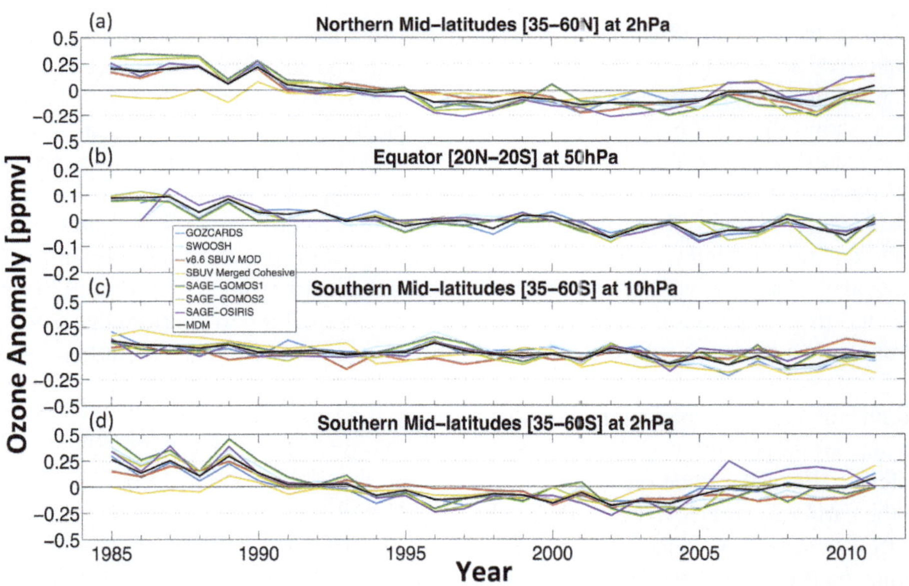

Figure 7. Annual mean anomalies for **(a)** the Northern Hemisphere mid-latitudes (35–60° N) at 2 hPa, **(b)** the tropics (20° N–20° S) at 50 hPa, and **(c)** the Southern Hemisphere mid-latitudes (35–60° S) at 10 hPa. Annual averages are only shown if data for more than 7 of 12 months per year were available. Note the different *y*-axis ranges.

similar to each other but show a smaller decrease than the two SAGE-GOMOS data sets. SAGE-OSIRIS is similar to SWOOSH and GOZCARDS, and is perhaps closest to the MDM in this region.

In the southern mid-latitude middle stratosphere, changes in anomalies tend to be small (Figs. 6c and 7c), with the SAGE-GOMOS and SAGE-OSIRIS data sets showing anomalies fluctuating around zero over the entire record. SWOOSH and GOZCARDS show some tendency towards

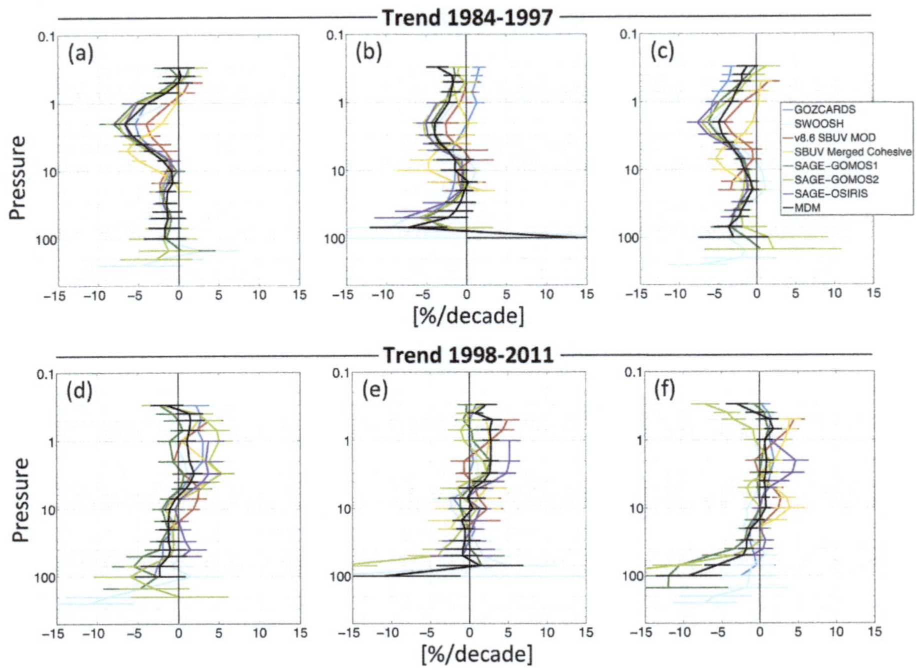

Figure 8. Latitudinal average ozone profile trends (in % $decade^{-1}$) for 1984–1997 (top row) and 1998–2011 (bottom row). Left column (**a** and **d**): Southern Hemisphere mid-latitudes (35–60° S); middle column: (**b** and **e**) tropics (20° N–20° S); right column: (**c** and **f**) Northern Hemisphere mid-latitudes (35–60° N). The horizontal error bars indicate $\pm 2\sigma$ uncertainties.

more negative anomalies from 2004 to 2009, but thereafter anomalies again become close to zero. The largest change in terms of annual mean anomalies is seen in the SBUV Merged Cohesive data set, which shows the most positive anomalies at the beginning of the record and then the most negative anomalies at the end of the record (Fig. 7c). This is also true in the northern mid-latitude middle stratosphere (not shown). The v8.6 SBUV MOD data set, in contrast, shows a more similar tendency to the other data sets, with anomalies becoming more negative and then more positive, although at the end of the record anomalies are more positive than any other data set.

In terms of both monthly and annual mean anomalies, differences between data sets are largest prior to about 1995 in all regions and at all levels. Even the three data sets based solely on SAGE-II in the earlier period (SAGE-OSIRIS and the two SAGE-GOMOS data sets) show relatively large differences from each other (particularly SAGE-OSIRIS from the two SAGE-GOMOS data sets), indicating that the different merging processes have an impact on the estimated trends. After 1995 the data sets are more similar to each other until about 2005, after which differences become larger again. This is likely because from 2005 onwards, when SAGE-II was turned off, the data sets are based on different instrument records (see Fig. 1). Overall, however, data sets based on the same or very similar instrument records (e.g. SWOOSH and GOZCARDS) tend to be more similar to each other than to those based on different underlying instrument records. The exceptions to this are the two SBUV data sets, which tend to differ notably from one another despite relying on the same input data. This is because the SBUV record is made up of several individual instrument records rather than the single SAGE-II record anchoring the other data sets. This results in a much larger choice of possibilities in terms of which records are used and how they are intercalibrated. Therefore, despite the similar input to the SBUV-based data sets, the merging approaches result in substantial differences in the anomalies derived from these two products. The result this has on derived trends is further discussed in Sect. 4.

4 Trends

The multiple linear regression model used to estimate the trends discussed in this section is described in Sect. 2.4. As above, an MDM is calculated, this time as the mean trend and uncertainty for all data sets available (again unweighted), i.e. the multiple linear regression is not applied separately to the monthly MDM constructed and shown in Figs. 6 and 7.

4.1 1984–1997 trends

For the 1984–1997 period, the data sets show relatively similar trend profiles in all latitude bands, most with maximum negative trends in the upper stratosphere (top row Fig. 8 and Tables 8–10). In the southern mid-latitudes (Fig. 8a), SWOOSH, SAGE-OSIRIS, and both SAGE-GOMOS data

Table 8. Trends in % decade^{-1} for the 1984–1997 period at 2 hPa. Values in brackets show the 2σ uncertainty estimates, while those in bold are significantly different from zero at the 2σ level.

Data set	Northern mid-latitudes (35–60° N)	Tropics (20° N–20° S)	Southern mid-latitudes (35–60° S)
v8.6 SBUV MOD	**−4.12 (1.11)**	**−2.82 (1.56)**	**−4.13 (1.04)**
SBUV Merged Cohesive	−0.58 (0.87)	0.26 (1.13)	**−2.12 (0.90)**
SAGE-GOMOS1	**−6.28 (0.88)**	**−4.35 (1.07)**	**−8.02 (1.82)**
SAGE-GOMOS2	**−7.08 (1.08)**	**−5.40 (1.28)**	**−7.44 (1.54)**
SAGE-OSIRIS	**−7.64 (1.36)**	**−5.10 (1.21)**	**−6.69 (1.65)**
GOZCARDS	**−6.09 (0.91)**	0.09 (1.02)	**−5.38 (1.31)**
SWOOSH	**−6.02 (0.86)**	**−3.52 (0.93)**	**−6.74 (1.35)**
MDM	**−5.40 (1.01)**	**−2.98 (1.17)**	**−5.79 (1.37)**

Table 9. As for Table 8 but for 10 hPa.

Data set	Northern mid-latitudes (35–60° N)	Tropics (20° N–20° S)	Southern mid-latitudes (35–60° S)
v8.6 SBUV MOD	−1.73 (2.09)	−0.58 (2.27)	**−1.43 (0.91)**
SBUV Merged Cohesive	**−4.07 (2.28)**	**−4.77 (3.97)**	**−2.46 (0.91)**
SAGE-GOMOS1	**−1.23 (1.08)**	−0.41 (1.60)	−0.20 (1.00)
SAGE-GOMOS2	**−1.52 (1.19)**	−0.05 (1.72)	**−1.20 (1.07)**
SAGE-OSIRIS	**−1.55 (0.93)**	−0.54 (1.51)	−0.80 (1.04)
GOZCARDS	**−1.32 (0.91)**	−1.39 (1.76)	−0.86 (1.15)
SWOOSH	0.58 (0.84)	1.22 (1.96)	0.30 (1.15)
MDM	**−1.54 (1.09)**	−0.93 (2.11)	**−0.95 (1.03)**

sets show largest negative trends at 2 hPa (between −8 and −6 % decade^{-1}), slightly more negative than GOZCARDS and v8.6 SBUV MOD, which indicate maximum negative trends of −5 and −4 % decade^{-1} at 2 hPa, respectively (see also Table 8). Trends in the northern mid-latitudes show a similar shape (Fig. 8c), with maximum negative values at 2 hPa, and are also of similar magnitude. All five SAGE-based data sets agree to within 2 %, with maximum negative trends ranging between −6 and −7 % decade^{-1}, while v8.6 SBUV MOD again shows less negative trends, peaking at −4 % decade^{-1}. In the equatorial region (Fig. 8b), trends are somewhat smaller, up to −3 to −5 % decade^{-1} at 2 hPa, with GOZCARDS and the SBUV Merged Cohesive data set even showing zero trend at this level. The large difference between GOZCARDS and SWOOSH, which are based on similar underlying instrument records, likely results from the different temperature record used to convert the different SAGE-II versions. In the tropical upper stratosphere, the NCEP temperatures used to convert the SAGE-II v6.2 profiles in GOZCARDS have different temporal variations than MERRA, which leads to some significant differences between the two data sets (Froidevaux et al., 2015); in most other regions, the GOZCARDS and SWOOSH results shown here tend to be quite similar. The MERRA reanalysis used to convert the SAGE-II v7.0 data set used in SWOOSH produces results more similar to the SAGE-OSIRIS and SAGE-GOMOS data sets, also converted using MERRA.

In all three latitude regions, SBUV Merged Cohesive shows a different trend profile structure from the other data sets, with maximum negative trends of nearly similar magnitude, but considerably lower down in the stratosphere (between 5 and 10 hPa rather than at 2 hPa). As seen in the annual mean anomalies in Fig. 7, this data set shows somewhat higher/lower mean values than the other data sets, particularly at the beginning of the record, which may influence the overall trend calculated for the 1984–1997 period. The difficulty lies in the lesser quality of the NOAA-9 and NOAA-14 data, which appear at the end of the 1984–1997 period. Due to this reduced quality, the SBUV Merged Cohesive data set does not tie the beginning of the time series to these data but instead uses the same adjustments for the ascending node of NOAA-11 as determined from the overlap of the descending node of NOAA-11 with NOAA-16 (Wild and Long, 2015). The assumption that NOAA-11 ascending and descending have the same properties may not be warranted, and this may influence the trends from this data set for the 1984–1997 period. The different trends highlight the influence of using different merging techniques, since the underlying data (SBUV v8.6) and the regression model used are the same as for the v8.6 SBUV MOD data set.

Table 10. As for Table 8 but for 50 hPa. Trends were not calculated for the SBUV data sets at pressures greater than 20 and 30hPa in the tropics and mid-latitudes, respectively; therefore they are not shown (NA).

Data set	Northern mid-latitudes (35–60° N)	Tropics (20° N–20° S)	Southern mid-latitudes (35–60° S)
v8.6 SBUV MOD	NA	NA	NA
SBUV Merged Cohesive	NA	NA	NA
SAGE-GOMOS1	**−3.05 (1.34)**	**−5.18 (3.19)**	−0.92 (1.81)
SAGE-GOMOS2	**−1.99 (1.93)**	**−5.88 (3.53)**	−1.32 (2.05)
SAGE-OSIRIS	**−3.34 (1.35)**	**−7.86 (3.68)**	−1.46 (2.25)
GOZCARDS	**−2.99 (1.48)**	**−5.58 (3.31)**	−1.25 (2.21)
SWOOSH	**−3.30 (1.60)**	**−7.33 (3.30)**	−1.80 (2.42)
MDM	**−2.93 (1.54)**	**−6.37 (3.40)**	−1.35 (2.15)

Trends in the middle stratosphere in all latitude regions tend to be small, mostly less than $-2\,\%\,\text{decade}^{-1}$, and in some cases even being slightly positive (see Table 9). The mean trend of all data sets (MDM) is, however, significant and negative in both mid-latitude regions but remains small. In the lower stratosphere the SAGE-based data sets show small negative trends in the northern mid-latitudes, ranging from -2 to $-3\,\%\,\text{decade}^{-1}$, and quite large negative trends in the tropics, almost of similar magnitude to the mid-latitude upper stratosphere (-5 to $-7\,\%\,\text{decade}^{-1}$), although uncertainties are significantly larger. Nonetheless, at 50 hPa in the northern mid-latitudes and tropics, all five data sets indicate significant negative trends (see Table 10). In contrast, in the southern mid-latitudes, trends are small and insignificant in all data sets.

The trends calculated for this period agree very well with a wide range of previous studies (e.g. WMO, 1999, and references therein; Harris et al., 1998). In particular, more recent studies using some of the same data sets as here show similar results. For example, using the same SAGE-GOMOS1 data set, Kyrölä et al. (2013) derive trends of the same magnitude in the upper stratosphere (ranging from -8 to $-6\,\%\,\text{decade}^{-1}$ in the mid-latitudes and up to $-4\,\%\,\text{decade}^{-1}$ in the tropics), and, using a more complex analysis technique, Laine et al. (2013) present very similar results as well. Using the SAGE-OSIRIS data set also used in this study, Bourassa et al. (2014) estimate significant trends of the same magnitude in the upper stratosphere (again between approximately -8 and $-6\,\%\,\text{decade}^{-1}$ in the mid-latitudes and up to $-4\,\%\,\text{decade}^{-1}$ in the tropics) but show no significant trends below 35 km. In comparison, in the tropics we estimate significant negative trends between 30 and 50 hPa for the SAGE-OSIRIS data set as well as SWOOSH, GOZCARDS, and the two SAGE-GOMOS data sets. These significant negative trends are in agreement with data from SAGE-II, GOZCARDS, and model results presented in WMO (2015), although they cover a slightly different period (1985–1995). Damadeo et al. (2014) show no significant or slightly positive trends in the same SAGE-II data (which underlies all data sets showing negative trends during this period). Obviously there is still some uncertainty regarding the trends in the tropics during this period, and further work is needed to fully resolve this issue.

4.2 1998–2011 trends

For the 1998–2011 period, calculated trends are less consistent among the data sets but overall show a general shift towards increasing ozone in the middle and upper stratosphere (bottom row Fig. 8 and Tables 11–13). The smaller change in trends is somewhat to be expected given that the lifetimes of most ODS species are long (several decades), and thus the removal of these species will occur over a considerably longer timescale than the relatively brief period during which their concentrations increased (WMO, 2015). Trends in the northern mid-latitude upper stratosphere are mostly positive (Fig. 8f), although relatively small in magnitude compared to the large negative trend seen in the 1984–1997 period. SAGE-OSIRIS shows the largest positive trends, peaking at $+5\,\%\,\text{decade}^{-1}$ at 2 hPa. This is unsurprising given the relatively large positive tendency in the annual mean values presented in Fig. 7a and agrees well with the findings of Bourassa et al. (2014), who show trends of similar magnitude in this location. Both SBUV data sets also show significant positive trends (up to $4\,\%\,\text{decade}^{-1}$) but at different pressures than SAGE-OSIRIS, above 1 hPa (where the SAGE-OSIRIS data are not available for comparison) as well as lower down in the middle stratosphere between 10 and 7 hPa. Both SAGE-GOMOS data sets, GOZCARDS, and SWOOSH all show almost no significant trend in the upper stratosphere. At pressures <1 hPa differences between calculated trends are large in all three latitude regions. For example, in the northern mid-latitudes, SAGE-GOMOS2 shows negative trends of up to $-7\,\%\,\text{decade}^{-1}$, whereas GOZCARDS and SAGE-GOMOS1 indicate no significant trend and the SBUV data sets indicate positive significant trends up to $4\,\%\,\text{decade}^{-1}$. It is very likely that the diurnal cycle in ozone plays a role at these altitudes, affecting the estimated trends quite strongly depending on how, and if, this is treated in the merging process. Whether and how the diurnal cycle of ozone is treated

Table 11. Trends in % decade^{-1} for the 1998–2011 period at 2 hPa. Values in brackets show the 2σ uncertainty estimates, while those in bold are significantly different from zero at the 2σ level.

Data set	Northern mid-latitudes (35–60° N)	Tropics (20° N–20° S)	Southern mid-latitudes (35–60° S)
v8.6 SBUV MOD	−0.61 (1.23)	−0.56 (1.66)	−0.36 (1.17)
SBUV Merged Cohesive	**1.98 (0.91)**	**2.12 (1.16)**	**3.47 (0.99)**
SAGE-GOMOS1	0.26 (1.16)	**2.72 (1.21)**	−0.65 (1.84)
SAGE-GOMOS2	**1.31 (1.30)**	**1.78 (1.45)**	**3.79 (1.90)**
SAGE-OSIRIS	**4.70 (1.73)**	**5.19 (1.36)**	**3.34 (2.13)**
GOZCARDS	**1.90 (1.01)**	0.53 (1.00)	**2.96 (1.44)**
SWOOSH	0.26 (0.96)	**0.99 (0.97)**	0.74 (1.52)
MDM	**1.40 (1.19)**	**1.83 (1.26)**	**1.90 (1.57)**

Table 12. As for Table 11 but for 10 hPa.

Data set	Northern mid-latitudes (35–60° N)	Tropics (20° N–20° S)	Southern mid-latitudes (35–60° S)
v8.6 SBUV MOD	**2.31 (0.96)**	1.57 (2.33)	**1.55 (0.98)**
SBUV Merged Cohesive	**3.91 (1.98)**	−0.18 (4.28)	**−1.55 (0.98)**
SAGE-GOMOS1	0.08 (1.02)	−1.18 (1.56)	**−2.02 (1.20)**
SAGE-GOMOS2	0.18 (1.28)	0.93 (1.71)	−0.24 (1.14)
SAGE-OSIRIS	−0.02 (1.02)	1.35 (1.53)	−0.44 (1.22)
GOZCARDS	−0.18 (0.93)	−1.04 (1.73)	**−1.64 (1.15)**
SWOOSH	**−1.71 (0.82)**	**−2.28 (1.85)**	**−2.36 (1.11)**
MDM	0.65 (1.14)	−0.12 (2.14)	−0.96 (1.11)

in all data sets is shown in Table 6. The differences between data sets are more strongly evident in this later period (1998–2011), particularly since a wide range of different instruments which may be measuring at different times of the day is used to extend the SAGE-II record after 2005.

In the southern mid-latitudes (Fig. 8d), all data sets, excluding SWOOSH and SAGE-GOMOS1, indicate significant positive trends in the upper stratosphere, although they vary in where maximum trends are located in the vertical. SAGE-OSIRIS and SAGE-GOMOS2 show trends ranging from +3 to +5 % decade^{-1} between 4 and 0.5 hPa, while GOZCARDS shows maximum positive trends slightly smaller in magnitude and somewhat higher up. SBUV Merged Cohesive shows trends of similar magnitude to SAGE-GOMOS2, but peaking very clearly at 3 hPa. The v8.6 SBUV MOD data set also shows significant positive trends but peaking between 5 and 10 hPa and only up to +3 % decade^{-1}. SWOOSH shows no significant trend over the entire profile, similar to the SAGE-GOMOS1 data set, which differs only in that it shows small but significant negative trends at pressures greater than 50 hPa. SAGE-GOMOS2 also indicates similar magnitude negative trends at these levels; however, no such significant trends are evident in either GOZCARDS and SWOOSH and overall the MDM shows insignificant trends in the southern mid-latitudes at pressures > 5 hPa. As in the northern mid-latitudes, at pressures < 1 hPa the trends vary widely between data sets, again likely due to the diurnal cycle in ozone.

The mid-latitude trends shown here agree well with WMO (2015), which, although considering just post-2000 trends, shows a similar range of estimates using some of the same data sets (e.g. GOZCARDS and v8.6 SBUV MOD) as well as several other sources, including ground-based observations. As for the 1984–1997 period, the trend profiles for 1998–2011 also agree with the results presented by Bourassa et al. (2014) using the same SAGE-OSIRIS data set, as well as Kyrölä et al. (2013) and Laine et al. (2013) using the same SAGE-GOMOS data set. Looking at a somewhat shorter period (August 2002–April 2012) Gebhardt et al. (2014) show similar magnitude positive trends in SCIAMACHY (SCanning Imaging Absorption spectroMeter for Atmospheric CHartographY) of approximately 4 % decade^{-1} in the southern mid-latitude upper stratosphere; however, they show almost no significant positive trend in the northern mid-latitude upper stratosphere. This tendency towards a stronger signal in the Southern Hemisphere is also seen in the MIPAS observations covering the same period (Eckert et al., 2014), although they show a difference in the altitude range, with a larger region of positive trends seen in the southern mid-latitude upper stratosphere.

In the equatorial region, trends are least consistent. In the upper stratosphere, between 5 and 2 hPa both SAGE-

Table 13. As for Table 11 but for 50 hPa. Trends were not calculated for the SBUV data sets at pressures greater than 20 and 30hPa in the tropics and mid-latitudes, respectively, therefore they are not shown (NA).

Data set	Northern mid-latitudes (35–60° N)	Tropics (20° N–20° S)	Southern mid-latitudes (35–60° S)
v8.6 SBUV MOD	NA	NA	NA
SBUV Merged Cohesive	NA	NA	NA
SAGE-GOMOS1	**−6.29 (1.31)**	0.91 (3.65)	**−4.42 (2.00)**
SAGE-GOMOS2	**−3.45 (2.07)**	**−4.03 (3.96)**	−1.08 (2.18)
SAGE-OSIRIS	−1.03 (1.54)	−1.00 (4.13)	0.40 (2.48)
GOZCARDS	−0.75 (1.55)	−0.94 (3.50)	−1.95 (2.18)
SWOOSH	0.62 (1.69)	1.79 (3.54)	0.20 (2.45)
MDM	**−2.18 (1.63)**	−0.65 (3.75)	−1.37 (2.26)

GOMOS data sets as well as SBUV Merged Cohesive suggest small positive trends of approximately +2 % decade^{-1} peaking near 2 hPa. SAGE-OSIRIS also shows positive trends, but much larger in magnitude (up to +5 % decade^{-1}). These estimates agree well with the work of Bourassa et al. (2014), although, with potential positive drift in the OSIRIS data taken into account, the trends they present become less significant in this region (being insignificant between 10° N and 10° S). SWOOSH, GOZCARDS, and v8.6 SBUV MOD all show no significant trends in this region, except for at pressures < 1 hPa, where the v8.6 SBUV MOD data set shows significant positive trends. In the middle stratosphere, between approximately 10 and 5 hPa, GOZCARDS, SWOOSH, and the two SAGE-GOMOS data sets show small negative trends up to −3 % decade^{-1}, but only the trends from SWOOSH at 10 hPa are significant (see Tables 11–13). For the same period and levels, Kyrölä et al. (2013) also show significant negative trends even larger in magnitude, up to −5 % decade^{-1}, but only between 10° N and 10° S. The larger latitude band used here (20° N–20° S) likely accounts for the difference in magnitude and significance in trends. Gebhardt et al. (2014) also found large negative trends in this region of the middle stratosphere (as large as −10 % decade^{-1}) for the 2002–2012 period in SCIAMACHY observations. However, in a comparison for just the 2004–2012 period they show that the SCIAMACHY trends are considerably more negative than either Aura MLS or OSIRIS. Using MIPAS data for the 2002–2012 period, Eckert et al. (2014) show negative trends of similar magnitude to Kyrölä et al. (2013) of approximately −5 % decade^{-1}, although somewhat lower in the middle stratosphere. While there is a clear negative trend in many data sets, the range of trend estimates remains large. Finally, in the tropical lower tropical stratosphere, trends are insignificant for nearly all the data sets shown here, largely as a result of the large uncertainties in measurements made in this part of the atmosphere.

5 Conclusions

This paper presents the first intercomparison of seven new merged satellite ozone profile data sets for the period 1984–2011, common to all data sets. We also present an overview of the methods, assumptions, and challenges involved in producing such merged data sets. The analysis focuses on the representation of the annual cycle, interannual variability, and, in particular, long-term trends. Overall, the data sets are most similar in the mid-latitude lower and middle stratosphere, remaining largely within ±5 % of the mean of all data sets (MDM). Larger differences are found in the tropical lower stratosphere, where the spread between data sets is ±10 % from the MDM, and in the upper stratosphere globally, where data sets for the most part are within ±8 % of the MDM. These results are in agreement with the inter-instrument comparison of Tegtmeier et al. (2013). For the data sets based on SAGE-II (SAGE-GOMOS1, SAGE-GOMOS2, SAGE-OSIRIS, SWOOSH, and GOZCARDS), the choice of instrument records to be merged was found to have a greater impact on differences than the choice of merging technique. For these data sets, those based on the same individual instrument records tend to be more similar to each other than those based on different instrument records. In general, it also appears that the inclusion of a greater number of instrument records in an individual merged data set and the potential this creates for bias cancellation tends to improve such long-term merged data sets. The SBUV v8.6 records, on the other hand, show some significant differences, despite being based on the same underlying data. This is in large part because the SBUV record is made up of several individual instrument records, allowing for more potential choices as to what records to use and, more importantly, how to intercalibrate those records.

Piecewise linear regression was used on the entire time period (1984–2011) to calculate trends for a "decrease" period (1984–1997) and a "recovery" period (1998–2011). Trends estimated for the first period are more similar between data sets, with most data sets showing significant negative trends in the mid-latitude upper stratosphere ranging between −4 and $-8\,\%\,\mathrm{decade}^{-1}$. Significant negative trends are also evident in the tropical upper stratosphere in most data sets, although smaller in magnitude. In the middle and lower stratosphere, trends are small, and for the most part insignificant, but depend critically on the uncertainties. The good agreement between data sets is, however, somewhat unsurprising given that most data sets are predominantly (or only) based on the SAGE-II record for the 1984–1997 period. The trends presented for this period agree well with both previous and more recent studies (e.g. Harris et al., 1998; WMO, 1999; Kyrölä et al., 2013; Bourassa et al., 2014; Damadeo et al., 2014; Eckert et al., 2014; WMO, 2015).

For the second period (1998–2011), calculated trends vary more between data sets, ranging from -1 to $+5\,\%\,\mathrm{decade}^{-1}$ in the upper stratosphere in all three latitude regions considered. There is, however, a clear shift from mostly negative to mostly positive trends, which is most evident in the southern mid-latitude and equatorial upper stratosphere. In the northern mid-latitude upper stratosphere, only two of the seven data sets show significant positive trends and the MDM trend is only just barely significantly positive. In the middle and lower stratosphere, trends are again mostly small and not significantly different from zero. The larger differences between data sets for the "recovery" period are perhaps to be expected since the five SAGE-based data sets use different instrument records to complete the SAGE-II record. Nevertheless, the trends calculated largely fall within the range of estimates presented in several recent analyses (Kyrölä et al., 2013; Bourassa et al., 2014; Damadeo et al., 2014; Eckert et al., 2014; Laine et al., 2013; Sioris et al., 2014; WMO, 2015).

Recent studies have also demonstrated that it remains difficult to identify stable, statistically significant positive trends in global ozone (Frith et al., 2014; Harris et al., 2015). The low-frequency variability in the ozone record (Frith et al., 2014) and the choice of start and end dates (Frith et al., 2014; Harris et al., 2015) are contributing factors, as well as uncertainty related to the proxies chosen for the regression model (de Laat et al., 2015). This study, using an "ensemble" of recently produced long-term, vertically resolved, merged satellite ozone data sets, shows a fairly wide range of estimates of long-term ozone trends; not all data sets unequivocally indicate significant positive trends in the upper stratosphere for the 1998–2011 period. However, the mean trend of all data sets (MDM) is significant and positive in the southern mid-latitudes between 5 and 0.5 hPa, in the tropics between 3 and 1 hPa, and in the northern mid-latitudes at just 2 hPa. The use of more complex regression techniques that take into account spatial and temporal bias, such as that used by Damadeo et al. (2014), or that allow a latitude- and altitude-varying change point (e.g. Laine et al., 2013), may help better constrain estimates of long-term ozone trends. As newer individual satellite data sets are better understood and merging techniques become more refined, differences between data sets and derived trends may be reduced. In this context, continued high-quality long-term ozone profile measurements are essential to unambiguously identify an ozone recovery in response to decreasing ODSs within a changing climate.

Acknowledgements. The authors thank all those involved in the SI2N initiative whose research underlies the results presented here. Fiona Tummon thanks the Swiss National Science Foundation for funding. Neil Harris thanks the UK Natural Environment Research Council for an Advanced Research Fellowship. Work at the Jet Propulsion Laboratory, California Institute of Technology, was performed under contract with the National Aeronautics and Space Administration. The use of MERRA data from GMAO is also acknowledged. We thank the two reviewers for helpful comments and suggestions.

Edited by: R. Eckman

References

Adams, C., Bourassa, A. E., Bathgate, A. F., McLinden, C. A., Lloyd, N. D., Roth, C. Z., Llewellyn, E. J., Zawodny, J. M., Flittner, D. E., Manney, G. L., Daffer, W. H., and Degenstein, D. A.: Characterization of Odin-OSIRIS ozone profiles with the SAGE II dataset, Atmos. Meas. Tech., 6, 1447–1459, doi:10.5194/amt-6-1447-2013, 2013.

Adams, C., Bourassa, A. E., Sofieva, V., Froidevaux, L., McLinden, C. A., Hubert, D., Lambert, J.-C., Sioris, C. E., and Degenstein, D. A.: Assessment of Odin-OSIRIS ozone measurements from 2001 to the present using MLS, GOMOS, and ozonesondes, Atmos. Meas. Tech., 7, 49–64, doi:10.5194/amt-7-49-2014, 2014.

Austin, J. and Butchart, N.: Coupled chemistry-climate model simulations for the period 1980 to 2020: Ozone depletion and the start of ozone recovery, Q. J. R. Meteorol. Soc., 129, 3225–3249, doi:10.1256/qj.02.203, 2003.

Austin, J. and Wilson, R. J.: Ensemble simulations of the decline and recovery of stratospheric ozone, J. Geophys. Res., 111, D16314, doi:10.1029/2005JD006907, 2006.

Austin, J., Tourpali, K., Rozanov, E., Akiyoshi, H., Bekki, S., Bodeker, G. E., Brühl, C, Butchart, N., Chipperfield, M., Deushi, M., Fomichev, V. I., Giorgetta, M. A., Gray, L., Kodera, K., Lott, F., Manzini, E., Marsh, D., Matthes, K., Nagashima, T., Shibata, K., Stolarski, R. S., Struthers, H., and Tian W.: Coupled chemistry climate model simulations of the solar cycle in ozone and temperature, J. Geophys. Res., 113, D11306, doi:11310.11029/12007JD009391, 2008.

Bhartia, P. K., McPeters, R. D., Flynn, L. E., Taylor, S., Kramarova, N. A., Frith, S., Fisher, B., and DeLand, M.: Solar Backscatter UV (SBUV) total ozone and profile algorithm, Atmos. Meas. Tech., 6, 2533–2548, doi:10.5194/amt-6-2533-2013, 2013.

Bhatt, P. P., Remsberg, E. E., Gordley, L. L., McInerney, J. M., Brackett, V. G., and Russell III, J. M.: An evaluation of the quality of Halogen Occultation Experiment ozone profiles in the lower stratosphere, J. Geophys. Res., 104, 9261–9275, 1999.

Bodeker, G. E., Boyd, I. S., and Matthews, W. A.: Trends and variability in vertical ozone and temperature profiles measured by ozonesondes at Lauder, New Zealand: 1986–1996, J. Geophys. Res.-Atmos., 103, 28661–28681, doi:10.1029/98JD02581, 1998.

Bodeker, G. E., Shiona, H., and Eskes, H.: Indicators of Antarctic ozone depletion, Atmos. Chem. Phys., 5, 2603–2615, doi:10.5194/acp-5-2603-2005, 2005.

Bosilovich, M. G., Robertson, F. R., and Chen, J.: Global Energy and Water Budgets in MERRA, J. Clim., 24, 5721–5739, doi:10.1175/2011JCLI4175.1, 2011.

Bourassa, A. E., Degenstein, D. A., Randel, W. J., Zawodny, J. M., Kyrölä, E., McLinden, C. A., Sioris, C. E., and Roth, C. Z.: Trends in stratospheric ozone derived from merged SAGE II and Odin-OSIRIS satellite observations, Atmos. Chem. Phys., 14, 6983–6994, doi:10.5194/acp-14-6983-2014, 2014.

Brasseur, G. P., and Solomon, S.: Aeronomy of the middle atmosphere, 3rd ed., Springer, New York, 644 pp., 2005.

Chehade, W., Weber, M., and Burrows, J. P.: Total ozone trends and variability during 1979–2012 from merged data sets of various satellites, Atmos. Chem. Phys., 14, 7059–7074, doi:10.5194/acp-14-7059-2014, 2014.

Damadeo, R. P., Zawodny, J. M., Thomason, L. W., and Iyer, N.: SAGE version 7.0 algorithm: application to SAGE II, Atmos. Meas. Tech., 6, 3539–3561, doi:10.5194/amt-6-3539-2013, 2013.

Damadeo, R. P., Zawodny, J. M., and Thomason, L. W.: Reevaluation of stratospheric ozone trends from SAGE II data using a simultaneous temporal and spatial analysis, Atmos. Chem. Phys., 14, 13455–13470, doi:10.5194/acp-14-13455-2014, 2014.

Davis, S. M. and Rosenlof, K. H.: A Multidiagnostic Intercomparison of Tropical-Width Time Series Using Reanalyses and Satellite Observations, J. Clim., 25, 1061–1078, doi:10.1175/JCLI-D-11-00127.1, 2012.

Davis, S. M., et al.: The Stratospheric Water and Ozone Satellite Homogenized (SWOOSH) database: A long-term database for climate studies, in preparation, 2015.

de Laat, A. T. J., van der A, R. J., and van Weele, M.: Tracing the second stage of ozone recovery in the Antarctic ozone-hole with a "big data" approach to multivariate regressions, Atmos. Chem. Phys., 15, 79–97, doi:10.5194/acp-15-79-2015, 2015.

DeLand, M. T., Taylor, S. L., Huang, L. K., and Fisher, B. L.: Calibration of the SBUV version 8.6 ozone data product, Atmos. Meas. Tech., 5, 2951–2967, doi:10.5194/amt-5-2951-2012, 2012.

Ebita, A., Kobayashi, S., Ota, Y., Moriya, M., Kumabe, R., Onogi, K., Harada, Y., Yasui, S., Miyaoka, K., Takahashi, K., Kamahori, H., Kobayashi, C., Endo, H., Soma, M., Oikawa, Y., and Ishimizu, T.: The Japanese 55-year Reanalysis "JRA-55": an interim report, SOLA, 7, 149–152, 2011.

Eckert, E., von Clarmann, T., Kiefer, M., Stiller, G. P., Lossow, S., Glatthor, N., Degenstein, D. A., Froidevaux, L., Godin-Beekmann, S., Leblanc, T., McDermid, S., Pastel, M., Steinbrecht, W., Swart, D. P. J., Walker, K. A., and Bernath, P. F.: Drift-corrected trends and periodic variations in MIPAS IMK/IAA ozone measurements, Atmos. Chem. Phys., 14, 2571–2589, doi:10.5194/acp-14-2571-2014, 2014.

Eyring, V., Waugh, D. W., Bodeker, G. E., Cordero, E., Akiyoshi, H., Austin, J., Beagley, S. R., Boville, B. A., Braesicke, P., Brühl, C., Butchart, N., Chipperfield, M. P., Dameris, M., Deckert, R., Deushi, M., Frith, S. M., Garcia, R. R., Gettelman, A., Giorgetta, M. A., Kinnison, D. E., Mancini, E., Manzini, E., Marsh, D. R., Matthes, S., Nagashima, T., Newman, P. A., Nielsen, J. E., Pawson, S., Pitari, G., Plummer, D. A., Rozanov, E., Schraner, M., Scinocca, J. F., Semeniuk, K., Shepherd, T. G., Shibata, K., Steil, B., Stolarski, R. S., Tian, W., and Yoshiki, M.: Multimodel projections of stratospheric ozone in the 21st century, J. Geophys. Res., 112, D16, doi:10.1029/2006JD008332, 2007.

Eyring, V., Shepherd, T. G., and Waugh, D. W. (Eds.): SPARC Report on the Evaluation of Chemistry-Climate Models, SPARC Report No. 5, WCRP-132, WMO/TD-No. 1526, 2010a.

Eyring, V., Cionni, I., Bodeker, G. E., Charlton-Perez, A. J., Kinnison, D. E., Scinocca, J. F., Waugh, D. W., Akiyoshi, H., Bekki, S., Chipperfield, M. P., Dameris, M., Dhomse, S., Frith, S. M., Garny, H., Gettelman, A., Kubin, A., Langematz, U., Mancini, E., Marchand, M., Nakamura, T., Oman, L. D., Pawson, S., Pitari, G., Plummer, D. A., Rozanov, E., Shepherd, T. G., Shibata, K., Tian, W., Braesicke, P., Hardiman, S. C., Lamarque, J. F., Morgenstern, O., Pyle, J. A., Smale, D., and Yamashita, Y.: Multimodel assessment of stratospheric ozone return dates and ozone recovery in CCMVal-2 models, Atmos. Chem. Phys., 10, 9451–9472, doi:10.5194/acp-10-9451-2010, 2010b.

Fioletov, V. E.: Ozone climatology, trends, and substances that control ozone, Atmos. Ocean., 46, 39–67, doi:10.3137/ao.460103, 2008.

Frith, S. M., Kramarova, N. A., Stolarski, R. S., McPeters, R. D., Bhartia, P. K., and Labow, G. J.: Recent changes in column ozone based on the SBUV version 8.6 merged ozone dataset, J. Geophys. Res.-Atmos., 119, 9735–9751, doi:10.1002/2014JD021889, 2014.

Froidevaux, L., Jiang, Y. B., Lambert, A., Livesey, N. J., Read, W. G., Waters, J. W., Browell, E. V., Hair, J. W., Avery, M. A., McGee, T. J., Tiwgg, L. W., Sumnicht, G. K., Jucks, K. W., Margitan, J. J., Sen, B., Stachnik, R. A., Toon, G. C., Bernath, P. F., Boone, C. D., Walker, K. A., Filipiak, M. J., Harwood, R. S., Fuller, R. A., Manney, G. L., Schwartz, M. J., Daffer, W. H., Drouin, B. J., Cofield, R. E., Cuddy, D. T., Jarnot, R. F., Knosp, B. W., Perun, V. S., Snyder, W. V., Stek, P. C., Thurstans, R. P., and Wagner, P. A.: Validation of Aura Microwave Limb Sounder stratospheric and mesospheric ozone measurements, J. Geophys. Res., 113, D15S20, doi:10.1029/2007JD008771, 2008.

Froidevaux, L., Anderson, J., Wang, H.-J., Fuller, R. A., Schwartz, M. J., Santee, M. L., Livesey, N. J., Pumphrey, H. C., Bernath, P., Russell III, J. M., and McCormick, M. P., Global Ozone Chemistry And Related Datasets for the Stratosphere (GOZCARDS): methodology and sample results with a focus on HCl, H2O, and O3, Atmos. Chem. Phys. Discuss., 15, 5849-5957, doi:10.5194/acpd-15-5849-2015, 2015.

Gebhardt, C., Rozanov, A., Hommel, R., Weber, M., Bovensmann, H., Burrows, J. P., Degenstein, D., Froidevaux, L., and Thompson, A. M.: Stratospheric ozone trends and variability as seen by SCIAMACHY from 2002–2012, Atmos. Chem. Phys., 14, 831–846, doi:10.5194/acp-14-831-2014, 2014.

Grooß, J.-U. and Russell III, James M.: Technical note: A stratospheric climatology for O_3, H_2O, CH_4, NO_x, HCl and HF derived from HALOE measurements, Atmos. Chem. Phys., 5, 2797–2807, doi:10.5194/acp-5-2797-2005, 2005.

Harris, N. R. P., Hudson, R., and Phillips, C. (Eds.): Assessment of trends in the vertical distribution of ozone, Stratospheric Processes and Their Role in Climate/International Ozone Commission/Global Atmospheric Watch (SPARC/IOC/ GAW) Report 1, World Meteorol. Organ. OzoneRes. and Monit. Proj. Rep. 43, 289 pp., Geneva, 1998.

Harris, N. R. P., Hassler, B., Tummon, F., Bodeker, G. E., Hubert, D., Petropavlovkikh, I., Steinbrecht, W., Anderson, J., Bhartia, P. K., Boone, C. D., Bourassa, A., Davis, S. M., Degenstein, D., DelCloo, A., Frith, S. M., Froidevaux, L., Godin-Beekmann, S., Jones, N., Kurylo, M. J., Kyrölä, E., Laine, M., Leblanc, S. T., Lambert, J.-C., Liley, B. Mahieu, E., Maycock, A., de Mazière, M., Parrish, A., Qerel, R., Rosenlof, K. H., Roth, C., Sioris, C., Staehelin, J., Stolarski, R. S., Stübi, R., Tamminen, J., Vigouroux, C., Walker, K., Wang, H. W., Wild, J., and Zawodny, J. M.: Past changes in the Vertical Distribution of Ozone – Part III: Analysis and Interpretation of Trends, Atmos. Chem. Phys. Discuss., submitted, 2015.

Hassler, B., Bodeker, G. E., and Dameris, M.: Technical Note: A new global database of trace gases and aerosols from multiple sources of high vertical resolution measurements, Atmos. Chem. Phys., 8, 5403–5421, doi:10.5194/acp-8-5403-2008, 2008.

Hassler, B., Young, P. J., Portmann, R. W., Bodeker, G. E., Daniel, J. S., Rosenlof, K. H., and Solomon, S.: Comparison of three vertically resolved ozone data sets: climatology, trends and radiative forcings, Atmos. Chem. Phys., 13, 5533–5550, doi:10.5194/acp-13-5533-2013, 2013.

Hassler, B., Petropavlovskikh, I., Staehelin, J., August, T., Bhartia, P. K., Clerbaux, C., Degenstein, D., Mazière, M. De, Dinelli, B. M., Dudhia, A., Dufour, G., Frith, S. M., Froidevaux, L., Godin-Beekmann, S., Granville, J., Harris, N. R. P., Hoppel, K., Hubert, D., Kasai, Y., Kurylo, M. J., Kyrölä, E., Lambert, J.-C., Levelt, P. F., McElroy, C. T., McPeters, R. D., Munro, R., Nakajima, H., Parrish, A., Raspollini, P., Remsberg, E. E., Rosenlof, K. H., Rozanov, A., Sano, T., Sasano, Y., Shiotani, M., Smit, H. G. J., Stiller, G., Tamminen, J., Tarasick, D. W., Urban, J., van der A, R. J., Veefkind, J. P., Vigouroux, C., von Clarmann, T., von Savigny, C., Walker, K. A., Weber, M., Wild, J., and Zawodny, J. M.: Past changes in the vertical distribution of ozone –Part 1: Measurement techniques, uncertainties and availability, Atmos. Meas. Tech., 7, 1395–1427, doi:10.5194/amt-7-1395-2014, 2014.

Hegglin, M. I., Plummer, D. A., Shepherd, T. G., Scinocca, J. F., Anderson, J., Froidevaux, L., Funke, B., Hurst, D., Rozanov, A., Urban, J., von Clarmann, T., Walker, K. A., Wang, H. J., Tegtmeier, S., and Weigel., K.: Vertical structure of stratospheric water vapour trends derived from merged satellite data, Nature: Geoscience, doi:10.1038/ngeo2236, 2014.

Hervig, M. and McHugh, M.: Cirrus detection using HALOE measurements, Geophys. Res. Lett., 26, 719–722, 1999.

Jones, A., Urban, J., Murtagh, D. P., Eriksson, P., Brohede, S., Haley, C., Degenstein, D., Bourassa, A., von Savigny, C., Sonkaew, T., Rozanov, A., Bovensmann, H., and Burrows, J.: Evolution of stratospheric ozone and water vapour time series studied with satellite measurements, Atmos. Chem. Phys., 9, 6055–6075, doi:10.5194/acp-9-6055-2009, 2009.

Kramarova, N. A., Frith, S. M., Bhartia, P. K., McPeters, R. D., Taylor, S. L., Fisher, B. L., Labow, G. J., and DeLand, M. T.: Validation of ozone monthly zonal mean profiles obtained from the version 8.6 Solar Backscatter Ultraviolet algorithm, Atmos. Chem. Phys., 13, 6887–6905, doi:10.5194/acp-13-6887-2013, 2013.

Kyrölä, E., Laine, M., Sofieva, V., Tamminen, J., Päivärinta, S.-M., Tukiainen, S., Zawodny, J., and Thomason, L.: Combined SAGE II-GOMOS ozone profile data set for 1984–2011 and trend analysis of the vertical distribution of ozone, Atmos. Chem. Phys., 13, 10645–10658, doi:10.5194/acp-13-10645-2013, 2013.

Labow, G. J., McPeters, R. D., Bhartia, P. K., and Kramarova, N.: A comparison of 40 years of SBUV measurements of column ozone with data from the Dobson/Brewer network, J. Geophys. Res.-Atmos., 118, 7370–7378, doi:10.1002/jgrd.50503, 2013.

Laine, M., Latva-Pukkila, N., and Kyrölä, E.: Analyzing time varying trends in stratospheric ozone time series using state space approach, Atmos. Chem. Phys. Discuss., 13, 20503–20530, doi:10.5194/acpd-13-20503-2013, 2013.

Lambert, J.-C., et al.: Past changes in the Vertical Distribution of Ozone - Part II: Measurement Intercomparisons, in preparation, 2015.

Livesey, N. J., Read, W. J., Froidevaux, L., Waters, J. W., Santee, M. L., Pumphrey, H. C., Wu, D. L., Shippony, Z., and Jarnot, R. F.: The UARS Microwave Limb Sounder version 5 dataset: Theory, characterization and validation, J. Geophys. Res., 108, 4378, doi:10.1029/2002JD002273, 2003.

McCormick, M. P., Zawodny, J. M., Veiga, R. E., Larsen, J. C. and Wang, P. H.: An overview of sage I and II ozone measurements, Planet. Space Sci., 37, 1567–1586, doi:10.1016/0032-0633(89)90146-3, 1989.

McLinden, C. A. and Fioletov, V.: Quantifying stratospheric ozone trends: Complications due to stratospheric cooling, Geophys. Res. Lett., 38, L03808, doi:10.1029/2010GL046012, 2011.

McPeters, R. D., Bhartia, P. K., Haffner, D., Labow, G. J., and Flynn, L.: The version 8.6 SBUV ozone data record: An overview, J. Geophys. Res.-Atmos., 118, 8032–8039, doi:10.1002/jgrd.50597, 2013.

McPeters, R. D. and Labow, G. J.: Climatology 2011: An MLS and sonde derived ozone climatology for satellite retrieval algorithms, J. Geophys. Res., 117, D10, doi:10.1029/2011JD017006, 2012.

Newchurch, M. J., Yang, E.-S., Cunnold, D. M., Reinsel, G. C., Zawodny, J. M., and Russell III, J. M.: Evidence for slowdown in stratospheric ozone loss: First stage of ozone recovery, J. Geophys. Res., 108, D16, doi10.1029/2003JD003471, 2003.

Parrish, A., Boyd, I. S., Nedoluha, G. E., Bhartia, P. K., Frith, S. M., Kramarova, N. A., Connor, B. J., Bodeker, G. E., Froidevaux, L., Shiotani, M., and Sakazaki, T.: Diurnal variations of stratospheric ozone measured by ground-based microwave remote sensing at the Mauna Loa NDACC site: measurement validation and GEOSCCM model comparison, Atmos. Chem. Phys., 14, 7255–7272, doi:10.5194/acp-14-7255-2014, 2014.

Penckwitt, A. A., Bodeker, G. E., Revell, L. E., Richter, L., Kyrölä, E., and Young,P.: Construction and analysis of a new merged SAGE II-GOMOS ozone profile data set for 1984–2012, Earth Syst. Sci. Data, in preparation, 2015.

Perliski, L. M., Solomon, S., and London, J.: On the interpretation of seasonal variations of stratospheric ozone, Planet. Space Sci., 37, 1527–1538, doi:10.1016/0032-0633(89)90143-8,1989.

Randel, W. J., Park, M., Wu, Fu., and Livesey, N.: A large annual cycle in ozone above the tropical tropopause linked to the Brewer-Dobson circulation, J. Atmos. Sci., 64, 4479–4488, doi:10.1175/2007JAS2409.1, 2007.

Reinsel, G. C., Tiao, G. C., Wuebbles, D. J., Kerr, J. B., Miller, A. J., Nagatani, R. M., Bishop, L., and Ying, L. H.: Seasonal trend analysis of published ground-based and TOMS total ozone data through 1991, J. Geophys. Res., 99, 5449–5464, 1994.

Rienecker, M. M., Suarez, M. J., Gelaro, R., Todling, R., Bacmeister, J., Liu, E., Bosilovich, M. G., Schubert, S. D., Takacs, L., Kim, G.-K., Bloom, S., Chen, J., Collins, D., Conaty, A., da Silva, A., Gu, W., Joiner, J., Koster, R. D., Lucchesi, R., Molod, A., Owens, T., Pawson, S., Pegion, P., Redder, C. R., Reichle, R., Robertson, F. R., Ruddick, A. G., Sienkiewicz, M. and Woollen, J.: MERRA: NASA's Modern-Era Retrospective Analysis for Research and Applications, J. Clim., 24, 3624–3648, doi:10.1175/JCLI-D-11-00015.1, 2011.

Rosenfield, J. E., Frith, S. M., and Stolarski, R. S.: Version 8 SBUV ozone profile trends compared with trends from a zonally averaged chemical model, J. Geophys. Res.-Atmos., 110, D12, doi:10.1029/2004JD005466, 2005.

Sakazaki, T., Fujiwara, M., Mitsuda, C., Imai, K., Manago, N., Naito, Y., Nakamura, T., Akiyoshi, H., Kinnison, D., Sano, T., Suzuki, M. and Shiotani, M.: Diurnal ozone variations in the stratosphere revealed in observations from the Superconducting Submillimeter-Wave Limb-Emission Sounder (SMILES) on board the International Space Station (ISS), J. Geophys. Res.-Atmos., 118, 2991–3006, doi:10.1002/jgrd.50220, 2013.

Schanz, A., Hocke, K., and Kämpfer, N.: Daily ozone cycle in the stratosphere: global, regional and seasonal behaviour modelled with the Whole Atmosphere Community Climate Model, Atmos. Chem. Phys., 14, 7645–7663, doi:10.5194/acp-14-7645-2014, 2014.

Simmons, A. J., Poli, P., Dee, D. P., Berrisford, P., Hersbach, H., Kobayashi, S., and Peubey, C.: Estimating low-frequency variability and trends in atmospheric temperature using ERA-Interim, Q. J. R. Meteorol. Soc., 140, 329–353, doi:10.1002/qj.2317, 2014.

Sioris, C. E., McLinden, C. A., Fioletov, V. E., Adams, C., Zawodny, J. M., Bourassa, A. E., Roth, C. Z., and Degenstein, D. A.: Trend and variability in ozone in the tropical lower stratosphere over 2.5 solar cycles observed by SAGE II and OSIRIS, Atmos. Chem. Phys., 14, 3479–3496, doi:10.5194/acp-14-3479-2014, 2014.

Steinbrecht, W., Claude, H., Schönenborn, F., McDermid, I. S., Leblanc, T., Godin, S., Song, T., Swart, D. P. J., Meijer, Y. J., Bodeker, G. E., Connor, B. J., Kämpfer, N., Hocke, K., Calisesi, Y., Schneider, N., de la Noë, J., Parrish, A. D., Boyd, I. S., Brühl, C., Steil, B., Giorgetta, M. A., Manzini, E., Thomason, L. W., Zawodny, J. M., McCormick, M. P., Russell, J. M., Bhartia, P. K., Stolarski, R. S. and Hollandsworth-Frith, S. M.: Long-term evolution of upper stratospheric ozone at selected stations of the Network for the Detection of Stratospheric Change (NDSC), J. Geophys. Res.-Atmos., 111, D10308, doi:10.1029/2005JD006454, 2006.

Studer, S., Hocke, K., Schanz, A., Schmidt, H., and Kämpfer, N.: A climatology of the diurnal variations in stratospheric and mesospheric ozone over Bern, Switzerland, Atmos. Chem. Phys., 14, 5905–5919, doi:10.5194/acp-14-5905-2014, 2014.

Tegtmeier, S., Hegglin, M. I., Anderson, J., Bourassa, A., Brohede, S., Degenstein, D., Froidevaux, L., Fuller, R., Funke, B., Gille, J., Jones, A., Kasai, Y., Krüger, K., Kyrölä, E., Lingenfelser, G., Lumpe, J., Nardi, B., Neu, J., Pendlebury, D., Remsberg, E., Rozanov, A., Smith, L., Toohey, M., Urban, J., von Clarmann, T., Walker, K. A. and Wang, R. H. J.: SPARC Data Initiative: A comparison of ozone climatologies from international satellite limb sounders, J. Geophys. Res.-Atmos., 118, 12229–12247, doi:10.1002/2013JD019877, 2013.

Thompson, D. J. W., Seidel, D. J., Randel, W. J., Zou, C.-Z., Butler, A. H. Mears, C., Osso, A., Long, C. and Lin, R.: The mystery of recent stratospheric temperature trends, Nature, 491, 692–697, doi:10.1038/nature11579, 2012.

Toohey, M. and von Clarmann, T.: Climatologies from satellite measurements: the impacts of orbital sampling on the standard error of the mean, Atmos. Meas. Tech., 6, 37–948, doi:10.5194/amtd-5-8241-2012, 2013.

Toohey, M., Hegglin, M. I., Tegtmeier, S., Anderson, J., Anel, J. A., Bourassa, A., Brohede, S., Degenstein, D., Froidevaux, L., Fuller, R., Funke, B., Gille, J., Jones, A., Kasai, Y., Krüger, K., Kyrölä, E., Neu, J. L., Rozanov, A., Smith, L., Urban, J., von Clarmann, T., Walker, K. A. and Wang, R. H. J.: Characterizing sampling biases in the trace gas climatologies of the SPARC Data Initiative, J. Geophys. Res.-Atmos., 118, 11847–11862, doi:10.1002/jgrd.50874, 2013.

Wang, H. J., Cunnold, D. M., Thomason, L. W., Zawodny, J. M., and Bodeker, G. E.: Assessment of SAGE version 6.1 ozone data quality, J. Geophys. Res., 107, D23, doi:10.1029/2002JD002418, 2002.

Wang, L., Zou, C.-Z., and Qian, H.: Construction of Stratospheric Temperature Data Records from Stratospheric Sounding Units, J. Clim., 25, 2931–2946, doi:10.1175/JCLI-D-11-00350.1, 2011.

Waugh, D. W., Oman, L., Kawa, S. R., Stolarski, R. S., Pawson, S., Douglass, A. R., Newman, P. A., and Nielsen, J. E.: Impacts of climate change on stratospheric ozone recovery, Geophys. Res. Lett., 36, doi:10.1029/2008GL036223, 2009.

Wild, J. D. and Long, C. S.: A Coherent Ozone Profile Dataset from SBUV, SBUV/2: 1979 to 2013, in preparation, 2015.

WMO, World Meteorological Organization: Scientific Assessment of Ozone Depletion: 1998, Global Ozone Research and Monitoring Project Report No. 44, Geneva, Switzerland, 1999.

WMO, World Meteorological Organization: Scientific Assessment of Ozone Depletion: 2010, Global Ozone Research and Monitoring Project Report No. 52, Geneva, Switzerland, 516 pp., 2011.

WMO, World Meteorological Organization: Scientific Assessment of Ozone Depletion: 2014, Global Ozone Research and Monitoring Project Report No. 55, Geneva, Switzerland, 416 pp., 2015.

Xu, J. and Powell Jr., A. M.: Uncertainty of the stratospheric/tropospheric temperature trends in 1979–2008: multiple satellite MSU, radiosonde, and reanalysis datasets, Atmos. Chem. Phys., 11, 10727–10732, doi:10.5194/acp-11-10727-2011, 2011.

11

Stratospheric and mesospheric HO_2 observations from the Aura Microwave Limb Sounder

L. Millán[1,2], S. Wang[2], N. Livesey[2], D. Kinnison[3], H. Sagawa[4], and Y. Kasai[4]

[1]Joint Institute for Regional Earth System Science and Engineering, University of California, Los Angeles, California, USA
[2]Jet Propulsion Laboratory, California Institute of Technology, Pasadena, California, USA
[3]National Center for Atmospheric Research, Boulder, Colorado, USA
[4]National Institute of Information and Communications Technology, Koganei, Tokyo, Japan

Correspondence to: L. Millán (lmillan@jpl.nasa.gov)

Abstract. This study introduces stratospheric and mesospheric hydroperoxyl radical (HO_2) estimates from the Aura Microwave Limb Sounder (MLS) using an offline retrieval (i.e. run separately from the standard MLS algorithm). This new data set provides two daily zonal averages, one during daytime from 10 to 0.0032 hPa (using day-minus-night differences between 10 and 1 hPa to ameliorate systematic biases) and one during nighttime from 1 to 0.0032 hPa. The vertical resolution of this new data set varies from about 4 km at 10 hPa to around 14 km at 0.0032 hPa. A description of the methodology and an error analysis are presented. Comparisons against the Whole Atmosphere Community Climate Model (WACCM), the Superconducting Submillimeter-Wave Limb-Emission Sounder (SMILES) and the Far Infrared Spectrometer (FIRS-2) measurements, as well as photochemical simulations, demonstrate the robustness of the retrieval and indicate that the retrieval is sensitive enough to detect mesospheric HO_2 layers during both day and night. This new data set is the first long-term HO_2 stratospheric and mesospheric satellite record and it provides needed constraints to help resolve the O_3 deficit problem and the "HO_x dilemma".

1 Introduction

Since 1985, when for the first time the now famous O_3 hole was reported (Farman et al., 1985), the stratospheric O_3 layer has received massive scientific attention. Although it peaks in the lower stratosphere, this layer extends well into the mesosphere where O_3 chemistry is controlled by catalytic cycles involving the HO_x (HO_2, OH and H) family (Brasseur and Solomon, 2005):

$$X + O_3 \rightarrow XO + O_2 \tag{R1}$$
$$O + XO \rightarrow O_2 + X, \tag{R2}$$

where the net effect of these two reactions is simply

$$O + O_3 \rightarrow 2O_2 \tag{R3}$$

which destroys O_3 without changing the abundance of the catalyst, X, which in this case is either OH or H.

The presence of the HO_x family in the middle atmosphere is a consequence of the transport of H_2O and CH_4 from the troposphere to higher altitudes. The production of HO_x species in the middle atmosphere is primarily due to O_3 photolysis:

$$O_3 + h\nu \rightarrow O_2 + O(^1D) \tag{R4}$$
$$O(^1D) + H_2O \rightarrow 2OH \tag{R5}$$
$$OH + O_3 \rightarrow O_2 + HO_2 \tag{R6}$$

and, above 60 km, by photodissociation of H_2O by absorption of UV radiation (particularly in the Lyman-alpha region and the Schumann–Runge bands),

$$H_2O + h\nu \rightarrow OH + H, \tag{R7}$$

where H turns into HO_2 due to the three-body reaction

$$H + O_2 + M \rightarrow HO_2 + M. \tag{R8}$$

The removal of HO_x is mainly through the self-reaction

$$HO_2 + OH \rightarrow H_2O + O_2 \tag{R9}$$

and the partitioning between OH and HO_2 above ~ 40 km is primarily driven by

$$HO_2 + O \rightarrow OH + O_2 \tag{R10}$$

$$H + O_3 \rightarrow O_2 + OH \tag{R11}$$

$$OH + O \rightarrow O_2 + H, \tag{R12}$$

where the H produced in Reaction (R12) is quickly converted to HO_2 by Reaction (R8).

Despite the apparent simplicity of the HO_x chemistry, models were not able to give a complete picture of its chemistry in the middle atmosphere: Summers et al. (1997) reported that OH satellite observations by the Middle Atmosphere High Resolution Spectrograph Investigation (MAHRSI) were 30 to 40 % lower than the values computed using standard photochemical models. Sandor et al. (1998) reported that mesospheric ground-based microwave measurements of HO_2 were 23 to 47 % higher than photochemical model predictions at midday, agreed with the model values prior to 09:00 LT, and were 70 to 100 % higher immediately after sunset. Jucks et al. (1998) using simultaneous OH and HO_2 balloon observations by the Far Infrared Spectrometer (FIRS-2) reported that OH agreed reasonably well with model estimates, a conclusion supported by Pickett et al. (2008), while HO_2 was 25 % higher than the model estimates. However, Canty et al. (2006) concluded that, at least between 25 and 60 km, the HO_x Microwave Limb Sounder (MLS) and FIRS-2 observations were reasonably well described by photochemical models. In addition to these modelling discrepancies (a problem known as the HO_x dilemma), models have consistently underpredicted the amounts of O_3 at such altitudes, an issue known as the O_3 deficit problem (Crutzen and Schmailzl, 1983; Solomon et al., 1983; Eluszkiewicz and Allen, 1993; Summers et al., 1997; Varandas, 2004; Siskind et al., 2013).

There have been many suggested solutions for the "HO_x dilemma" and the O_3 deficit problem in the literature (e.g. Miller et al., 1994; Jucks et al., 1998; Varandas, 2004); however, difficulties persist, in part because there are very few observations of HO_2 in the mesosphere. To date, there are four mesospheric data sets available: (1) 6 days spread between April 1992 and December 1996 measured by the Kitt Peak National Radio Astronomy Observatory (NRAO) (Sandor et al., 1998), (2) the Sub-Millimeter Radiometer (SMR) aboard the Odin satellite data set (Baron et al., 2009), which consists of one observation period of 24 h each month between October 2003 and December 2005, (3) the Superconducting Submillimeter-Wave Limb-Emission Sounder (SMILES) data set (Kikuchi et al., 2010), which provides daily coverage from 38° S to 65° N between October 2009 and April 2010, and (4) the standard MLS data set (Pickett et al., 2006, 2008), which provides coverage mostly from 55° S to 55° N, up to 0.046 hPa and only during daytime.

In this study, we introduce a new data set of global observations of stratospheric and mesospheric HO_2 from the MLS instrument. This new offline (i.e. run separately from the standard MLS algorithm) retrieval extends the HO_2 vertical range well into the mesosphere (up to 0.003 hPa or ~ 90 km), which, in addition to the standard MLS H_2O, OH and O_3, potentially allows the MLS data set to study the HO_x–O_3 chemical system to provide insights for the long-standing O_3 deficit problem. To date, this data set provides 10 years of data and, in the near future, it will be publicly available for download in a daily-based hierarchical data format (HDF). In the next section we give an overview of the MLS instrument as well as the MLS HO_2 measurements. Section 3 presents the offline retrieval approach and describes its vertical resolution, precision, and systematic uncertainties. Comparisons between the offline MLS HO_2 data set and global climate model simulations, balloon-borne and satellite data sets are shown in Sect. 4. Lastly, the offline HO_2 data are compared to the results of a 1-D photochemical model to investigate whether standard reaction rates can model the extended HO_2 vertical range. Section 5 provides a summary.

2 MLS HO_2 observations

The Microwave Limb Sounder (MLS) is, in essence, a small radio telescope on board the Aura satellite which was launched into a polar sun-synchronous orbit in July 2004. MLS measures limb millimetre and submillimetre atmospheric thermal emission at 120 different tangent altitudes from the ground to about 95 km every 24.7 s. It covers latitudes between 82° S and 82° N, providing near-global observations with roughly half of these measurements during daytime ($\sim$ 13:45 LT), and the other half during nighttime ($\sim$ 01:45 LT), except near the poles where the observations transition between daytime and nighttime conditions and vice versa. The incoming radiance is collected by a 1.6 m antenna which directs it onto four heterodyne radiometers covering spectral regions near 118, 191, 240 and 640 GHz (a fifth radiometer located at 2.5 THz is fed by a separate antenna). The radiant flux measured by the GHz radiometers is then analysed by 22 filter banks and four digital autocorrelator spectrometers. Most of these filter banks were designed to measure essentially a single spectral line, despite the double side band nature of the radiometers (i.e. each filter bank measures two different spectral regions, one on each side of the local oscillator). Furthermore, these banks have narrower filters at the band centre than at the extremes allowing them to measure the strong pressure broadening at microwave frequencies. A more detailed description of the Aura MLS instrument is given by Waters et al. (1999, 2006).

HO_2 is measured by two of these filter banks, known as band 28 and band 30, centred at the 649.72 and 660.50 GHz HO_2 lines. These filter banks consist of 11 channels with widths varying from 6 to 32 MHz, giving a total width of

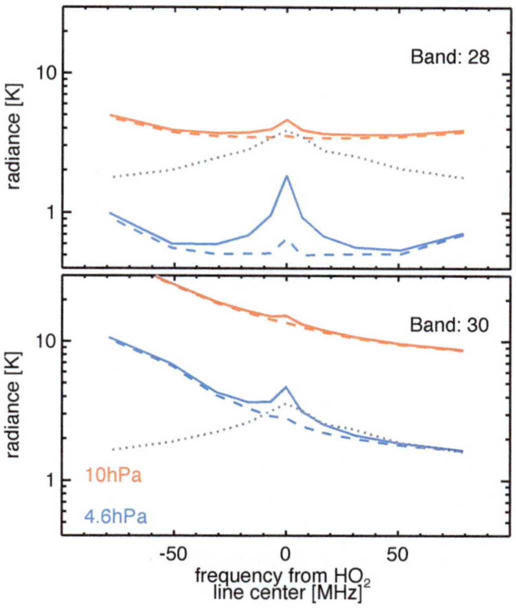

Figure 1. Average MLS radiance as detected by bands 28 (top) and 30 (bottom), separated into daytime (solid line) and nighttime (dashed line) measurements. Average is from 55° S and 55° N and for limb tangents at 10 and 4.6 hPa for January 2005. The dotted grey line is the expected single-scan noise.

around 200 MHz. An example of the observed radiances is shown in Fig. 1. The ~ 1 K HO_2 signal is relatively small compared to the individual limb radiance precision which varies from 2 K at the bands edges to 4 K at the band centre (grey dotted line), hence some averaging is required to obtain HO_2 abundances with a useful signal to noise ratio. The clear tilt in band 30 is due to the proximity of an O_3 line.

As mentioned above, HO_2 is one of the MLS products retrieved with the standard algorithm described in Livesey et al. (2006a). Briefly, the algorithm uses the optimal estimation technique (Rodgers, 2000) retrieving one profile for each scan using a two-dimensional approach. The smallness of the HO_2 signal translates to a retrieved product usable only between 10 and 0.046 hPa. For greater pressures, the signal is lost due to pressure broadening (Fig. 1 shows how the lines broaden rapidly as the pressure increases) and stronger emissions from O_3. For smaller pressures the signal is indistinguishable from the noise.

The latest version (V3.3) of the MLS HO_2 standard product (Pickett et al., 2008; Livesey et al., 2011) produces ~3500 abundance profiles daily with typical precisions varying from 0.15 ppbv (52×10^6 cm^{-3}) at 10 hPa to 3 ppbv (5×10^6 cm^{-3}) at 0.046 hPa for measurements zonally averaged and binned in a 10° latitude bin. Due to the seasonality of HO_2, these values can change up to 10 % depending on the pressure level observed. Its vertical resolution varies from 4 to 10 km between 10 and 0.046 hPa. In this pressure range, since negligible HO_2 is expected at night (~ 1:45 a.m. LT for MLS at the equator), it is recommended to use the non-zero nighttime abundances as an indication of systematic biases (i.e. affecting both day and night retrievals). Hence, the HO_2 day-minus-night difference should be used as a better estimate of the daytime HO_2 than taking the daytime measurements at face value. During summer and winter, this restricts the usable HO_2 standard product data to between roughly 55° S and 55° N, where MLS observes both daytime and nighttime data. In addition to the MLS HO_2 product the MLS HO_2 product, this day-minus-night difference approach to ameliorate biases has been used successfully for the BrO and OH MLS products (Livesey et al., 2006b; Pickett et al., 2008; Millán et al., 2012).

3 Offline retrieval

For this study, an HO_2 offline retrieval has been developed similar to that described by Millán et al. (2012). In essence, we compute a daily zonal mean of the radiances, for band 28 and 30, from which we retrieve daily HO_2 concentrations. The daily zonal mean radiances are formed by collocating them into 10° latitude bins, sorting them into daytime and nighttime using solar zenith angles (SZAs) lower than 90° and greater than 100°. Those radiances having SZAs between 90 and 100° are disregarded to avoid twilight measurements. These sorted radiances are then averaged onto a vertical grid with six pressure levels per decade (spacing of ~ 3 km) using the tangent pressure retrieved by the standard algorithm.

The retrieval uses the optimal estimation technique as described by Rodgers (2000) assuming an HO_2 a priori of zero, an a priori precision of 20 ppbv and using the corresponding daily zonal means of temperature, O_3, HNO_3 and HCl from the standard MLS algorithm (version 3.3) as part of the atmospheric state. In addition to this constraint, a Twomey–Tikhonov regularization (Tikhonov, 1963; Twomey, 1963) is used to reduce noise and smooth the profiles at the expense of some vertical resolution. Furthermore, at each pressure level, a constant baseline is retrieved for each band to correct any instrument baselines as well as to take care of the water vapour continuum contribution. The best retrievals were found when doing a joint band 28 and 30 retrieval as opposed to doing retrievals using only band 28 or only band 30, even despite the O_3 line influencing band 30.

The offline daytime HO_2 estimates are confined to a pressure range between 10 and 0.0032 hPa with day-minus-night differences used as a measure of daytime HO_2 for pressures between 10 and 1 hPa where the nighttime values exhibit non-zero values indicative of biases. Note that a visual inspection of the 10° bin monthly average profiles has shown, overall, no sings of a discontinuity at 1 hPa when using this approach. The offline nighttime HO_2 estimates are confined to a pressure range between 1 and 0.0032 hPa.

Figure 2 shows monthly (January 2005) zonal means for the standard and offline HO_2 data sets in volume mixing ra-

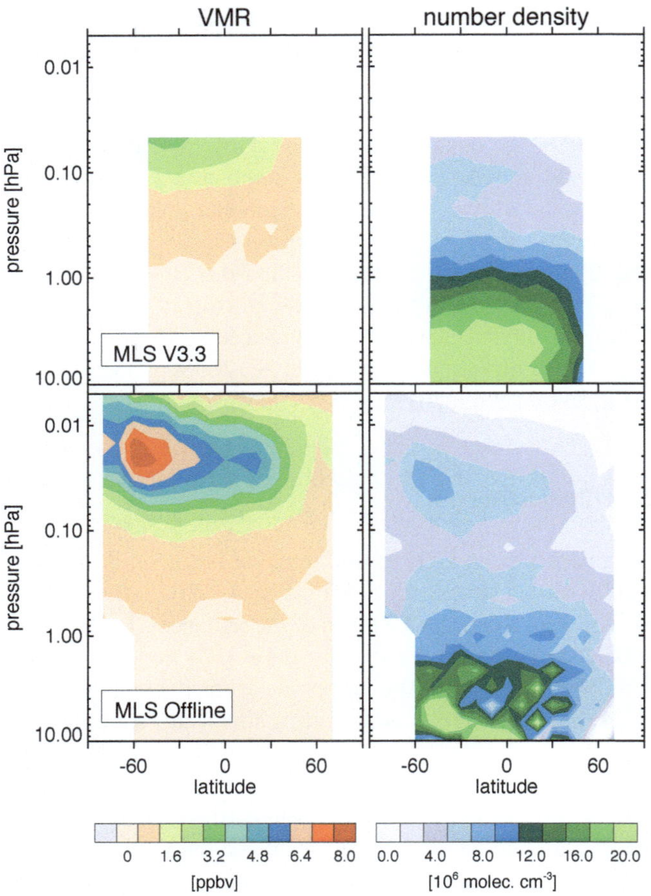

Figure 2. January 2005 zonal mean for MLS HO_2 data sets. The MLS V3.3 data set (top) is the standard product described by Pickett et al. (2008), here shown as the day-minus-night difference as recommended by the MLS data guidelines (Livesey et al., 2011). The MLS offline (bottom) is the data set introduced here. In this case, the day-minus-night difference has been taken only between 10 and 1 hPa, which is the reason why this data set can be extended to 10 hPa. The left column shows the data in VMR and the right one in number density.

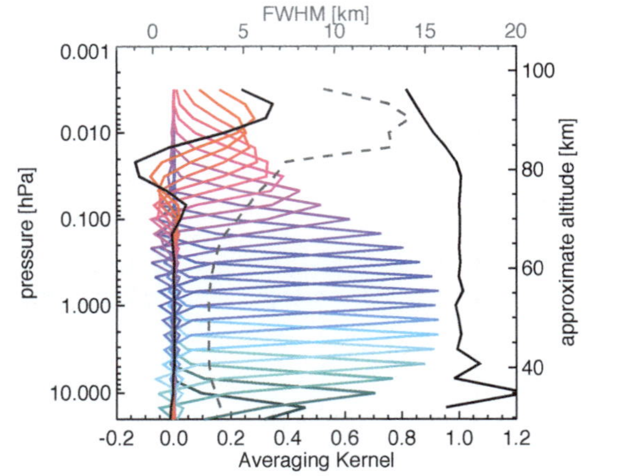

Figure 3. Daytime averaging kernels for the HO_2 offline retrieval at the equator (those at other latitudes are very similar). The black line is the integrated area under each kernel: values near unity indicate that most information was provided by the measurements while lower values indicate that the retrieval was influenced by the a priori. The dashed grey line is a measure of the vertical resolution of the retrieved profile (derived from the FWHM of the averaging kernels approximately scaled into kilometres).

tio (VMR) and density units. The temperature used to convert the data from VMR to number density is the same zonal mean temperature from the standard MLS algorithm that is used as part of the atmospheric state. As can be seen, the offline retrieval has two distinct improvements over the standard product: (1) an extended pressure range, which enables the measurements of the mesospheric local maxima that occur at around 0.02 hPa at most latitudes, and (2) an extended latitudinal coverage, making it possible to measure the polar regions, where in the summer one the HO_2 maximum lies. Furthermore, as shown in the following sections, the offline retrieval also estimates HO_2 during night between 1 and 0.0032 hPa. Note that the standard MLS product is smoother because it is highly constrained due to the poor signal-to-noise ratio of the individual radiance profiles.

3.1 Vertical resolution

Figure 3 shows typical averaging kernels for the HO_2 offline retrieval. These kernels delimitate the region of the atmosphere from which the information is contributing to the retrieved values at a given pressure level. As such, their full width at half maximum (FWHM) is a measure of the vertical resolution. The offline HO_2 product has a vertical resolution of about 4 km between 10 and 0.1 hPa, 8 km at 0.02 hPa and around 14 km for smaller pressures. The vertical resolution of this offline retrieval is similar to that of the HO_2 standard product for the pressure range where they overlap. The integrated kernel shows that most of the information arises from the measurements.

3.2 Error assessment

The total error in the retrieved product is a combination of the random noise in the measurements, the smoothing error, and the errors due to systematic uncertainties, such as instrumental and calibration errors and forward model and retrieval approximations.

Figure 4 displays the offline HO_2 expected precision for daily, monthly and yearly profiles of measurements averaged over a 10° latitude bin. The expected precision is the error due to the combination of the random noise in the measurements and the a priori uncertainty and is given by the diagonal elements of the covariance matrix of the retrieved state (Rodgers, 2000). For measurements averaged over a 10° latitude bin, for both the daytime and nighttime data, the daily HO_2 precision ranges from 0.09 ppbv (29×10^6 molec cm^{-3}) at 10 hPa to 1.4 ppbv (2.2×10^6 molec cm^{-3}) at 0.046 hPa,

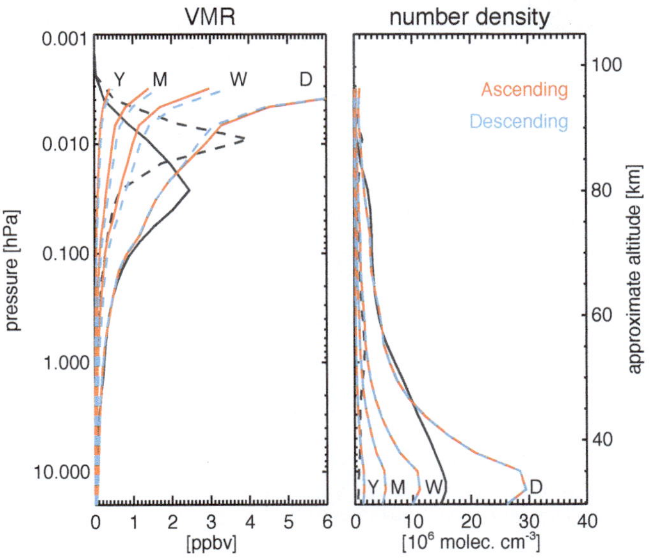

Figure 4. Expected precision for daily (D), weekly (W), monthly (M) and yearly (Y) MLS HO_2 offline data averaged over a 10° latitude bin. Precision expected in MLS HO_2 offline data averaged over a 10° latitude bin for a day (D), a week (W), a month (M) and a year (Y). The left column shows the data in VMR and the right one in number density. The black lines show typical HO_2 profiles, daytime in solid and nighttime dashed. These profiles are a yearly average over all latitudes of the SD-WACCM model (see Sect. 4.3 for the model description).

and up to 7.7 ppbv (1×10^6 molec cm^{-3}) at 0.003 hPa. Due to the temporal variability of HO_2, these precision values can change seasonally up to 40 % depending on the pressure level observed. This variability is greater than the one found in the standard MLS product because the standard product is highly constrained due to the poor signal-to-noise ratio. Although significant averaging such as monthly means is needed to achieve usable HO_2 estimates, the retrieval algorithm presented in this study uses daily zonal mean radiances, instead of weekly or monthly, in order to enable averaging different combination of days as needed.

Figure 5 summarizes the impact of the dominant systematic uncertainties for the offline MLS HO_2 product. These arise from instrumental issues such as uncertainties in the radiometric calibration, the field of view (FOV) characterization, the spectroscopy parameters, the pointing knowledge, the temperature profile used, the contaminant species errors, such as the O_3 line influencing band 30, and the retrieval approximations (a complete list and a more detailed discussion of the systematic error analysis is given by Read et al., 2007, Appendix A). The contribution of these uncertainties to the total HO_2 error was estimated using end-to-end calculations. For each systematic error a full day ($\sim$ 3500 profiles) of perturbed radiances was generated and binned into 10° zonal bins and processed by the offline algorithm. Each perturbation corresponds to either 2σ estimates of uncertainties in the relevant parameter, or an estimate of their maximum reasonable error based on instrument knowledge. Comparisons of these results with those using unperturbed radiances are a measure of the impact of each systematic error source. The comparison between the unperturbed noise-free radiances run, and the "true" model atmosphere estimates the errors due to the retrieval numerics, which, in other words, is a measure of error due to the retrieval formulation itself, in this case, mostly an smoothing error.

The impact of typically small error sources, such as (but not limited to) errors due to the spectrometer nonlinearities, uncertainties in the MLS spectral filter position, and the antenna transmission losses, has been quantified with a simple analytical model of the MLS measurement system (Read et al., 2007, Auxiliary material). Unlike the end-to-end estimates, these calculations only provide a multiplicative error.

Between 10 and 0.1 hPa, both for the daytime and nighttime case, the total systematic error is around 0.04 ppbv (up to $\sim 10 \times 10^6$ molec cm^{-3}). In this region, the main source of systematic bias arises from radiometric and spectroscopy uncertainties, in particular due to standing waves. Standing waves are a consequence of multiple reflections in the MLS optics of the hot and cold targets used as part of the radiometric calibration (Jarnot et al., 2006). For pressures smaller than 0.1 hPa, the main source of bias and scatter are retrieval numerics, which, although unsatisfactory, is understandable given the $\sim$ 14 km vertical resolution in this region. Around 0 0032 hPa, the total error is as large as 1.2 ppbv ($\sim 0.2 \times 10^6$ molec cm^{-3}) .

As already mentioned, between 10 and 1 hPa, the effects of the systematic biases can be diminished by subtracting the nighttime retrieved values from the daytime, taking advantage of the pronounced HO_2 diurnal variation below $\sim$ 1 hPa where negligible HO_2 is expected during night.

4 Results

In this section we compare the offline HO_2 data set with balloon-borne and other satellite measurements, as well as with global climate and photochemical model simulations. In making these comparisons, i.e. when showing the absolute or percentage differences between the data sets, the MLS averaging kernels has been applied to properly compare them. Furthermore, when comparing against the global climate or the photochemical model simulations, their high vertical resolution has been reduced to the MLS one using a least square fit as described by Livesey et al. (2011, Sect. 1.9). In these comparisons, no altitude extrapolation has been applied to any data set. To alleviate biases in the MLS offline HO_2 data, the day-minus-night differences are used as a measure of daytime HO_2 for pressures between 10 and 1 hPa.

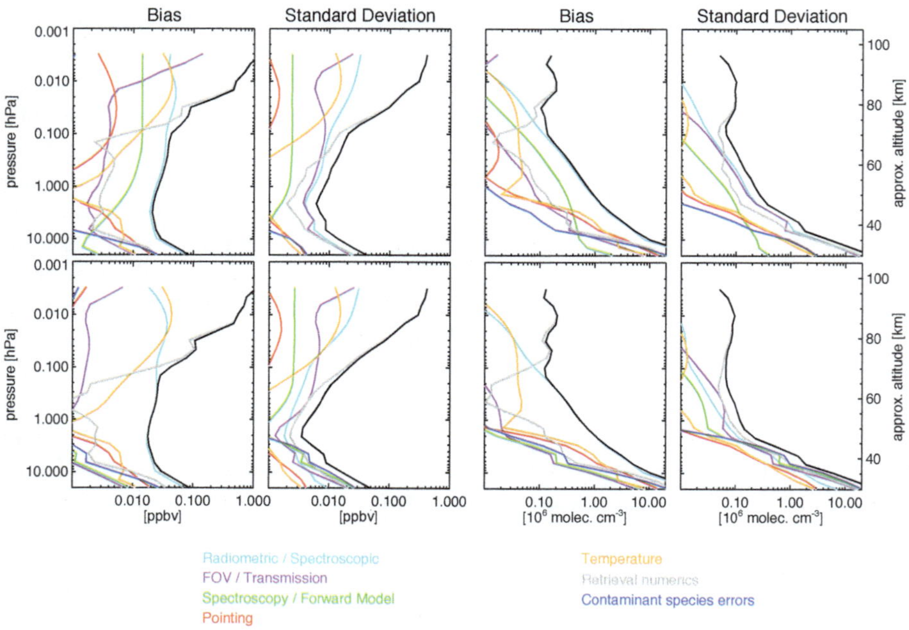

Figure 5. Estimated impact of various families of systematic error sources on the MLS offline HO_2 observations. Cyan lines correspond to uncertainties in the MLS radiometric and spectral calibration. Purple lines denote uncertainties associated with the MLS FOV and antenna transmission efficiency. Green lines denote errors in the spectroscopic databases and due to the forward model approximations. The impact of the uncertainties in the MLS pointing are depicted by the red lines. The yellow lines correspond to errors in the retrieved MLS temperature, while the blue lines show the impact of the retrieved errors in contaminant species. Errors due to retrieval formulation are shown in grey. The black lines are the root sum squares of all the biases or the scatters shown. The top panel corresponds to the daytime part of the orbit while the lower panel corresponds to the nighttime part. The left panels show the biases and additional scatter introduced by each family of systematic errors in VMR while the right panels show them in number density.

4.1 Comparison with balloon-borne instruments

Figure 6 compares the MLS offline and the Far Infrared Spectrometer (FIRS-2) HO_2 data. The MLS HO_2 profile corresponds to day-minus-night differences averaged over a 20° latitude bin centred at 30° N averaged over 10 days centred on the day of the balloon flight. The mean SZA of this profile is $\sim 32°$. The FIRS-2 profile shown was taken on the 20 September 2005 in Fort Sumner, New Mexico, USA (34.5° N). In that flight, the balloon stayed aloft at around 38 km for nearly 24 h. The FIRS-2 profile with the closest SZA to the MLS one is displayed, in this case $\sim 31°$.

FIRS-2 is a thermal emission far-infrared Fourier transform spectrometer developed at the Smithsonian Astrophysical Observatory. It measures emission spectra between 75 and 1000 cm^{-1} with high interferometric efficiency. It retrieves HO_2 from 43 rotational transitions between 110 and 220 cm^{-1} (Jucks et al., 1998). The retrieval algorithm first estimates slant columns and then, in an onion-peeling fashion, a singular value decomposition routine is used to retrieve mixing ratios on a 1 km vertical grid (Johnson et al., 1996). The total systematic errors for the retrieved HO_2 are estimated to be 3 %.

Overall, the two instruments agree on the HO_2 vertical structure, with increasing HO_2 with height (in the VMR representation). However, there seems to be an bias between them, with the MLS data in the lower bound. This might be due to the differences between the FIRS-2 single profile and the MLS zonal mean profile. In SD-WACCM (see Sect. 4.3 for the model description), these differences (i.e. comparing a single profile versus a zonal mean profile at this location) are around 30 %. Quantitatively, the MLS offline data agree with the FIRS-2 data within their uncertainties – a result also found by Pickett et al. (2008) using the standard MLS HO_2 product.

4.2 Satellite intercomparison

Comparisons of monthly means were made with those from the Superconducting Submillimeter Wave Limb Emission Sounder (SMILES). SMILES is a 4 K cooled radiometer on board the Japanese Experiment Module (JEM) on the International Space Station (ISS) that performed atmospheric observations from October 2009 to April 2010. It measured, on a time-sharing basis, two of three frequency bands: 624.32–625.52 GHz (band A), 625.12–626.32 GHz (band B) and 649.12–650.32 GHz (band C) covering tangent heights between 10 and at least 60 km about 1600 times per day covering mostly from 65° N to 38° S. HO_2 retrievals are based on radiances from band C which measures an HO_2 line centred

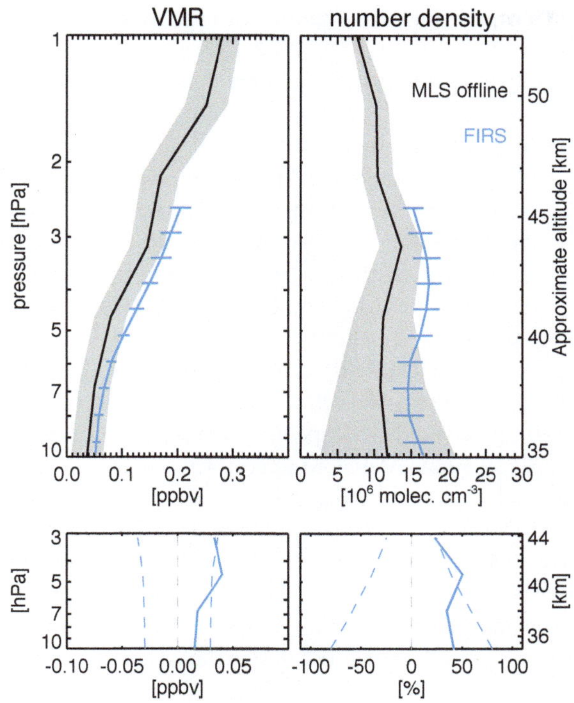

Figure 6. Comparison of MLS offline and FIRS-2 HO_2 data from 20 September 2005 at 34.5° N. The MLS data correspond to the day-minus-night average of 15–25 September 2005 for the 20° latitude bin centred at 30° N. The FIRS profile corresponds to the one with the closest SZA to the MLS (daytime only) data. The MLS (grey shaded region) and FIRS-2 errors represent a combination of precision and accuracy. The left column shows the data in VMR and the right one in number density. The differences shown in the bottom panels are FIRS-2 – MLS or (FIRS-2 – MLS) 100 / MLS (both in VMR units). The MLS averaging kernels were applied to the FIRS-2 data to fairly compare the two. The error margins (dashed lines) in the difference subplots correspond to the MLS and FIRS-2 errors added in quadrature.

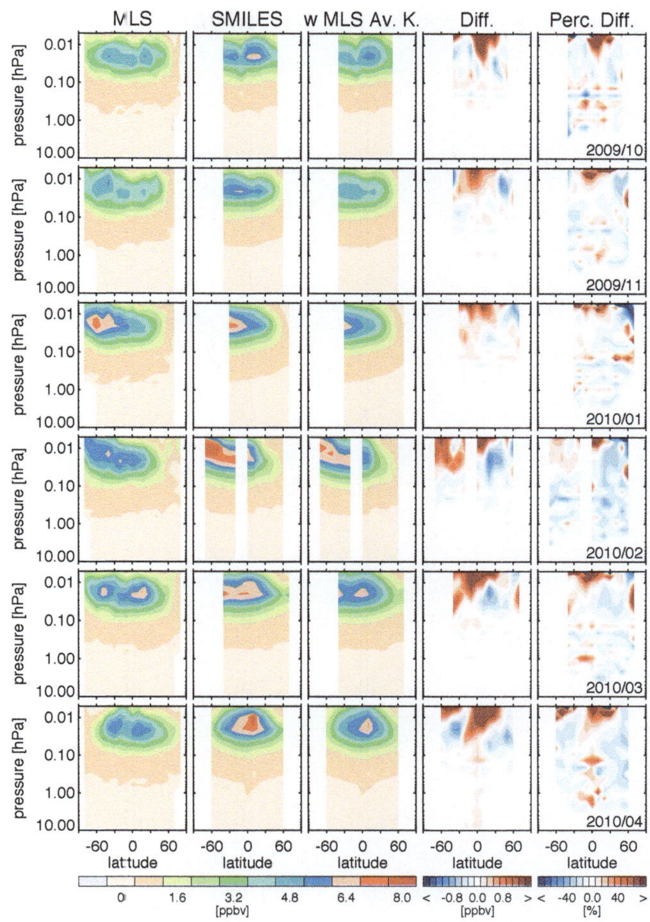

Figure 7. Monthly daytime zonal mean comparisons between MLS offline and SMILES HO_2 data for October 2009 to April 2011. The MLS averaging kernels were applied to the SMILES data to fairly compare the two. The differences shown are SMILES – MLS or (SMILES – MLS) 100 / MLS (both in VMR units). Note that there are no HO_2 SMILES data available for December 2009. As before, to alleviate biases in the MLS HO_2 data, the day-minus-night differences are used as a measure of daytime HO_2 for pressures between 10 and 1 hPa. From left to right, the data sets shown are: MLS, SMILES, SMILES with the MLS average kernels, the absolute difference and the percentage difference.

at 649.7 GHz (the same line measured by the upper side band of MLS band 28). The ISS follows a non-sun-synchronized circular orbit with an inclination of 51.6° which allowed SMILES to make measurements at local times that drifted ~ 20 min earlier each day covering the entire diurnal cycle in a period of about 2 months (Kikuchi et al., 2010). In this study we use the HO_2 retrievals from the research processor (version 3.0.0) developed by NICT (National Institute of Information and Communications Technology). This version is an outcome of the latest calibrated radiances which includes an improved determination of the tangent height (e.g. Ochiai et al., 2013). These HO_2 retrievals have been used for inter-satellite comparisons (Khosravi et al., 2013) and for a reaction rate estimation (Kuribayashi et al., 2014). They have a vertical resolution varying from 4 to 5 km at 35 and 55 km, respectively. The estimated precision of a single profile was estimated to be better than 30 % in the vertical range 20–86 km. In this study, only retrieval levels with a measurement response greater than 0.8 and lower than 1.2 have been used to avoid altitudes influenced too much by the a priori (Baron et al., 2011).

Figures 7 and 8 show daytime and nighttime comparisons (monthly means and the differences of the monthly means), respectively. To alleviate biases in the MLS HO_2 data, the day-minus-night differences are used as a measure of daytime HO_2 for pressures between 10 and 1 hPa. Note that only SMILES measurements made within half an hour local time of the MLS measurements were used in this comparison. During daytime, the retrieval top-level difference is as much as 80 %; however, at around 0.02 hPa (the mesospheric local maxima height), the difference is overall less

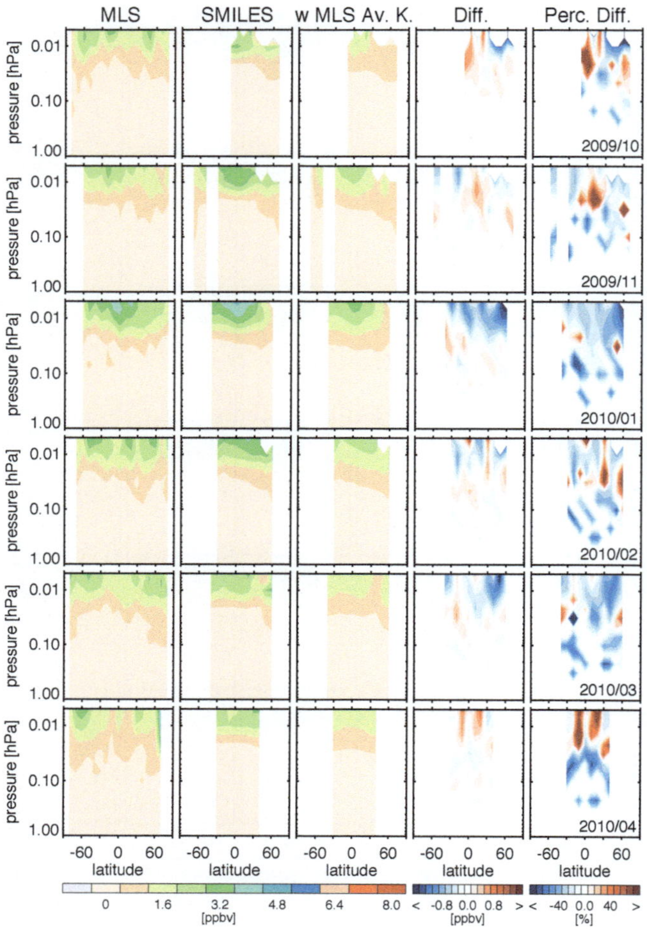

Figure 8. As Fig. 7, but for nighttime data. Note that the large positive percentage differences shown for April 2010 close to 20° S and 20° N are due to underestimations by the MLS retrieval.

than ±30 %. The retrieval top-level differences will need to be explored further, to investigate whether they are due to retrieval artifacts (both retrievals are more sensitive to the a priori at these levels), calibration uncertainties or sampling differences (unlike MLS, SMILES data are not regularly distributed); this will require a joint effort from the MLS and SMILES teams. During nighttime, the overall differences are around 30 % with localized spikes due to the small nighttime values. Overall, below around 0.1 hPa these HO_2 estimates agree within their uncertainties.

Comparisons against SMR were not performed for two reasons: (1) the SMR mesospheric data mentioned in Sect. 1 are not publicly available and (2) SMR ascending- and descending-node times are about 6.00 a.m. and 6.00 p.m., just during sunrise and sunset, which complicates the comparison.

4.3 Comparison with a global climate model

The Whole Atmosphere Community Climate Model (WACCM Version 4) is a fully interactive chemistry climate model in which the radiatively active gases affect heating and cooling rates and therefore dynamics (Garcia et al., 2007). It simulates the atmosphere from the Earth's surface to the thermosphere. For this analysis, the model was run with a horizontal resolution of $1.9° \times 2.5°$ in latitude and longitude, and a vertical coordinate purely isobaric in the stratosphere with a variable spacing of 1.1 to 1.75 km using meteorological fields derived from the Goddard Earth Observing System 5 (GEOS-5) analyses. The use of offline meteorological fields, a capability described by Lamarque et al. (2012), allows WACCM to perform as a chemical transport model, thus facilitating comparisons with observations. The WACCM chemical module is based on the 3-D chemical transport Model of Ozone and Related Tracers (MOZART), Version 4 (Kinnison et al., 2007). The results of this Specified Dynamics WACCM (SD-WACCM/GEOS-5) run were then sampled to the corresponding MLS observation time.

Figure 9 shows a daytime monthly (January 2005) mean comparison between the offline MLS data and the SD-WACCM simulations both in VMR and density units. To alleviate biases in the MLS HO_2 data, the day-minus-night differences are used as a measure of daytime HO_2 for pressures between 10 and 1 hPa. As can be seen, MLS and SD-WACCM display similar VMR structures with a gradient from the winter pole towards the summer pole. Note that, to properly compare the data with the model simulations, Fig. 9 also shows the SD-WACCM simulations convolved with the offline MLS averaging kernels reducing the SD-WACCM high vertical resolution using a least square fit as described by Livesey et al. (2011, Sect. 1.9).

In Fig. 9, the HO_2 VMR peak between 0.05 and 0.01 hPa reflects the mesospheric source – Reaction (R8) following Reaction (R7). The latitudinal gradient is a consequence of the varying SZA and the H_2O distribution, which also shows a latitudinal gradient due to the meridional circulation. The height of this peak is set by the balance between the thin air density at higher altitudes and the weakening of the UV irradiance responsible for the H_2O photolysis at lower altitudes, which leads to smaller HO_2 production.

In the stratosphere, as H_2O photolysis becomes less important and the H abundance decreases, HO_2 mainly forms from the transformation of OH through reactions with O_3 – Reaction (R6). This source is reflected in the HO_2 number density peak in the stratosphere and has a similar shape to the peak of OH (e.g. Pickett et al., 2008) but at lower altitudes due to the O_3 maximum height in the lower stratosphere.

Even though the offline MLS data set and the SD-WACCM simulations display similar structures (see Fig. 9), they differ in magnitude. The offline MLS data suggest that there is more mesospheric HO_2 than predicted by the model, particularly at ~ 0.02 hPa, which requires further investigation.

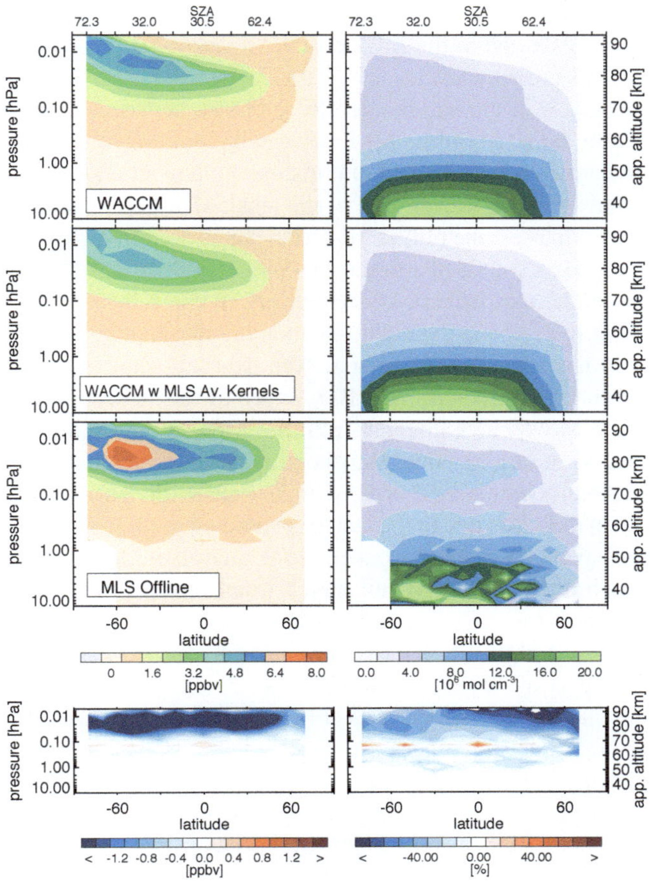

Figure 9. January 2005 monthly daytime zonal mean for the MLS offline HO_2 observations and the SD-WACCM model. To alleviate biases in the MLS HO_2 data, the day-minus-night differences are used as a measure of daytime HO_2 for pressures between 10 and 1 hPa. The MLS averaging kernels were applied to the SD-WACCM data set to fairly compare the two. The left column shows the data in VMR and the right one in number density. From top to bottom, the data sets shown are: SD-WACCM, SD-WACCM with the MLS averaging kernels, MLS and the absolute and percentage differences. The differences shown are SD-WACCM – MLS or (SD-WACCM – MLS) 100 / MLS (both in VMR units).

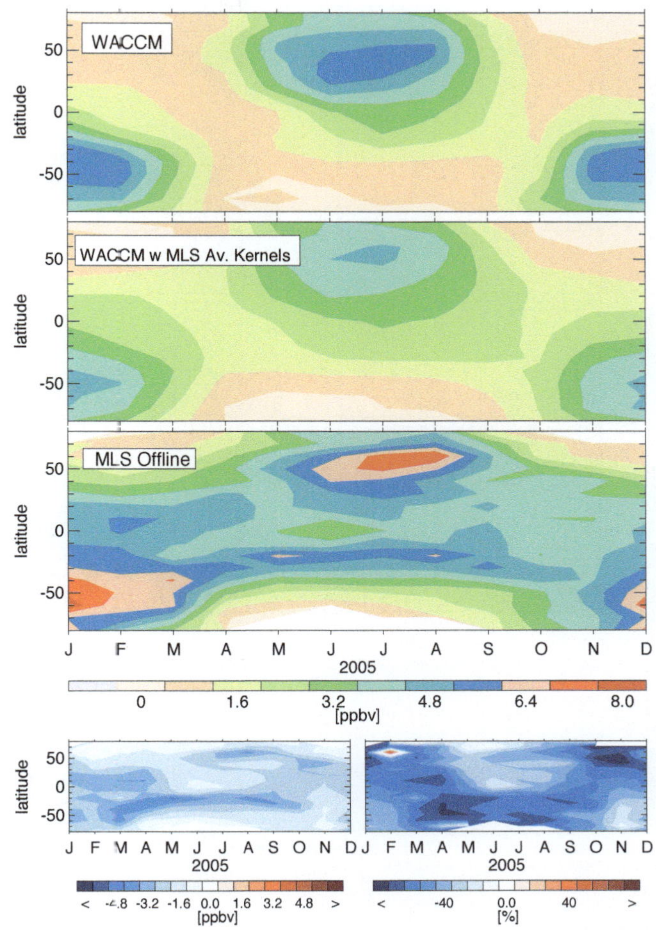

Figure 10. Daytime latitude/time cross-sections of the MLS offline HO_2 observations and the SD-WACCM model for 2005 at 0.014 hPa. As in Fig. 9, the MLS averaging kernels were applied to the SD-WACCM data set to fairly compare the two. From top to bottom, the data sets shown are SD-WACCM, SD-WACCM with the MLS averaging kernels, MLS and the absolute and percentage differences. The differences shown are SD-WACCM – MLS or (SD-WACCM – MLS) 100 / MLS (both in VMR units).

These differences correspond to SD-WACCM values smaller than the retrieved values by 50 % near the summer pole and smaller than 80 % near the winter pole (where SD-WACCM estimates near zero values).

To investigate whether these discrepancies, particularly the ∼ 4 ppbv difference at around 50° S, were due to measurement errors or due to assumptions in the SD-WACCM model, we compared measured MLS radiances to synthetic radiances computed using the offline MLS data and the SD-WACCM values. The SD-WACCM simulated radiances were smaller than the measured radiances by ∼ 40 %, outside the radiance error. Although MLS calibration errors cannot be ruled out, this is unlikely due to the magnitude of the offset and because this type of discrepancy cannot be found in the nighttime monthly mean comparison (Fig. 11). Hence, these simulations suggest that there is more mesospheric HO_2 than modelled at around 0.02 hPa. Furthermore, as can be seen in the latitude/time cross-section shown in Fig. 10, the behaviour found in Fig. 9 (a 50 % difference near the summer pole and more than 80 % near the winter one) repeats itself. These discrepancies might be due to a variety of reasons, for example: (1) our understanding of middle atmospheric chemistry may not be complete, and (2) there might be due to differences between recent solar spectral irradiance (SSI) satellite measurements (Snow et al., 2005; Harder, 2010) and most parametrizations. These SSI measurements display a larger variability in solar UV irradiance which cannot be reconstructed with SSI models, including the model of Lean et al. (2005), used in this SD-WACCM run (Marsh et al.,

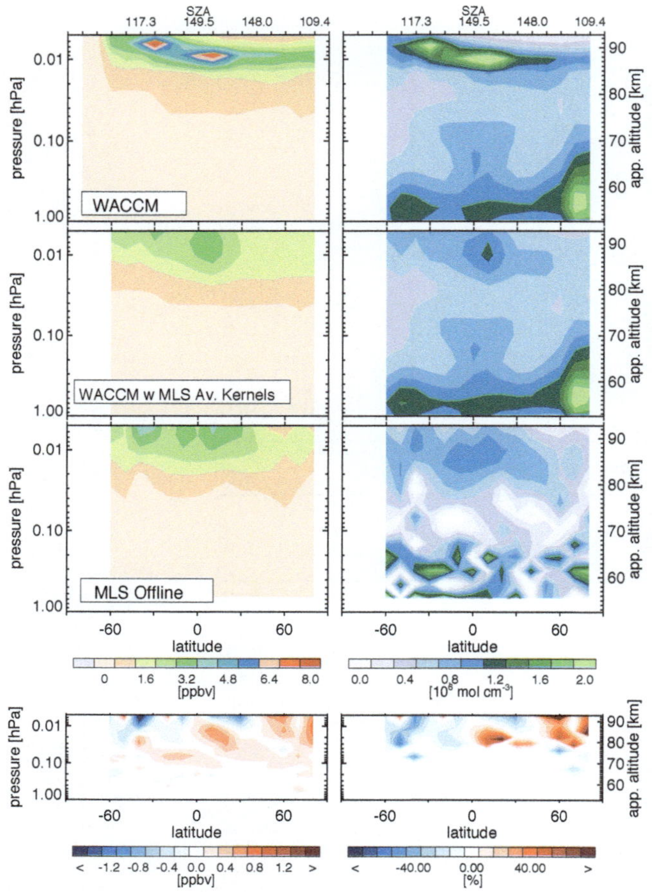

Figure 11. As Fig. 9, but for nighttime.

2013). These SSI measurement–model differences have been proven to affect the HO_x photochemistry (Haigh et al., 2010; Merkel et al., 2011; Ermolli et al., 2013); more UV irradiance leads to an enhancement of O_3 photolysis as well as H_2O photodissociation, which leads to more HO_x production through Reactions (R4)–(R8). Further, Wang et al. (2013) showed that using a solar forcing derived from these SSI measurements the modelled OH variability agrees much better with observations. Lastly, (3) these discrepancies might be related to the WACCM representation of the mean meridional circulation which has been shown to have some deficiencies (Smith et al., 2011; Smith, 2012), suggesting that the gravity wave parametrization needs to be modified. In addition, Garcia et al. (2014) have shown that adjusting the Prandtl number, used to calculate the diffusivity due to gravity waves, significantly alters the CO_2 SD-WACCM simulations, improving the agreement with satellite measurements. Such adjustment should also affect the H_2O and hence the HO_x chemistry.

Figure 9 also shows that SD-WACCM underpredicts HO_2 by about 20–30 % between 1 and 0.1 hPa. This result agrees with previous studies (Sandor et al., 1998; Khosravi et al., 2013) but contradicts the result of the study by Canty et al. (2006). The apparent model overestimation near 0.2 hPa is probably related to the MLS change in vertical scan pattern near this pressure level.

In Fig. 9 in the number density subplots, between 10 and 0.1 hPa, both the offline MLS data set and the SD-WACCM simulations behave in a similar manner both in structure and in magnitude; however, due to the small HO_2 signal in the MLS radiances, the offline MLS retrieval is noisier. The lack of a peak at ~0.02 hPa in the SD-WACCM data set reflects the smaller mesospheric concentrations in this data set.

Figure 11 shows the nighttime monthly mean comparison. As in the daytime comparison, both data sets show similar structures with a narrower HO_2 layer in the upper mesosphere as well as a gradient from the winter to the summer pole. As during daytime, HO_2 is principally formed through the three-body reaction of H with O_2 – Reaction (R8); however in this case, due to the lack of photodissociation of H_2O, the H available is the one generated at sunlit latitudes (in this case, in the Northern Hemisphere), transported at high altitudes towards the winter pole where it descends and reacts with O_2 at night (in this case, in the Southern Hemisphere) (Pickett et al., 2006). As in daytime, despite having similar structures, both data sets differ in magnitude. Overall, the nighttime differences are smaller than the daytime differences with SD-WACCM underpredicting the midlatitude summer regions by as much as 30 % and overestimating by as much as 50 % over the polar winter regions.

The strong contrast, around 6 ppbv, between the daytime and the nighttime HO_2 mesospheric peaks demonstrates the capability of the new offline retrieval to invert large mesospheric variations.

4.4 Photochemical model comparisons

Figure 12 shows a comparison between the HO_2 estimates from MLS offline and model simulations using the Caltech/JPL-Kinetics 1-D photochemical model. This model covers from the surface to 130 km in 66 layers. It has vertical transport (including eddy, molecular and thermal diffusion) and coupled radiative transfer (Allen et al., 1981, 1984). The kinetic parameters for the calculation of rate constants and photolysis rates were specified according to JPL 2011 recommendations (Sander et al., 2011). The model was run in a diurnally varying mode with no transport until the HO_x concentrations were repetitive – which means that a steady state was reached. Two model runs are shown: Kinetics 1 which constrains the model using MLS measurements of H_2O, O_3, and temperature to test the HO_x production and loss balance as well as the HO_x partitioning, and Kinetics 2 which, in addition, constrains the model using MLS OH measurements to mostly test the HO_x partitioning – Reactions (R10)–(R12).

As shown in Fig. 12, in the upper mesosphere (pressures smaller than 0.1 hPa), the Kinetics 1 simulations do not reproduce the magnitude of the measured peak, underestimating it by as much as 60 %. On the other hand, Kinetics 2

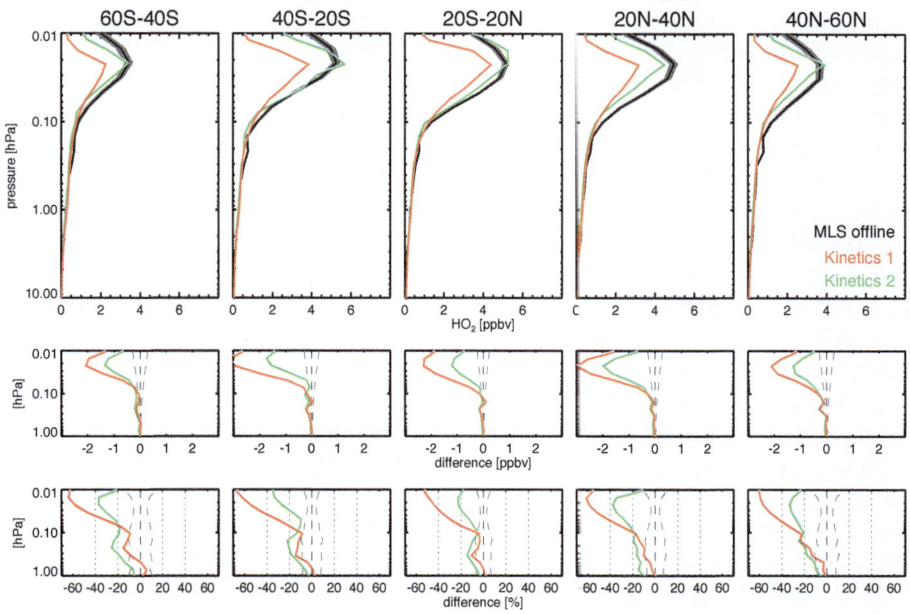

Figure 12. Daytime MLS offline HO_2 estimates for April 2005 and model results using JPL11 kinetics. Kinetics 1 is a photochemical model run constraining H_2O, O_3 and temperature while Kinetics 2 constrains H_2O, O_3, temperature and OH. The differences shown are Kinetics – MLS or (Kinetics – MLS) 100 / MLS. In the difference subplots, to properly compare the model runs with MLS, a least square fit has been used to reduce the model resolution and the corresponding averaging kernels has been applied. The dashed grey lines show the MLS precision as well as the 20 and 40 % percentage regions.

shows an improvement in the modelling of this peak, reducing the underestimation to less than 40 %. These discrepancies coincide with the ones discussed in the previous section, strongly suggesting that they are related to the model assumptions rather than to measurement errors. As with the SD-WACCM simulations, several factors could be the reason for this discrepancy: it might be due to limitations in our current understanding of middle atmospheric chemistry and/or due to the deficiencies in the model solar spectral irradiance used, in this case Rottman (1982).

Also, considering that Kinetics 2 (the run testing the HO_x partitioning) represents the measured HO_2 better, these simulations might suggest that the modelling problems are related to the HO_x production and loss balance rather than to the HO_x partitioning. In the upper stratosphere and lower mesosphere (between 1 and 0.1 hPa) for the most part the photochemical model underpredicts HO_2 by around 20 %, concurring with the SD-WACCM simulations as well as with previous studies (Sandor et al., 1998; Khosravi et al., 2013) but contradicting the result of the study by Canty et al. (2006).

5 Summary

We have introduced a stratospheric and mesospheric HO_2 data set derived from the Aura MLS using an offline retrieval algorithm. This offline HO_2 data set has three distinct improvements upon the standard MLS HO_2: (1) an extended pressure range, allowing measurements of the mesospheric peak local maxima that occur at around 0.02 hPa at most latitudes, (2) an extended latitudinal coverage, which allows to measure the poles, where the HO_2 maximum lies, and (3) nighttime HO_2 estimates.

The offline retrieval uses zonal mean MLS radiance spectra divided into daytime ($SZA \geq 90°$) and nighttime ($SZA \leq 100°$) which are then inverted using the optimal estimation technique to produce daily zonal mean HO_2 profiles from 10 to 0.0032 hPa during daytime and from 1 to 0.0032 hPa during nighttime. The MLS non-zero nighttime HO_2 abundances between 10 and 1 hPa were used as measure of systematic biases in the retrievals. Assuming that these biases are constant throughout day and night, we used the day-minus-night differences, only at these pressure levels, as a more accurate daytime HO_2 estimate. The vertical resolution of this data set is about 4 km between 10 and 0.1 hPa, 8 km at 0.02 hPa, and around 14 km for smaller pressures. Daily precision ranges from 0.1 ppbv in the upper stratosphere to up to 8 ppbv in the upper mesosphere, dropping to ~ 1.4 ppbv and ~ 0.5 ppbv for monthly and yearly averages, respectively. Between 10 and 0.1 hPa, for both daytime and nighttime cases, the total systematic error is around 0.04 ppbv (up to $\sim 10 \times 10^6$ molec cm^{-3}), while for smaller pressure levels the systematic error is as big as 1.2 ppbv ($\sim 0.2 \times 10^6$ molec cm^{-3}).

Comparison with the balloon-borne FIRS-2 measurements revealed that both data sets agree within their uncertainties, however there seems to be an offset with MLS on the low side. Comparisons with SMILES were found to agree both in structure and magnitude within the uncertainties below 0.02 hPa. Qualitatively, the offline MLS HO_2 agrees well with the SD-WACCM model. Quantitatively, however, the offline MLS HO_2 exceeds the model by up to 100 % in the mesosphere during day (in regions where SD-WACCM estimates near-zero values) and up to 40 % at night. Also, between 1 and 0.1 hPa, SD-WACCM underpredicts HO_2 by about 20–30 %, agreeing with previous studies (Sandor et al., 1998; Khosravi et al., 2013) but contradicting the result of the study by Canty et al. (2006).

Using the Caltech/JPL-Kinetics 1-D photochemical model we found similar results. In the upper mesosphere, we found an underestimation by the model by as much as 60 %, and in the upper stratosphere/lower mesosphere an underestimation by about 20 %. These results strongly suggest that these discrepancies are related to the model assumptions rather than to measurement errors.

The results presented in this study show that this new data set, in addition to the standard MLS OH, H_2O and O_3 measurements, offers the possibility to study the impact of the HO_x family upon the mesospheric O_3 as well as the HO_x dilemma. Furthermore, this retrieval, or a similar one using geomagnetic latitudes to sort the radiances, may help to understand the impact of solar proton events and energetic electron particles upon the HO_x family by comparing averages of days impacted by these events with averages of non-impacted days.

Acknowledgements. We thank M. Allen and K. Willacy for their help with setting up and running the Caltech/JPL-Kinetics 1-D photochemical model. FIRS-2 data were funded by the NASA Upper Atmosphere Research programme. JEM/SMILES mission is a joint project of Japan Aerospace Exploration Agency (JAXA) and National Institute of Information and Communications Technology (NICT). The WACCM modelling work was sponsored by the National Science Foundation and by the NASA Atmospheric Composition: Modeling and Analysis, solicitation NNH10ZDA001N-ACMAP. The research described in this paper was carried out by the Jet Propulsion Laboratory, California Institute of Technology, under contract with the National Aeronautics and Space Administration.

Edited by: G. Stiller

References

Allen, M., Yung, Y., and Waters, J.: Vertical transport and photochemistry in the terrestrial mesosphere and lower thermosphere (50–120 km), J. Geophys. Res., 86, 3617–3627, 1981.

Allen, M., Lunine, J., and Yung, Y.: The vertical distribution of ozone in the mesosphere and lower thermosphere, J. Geophys. Res., 89, 4841–4872, 1984.

Baron, P., Dupuy, E., Urban, J., Murtagh, D. P., Eriksson, P., and Kasai, Y.: HO_2 measurements in the stratosphere and the mesosphere from the sub-millimetre limb sounder Odin/SMR, Int. J. Remote. Sens., 30, 4195–4208, 2009.

Baron, P., Urban, J., Sagawa, H., Möller, J., Murtagh, D. P., Mendrok, J., Dupuy, E., Sato, T. O., Ochiai, S., Suzuki, K., Manabe, T., Nishibori, T., Kikuchi, K., Sato, R., Takayanagi, M., Murayama, Y., Shiotani, M., and Kasai, Y.: The Level 2 research product algorithms for the Superconducting Submillimeter-Wave Limb-Emission Sounder (SMILES), Atmos. Meas. Tech., 4, 2105–2124, doi:10.5194/amt-4-2105-2011, 2011.

Brasseur, G. P. and Solomon, S.: Aeronomy of the Middle Atmosphere, Chemistry and Physics of the Stratosphere and Mesosphere, 3rd revised and enlarged edn., Springer, the Netherlands, 307–326, 2005.

Canty, T., Pickett, H. M., Salawitch, R. J., Jucks, K. W., Traub, W. A., and Waters, J. W.: Stratospheric and mesospheric HO_X: results from Aura MLS and FIRS-2, Geophys. Res. Lett., 33, L12802, doi:10.1029/2006GL025964, 2006.

Crutzen, P. J. and Schmailzl, U.: Chemical budgets of the stratosphere, Planet. Space. Sci., 31, 1009–1032, doi:10.1016/0032-0633(83)90092-2, 1983.

Eluszkiewicz, J. and Allen, M.: A global analysis of the ozone deficit in the upper stratosphere and lower mesosphere, J. Geophys. Res., 98, 1069, doi:10.1029/92JD01912, 1993.

Ermolli, I., Matthes, K., Dudok de Wit, T., Krivova, N. A., Tourpali, K., Weber, M., Unruh, Y. C., Gray, L., Langematz, U., Pilewskie, P., Rozanov, E., Schmutz, W., Shapiro, A., Solanki, S. K., and Woods, T. N.: Recent variability of the solar spectral irradiance and its impact on climate modelling, Atmos. Chem. Phys., 13, 3945–3977, doi:10.5194/acp-13-3945-2013, 2013.

Farman, J. C., Gardiner, B. G., and Shanklin, J. D.: Large losses of total ozone in Antartica reveal seasonal ClO_x/NO_x interaction, Nature, 315, 2007–2010, 1985.

Garcia, R. R., Marsh, D., Kinninson, D. E., Boville, B., and Sassi, F.: Simulations of secular trends in the middle atmosphere, J. Geophys. Res., 112, D09301, doi:10.1029/2006JD007485, 2007.

Garcia, R. R., López-Puertas M., Funke D., Marsh D. R., Kinnison D. E., Smith A. K., and González-Galindo F.:, On the distribution of CO2 and CO in the mesosphere and lower thermosphere, J. Geophys. Res., 119, 5700–5718, doi:10.1002/2013JD021208, 2014.

Haigh, J. D, Winning, A. R., Toumi, R., and Harder, J. W.: An influence of solar spectral variations on radiative forcing of climate, Nature, 467, 696–699, doi:10.1038/nature09426, 2010.

Harder, J. W., Thuillier, G., Richard, E. C., Brown, S. W., Lykke, K. R., Snow, M., McClintock, W. E., Fontenla, J. M., Woods, T. N., and Pilewskie, P.: The SORCE SIM solar spectrum: comparison with recent observations, Sol. Phys., 263, 3–24, doi:10.1007/s11207-010-9555-y, 2010.

Jarnot, R. F., Perun, V. S., and Schwartz, M. J.: Radiometric and spectral performance and calibration of the GHz bands of EOS MLS, IEEE T. Geosci. Remote., 44, 1131–1143, 2006.

Johnson, D. G., Orphal, J., Toon, G. C., Chance, K. V., Traub, W. A., Jucks, K. W., Guelachvili, G., and Morillon-Chapey, M.: Measurements of chlorine nitrate in the stratosphere using the ν_4 and ν_5 bands, Geophys. Res. Lett., 23, 1745–1748, 1996.

Jucks, K. W., Johnson, D. G., Chance, K. V., and Traub, W. A.: Observations of OH, HO_2, H_2O and O_3 in the upper stratosphere:

implications for HO_x photochemistry, Geophys. Res. Lett., 25, 3935–3938, 1998.

Khosravi, M., Baron, P., Urban, J., Froidevaux, L., Jonsson, A. I., Kasai, Y., Kuribayashi, K., Mitsuda, C., Murtagh, D. P., Sagawa, H., Santee, M. L., Sato, T. O., Shiotani, M., Suzuki, M., von Clarmann, T., Walker, K. A., and Wang, S.: Diurnal variation of stratospheric and lower mesospheric HOCl, ClO and HO_2 at the equator: comparison of 1-D model calculations with measurements by satellite instruments, Atmos. Chem. Phys., 13, 7587–7606, doi:10.5194/acp-13-7587-2013, 2013.

Kikuchi, K., Nishibori, T., Ochiai, S., Ozeki, H., Irimajiri, Y., Kasai, Y., Koike, M., Manabe, T., Mizukoshi, K., Murayama, Y., Nagahama, T., Sano, T., Sano, R., Seta, M., Takahashi, C., Takayanagi, M., Masuko, H., Inatani, J., Susuki, M., and Shiotani, M.: Overview and early results of the Superconducting Submillimeter Wave Limb Emission Sounder (SMILES), J. Geophys. Res., 115, D23306, doi:10.1029/2010JD014379, 2010.

Kinnison, D. E., Brasseur, G. P., Walters, S., Garcia, R. R., Marsh, D. R., Sassi, F., Harvey, V. L., Randall, C. E., Emmons, L., Lamarque, J. F., Hess, P., Orlando, J. J., Tie, X. X., Randel, W., Pan, L. L., Gettelman, A., Granier, C., Diehl, T., Niemeier, U., and Simmons, A. J.: Sensitivity of chemical tracers to meteorological parameters in the MOZART-3 chemical transport model, J. Geophys. Res., 112, D20302, doi:10.1029/2006JD007879, 2007.

Kuribayashi, K., Sagawa, H., Lehmann, R., Sato, T. O., and Kasai, Y.: Direct estimation of the rate constant of the reaction $ClO + HO_2 \rightarrow HOCl + O_2$ from SMILES atmospheric observations, Atmos. Chem. Phys., 14, 255–266, doi:10.5194/acp-14-255-2014, 2014.

Lamarque, J.-F., Emmons, L. K., Hess, P. G., Kinnison, D. E., Tilmes, S., Vitt, F., Heald, C. L., Holland, E. A., Lauritzen, P. H., Neu, J., Orlando, J. J., Rasch, P. J., and Tyndall, G. K.: CAM-chem: description and evaluation of interactive atmospheric chemistry in the Community Earth System Model, Geosci. Model Dev., 5, 369–411, doi:10.5194/gmd-5-369-2012, 2012.

Lean, J., Rottman, G., Harder, J., and Kopp, G.: SORCE contributions to new understanding of global change and solar variability, Sol. Phys., 230, 27–53, 2005.

Livesey, N., Snyder, W. V., Read, W. G., and Wagner, P.: Retrieval algorithms for the EOS Microwave Limb Sounder (MLS), IEEE T. Geosci. Remote., 44, 1144–1155, 2006a.

Livesey, N., Kovalenko L., Salawitch, R., MacKenzie, I., Chipperfield, M., Read, W., Jarnot, R., and Waters, J.: EOS microwave limb sounder observations of upper stratospheric BrO: implications for total bromine, Geophys. Res. Lett., 33, L20817, doi:10.1029/2006GL026930, 2006b.

Livesey, N., Read, W. G., Frovideaux, L., Lambert, A., Manney, G. L., Pumphrey, H. C., Santee, M. L., Schwartz, M. J., Wang, S., Cofield, R. E., Cuddy, D. T., Fuller, R. A., Jarnot, R. F., Jiang, J. H., Knosp, B. W., Stek, P. C., Wagner, P. A., and Wu, D. L.: Earth Observing System (EOS) Aura Microwave Limb Sounder (MLS) Version 3.3 Level 2 data quality and description document, JPL D-33509, JPL publication, USA, 2011.

Marsh, D. R., Mills, M. J., Kinnison, D. E., Lamarque, J.-F., Calvo, N., and Polvani, L. M.: Climate Change from 1850 to 2005 Simulated in CESM1(WACCM), J. Climate, 26, 7372–7391, 2013.

Merkel, A. W., Harder, J. W., Marsh, D. R., Smit, A. K., Fontenla, J. M., and Woods, T. N.: The impact of solar spectral irradiance variability on middle atmospheric ozone, Geophys. Res. Lett., 38, L13802, doi:10.1029/2011GL047561, 2011.

Miller, R. L., Suits, A. G., Houston, P. L., Toumi, R., Mack, J. A., and Wodtke, J. A.: The "Ozone Deficit" Problem: $O_2(X, \upsilon \geq 26) + O(^3P)$ from 226-nm Ozone Photodissociation, Science, 265, 1831–1837, 1994.

Millán, L., Livesey, N., Read, W., Froidevaux, L., Kinnison, D., Harwood, R., MacKenzie, I. A., and Chipperfield, M. P.: New Aura Microwave Limb Sounder observations of BrO and implications for Br_y, Atmos. Meas. Tech., 5, 1741–1751, doi:10.5194/amt-5-1741-2012, 2012.

Ochiai, S., Kikuchi, K., Nishibori, T., Manabe, T., Ozeki, H., Mizobuchi, S., and Irimajiri, Y.: Receiver performance of the superconducting submillimeter-wave limb-emission sounder (SMILES) on the international space station, IEEE T. Geosci. Remote., 51, 3791–3801, doi:10.1109/TGRS.2012.2227758, 2013.

Pickett, H. M., Read, W. G., Lee, K. K., and Lung, Y. L.: Observation of night OH in the mesosphere, Geophys. Res. Lett., 33, L19808, doi:10.1029/2006GL026910, 2006.

Pickett, H. M., Drouin, B. J., Canty, T., Salawitch, R. J., Fuller, R. A., Perun, V. S., Livesey, N. J., Waters, J. W., Stachnik, R. A., Sander, S. P., Traub, W. A., Jucks, K. W., and Minschwaner, K.: Validation of Aura Microwave Limb Sounder OH and HO_2 measurements, J. Geophys. Res., 113, D16S30, doi:10.1029/2007JD008775, 2008.

Read, W., Lambert, A., Bacmeister, J., Cofield, R. E., Christensen, L. E., Cuddy, D. T., Daffer, W. H., Drouin, B. J., Fetzer, E., Froidevaux, L., Fuller, R., Herman, R., Jarnot, R. F., Jiang, J. H., Jiang, Y. B., Kelly, K., Knosp, B. W., Kovalenko, L. J., Livesey, N. J., Liu, H.-C., Manney, G. L., Pickett, H. M., Pumphrey, H. C., Rosenlof, K. H., Sabounchi, X., Santee, M. L., Schwartz, M. J., Snyder, W. V., Stek, P. C., Su, H. S., Takacs, L. L., Thurstans, R. P., Vömel, H., Wagner, P. A., Waters, J. W., Webster, C. R., Weinstock, E. M., and Wu, D. L.: Aura microwave limb sounder upper tropospheric and lower stratospheric H_2O and relative humidity with respect to ice validation, J. Geophys. Res., 112, D24S35, doi:10.1029/2007JD008752, 2007.

Rodgers, C.: Inverse Methods for Atmospheric Sounding: Theory and Practice, Vol. 2 of Series on Atmospheric, Oceanic and Planetary Physics, World Scientific, Singapore, 2000.

Rottman, G. J., Barth, C. A., Thomas, R. J., Mount, G. H., Lawrence, G. M., Rusch, D. W., Sanders, R. W., Thomas, G. E., London, J.:, Solar spectral irradiance, 120 to 190nm, October 13, 1981–January 3, 1982, Geophys. Res. Lett., 9, 587–590, doi:10.1029/GL009i005p00587, 1982.

Sander, S. P., Friedl, R. R., Barker, J. R., Golden, D. M., Burkholder, J. B., Kolb, C. E., Kurylo, M. J., Moortgat, G. K., Wine, P. H., Abbatt, J. P. D., Huie, R. E., and Orkin, V. L.: Chemical kinetics and photochemical data for use in atmospheric studies, Tech. rep., JPL, evaluation number 17, JPL Publ., 10–6, JPL publication, USA, 2011.

Sandor, B. J. and Clancy, R. T.: Mesospheric HO_x chemistry from diurnal microwave observations of HO_2, O_3 and H_2O, J. Geophys. Res., 103, 13337–13351, 1998.

Smith, A. K., Garcia R. R., Marsh D. R., and Richter J. H.:, WACCM simulations of the mean circulation and trace species transport in the winter mesosphere, J. Geophys. Res., 116, D20115, doi:10.1029/2011JD016083, 2011.

Smith, A. K.: Global Dynamics of the MLT, Surv. Geophys. 33, 1177–1230, doi:10.1007/s10712-012-9196-9, 2012.

Snow, M., McClintock, W. E., Rottman, G., and Woods, T. N.: Solar–stellar irradiance comparison experiment II (Solstice II): examination of the solar–stellar comparison technique, Sol. Phys., 230, 295–324, doi:10.1007/s11207-005-8763-3, 2005.

Solomon, S., Rusch, D. W., Thomas, R. J., and Eckman, R. S.: Comparison of mesospheric ozone abundances measured by the Solar Mesosphere Explorer and model calculations, Geophys. Res. Lett., 10, 249, doi:10.1029/GL010i004p00249, 1983.

Siskind, D. E., Stevens, M. H., Englert, C. R., and Mlynczak, M. G.: Comparison of a photochemical model with observations of mesospheric hydroxyl and ozone, J. Geophys. Res., 118, 195–207, doi:10.1029/2012JD017971, 2013.

Summers, M. E., Conway, R. R., Siskind, D. E., Stevens, M. H., Offerman, D., Riese, M., Preusse, P., Strobel, D. F., and Russell, III J. M.: Implications of satellite OH observations for middle atmospheric H_2O and ozone, Science, 227, 1967–1970, 1997.

Tikhonov, S.: On the solution of incorrectly stated problems and a method of a regularization, Dokl. Acad. Nauk. SSSR, 151, 501–504, 1963.

Twomey, S.: On the numerical solution of Fredholm integral equation of the first kind by the inversion of linear system produced by quadrature, J. Assoc. Comput. Mach, 10, 97–101, 1963.

Varandas, A. J. C.: Are vibrationally excited molecules a clue for the "O_3 deficit problem" and "HO_x dilemma" in the middle atmosphere?, J. Phys. Chem. A, 108, 758–769, 2004.

Wang, S., Li, K.-F., Pongetti, T. J., Sander, S. P., Yung, Y. L., Liang, M.-C., Livesey, N., Santee, M., Harder, J., Snow, M., and Mills, F.: Midlatitude atmospheric OH response to the most recent 11-y solar cycle, P. Natl. Acad. Sci. USA, 110, 2023–2028, doi:10.1073/pnas.1117790110, 2013.

Waters, J., Read, W., Froidevaux, L., Jarnot, R., Cofield, R., Flower, D., Lau, G., Pickett, H., Santee, M., Wu, D., Boyles, M., Burke, J., Lay, R., Loo, M., Livesey, N., Lungu, T., Manney, G., Nakamura, L., Perun, V., Ridenoure, B., Shippony, Z., Siegel, P., Thurstans, R., Harwood, R., and Filipiak, M.: The UARS and EOS microwave limb sounder experiments, J. Atmos. Sci., 56, 194–218, 1999.

Waters, J., Froidevaux, L., Harwood, R., Jarnot, R., Pickett, H., Read, W., Siegel, P., Cofield, R., Filipiak, M., Flower, D., Holden, J., Lau, G., Livesey, N., Manney, G., Pumphrey, H., Santee, M., Wu, D., Cuddy, D., Lay, R., Loo, M., Perun, V., Schwartz, M., Stek, P., Thurstans, R., Boyles, M., Chandra, S., Chavez, M., Chen, G.-S., Chudasama, B., Dodge, R., Fuller, R., Girard, M., Jiang, J., Jiang, Y., Knosp, B., LaBelle, R., Lam, J., Lee, K., Miller, D., Oswald, J., Patel, N., Pukala, D., Quintero, O., Scaff, D., Snyder, W., Tope, M., Wagner, P., and Walch, M.: The Earth Observing System Microwave Limb Sounder (EOS MLS) on the Aura satellite, IEEE T. Geosci. Remote., 44, 1075–1092, doi:10.1109/TGRS.2006.873771, 2006.

12

Mesoscale modeling study of the interactions between aerosols and PBL meteorology during a haze episode in Jing–Jin–Ji (China) and its nearby surrounding region – Part 1: Aerosol distributions and meteorological features

H. Wang[1,2], M. Xue[1], X. Y. Zhang[1], H. L. Liu[1], C. H. Zhou[1], S. C. Tan[3], H. Z. Che[1], B. Chen[3], and T. Li[4]

[1]Institute of Atmospheric Composition (IAC), Key Laboratory of Atmospheric Chemistry (LAC) of China Meteorological Administration (CMA), Chinese Academy of Meteorological Sciences (CAMS), Beijing, 100081, China
[2]Collaborative Innovation Center on Forecast and Evaluation of Meteorological Disasters, Nanjing University of Information Science & Technology, Nanjing, 210044, China
[3]State Key Laboratory of Numerical Modeling for Atmospheric Sciences and Geophysical Fluid Dynamics (LASG), Institute of Atmospheric Physics, Chinese Academy of Sciences, Beijing, 100029, China
[4]School of Atmospheric Physics, Nanjing University of Information Science & Technology, Nanjing, 210044, China

Correspondence to: H. Wang (wangh@cams.cma.gov.cn) and M. Yue (xm@cams.cma.gov.cn)

Abstract. The urbanized region ofJing(Beijing)-Jin(Tianjin)-Ji (alias of Hebei province) and its nearby surrounding region (3JNS) is becoming China's most polluted area by haze, exceeding even the Yangtze and Pearl river deltas. Aside from pollutant emission, the meteorology of the planetary boundary layer (PBL) is the most important factor affecting haze pollution. Focusing on July 2008, the aerosol optical properties and PBL meteorology features closely related to haze formation were simulated in the 3JNS region using an online atmospheric chemical transport model. The relationship between regional PBL meteorology, $PM_{2.5}$, and haze is discussed. Model results accurately simulated the aerosol optical depth (AOD), single scattering albedo (SSA) and asymmetry parameter (ASY), validated by comparison with observations from the MODerate Resolution Imaging Spectroradiometer (MODIS), the China Aerosol Remote Sensing NETwork (CARSNET) and the Aerosol Robotic NETwork (AERONET). Modeled PBL wind speeds showed reasonable agreement with those from the National Centers for Environmental Prediction (NCEP) Reanalysis 2. A monthly mean AOD value as high as 1.2 was found from both model and observations, with a daily mean larger than 2.0 during haze episodes in the 3JNS region. Modeled and observed SSA values of 0.90–0.96 and ASY values of 0.72–0.74 demonstrated the high scattering characteristic of summer aerosols in this region. PBL wind speeds from modeled and NCEP data both showed a reversing trend of $PM_{2.5}$ variation, illustrating the importance of the "PBL window shadow" in haze formation. Turbulence diffusion and PBL height had opposite phases to surface $PM_{2.5}$, indicating that lower PBL height and weaker PBL turbulence diffusion are essential to haze formation. It is noted that homogeneous air pressure does not occur at the surface, but at an 850–950 hPa height during the haze episode. The momentum transmitting downward of the cold air from above the PBL to the low PBL and surface lead to an increase in surface wind speeds and haze dispersal.

1 Introduction

With its rapidly expanding urbanization, and both economic and industrial developments, China is faced with increasingly poor air quality and haze pollution. There are three main haze pollution regions in eastern China: the Yangtze River delta, the Pearl River (Zhu Jiang) delta, and Beijing–Tianjin–Hebei (shortened to Jing–Jin–Ji) and its nearby surrounding region (3JNS). These are all areas of high population, rapid economic growth, urbanization and energy consumption (Zhang et al., 2004; Chak et al., 2008; Che et al., 2009; Wu et al., 2010). The Yangtze River delta region consists of Shanghai, the urban agglomeration of southeastern Jiangsu Province and northeastern Zhejiang Province. The Pearl River delta metropolitan area includes Guangzhou, Shenzhen, Dongguan, Zhuhai and other nearby cities. The 3JNS region includes Beijing (Jing), Tianjin (Jin) and Hebei Province (Ji) and their near surroundings, including eastern Shanxi Province, western Shandong Province and northern Henan Province.

Many observational and model studies have focused on pollution in the Yangtze River delta (Zhang et al., 2008; Fu et al., 2008; Yin et al., 2009; Gao et al., 2009, 2011; Wang et al., 2012; Kang et al., 2013) and Pearl River delta regions around the time of their initial, rapid economic development (Zheng et al., 2000; Lee et al., 2001; Cao et al., 2004; Wu et al., 2006; Chan et al., 2006; Chak et al., 2008; Huang et al., 2008a; Tan et al., 2009, 2011). However, the 3JNS region has recently become the most polluted area of the three regions, and is now attracting serious concern (Wang et al., 2006, 2012, 2014a, b; Chen et al., 2007; Wu et al., 2008; Wei et al, 2010; W. Liu et al., 2010; Duan et al., 2012; Che et al., 2014). According to the China Environmental Condition Report by the Ministry of Environmental Protection of The People's Republic of China (MEPPRC), seven of the top ten polluted cities in China in the first 6 months of 2013 – Xingtai, Shijiazhuang, Handan, Baoding, Tangshan, Hengshui and Langfang – are located in this region. However, haze pollution and air quality studies in this region, especially modeling and simulation studies, are rare (Wang et al., 2008; Xing et al., 2011) and inadequate compared to the Yangtze River delta and Pearl River delta regions (Westerdahl, et al., 2009; Zhang et al., 2009, 2011, 2013; Quan et al., 2014; Wang et al., 2014a).

When haze occurs, local meteorological patterns strongly affect the transport and mixing of gases and aerosols, pollutant loading, spatiotemporal distributions and pollution strength. In particular, the meteorological conditions of the local planetary boundary layer (PBL), e.g., wind fields, turbulence diffusion, PBL height and atmospheric circulation patterns, are all key to hazy weather, and dominate whether the haze occurs or not, since emissions can remain stable within a defined period in a certain area. A PBL parameterization scheme and describing local PBL meteorological conditions in mesoscale atmospheric chemistry models form the basis of $PM_{2.5}$ and haze forecasting. In turn, high particle concentrations suspended in the PBL atmosphere during hazy weather may exert a remarkable influence on local PBL meteorology and circulation patterns by reforming the regional solar and thermal radiative budgets.

Focusing on July 2008 over the 3JNS region, this paper outlines the methodology for the online calculation of aerosol optical features of different species based on an external mixing scheme, introduced into the GRAPES-CUACE atmospheric chemical model to simulate the aerosol optical features and PBL meteorology condition. The local aerosol optical, meteorological and circulation patterns in the PBL as related to haze are also discussed, with a particular focus on the haze episode of 7–11 July 2008. The relationship between key PBL meteorological factors, $PM_{2.5}$ and the haze episode is analyzed. The impact of aerosols on local PBL is presented in a companion paper (Part 2, Wang et al., 2015).

2 Model description

The Chinese Unified Atmospheric Chemistry Environment (CUACE) has been integrated into the mesoscale version of Global/Regional Assimilation and PrEdiction System (GRAPES_Meso) developed by the Chinese Academy of Meteorological Sciences, China Meteorological Administration (CMA), to build an online chemical weather forecasting model, GRAPES-CUACE/haze, focusing especially on haze pollution forecasting in China and East Asia. An aerosol radiative parameterization scheme was incorporated into the GRAPES-CUACE model. The aerosol optical depth (AOD), single scattering albedo and asymmetry factor (ASY) are calculated online using this model. The following sections offer a brief introduction to the model.

2.1 GRAPES_Meso

GRAPES_Meso is a real-time operational weather forecasting model used by the CMA, which includes 3-D meteorological field data assimilation, a fully compressible nonhydrostatic model core and a modularized physics package (Chen et al., 2003, 2008; Zhang and Shen, 2008; Yang et al., 2007). The model's time integration discretization uses a semi-implicit and semi-Lagrangian temporal advection scheme. The model's horizontal discretization adopts an Arakawa-C staggered grid arrangement and a central finite-difference scheme with second-order accuracy, while the model's vertical discretization adopts a nonhydrostatic approximation scheme. Height-based, terrain-following coordinates are used. The model's vertical discretization adopts the vertically staggered variable arrangement proposed by Charney-Phillips. The large-scale horizontal and vertical transportation and diffusion processes for all gases and aerosols are processed in the dynamic framework of GRAPES_Meso. The GRAPES_Meso3.3 model was released in July 2013 and was used in the GRAPES-CUACE/haze in this study.

The physical processes principally involve large-scale condensation, cumulus convection, micro-physical precipitation, radiative transfer, land surface and boundary layer processes. Each physical process incorporates the use of several schemes (Xu et al., 2008). The model physics schemes and the related lead references used in this study are summa-

Table 1. Configured GRAPES_CUACE options for physical processes.

Physical process	Configured options	References
Longwave radiation	Goddard	Chou et al. (2001)
Shortwave radiation	Goddard	Chou et al. (1998)
Cumulus clouds	KFETA scheme	Kain (2004)
Surface layer	SFCLAY scheme	Pleim (2007)
Boundary layer	MRF scheme	Hong and Pan (1996)
Land surface	SLAB scheme	Kusaka et al. (2001)
Cloud microphysics	KESSLER scheme	Kessler (1969)
Gas-phase chemistry	RADM II	Stockwell et al. (1990)
Aerosol chemistry	CUACE	Gong and Zhang (2008)

rized and listed in Table 1. The PBL scheme is very important for correctly modeling and providing accurate weather forecasts (Vogelezang et al., 1996; Santanello et al., 2005), especially for accurate air pollution forecasts (Cheng et al., 2002; Pleim, 2007). The PBL is the lower tropospheric layer, with its height (PBLH) ranging from several hundred meters to a few kilometers, which is one basic feature of the accurate and realistic modeling. The processes of heat, moisture and momentum exchange between the Earth's surface and the rest of the atmosphere all occur within the PBL. The wind speed near the surface, turbulence diffusion, and stability are also calculated in the PBL scheme in the air quality model. Among the different definitions of PBL, there are still some general agreements. The Richardson number (Ri) is usually used by PBLH calculation. The height at which the local Richardson number exceeds a critical value is used to separate stable from turbulent flow. The Hong and Pan Medium Range Forecast (MRF) PBL scheme (Hong and Pan, 1996) was selected for this study (Table 1). The MRF PBL scheme uses nonlocal closure and relies heavily on Ri to compute PBLH. It defines PBLH as the height at which a critical Ri is reached (0.5). All the PBL parameters discussed in the following sections are based upon this PBL scheme.

2.2 CUACE

Components of the CUACE atmospheric chemistry model include an emission inventory and process system; gaseous, physical aerosol and chemistry processes; and related thermodynamic equilibrium modules for processing the transformation between gas and particle matter (Gong and Zhang, 2008; Wang et al., 2009, 2010). The CUACE module tracer consists of 66 gas species and seven species of aerosols, with 12 particle size bins.

2.2.1 Emissions

Based on official information about national emission sources in 2006 (Cao et al., 2006), the detailed high-resolution emission inventories of reactive gases, i.e., SO_2, NO_x, CO, NH_3 and VOCs, from emissions over China in 2007 were updated to form the current emission data (Cao et al., 2010). The Sparse Matrix Operator Kernel Emissions (SMOKE) system was used to transform these emission data into hourly gridded data as required by the GRAPES_CUACE model, which includes five aerosols species (black carbon (BC), organic carbon (OC), sulfate, nitrate and fugitive dust particles) and 27 gases including VOCs, NH_3, CO, CO_2, SO_x and NO_x (An et al., 2013).

2.2.2 CUACE/Gas

CUACE/Gas is based on the Regional Acid Deposition Model (RADM) (Stockwell et al., 1990), which consists of 66 gaseous species, including five second-order organic aerosols (SOA); 21 photochemical reactions and 121 gas phase reactions are also involved. Wet and dry deposition processes, simple SOA reactions and a liquid-phase chemical balance are also included. The gas-to-aerosol particle transformation process is described using a thermodynamic equilibrium equation.

2.2.3 CUACE/Aero

There are seven species of aerosol considered in CUACE/Aero: sulfates (SF), soil dust (SD), black carbon (BC), organic carbon (OC), sea salts (SS), nitrates (NI) and ammonium salts (AM). The model divides all the aerosol particles into 12 particle size bins with diameter ranges of 0.01–40.96 μm (excluding AM). CUACE/Aero includes the major aerosol processes in the atmosphere: hygroscopic growth, coagulation, nucleation, condensation, dry deposition and sedimentation, below-cloud scavenging, aerosol activation, aerosol–cloud interactions and chemical transformation of sulfur species (Gong and Zhang, 2008).

2.3 Online calculation of optical properties of externally mixed aerosols

Aerosol chemical properties and sizes are used to calculate aerosol optical and radiative properties. Each chemical constituent of an aerosol is associated with a set of complex refractive index (CRI) data as a function of wavelength. The CRI data of the seven species of aerosols are derived mainly from the HITRAN 2008 database (Rothman et al., 2009), and the Optical Properties of Aerosols and Clouds (OPAC) database (Hess et al., 1998). Optical model data, accounting for East Asian dust using both theory calculation and composition analysis of aerosol samples collected in the Chinese desert during the international project, Studies on the Origin and Transport of Aeolian Dust and its Effects on Climate (ADEC), is used to account for Chinese mineral dust CRI data (Wang et al., 2004, 2006). Based on these CRI data and particle sizes in GRAPES_CUACE, Mie theory is used to calculate the key optical parameters of dry aerosol particles in determining aerosols' direct radiative effects, i.e., the extinction coefficient (Q_e), the SSA and the ASY. The mass extinction coefficient (K_{ext} in $m^2 g^{-1}$) is calculated according to the following formula:

$$K_{\text{ext}_{m,n}}(\lambda) = 3Q_{\text{e}_{m,n}}(\lambda)/4r_n\rho_m, \tag{1}$$

where n represents the particle bin from size 1 to 12 and r_n is the corresponding effective radius of the nth aerosol size, ρ is the particle mass density of the particular aerosol calculated, and m is the aerosol species, i.e., BC, SF, SD, BC, OC, SS, NI or AM, and λ is the wavelength. Figure 1 shows the K_{ext} (Fig. 1a), SSA (Fig. 1b) and ASY (Fig. 1c) for six radii of the model's 12 dry particle size bins (*R*_dry) from small to large particles of seven species of aerosol. The red line in Fig. 1 indicates the representative particle size bin for the species of aerosol with the highest concentration in the model. For SD aerosol particles, the size range is relatively large and all particle size bins are used, while for AM aerosol particles, only the particle size bin with a radius of 0.06 μm is employed. It can be seen from Fig. 1 that the optical features of dry aerosol particles change with chemical composition, particle size and wavelength. These changes are described in the following study of the aerosol radiative parameterization scheme. The AOD of any of the 12 particle bin sizes for any seven of the aerosol species is calculated using:

$$\text{AOD}_{m,n}(\lambda) = \sum_{i=1}^{k} K_{\text{ext}_{m,n}}(\lambda) C_{m,n} \Delta z_i \tag{2}$$

where $C_{m,n}$ is the mass concentration of the aerosol, n stands for the nth particle size bin and m indicates the aerosol type, i is the index of vertical layers, K is the total number of layers in the model, and Δz_i is the thickness of the model layer. For hygroscopic aerosols, i.e., SF, OC, SS, NI and AM, aerosol sizes of wet particles are calculated as a function of relative humidity (RH) using the Kola equation. A total of 10 RH values, 0, 45, 50, 60, 70, 80, 90, 95, 98, and 99 %, are considered in the model. The AOD, SSA and ASY of wet particles are a function of chemical composition, RH, particle size and wavelength, which are described as $\text{AOD}_{m,n}$ (rh,λ), $\text{SSA}_{m,n}$ (rh,λ) and $\text{ASY}_{m,n}$ (rh,λ). Figure 2 shows the typical particle size of the above three optical factors (the red line in Fig. 1) for five hygroscopic aerosol species at six different RHs. It can be seen from Fig. 2a that the extinction efficiencies of different aerosol species, and their changing trends along with wavelength and RH, are distinctly diverse. Wet aerosol SSA (Fig. 2b) and ASY (Fig. 2c) also show similar dependencies on RH, particle size and chemical composition. Figure 2 indicates a detailed aerosol radiation parameterization scheme considering particle size, atmospheric RH and the particle chemical aerosol composition is essential to evaluate aerosol radiative feedback. An external mixing scheme is used for the different particle size bins for one aerosol type and for different aerosol species to calculate composite aerosol optical properties for each model grid, according the following formulae:

$$\text{AOD}(\text{rh}, \lambda) = \sum_{m=1}^{7} \sum_{n=1}^{1} 2\text{AOD}_{m,n}(\text{rh}, \lambda) \tag{3}$$

$$\text{SSA}(\text{rh}, \lambda) = \frac{\sum\limits_{m=1}^{7} \sum\limits_{n=1}^{12} \text{SSA}_{m,n}(\text{rh}, \lambda) \times \text{AOD}(\text{rh}, \lambda)_{m,n}}{\text{AOD}(\text{rh}, \lambda)} \tag{4}$$

$$\text{ASY}(\text{rh}, \lambda) = \frac{\sum\limits_{m=1}^{7} \sum\limits_{n=1}^{12} \text{AOD}_{m,n}(\text{rh}, \lambda) \times \text{SSA}_{m,n}(\text{rh}, \lambda) \times \text{ASY}_{m,n}(\text{rh}, \lambda)}{\sum\limits_{m=1}^{7} \sum\limits_{n=1}^{12} \text{SSA}_{m,n}(\text{rh}, \lambda) \times \text{AOD}(\text{rh}, \lambda)} \tag{5}$$

Composite aerosol optical properties change with particle size bin (n) and concentration (C_m); RH forecast according to the GRAPES_CUACE model; and wavelength (λ) according to Eqs. (1)–(5).

3 Experiment design

A simulation experiment was achieved in this study by treating any composite aerosol only as a dynamic tracer: the aerosol's radiation feedback to its dynamic process was not calculated in this model experiment.

The model run for this study commenced on 25 June 2008 and the simulated results for 1–31 July 2008 serve as the base simulations for this research. The GRAPES_CAUCE3.0 model adopts an alterable horizontal resolution, a time step and a forecasting time. There are 31 model layers ascending vertically from the Earth's surface to 31 km in height. For the purposes of this study, the horizontal resolution was set to 0.15° × 0.15°, the time step to 100 s and the forecasting time to 72 h. The model domain was set to 90–140° E, 20–55° N. NCEP 1 × 1° Reanalysis data were used for the model's initial and 6 h meteorological lateral boundary input fields. The monthly mean values of all tracers from observation data are used for initialization at the very beginning of the model run. The initial values of all gases in RADM2 and aerosol concentrations are based on the 24 h forecast made by the previous day's model run. The simulation results after the first 3 days' model runs are used in this study to eliminate the model errors from the chemical tracer initialization.

4 Results

4.1 Optical properties of aerosols

AOD is a good parameter for elucidating aerosol column loading in the atmosphere. Aerosol optical properties contributing to AOD, SSA and ASY are the most direct and critical parameters for aerosol direct radiative forcing, radiative heating effects, and feedback to atmospheric circulation (Wang et al., 2006; Huang et al., 2006, 2009). AOD data from the Moderate Resolution Imaging Spectroradiometer

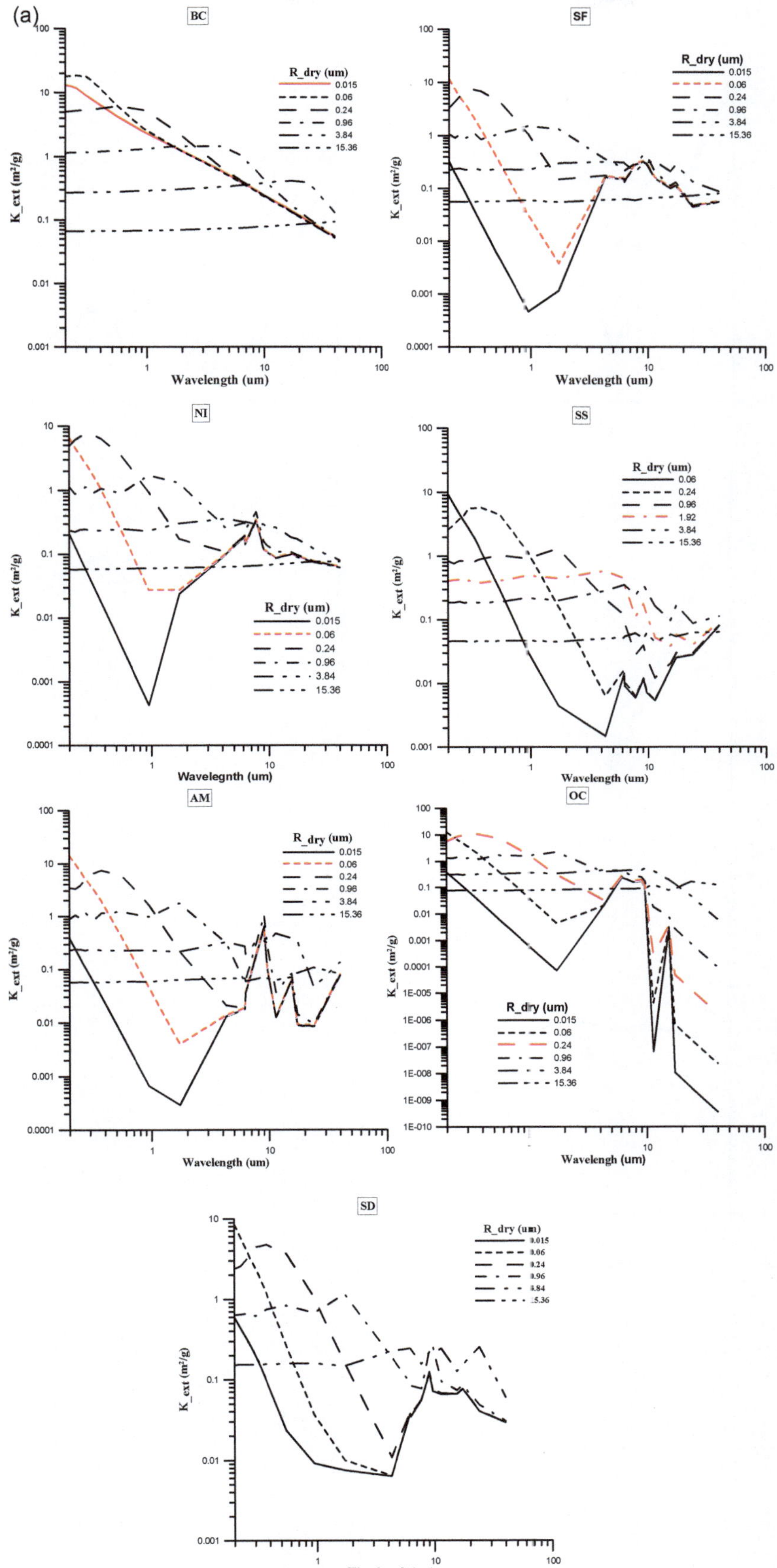

(a)
BC
SF
NI
SS
AM
OC
SD
R_dry (um)
0.015
0.06
0.24
0.96
1.92
3.84
15.36
K_ext (m²/g)
Wavelength (um)
Wavelengh (um)

Figure 1.

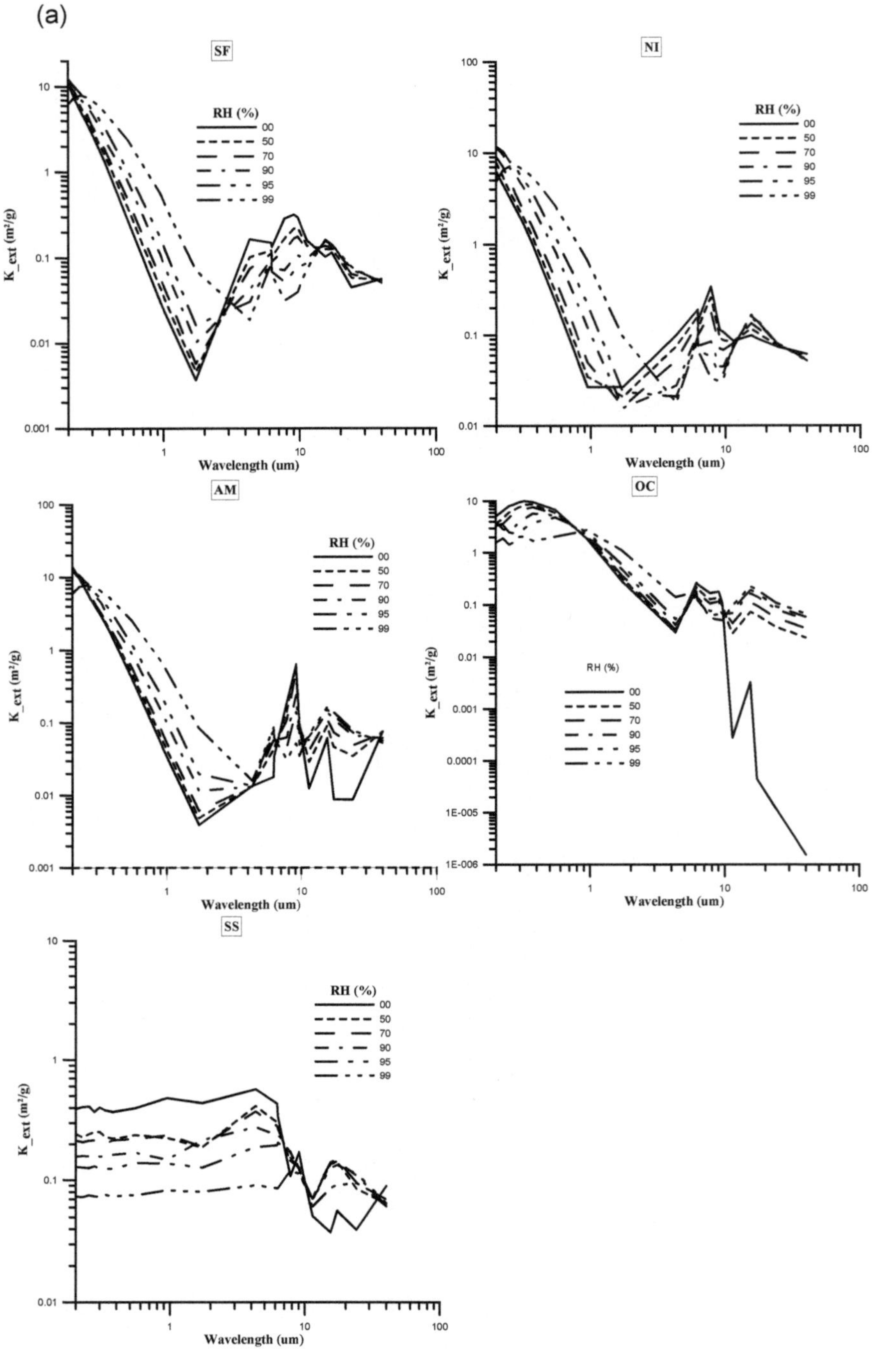
(a)
SF
NI
AM
OC
SS
RH (%)
00
50
70
90
95
99
K_ext (m²/g)
Wavelength (um)
0.001
0.01
0.1
1
10
100
0.0001
1E-005
1E-006

Figure 1.

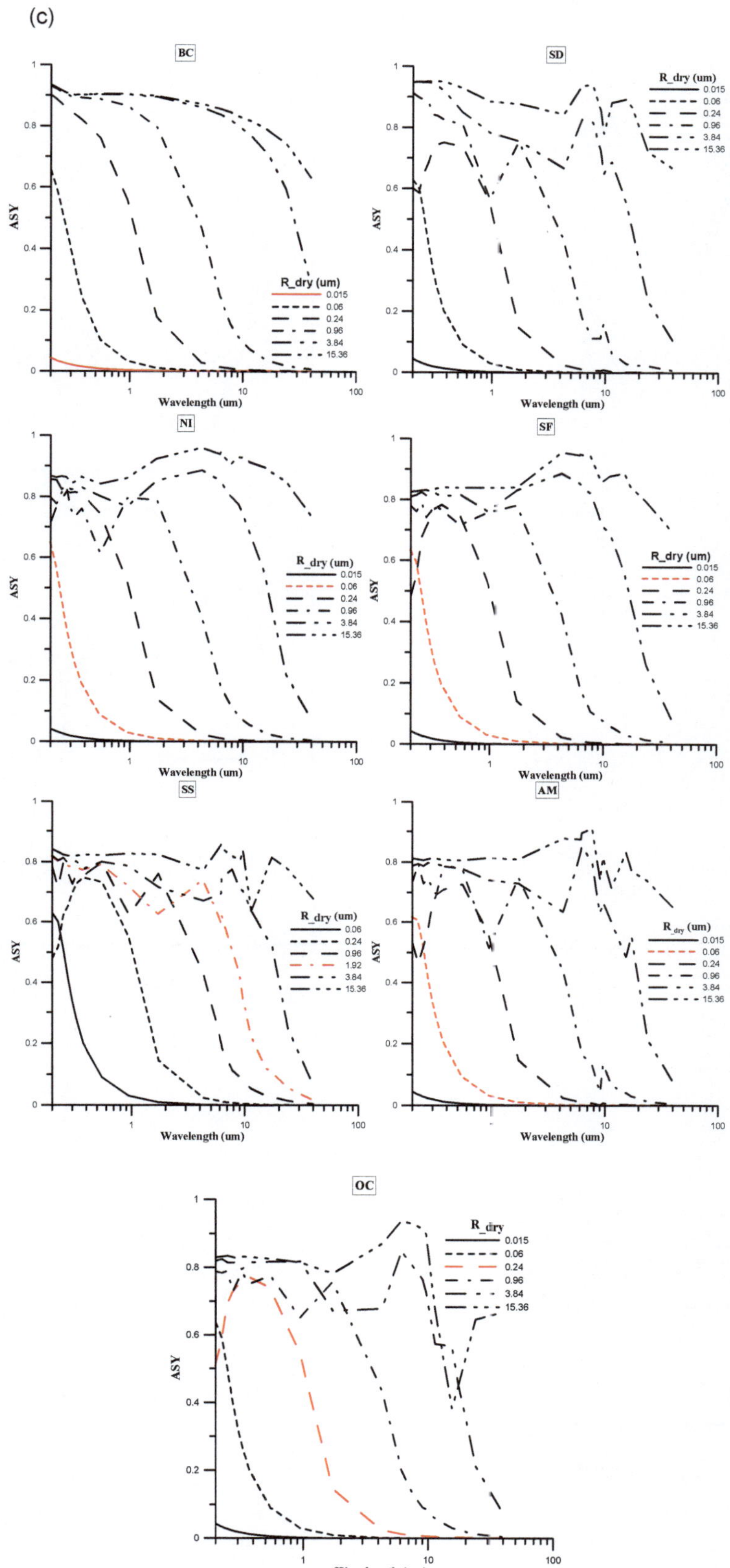

Figure 1. The **(a)** mass extinction coefficient, K_{ext} ($\text{m}^2\ \text{g}^{-1}$), **(b)** SSA, and **(c)** ASY for six of the 12 model size bins for seven species of dry aerosols.

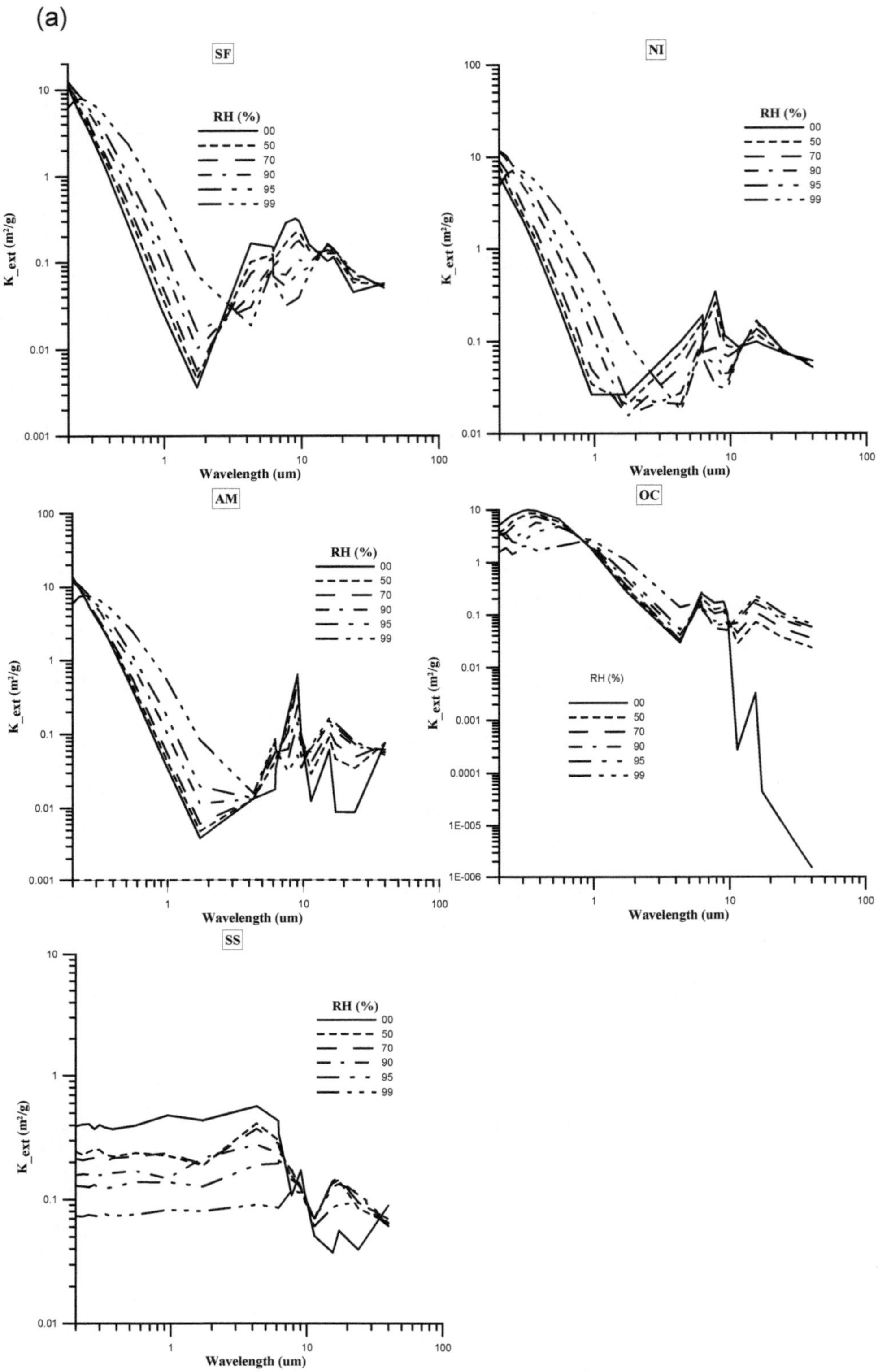
(a)
SF
RH (%)
00
50
70
90
95
99
K_ext (m²/g)
0.001
0.01
0.1
1
10
Wavelength (um)
1
10
100
NI
RH (%)
00
50
70
90
95
99
K_ext (m²/g)
0.01
0.1
1
10
100
Wavelength (um)
AM
RH (%)
00
50
70
90
95
99
K_ext (m²/g)
0.001
0.01
0.1
1
10
100
Wavelength (um)
OC
RH (%)
00
50
70
90
95
99
K_ext (m²/g)
1E-006
1E-005
0.0001
0.001
0.01
0.1
1
10
Wavelength (um)
SS
RH (%)
00
50
70
90
95
99
K_ext (m²/g)
0.01
0.1
1
10
Wavelength (um)

Figure 2.

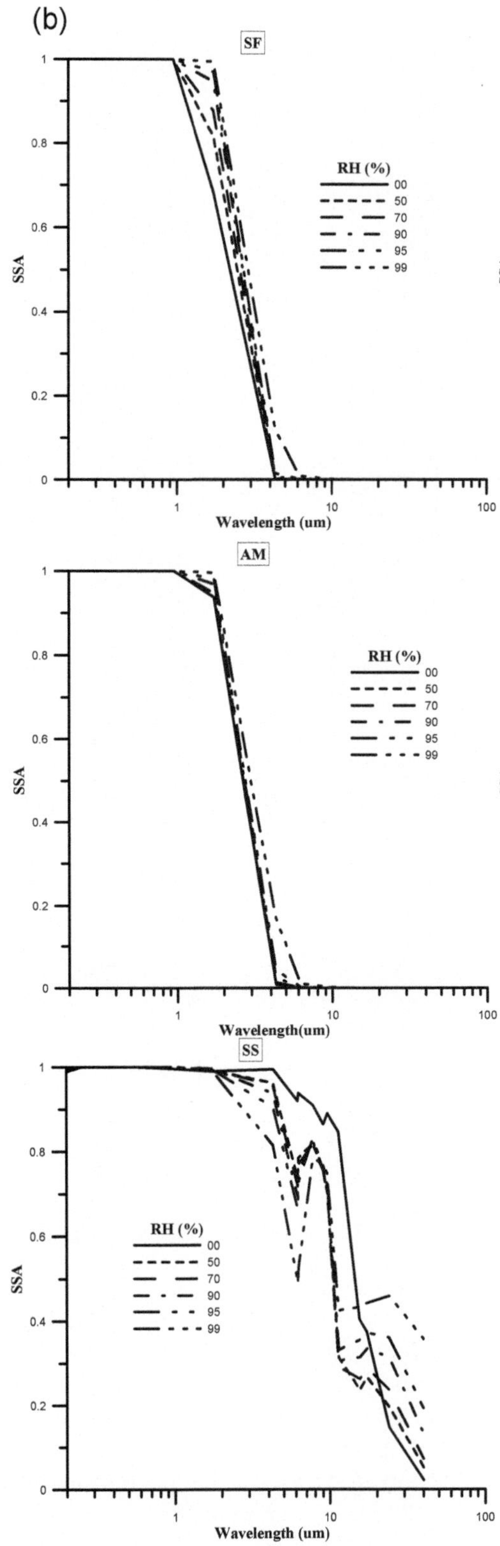
(b)
SF
NI
AM
OC
SS
RH (%)
00
50
70
90
95
99
SSA
Wavelength (um)
1
10
100
0
0.2
0.4
0.6
0.8

Figure 2.

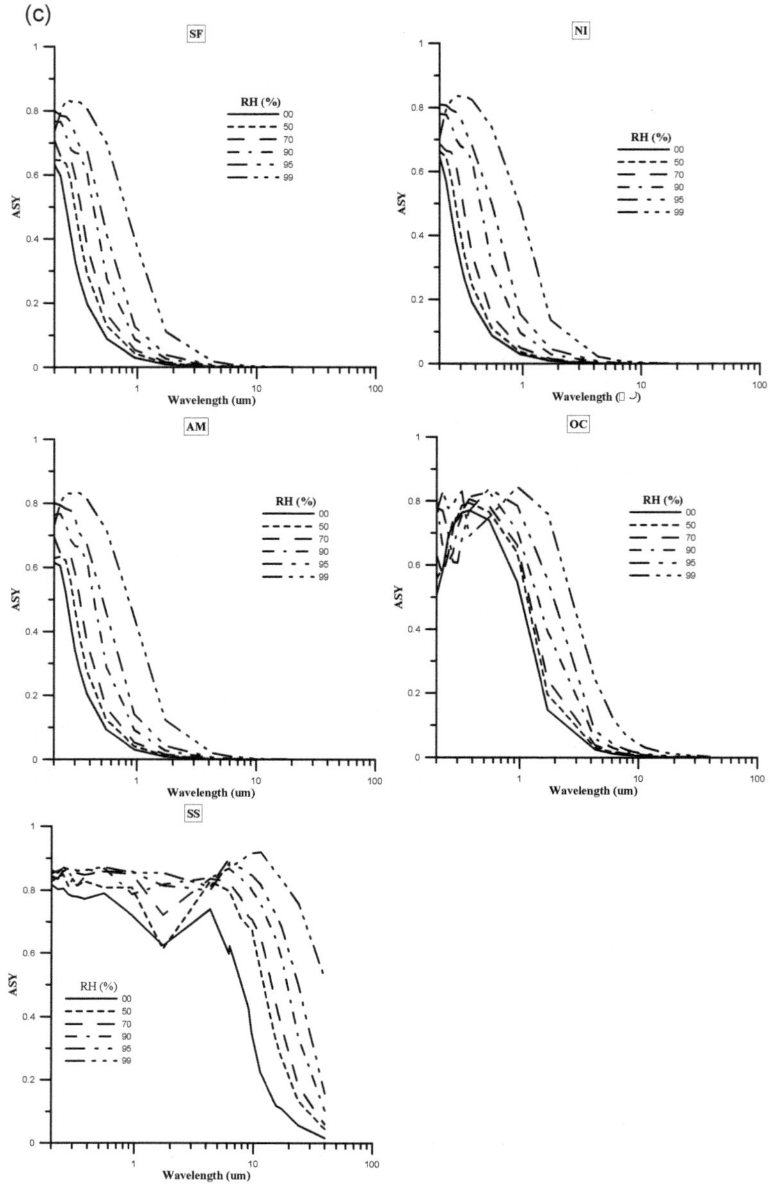

Figure 2. The **(a)** mass extinction coefficient, K_{ext} ($m^2\ g^{-1}$), **(b)** SSA, and **(c)** ASY of a typical particle size (the red line in Fig. 1) for five hygroscopic aerosol species at six different RHs.

(MODIS), onboard the Aqua satellite, have been widely used in evaluating and investigating the aerosol burden (Ichoku et al., 2002; Kahn et al., 2007; Zhang and Reid, 2010). The daily MODIS AOD by the Deep Blue algorithm, which fills in data gaps remaining in the dark target land aerosol retrieval over bright surfaces (Hsu et al., 2006) at 550 nm (MODIS/Aqua Collection 5.1 MYD08_D3 product) with a spatial resolution of $1° \times 1°$, is used in this paper to evaluate the modeled AOD.

Figure 3 compares the modeled monthly averaged AOD for July with MODIS Deep Blue AOD at 550 nm. It can be seen from Fig. 3 that both the MODIS and modeled AOD results show that the highest AOD values are in the 3JNS region, reaching 1 for most of this region, and even 2. As a general rule, the modeled AOD results seem a little higher than the MODIS AOD data. Considering the uncertainties of MODIS Deep Blue AOD over land (Remer et al., 2005), especially in spring and summer time in eastern China (Yang et al., 2011), its time-limited scans of China, and the different integrating times of the monthly averaged MODIS and modeled AOD, the resulting consistency of the horizontal distribution, the AOD's central location, and the values of both AOD data sets are both acceptable and reasonable. These results also prove the model's performance in July 2008 in describing aerosol column loading and the extinction effects by the composite atmospheric aerosol in this region.

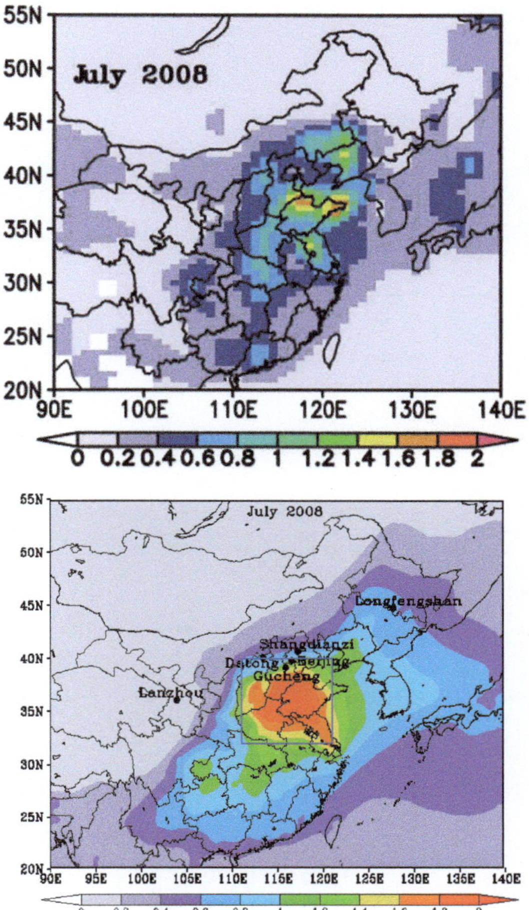

Figure 3. Monthly mean MODIS (top) and modeled AOD (bottom) for July 2008.

The ground-based observed AOD from the China Aerosol Remote Sensing NETwork (CARSNET) (Che et al., 2008) is also employed to evaluate the modeled AOD. CARSNET AODs are retrieved in the 440, 675, 870, and 1020 nm bands using the automatic Cimel sun and sky scanning radiometer (Cimel-318, Cimel Electronique). Taking the data continuity and the locations of the stations into account, the AOD at 440 nm at six surface CRASNET stations are used here. The locations and altitudes of these stations are listed in Table 2. The Beijing, Xianghe and Shangdianzi stations are all located in the Beijing metropolitan region. The Beijing station is located on CMA premises, and was taken as being representative of urban Beijing, while the Xianghe and Shangdianzi stations are located in the rural areas around Beijing. The city of Lanzhou is taken as having air pollution conditions typical of western China. The Lanzhou CARSNET station is located in urban Lanzhou; the Semi-Arid Climate and Environment Observatory of Lanzhou University (SACOL) station located on the Lanzhou University campus in Yuzhong, outside the city, represents the rural region of Lanzhou (Huang et al., 2008b). The Gucheng station located in Hebei Province represents Beijing's periphery. Datong is a medium-sized city in Shanxi Province, westward and windward of Beijing. All the stations are located in eastern central China, except for Lanzhou.

Table 2. CARSNET and AERONET station locations.

Station	Lat.	Long.	Altitude (m)
Beijing	39.80	116.47	31.3
Xianghe (AERONET)	39.76	117.00	
Datong	40.10	113.33	1067.2
Lanzhou	36.05	103.88	1517.2
SOCAL (AERONET)	35.57	104.08	
Shangdianzi	40.65	117.12	293.3
Longfengshan	44.73	127.60	330.5
Gucheng	39.13	115.80	15.1

Figure 4 shows a comparison between daily averaged CRASNET and modeled AOD data in July 2008. It can be seen that both the simulated and the observed AOD data congruously indicate two pollution episodes in the 3JNS region, one from 7 to 11 July 2008, in Beijing, Shangdianzi, Gucheng and Datong, and another from 23 to 29 July 2008 in Beijing, Shangdianzi and Gucheng. Both modeled and observed AOD data show that the daily averaged AOD values reached 1.5–3 during the episode on 7–11 July 2008, indicating that the pollution affects not only Beijing and its environs, but also Shanxi Province (i.e., Datong) to the west. The 23–29 July 2008 pollution episode appears weaker, but lasted longer than the episode on 7–11 July 2008, and was not observed at the Datong station. The modeled and CARSNET AOD data from 1 to 31 July 2008 show fairly consistent diurnal trends, validating the modeled AOD data at these four stations. In Lanzhou, CARSNET AOD values remained low from 1 to 31 July 2008, and almost all were <0.5, indicating clear air conditions in this city in July. Modeled AOD values show very similar results with the observed CARSNET data, corroborating the model's validity for western China. For Longfengshan station in northeastern China, the model's performance is not as good as for other stations as compared to CARSNET AOD data.

SSA and ASY values also have a substantial impact on aerosol radiative effects and feedback to atmospheric circulation, which help determine the existence of aerosol radiative forcing, i.e., the heating or cooling of the atmosphere, and the negative or positive radiative feedback from the haze episode itself. The SSA and ASY observation data set from the Aerosol Robotic Network (AERONET) at the Xianghe and SOCAL stations were used to evaluate the model's performance. Monthly averaged AERONET and modeled SSA and ASY were calculated, together with the SSA and ASY bias for the model (Table 3). It can be seen from Table 3 that observed SSA at the Xianghe station was 0.96, and 0.95 at the SOCAL station; the modeled values were 0.93 and 0.90 for the two stations, respectively. AERONET and modeled

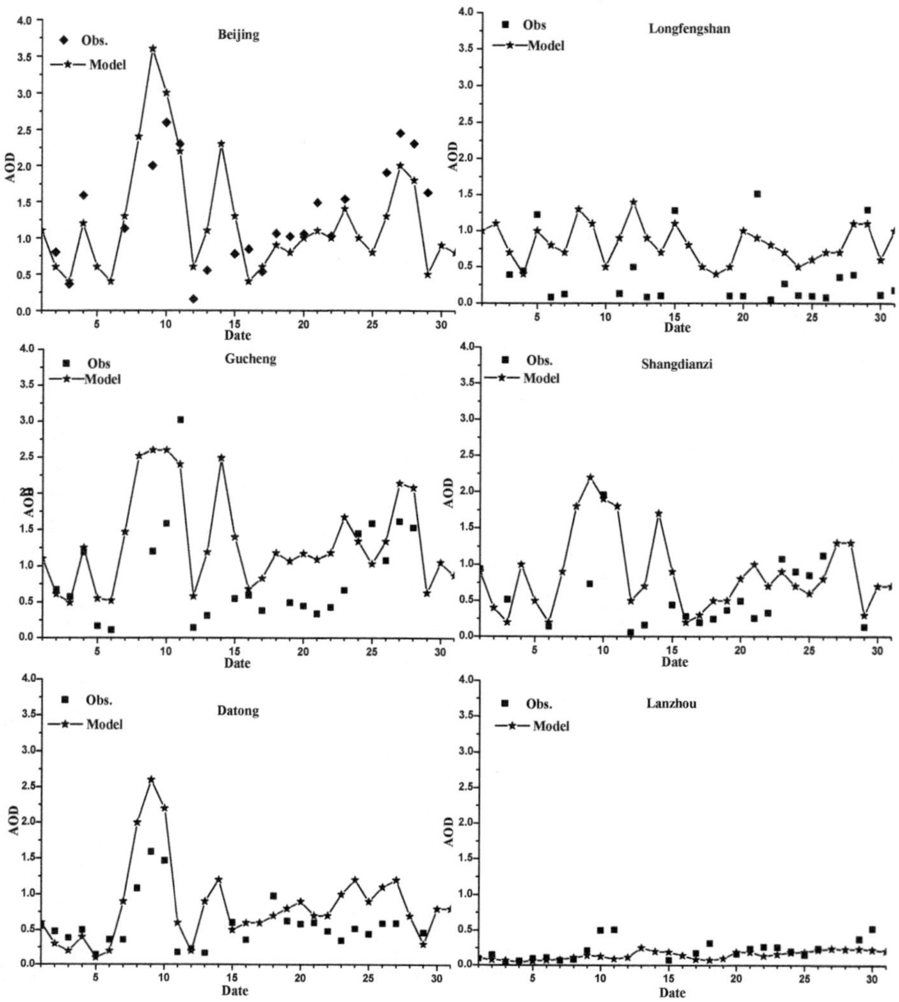

Figure 4. Daily variation of CRASNET observations (Obs.) and model-derived (Model) AOD from 1 to 31 July 2008.

Table 3. Observed and modeled SSA and ASY, and model bias.

Station	Parameters	Obs. (times)	Model	Model bias
Xianghe	SSA	0.96 (10)	0.93	−3 %
	ASY	0.74 (10)	0.78	5 %
SOCAL	SSA	0.95 (7)	0.90	−5%
	ASY	0.72 (17)	0.77	7 %

SSA data for the two stations consistently indicate a high aerosol scattering ratio in western and eastern China. The SSA bias is −3 at the Xianghe station and −5 % at the SOCAL station. The AERONET ASY is 0.74 at the Xianghe station and 0.72 at the SOCAL station, while the modeled values are 0.78 and 0.77, respectively. The ASY bias is +5 at the Xianghe station and

+7 % at the SOCAL station. The SSA and ASY bias for the model is therefore both reasonable and acceptable, taking the experimental sensitivity of the impact of optical properties on radiative forcing into account (Wang et al., 2006). It is worth noting that both the SSA and ASY observed data sets for the Chinese mainland are sparse. Only 10 days' SSA and ASY data for the Xianghe station, and 7 days' SSA data and 17 days' ASY data for the SOCAL station were available for evaluating this study's modeled results.

The vertical distribution of the aerosol layer is the other key factor affecting aerosol radiation besides chemical composition and optical characteristics. Diurnal changes in the vertical distribution of $PM_{2.5}$ for the 3JNS region from 1 to 31 July 2008 were calculated and are displayed in Fig. 5a. It can be seen that the $PM_{2.5}$ pollutants were generally concentrated in the surface and near-surface atmosphere in July 2008. A $PM_{2.5}$ concentration $> 120\,\mu g\,m^{-3}$ generally occurs below 800 hPa, overlapping the PBL height, or a little higher. The greatest height of the concentrated $PM_{2.5}$ layer varies between 700 and 900 hPa on different days. Figure 5b shows the vertical distribution of averaged $PM_{2.5}$ and K_{ext} for 7–11 July 2008. A $PM_{2.5}$ concentration $> 150\,\mu g\,m^{-3}$ occurs below 900 hPa and a concentration $> 110\,\mu g\,m^{-3}$ is found

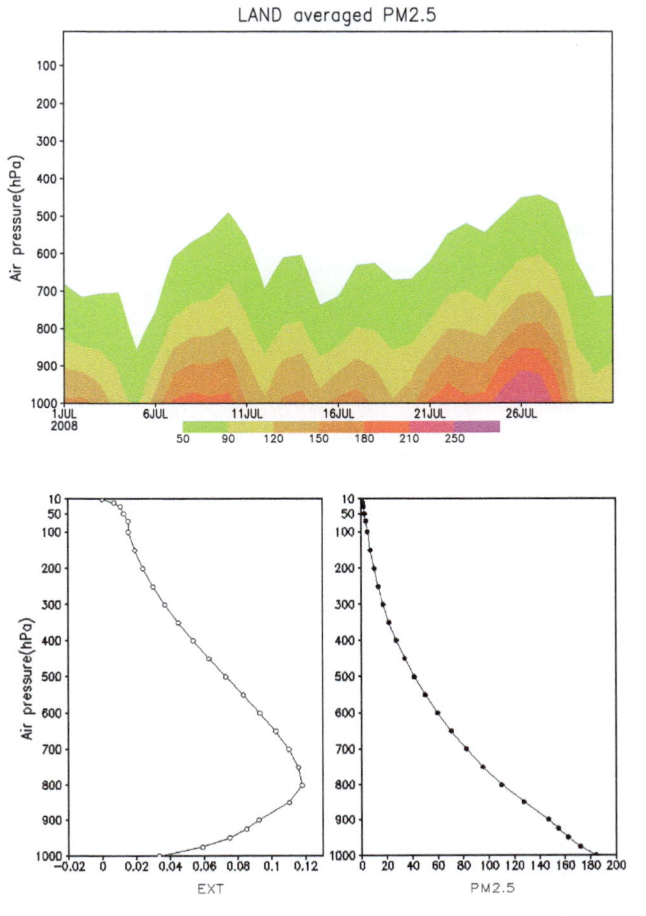

Figure 5. Vertical distribution of atmospheric particle pollutants: **(a)** daily variations of 3JNS – the mean $PM_{2.5}$ ($\mu g\,m^{-3}$) from 1 to 31 July 2008; **(b)** the averaged extinction coefficient (Kext) ($m^2\,g^{-1}$, left) and $PM_{2.5}$ (right) for 7–11 July 2008 for the 3JNS region.

below 800 hPa. The highest K_{ext} value appears at 800 hPa, suggesting that the greatest aerosol extinction occurs in the upper PBL, or above it, aiding the cooling of the lower PBL and the heating of the upper PBL.

4.2 PBL meteorological features corresponding to the haze episode

The PBL wind speed, diffusion coefficient and PBL height are the most important parameters representing the PBL characteristics affecting air and haze pollution. The terms "window shadow", "homogeneous air pressure" and "stable and steady" have been used to describe the typical patterns of local air circulation and surface meteorological fields for haze episodes. In the following section, these three factors and their correlations with particulate $PM_{2.5}$ pollutants are discussed. Air pressure patterns are also used to discuss the strength of, and changes in, PBL cold air; these patterns have an important effect on air pollution episodes, especially on the weakening and collapsing of haze episodes.

4.2.1 Planetary boundary layer winds

The surface winds not only directly impact haze impacts, but also the winds in the whole PBL. The accuracy of the model's forecasting of surface and PBL winds is critical for an accurate haze prediction. The conventional meteorology observation data from sounding balloons are only available at 00:00 and 12:00 UTC in China (early morning or dusk in local time). The NCEP/NCAR Reanalysis-2 meteorology data are available for numerous meteorological parameters including geopotential height, air temperature, vertical velocity, wind field, etc., at a time resolution of 6 h (00:00, 06:00, 12:00, and 18:00 UTC) at the height of sea level pressure 1000, 950, 850, 700, 500, and 300 hPa, etc. In view of the importance of daytime PBL meteorology conditions to the haze episode, Reanalysis-2 data are used to evaluate the model's results and study the PBL features instead of sounding balloon observations, due to the lack of daytime PBL meteorology data from the latter.

Figure 6 shows the model's daily averaged wind speeds and the NCEP analysis for the 3JNS region for 1–31 July 2008, together with $PM_{2.5}$ at the surface (Fig. 6a) and at 850–950 hPa (Fig. 6b). It can be seen that the modeled wind speed at 850–950 hPa agrees well with the NCEP wind speed, showing the model's fair ability to predict wind speed at this height. Averaged modeled $PM_{2.5}$ for the same region at 850–950 hPa is also shown in Fig. 6a. It can be seen that the diurnal variations in $PM_{2.5}$ in the PBL evince a completely contrary phase, shifting with the wind speeds of both modeled and NCEP Reanalysis; i.e., the peak values of $PM_{2.5}$ correspond to the trough values of wind speed, showing that low wind speed in the PBL may be the most important factor leading to haze pollution. Modeled surface wind values also correspond fundamentally to NCEP data, though not at 850–950 hPa. Nonetheless, the modeled surface wind speed and $PM_{2.5}$ show opposing trends, similar to the 850–950 hPa findings. Corresponding wind speeds are as low as 0.5 m s^{-1} at the surface and are <2 m s^{-1}, rising to 4–5 m s^{-1} at 850–950 hPa separately during the severe haze pollution episodes of 7–11 and 25–28 July 2008.

4.2.2 Planetary boundary layer turbulence diffusion

Turbulence diffusion is another important process significantly affecting surface pollutant concentration, since it defines horizontal transportation distance by affecting the vertical heights that pollutants may reach. The turbulence diffusion coefficient (fktm) parameterizes the PBL turbulence diffusion process (Wang et al., 2010). Figure 7 shows the averaged surface $PM_{2.5}$ and f_{ktm} horizontal distribution for 7–11 July 2008 (Fig. 7a), together with the diurnal changes of the 3JNS averaged $PM_{2.5}$ and f_{ktm} (Fig. 7b) from 1 to 31

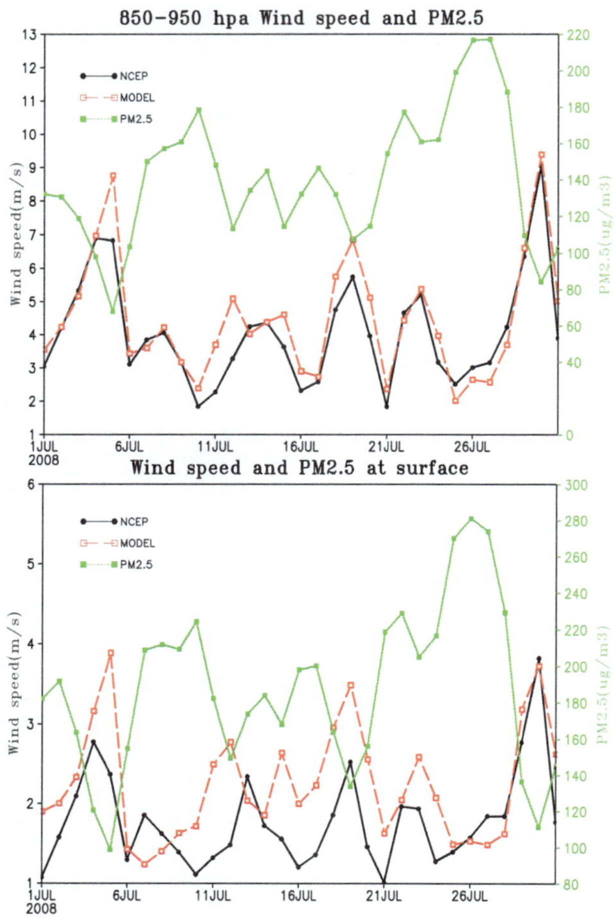

Figure 6. Daily changes in the 3JNS mean wind speeds ($m\,s^{-1}$) derived from the model and NCEP Reanalysis 2, together with simulated $PM_{2.5}$ ($\mu g\,m^{-3}$) at 850–950 hPa (Fig. 6a) and at the surface (Fig 6b).

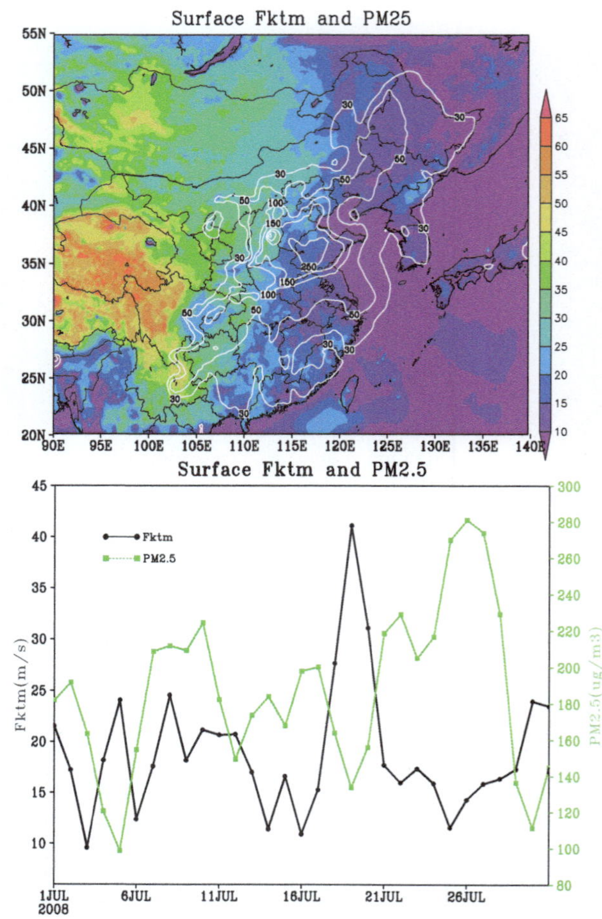

Figure 7. Mean surface $PM_{2.5}$ ($\mu g\,m^{-3}$, contour) and fktm ($m^2\,s^{-1}$, shaded) for 7–11 July 2008 (top) and the daily changes in the 3JNS mean $PM_{2.5}$ and fktm for 1–31 July 2008 (bottom).

July 2008. Figure 7a shows that lower f_{ktm} and higher $PM_{2.5}$ values occurred throughout middle and eastern China during 7–11 July 2008. The lowest f_{ktm} values and highest 400 $PM_{2.5}$ values appeared in the 3JNS region, showing the importance of turbulence diffusion in the strength of severe air pollution. It can also be seen from Fig. 7a that $PM_{2.5}$ concentrations exhibit higher values when f_{ktm} are lower, and that this is more marked over the 3JNS region than over southeastern China (22–30° N, 110–120° E). The daily averaged f_{ktm} and $PM_{2.5}$ values for the 3JNS region (Fig. 7b) show a marked correlation over the severely polluted parts of this region. There is a basic reverse trend in daily f_{ktm} and $PM_{2.5}$ values.

4.2.3 Planetary boundary layer height

PBL height is one of the most important variables in any PBL scheme. PBL height can be variously defined by reference to the local Richardson number, the height of the capping inversion, the proportionality of PBL height to the ratio between the friction velocity and the local Coriolis force parameter, the height of the minimum sensible heat flux, turbulent kinetic energy, or a specific vertical potential temperature gradient (Cheng et al., 2002; Santanello, et al., 2005; Hong et al., 2006; Pleim et al., 2007). Despite all these different definitions, some commonality of agreement on the definition of PBL height exists, e.g., the Richardson number, the capping inversion or the height where the potential temperature lapse rate becomes too positive, and so on. The medium-range forecast (MRF) PBL scheme uses non-local closure and relies heavily on the Richard index (Ri) to compute the PBL height for different regimes (Hong and Pan, 1996).

Figure 8 shows the horizontal distribution of the averaged surface $PM_{2.5}$ and PBL height for 7–11 July 2008 (Fig. 8a), together with diurnal regional trends in the 3JNS averaged $PM_{2.5}$ and PBL height (Fig. 8b). The PBL height values (Fig. 8a) are as low as 300–900 m, with high $PM_{2.5}$ values over eastern China, consistent with observational studies of this region (Wang et al., 2012). The lowest PBL height

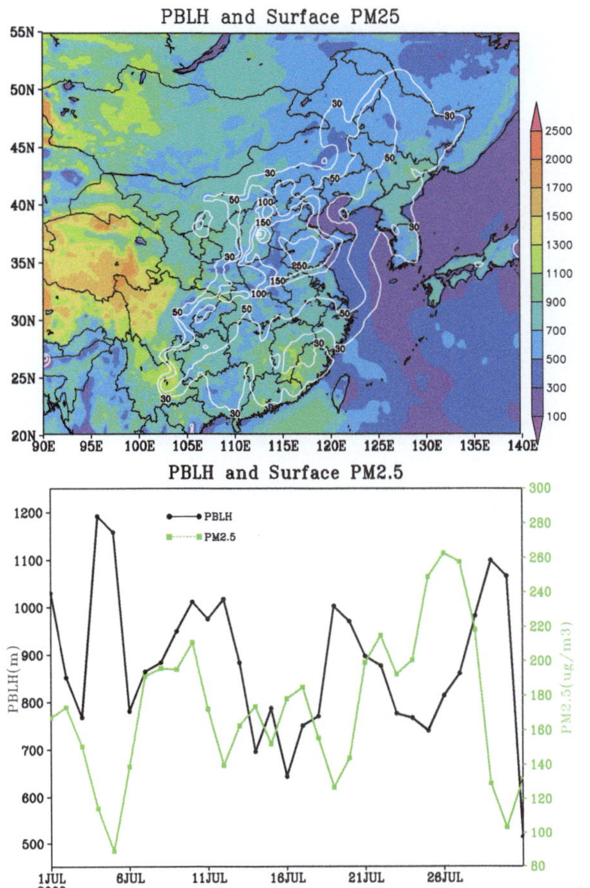

Figure 8. The mean surface $PM_{2.5}$ ($\mu g\,m^{-3}$, contours) and PBL height (m, shaded) for 7–11 July 2008 (top), and daily changes in the 3JNS mean surface $PM_{2.5}$ and PBL height (bottom).

almost overlaps with the regions of lowest f_{ktm} and highest $PM_{2.5}$ over eastern central China for this period. Comparing the similarly economically developed areas of eastern central and southeastern China (22–30° N, 110–120° E), eastern central China is much more severely polluted. This would suggest that a lower PBL height and weak turbulence diffusion (Fig. 7a) may be two of the main factors leading to higher pollution over this area. Diurnal changes in averaged PBL height and $PM_{2.5}$ over the central polluted area (Fig. 8b) also display a generally contrary correlation during July 2008, indicating the important impact of the PBL height on the pollution strength ($PM_{2.5}$ concentration) of the surface air; i.e., low PBL height and f_{ktm} are critical to the degree of haze pollution.

4.2.4 Patterns in planetary boundary layer air pressure fields

Surface "homogeneous air pressure" has been regarded as a typical surface air pressure feature associated with haze pollution (X. Liu et al., 2010). Detailed comparisons of PBL air pressure patterns between hazy and clear conditions for different seasons are, at present, limited. Modeled $PM_{2.5}$, AOD and CARSNET AOD data (Fig. 4) all show that 10 July 2008 was the severest haze day, while 12 July 2008 was the first clear day after this haze episode. Figure 9 displays the air pressure pattern, wind speed vector at the surface (bottom), geopotential height and wind speed vector at 950 (middle) and 850 hPa (top) on 10 July 2008 (left) and 12 July 2008 (right). July is midsummer in China and a subtropical high with high pressure controlled the surface in eastern China and the sea region east of China on 10 July 2008. No obvious difference was observed between the surface air pressure pattern on 12 July 2008 and on 10 July 2008 due to the effect of the strong subtropical high to the east. The patterns of geopotential heights for 850–950 hPa heights are quite different from those for surface pressure, showing a "homogeneous air pressure pattern" over the whole of eastern China on 10 July 2008. The southeasterly weakening and retrenchment of the subtropical high from 10 to 12 July 2008 is clear at 850–950 hPa height. The main area of subtropical high pressure withdrew across the sea west of the Korean Peninsula on 12 July 2008. At the same time, cold air from Mongolia moved to northern central China, and the air pressure over this region strengthened at 850–950 hPa during 10–12 July 2008. Eastern China was controlled by continental high pressure from the northwest, increasing the northerly wind speed over eastern central China at 850–950 hPa on 12 July 2008. The surface wind vector shows that, over eastern central China, the northerly wind speed on 12 July 2008 was obviously higher than that on 10 July 2008. This was not caused by any pressure-gradient force because the air pressure patterns on 10 and 12 July 2008 were very similar, but the transfer of cold air and momentum from 850 to 950 hPa to the surface led to an increase in the northerly wind speed at the surface, directly bringing the haze episode to an end over middle and eastern China, including the 3JNS region.

5 Conclusions

The mesoscale chemical weather forecasting model GRAPES-CUACE/haze, integrated with an online radiative parameterization scheme, based on an external mixing scheme of black carbon, organic carbon, soil dust, nitrates, sulfate, sea salt and ammonia aerosols, was employed to simulate the optical characteristics of aerosols and the PBL meteorological features related to haze in July 2008. The aerosols' AOD, SSA and ASY features were simulated and evaluated. The PBL characteristics of aerosol loading, PBL heights, turbulence diffusion, wind speed, air pressure, geopotential height patterns, and their relation to haze and $PM_{2.5}$, were evaluated, and the results are summarized below.

The comparison between modeled AOD, SSA and ASY data and MODIS, CARSNET and AERONET observational

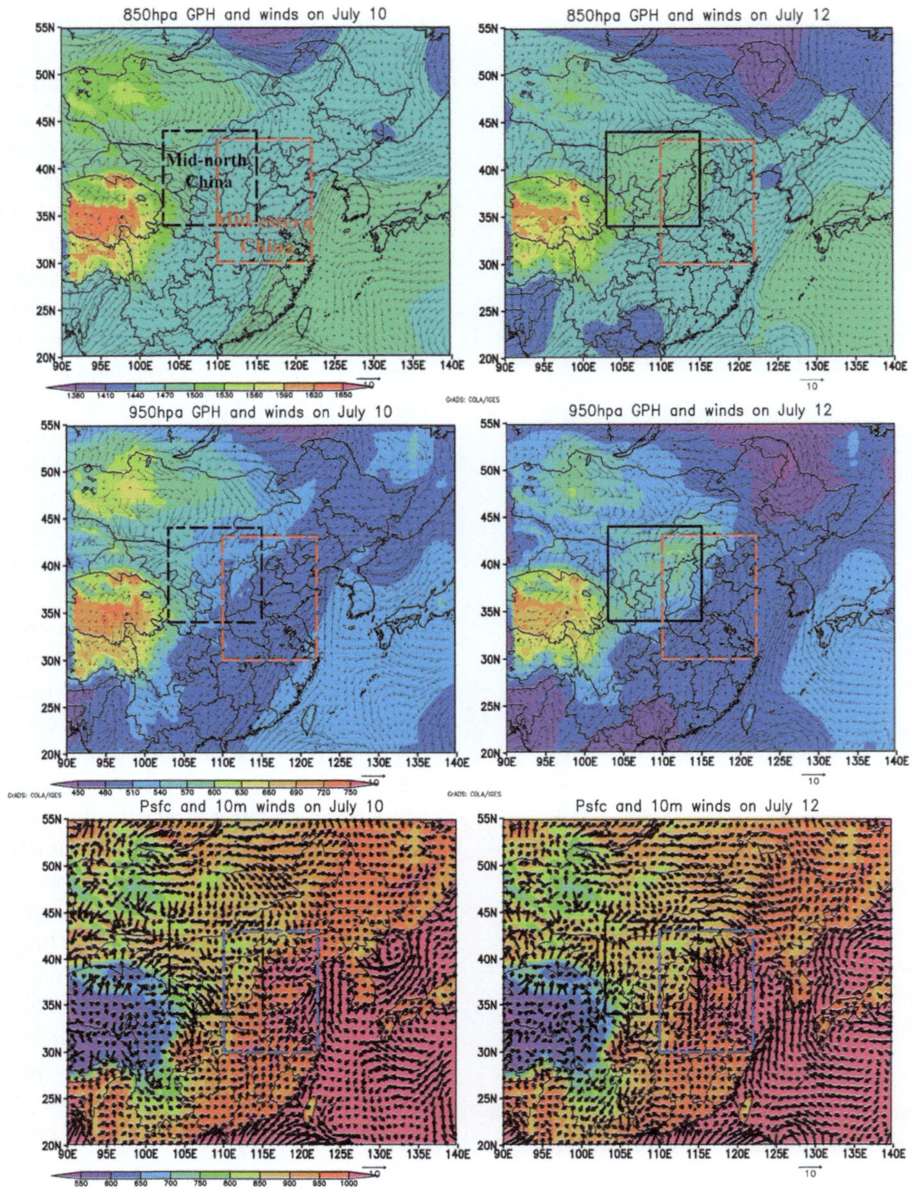

Figure 9. Air pressure patterns (hPa, shaded), wind vectors at the surface ($m\,s^{-1}$, bottom), geopotential height (gph, shaded), and wind vectors at 950 (middle) and 850 hPa (top) on 10 July (left) and 12 July 2008 (right).

data confirms the model's ability to predict aerosol column loading and aerosol optical features, fundamental to studying and evaluating the aerosols' radiative feedback to regional and local PBL circulation. The modeled PBL wind speed correlates closely with the NCEP analysis data, showing the model's ability to predict PBL wind speeds as a basis for haze forecasting. Modeled and observed AOD showed that the monthly mean AOD values may exceed 1.2 and that the daily mean value was likely larger than 2.0 for the period 7–11 July 2008 in the 3JNS region. SSA is as high as 0.90–0.96 and ASY is 0.72–0.74, showing the high scattering properties of aerosols in summer. The PBL wind speed from modeled and NCEP Reanalysis data both show a completely inverse trend with $PM_{2.5}$, changing diurnally. This illustrates the importance of the "PBL window shadow" to the haze episode. Low turbulence diffusion and PBL height are also important meteorological factors affecting haze episodes, as is evident from their correlation with near-surface $PM_{2.5}$ either on a regional or diurnal basis.

Contrary to the findings of most previous studies, a "homogeneous air pressure pattern" appears not to be typical of the surface pressure field in summer in middle and eastern China: surface pressure patterns are almost identical for both a hazy day (10 July 2008) and a clear day (12 July 2008) after haze. The geopotential height patterns of the PBL (850–950 hPa) show a "homogeneous air pressure" field in middle and eastern China between the land-based, cold high over northwestern China and Mongolia and the subtropical high over the East China Sea on a hazy day (10 July 2008). The haze ended on 12 July 2008 over eastern China, accompanied by an extension into northern central China of the land-based cold high from the northwest, and a weakening and easterly retrenchment to the east of the subtropical high. The PBL "homogeneous air pressure" over middle and eastern China was disrupted and cold air from Mongolia controlled northern central China at 850–950 hPa. This resulted in the wind speed increasing in middle and eastern China at this height. A downward momentum from the PBL (850–950 hPa) to the surface directly led to the surface wind increasing, and, combined with the PBL cold air, this resulted in the collapse of the hazy weather over eastern China.

Since all discussions above are based on the results from a haze episode in July 2008, any conclusions drawn concerning PBL meteorological and aerosol optical features during hazy weather may only apply to summer haze episodes. Haze episodes in other seasons need to be studied: the results may be different because of the different prevalent meteorological conditions in different seasons. In this study, aerosol optical parameters were calculated online, but their radiative effects were not entered online into the model's dynamical process. In the companion paper, aerosol radiative feedback and its impact on PBL meteorology and the haze episode itself will be implemented and discussed in detail.

Acknowledgements. This work is supported by the National Basic Research Program (973) (grant no. 2011CB403404), the National Natural Scientific Foundation of China (grant nos. 41275007, 41130104, and 41475136), and CAMS Key Projects (grant no. 2013Z007).

Edited by: S. Gong

References

An, X. Q., Sun, Z. B., Lin, W. L., Jin, M., and Li, N.: Emission inventory evaluation using observations of regional atmospheric background stations of China, J. Environ. Sci., 25, 537–546, 2013.

Cao, J. J., Lee, S. C., Ho, K. F., Zou, S. C., Fung, K., Li, Y., Watson, J. G., and Chow, J. C.: Spatial and seasonal variations of atmospheric organic carbon and elemental carbon in Pearl River Delta Region, Chin. Atmos. Environ., 38, 4447–4456, 2004.

Cao, G., Zhang, X., and Zheng, F.: Inventory of black carbon and organic carbon 446 emissions from China, Atmos. Environ., 40, 6516–6527, 2006.

Cao, G. L., An, X. Q., Zhou, C. H., Ren, Y. Q., and Tu, J.: Emission inventory of air pollutants in China, Chin. Environ., Sci., 30, 900–906, 2010.

Chak, K. C. and Yao, X. H.: Review Air pollution in mega cities in China, Atmos. Environ., 42, 1–42, 2008.

Chan, C. Y., Tang, J. H., Li, Y. S., and Chan, L. Y.: Mixing ratios and sources of halocarbons in urban, semi-urban and rural sites of the Pearl River Delta, South China, Atmos. Environ., 40, 7331–7345, 2006.

Che, H., Zhang, X., Chen, H., Damiri, B., Goloub, P., Li, Z., Zhang, X., Wei, Y., Zhou, H., Dong, F., Li, D., and Zhou, T.: Instrument calibration and Aerosol Optical Depth (AOD) validation of the China Aerosol Remote Sensing Network (CARSNET), J. Geophys. Res., 114, D03206, doi:10.1029/2008JD011030, 2008.

Che, H. Z., Zhang, X. Y., and Li, Y.: Haze trends over the capital cities of 31 provinces in China, 1981–2005, Theor. App. Clim., 97, 235–242, 2009.

Che, H., Xia, X., Zhu, J., Li, Z., Dubovik, O., Holben, B., Goloub, P., Chen, H., Estelles, V., Cuevas-Agulló, E., Blarel, L., Wang, H., Zhao, H., Zhang, X., Wang, Y., Sun, J., Tao, R., Zhang, X., and Shi, G.: Column aerosol optical properties and aerosol radiative forcing during a serious haze-fog month over North China Plain in 2013 based on ground-based sunphotometer measurements, Atmos. Chem. Phys., 14, 2125–2138, doi:10.5194/acp-14-2125-2014, 2014.

Chen, D. H., Xue, J. S., and Yang, X. S.: The new generation of hydrostatic/nonhydrostatic multi-scales numerical prediction model: Scientific design and experiments, in: CAMS Technical Report 1, 2003 (in Chinese).

Chen, D. H, Xue, J. S., Yang, X. S., Zhang, H. L., Shen, X. S., Hu, J. L., Wang, Y., Ji, L. R., and Chen, J. B.: (New generation of multi-scale NWP system (GRAPES): general scientific design, Chin. Sci. Bull., 53, 3433–3445, 2008.

Chen, Z. H , Cheng, S. Y., and Su, F. Q.: Analysis of large-scale weather pattern during heavy air pollution process in Beijing), Res. Environ. Sci., 20, 99–105, 2007 (in Chinese).

Cheng, Y., Canuto, V. M., and Howard, A. M.: An improved model for the turbulent PBL, J. Atmos. Sci., 59, 1550–1565, 2002.

Chou, M. D., Suarez, M. J., Ho, C. H., Yan, M. M. H., and Lee, K. T.: Parameterizations for Cloud Overlapping and Shortwave Single-Scattering Properties for Use in General Circulation and Cloud Ensemble Models, J. Clim., 11, 202–214, 1998.

Chou, M. D., Suarez, M. J., Liang, X. Z., and Michael M.-H. Y.: A Thermal Infrared Radiation Parameterization for Atmospheric Studies, Technical Report Series on Global Modeling and Data Assimilation, NASA/TM-2001-104606, 19, America, Goddard Space Flight Center, Greenbelt, Maryland, 55, 2001.

Duan, J. C., Guo, S. J., Tan, J. H., Wang, S., and Chai, F. H.: A Characteristics of atmospheric carbonyls during haze days in Beijing, China, Atmos. Res., 114–115, 17–27, doi:10.1016/j.atmosres.2012.05.010, 2012.

Fu, Q. Y., Zhuang, G. S., Wang, J., Xu, C., Huang, K., Li, J., Hou, B., Lu, T., and Streets, D. G.: Mechanism of formation of the heaviest pollution episode ever recorded in the Yangtze RiverDelta, China, Atmos. Environ., 42, 2023–2036, 2008.

Gao, J., Wang, T., Zhou, X. H., Wu, W. S., and Wang, W. X.: Measurement of aerosol number size distributions in the Yangtze River delta in China: formation and growth of particles under polluted conditions, Atmos. Environ., 43, 829–836, 2009.

Gao, L. N., Jia, G. S., Zhang, R. J., Che, H. Z., Fu, C. B., Wang, T. J., Zhang, M. G., and Jiang, H.: Visibility trends in the Yangtze River Delta of China during 1981–2005, J. Air. Waste. Manage., 61, 843–849, 2011.

Gong, S. L. and Zhang, X. Y.: CUACE/Dust – an integrated system of observation and modeling systems for operational dust forecasting in Asia, Atmos. Chem. Phys., 8, 2333–2340, doi:10.5194/acp-8-2333-2008, 2008.

Hess, M., Koepke, P., and Schult, I.: Optical Properties of Aerosols and Clouds: The software package OPAC, B. Am. Meteorol. Soc., 79, 831–844, 1998.

Hong, S. Y. and Pan, H. L.: Nonlocal boundary layer vertical diffusion in a Medium-Range Forecast model, Mon. Weather Rev., 124, 2322–2339, 1996.

Hong, S. Y., Noh, Y., and Dudhia, J.: A new vertical diffusion package with an explicit treatment of entrainment processes, Mon. Weather Rev., 134, 2318–2341, 2006.

Huang, J., Lin, B., Minnis, P., Wang, T., Wang, X., Hu, Y., Yi, Y., and Ayers, J.: Satellite-based assessment of possible dust aerosols semi-direct effect on cloud water path over East Asia, Geophys. Res. Lett., 33, L19802, doi:10.1029/2006GL026561, 2006.

Huang, J., Wu, D., Huang, M. H., Li, F., Bi, X. Y., Tan, H. B., and Deng, X. J.: Visibility variations in the Pearl River delta of China during the period of 1954–2004, J. App. Meteor., 19, 61–70, 2008a.

Huang, J. P., Zhang, W., Zuo, J. Q., Bi, J. R., Shi, J. S., Wang, X., Chang, Z. L., Huang, Z. W., Yang, S., Zhang, B. D., Wang, G. Y., Feng, G. H., Yuan, J. Y., Zhang, L., Zuo, H. C., Wang, S. G., Fu, C. B., and Chou, J. F. : An overview of the semi-arid climate and environment research observatory over the Loess Plateau, Adv. Atmos. Sci., 25, 906–921, 2008b.

Huang, J., Fu, Q., Su, J., Tang, Q., Minnis, P., Hu, Y., Yi, Y., and Zhao, Q.: Taklimakan dust aerosol radiative heating derived from CALIPSO observations using the Fu-Liou radiation model with CERES constraints, Atmos. Chem. Phys., 9, 4011–4021, doi:10.5194/acp-9-4011-2009, 2009.

Hsu, N. C., Si-Chee, T., King, M. D., and Herman, J. R.: Deep Blue Retrievals of Asian Aerosol Properties During ACE-Asia, IEEE T. Geosci. Remote, 44, 3180–3195, 2006.

Ichoku, C., Chu, D. A., Mattoo, S., Kaufman, Y. J., Remer, L. A., Tanré, D., Slutsker, I., and Holben, B. N.: A spatio-temporal approach for global validation and analysis of MODIS aerosol products, Geophys. Res. Lett., 29, 8006, doi:10.1029/2001GL013206, 2002.

Kang H. Q., Zhu, B., Su, J. F., Wang, H. L., Zhang, Q. C., and Wang, F.: Analysis of a long-lasting haze episode in Nanjing, China, Atmos. Res., 78–87, 2013.

Kahn, R., Garay, M., Nelson, D., Yau, K., Bull, M., and Martonchik, J.: Satellite derived aerosol optical depth over dark water from MISR and MODIS: comparisons with AERONET and implications for climatological studies, J. Geophys. Res., 112, D18205, doi:10.1029/2006JD008175, 2007.

Kain, J. S.: The Kain-Fritsch convective parameterization: An update, J. Appl. Meteor., 43, 170–181, 2004.

Kessler, E.: On the distribution and continuity of water substance in atmospheric circulation, Meteor. Monogr., 32, Amer. Meteor. Soc., 84 pp., 1969.

Kusaka, H., Kondo, H., Kikegawa, Y., and Kimura, F.: A simple single-layer urban canopy model for atmospheric models: Comparison with multi-layer and slab models, Bound.-Layer Meteor., 101, 329–358, 2001.

Lee, Y. L. and Sequeira, R.: Visibility degradation across Hong Kong: its components and their relative contributions, Atmos. Environ., 34, 5861–5872, 2001.

Liu, W. D., Jiang, Y. H., and Li, J.: Characteristics of aerosol distribution and transmission of a heavy air pollution process in Beijing area, Clim. Environ. Res., 15, 152–160, 2010 (in Chinese).

Liu, X. H., Zhang, Y., Cheng, S. H., Xing, J., Zhang, Q., Streets, D. G., Jang, C., Wang, W. X., and Hao, J. M.: Understanding of regional air pollution over China using CMAQ, Part 1: Performance evaluation and seasonal variation, Atmos. Environ., 44, 2415–2426, 2010.

Pleim, J.: A combined local and non-local closure model for the atmospheric boundary layer. Part II: Application and evaluation in a mesoscale meteorological model, J. Appl. Meteor. Clim., 46, 1396–1409, 2007.

Quan, J., Tie, X., Zhang, Q., and Liu, Q.: Characteristics of heavy pollution during the 2012–2013 winter in Beijing, China., Atmos. Envoirn., 88, 83–89, 2014.

Rothman, L. S., Gordon, I. E., Barbe, A., Chris Benner, D., Bernath, P. F., Birk, M., Boudon, V., Brown, L. R., Campargue, A., Champion, J. P., Chance, K., Coudert, L. H., Dana, V., Devi, V. M., Fally, S., Flaud, J.-M., Gamache, R. R., Goldman, A., Jacquemart, D., Kleiner, I., Lacome, N., Lafferty, W. J., Mandin, J.-Y., Massie, S. T., Mikhailenko, S. N., Miller, C. E., Moazzen-Ahmadi, N., Naumenko, O. V., Nikitin, A. V., Orphal, J., Perevalov, V. I., Perrin, A., Predoi-Cross, A., Rinsland, C. P., Rotger, M., Šimečková, M., Smith, M. A. H., Sung, K., Tashkun, S. A., Tennyson, J., Toth, R. A., Vandaele, A. C., and Vander Auwera, J.: The HITRAN 2008 molecular spectroscopic database, J. Quant. Spect. Rad. Trans., 110, 533–572, 2009.

Remer, L. A., Kaufman, Y. J., Tanre, D., Mattoo, S., Chu, D. A., Martins, J. V., Li, R. R., Ichoku, C., Levy, R. C., Kleidman, R. G., Eck, T. F., Vermote, E., and Holben, B. N.: The MODIS Aerosol Algorithm, Products, and Validation, J. Atmos. Sci., 66, 947–973, 2005.

Santanello, J. A., Friedl, M. A., and Kustas, W. P.: An empirical investigation of convective planetary boundary layer evolution and its relationship with the land surface, J. Appl. Meteor., 44, 917–932, 2005.

Stockwell, W. R., Middleton, P., Chang, J. S., and Tang, X.: The Second Generation Regional Acid Deposition Model Chemical Mechanism for Regional Air Quality Modeling, J. Geophys. Res., 95, 16343–16376, 1990.

Tan, J., Duan, J., Chen, D., Wang, X., Guo, S., Bi, X., Sheng, G., He, K., and Fu, J. : Chemical characteristics of haze during summer and winter in Guangzhou, Atmos. Res., 94, 238–245, 2009.

Tan, J., Guo, S. J., Ma, Y. L., Duan, J. C., Cheng, Y., He, K. B., and Yang, F. M.: Characteristics of particulate PAHs during a typical haze episode in Guangzhou, China, Atmos. Res., 102, 91–98, 2011.

Vogelezang, D. H. P. and Holtslag, A. A. M.: Evaluation and model impacts of alternative boundary-layer height formulations, Bound. Layer Meteor., 81, 245–269, doi:10.1007/BF02430331, 1996.

Wang, L. T., Hao, J. M., and He, K. B.: A modeling study of coarse particulate matter pollution in Beijing: regional source contributions and control implications for the 2008 Summer Olympics, J. Air. Waste. Manage., 58, 1057–1069, doi:10.3155/1047-3289.58.8.1057, 2008.

Wang, L. T., Xu, J., Yang, J., Zhao, X. J., Wei, W., Cheng, D. D., Pan, X. M., and Su, J.: Understanding haze pollution over the southern Hebei area of China using the CMAQ model, Atmos. Environ., 56, 69–79, 2012.

Wang, H., Shi, G. Y., Aoki, T., Wang, B., and Zhao, T. L.: Radiative forcing due to dust aerosol over east Asia-north Pacific region during spring 2001, Chin. Sci. Bull., 49, 2212–2219, 2004.

Wang, H., Shi, G. Y., Li., W., and Wang, B.: The impacts of optical properties on radiative forcing due to dust aerosol, Adv. Atmos. Sci., 23, 431–441, 2006.

Wang, H., Gong, S. L., Zhang, H. L., Chen, Y., Shen, X. S., Chen, D. H., Xue, J. S., Shen, Y. F., Wu, X. J., and Jin, Z. Y.: A new-generation sand and dust storm forecasting system GRAPES_CUACE/Dust: Model development, verification and numerical simulation, Chin. Sci. Bull., 55, 635–649, doi:10.1007/s11434-009-0481-z, 2009.

Wang, H., Zhang, X. Y., Gong, S., Chen, Y., Shi, G., and Li, W.: Radiative feedback of dust aerosols on the East Asian dust storms, J. Geophys. Res., 115, D23214, doi:10.1029/2009JD013430, 2010.

Wang, H., Tan, S. C., Wang, Y., Jiang, C., Shi, G. Y., Zhang M., and Che, H. Z.: A multisource observation study of the severe prolonged regional haze episode over eastern China in January 2013, Atmos. Environ., 89, 807–815, 2014a.

Wang, H., Xu, J. Y., Zhang, M., Yang, Y. Q., Shen, X., J., Wang, Y. Q., Chen, D., and Guo, J. P.: A study of the meteorological causes of a prolonged and severe haze episode in January 2013 over central-eastern China, Atmos. Environ., 98, 146–157, 2014b.

Wang, T. J., Jiang, F., Deng, J. J., Shen, Y., Q., Fu, Y., Wang, Q., Fu, Y., and Xu, J. H.: Urban air quality and regional haze weather forecast for Yangtze River Delta region, Atmos. Environ., 58, 70–83, 2012.

Wang, Y., Zhuang, G., Sun, Y., and An, Z.: The variation of characteristics and formation mechanisms of aerosols in dust, haze, and clear days in Beijing, Atmos. Environ., 40, 6579–6591, 2006.

Wang, Y. T., Li, W., Zhang. X. L., and Meng, W.: Relationship between atmospheric boundary layer and air pollution in summer stable weather in the Beijing urban area, Res. Environ. Sci., 25, 1092–1098, 2012 (in Chinese).

Wang, H., Shi, G. Y., Zhang, X. Y., Gong, S. L., Tan, S. C., Chen, B., Che, H. Z., and Li, T.: Mesoscale modelling study of the interactions between aerosols and PBL meteorology during a haze episode in China Jing-Jin-Ji and its near surrounding region – Part 2: Aerosols' radiative feedback effects, Atmos. Chem. Phys., 15, 3277–3287, doi:10.5194/acp-15-3277-2015, 2015.

Wei, W. X., Zhang, X., and Tian, G. Q.: Analysis of relation between haze distribution and terrain and wind speed in Hebei Province, J. Nat. Disasters, 19, 49–52, 2010 (in Chinese).

Westerdahl, D., Wang, X., Pan, X., and Zhang, K. M.: Characterization of on- road vehicle emission factors and microenvironmental air quality in Beijing, China, Atmos. Environ., 43, 697–705, 2009.

Wu, D., Tie, X. X., Li, C. C., Ying, Z. M., Lau, A. K., Huang, J., Deng, X. J., and Bi, X. Y.: An extremely low visibility event over the Guangzhou region: a case study, Atmos. Environ., 39, 6568–6577, 2006.

Wu, D., Wu, X. J., and Li, F.: Temporal and spatial variation of haze during 1951–2005 in Chinese mainland, Acta. Meteor. Sinica., 68, 680–688, 2010 (in Chinese).

Wu, Z. J., Hu, M., Lin, P., Liu, S., Wehner, B., and Wiedensohler, A.: Particle number size distribution in the urban atmosphere of Beijing, China, Atmos. Environ., 42, 7967–7980, 2008.

Xing, J., Zhang, Y., and Wang, S. X.: Modeling study on the air quality impacts from emission reductions and a typical meteorological conditions during the 2008 Beijing Olympics, Atmos. Environ., 45, 1786–1798, 2011.

Xu, G. Q., Chen, D. H., and Xue, J. S.: The program structure designing and optimizing tests of GRAPES physics, Chin. Sci. Bull., 53, 3470–3476, 2008.

Yang, S., Guang, Y. S., Chen, L., Wang, B., and Yang, H. L.: Evaluation of Moderate-Resolution Imaging Spectroradiometer (MODIS) Deep Blue Aerosol Products Using Ground-Based Measurements over Beijing, SOLA, 7, 133–136, doi:10.2151/sola.2011-034, 2011.

Yang, X. S., Chen, J. B., and Hu, J. L.: A semi-implicit semi-Lagran global nonhydrostatic model and the polar discretization scheme, Sci. China Ser. D, 50, 1885–1891, 2007.

Yin, Y., Tong, Y. Q., Wei, Y. X., Wang, T. J., Li, J. P., Yang, W. F., and Fan, S. X.: The analysis of chemistry composition of fine-mode particles in Nanjing, Trans. Atmos. Sci., 32, 723–733, 2009 (in Chinese).

Zhang, J. and Reid, J. S.: A decadal regional and global trend analysis of the aerosol optical depth using a data-assimilation grade over-water MODIS and Level 2 MISR aerosol products, Atmos. Chem. Phys., 10, 10949–10963, doi:10.5194/acp-10-10949-2010, 2010.

Zhang, M., Uno, I., and Yoshi, Y.: Transport and transformation of sulfur compounds over East Asia during the TRACE-P and ACE-Asia campaigns, Atmos. Environ., 38, 6947–6959, 2004.

Zhang, R. H. and Shen, X. S.: On the development of the GRAPES – A new generation of the national operational NWP system in China, Chin. Sci. Bull., 53, 3429–3432, 2008.

Zhang, X. Y., Wang, Y. Q., and Lin, W. L.: Changes of atmospheric composition and optical properties over Beijing 2008 Olympic monitoring Campaign, B. Am. Meteorol. Soc., 1634–1651, 2009.

Zhang, X. Y., Zhang, Y. F., Feng, Y. C., Han, S. Q., Han, B., Li, L., and Xu., H.: Influence of Synoptic Patterns on the Concentrations of PM_{10} in Tianjin, Res. Environ. Sci., 23, 1115–1116, 2013 (in Chinese).

Zhang, Y. M., Zhang, X. Y, and Sun, J. Y.: Characterization of new particle and secondary aerosol formation during summertime in Beijing, China, Tellus, 63, 382–394, 2011.

Zheng, M. and Fang, M.: Particle-associated polycyclic aromatic hydrocarbons in the atmosphere of Hong Kong, Water Air Soil Poll., 117, 175–189, 2000.

13

Mesoscale modelling study of the interactions between aerosols and PBL meteorology during a haze episode in China Jing–Jin–Ji and its near surrounding region – Part 2: Aerosols' radiative feedback effects

H. Wang[1,2], G. Y. Shi[3], X. Y. Zhang[1], S. L. Gong[1], S. C. Tan[3], B. Chen[3], H. Z. Che[1], and T. Li[4]

[1]Institute of Atmospheric Composition, Key Laboratory of Atmospheric Chemistry of the China Meteorological Administration, Chinese Academy of Meteorological Sciences (CAMS), Beijing 100081, China
[2]Collaborative Innovation Center on Forecast and Evaluation of Meteorological Disasters, Nanjing University of Information Science & Technology, Nanjing 210044, China
[3]State Key Laboratory of Numerical Modeling for Atmospheric Sciences and Geophysical Fluid Dynamics, Institute of Atmospheric Physics, Chinese Academy of Sciences, Beijing 100029, China
[4]School of Atmospheric Physics, Nanjing University of Information Science & Technology, Nanjing 210044, China

Correspondence to: H. Wang (wangh@cams.cma.gov.cn, wangh@rays.cma.gov.cn)

Abstract. Two model experiments, namely a control (CTL) experiment without aerosol–radiation feedbacks and a experiment with online aerosol–radiation (RAD) interactions, were designed to study the radiative feedback on regional radiation budgets, planetary boundary layer (PBL) meteorology and haze formation due to aerosols during haze episodes over Jing–Jin–Ji, China, and its near surroundings (3JNS region of China: Beijing, Tianjin, Hebei, East Shanxi, West Shandong and North Henan) with a two-way atmospheric chemical transport model. The impact of aerosols on solar radiation reaching Earth's surface, outgoing long-wave emission at the top of the atmosphere, air temperature, PBL turbulence diffusion, PBL height, wind speeds, air pressure pattern and $PM_{2.5}$ has been studied focusing on a haze episode during the period from 7 to 11 July 2008. The results show that the mean solar radiation flux that reaches the ground decreases by about 15 % in 3JNS and 20 to 25 % in the region with the highest aerosol optical depth during the haze episode. The fact that aerosol cools the PBL atmosphere but warms the atmosphere above it leads to a more stable atmospheric stratification over the region, which causes a decrease in turbulence diffusion of about 52 % and a decrease in the PBL height of about 33 %. This consequently forms a positive feedback on the particle concentration within the PBL and the surface as well as the haze formation. Additionally, aerosol direct radiative forcing (DRF) increases PBL wind speed by about 9 % and weakens the subtropical high by about 14 hPa, which aids the collapse of haze pollution and results in a negative feedback to the haze episode. The synthetic impacts from the two opposite feedbacks result in about a 14 % increase in surface $PM_{2.5}$. However, the persistence time of both high $PM_{2.5}$ and haze pollution is not affected by the aerosol DRF. On the contrary over offshore China, aerosols heat the PBL atmosphere and cause unstable atmospheric stratification, but the impact and its feedback on the planetary boundary layer height, turbulence diffusion and wind is weak, with the exception of the evident impacts on the subtropical high.

1 Introduction

Aerosol direct radiative forcing (DRF) arises from the reforming of the Earth-atmosphere radiation budget by the absorption and scattering of solar radiation and the absorption and the emission of thermal radiation. This may cool or heat the Earth-atmosphere system, leading to the reforming of the

Earth-atmosphere temperature profile and impacts on global and regional climate, which has been widely noted and studied (Hansen et al., 1997; Ramanathan et al., 2001; Liao et al., 2006; Yu et al., 2006; Huang et al., 2006a, b, 2009; Che et al., 2014).

Considering the short lifetime of most aerosol particles (about 1 week) and their sharp uneven local and regional distribution and high dependence on emission sources and local meteorological conditions in the lower atmosphere (Che et al., 2007, 2009; Huang et al., 2007, 2008; Wang et al., 2014), aerosol effects on smaller spatial and temporal atmospheric scales may be worthy of greater attention. Studies at regional or local scales have shown that the DRF due to aerosols can exceed, in terms of intensity, the DRF attributable to greenhouse gases and lead to complex and important feedback mechanisms at such scales (Ramanathan, 2001; Li et al., 2007; Shindell and Faluvegi, 2009). The radiative feedback and impacts on mesoscale weather due to aerosol DRF have caused widespread concern in recent years. Certain studies have been conducted to simulate the impact on mesoscale weather circulation and to evaluate the possible feedback on short- and medium-range weather and numerical prediction in different regions of the world (Grell et al., 2005; Fast et al., 2006; Perez et al., 2006; Wang and Christopher, 2006; Wang et at al., 2010a, b; Heinold et al., 2008; Chapman et al., 2009). However, the current understanding of aerosol effects on weather contains major uncertainties because the interactions among aerosols, meteorology, radiation and chemistry are very complex and require further study using the online coupled models.

Aerosols are the main pollutants when haze episodes occur in China and PM_{10} may reach up to $1000\,\mu m^{-3}$ in the 3JNS (Beijing, Tianjin, Hebei, East Shanxi, West Shandong and North Henan) region of China (Zhang et al., 2013; Wang et al., 2014) during severe, long-lasting hazy weather. Aerosol particles suspended in local atmosphere lead to significant DRF and impacts on local or regional circulation as well as on the developing process of hazy weather. The meteorological condition of the planetary boundary layer (PBL) has important impacts on the occurrence, persistence, dissipation and pollution density of the haze (Vogelezang et al., 1996; Santanello et al., 2005; Cheng et al., 2002; Pleim, 2007). Substantial aerosols may also influence PBL meteorology and circulation and, in turn, evidently affect the haze and air pollution process by its DRF since most aerosol particles concentrate in PBL during haze events.

Focusing on July 2008 and a haze episode from 7 to 11 July in 3JNS, an external mixing scheme of seven kinds of aerosols has been introduced into the GRAPES/CUACE model to evaluate the optical features of composite aerosols and discuss PBL aerosol loading, the PBL meteorological properties closely related to haze and their relationship to haze episodes (Part 1 of the companion paper, Wang et al., 2015). In this article, the aerosol optical properties are used as input parameters in a radiative transfer scheme where the radiative heating rates are fed back to the online dynamic frame of the GRAPES/CUACE. This allows us to evaluate aerosol DRF and its impact on the local radiation budget and the PBL meteorological features including air temperature, heating/cooling profile rates, wind intensity, planetary boundary layer height (PBLH), turbulence diffusion and air pressure pattern over 3JNS.

2 Model introduction

The dynamic core, physics processes option, chemical frame including emission sources, gas and aerosol processes, and interaction between gas and aerosols in the GRAPES/CUACE model have been introduced in Part 1 (Wang et al., 2015). This section provides a brief description of the radiative transfer scheme used in this research.

Several radiative transfer modes can be selected in the GRAPES/CUACE model. The short-wave (SW) and long-wave (LW) radiative transfer models developed by the Climate and Radiation Branch, NASA/Goddard Space Flight Center (CLIRAD-SW and CLIRAD-LW) (Chou et al., 1998, 2001), are used in this work for their convenience and fine capacity in processing aerosols (Wang et al., 2010, 2013). The CLIRAD includes the absorption due to water vapour, O_3, O_2, CO_2, clouds and aerosols. Interactions among the absorption and scattering by clouds and aerosols are considered. The solar spectrum in the CLIRAD is divided into 11 bands and the thermal infrared spectrum into 10 bands from 3.333 to $40\,\mu m$. For each atmospheric layer and spectral band, the effective optical thickness, single scattering albedo and asymmetry factor are summered up over all gases and particles:

$$\tau = \sum_i \tau_i, \tag{1}$$

$$\overline{\omega} = \sum_i \omega_i \tau_i \Big/ \sum_i \tau_i, \tag{2}$$

$$\overline{g} = \sum_i g_i \omega_i \tau_i \Big/ \sum_i \tau_i \omega_i, \tag{3}$$

where i denotes ozone, water vapour, clouds, aerosols and atmospheric gases. Aerosol optical depth (AOD) (τ_a), single scattering albedo (ω_a) and Asymetry factor (g_a) are calculated by an external mixing scheme of different types of aerosols as described in the companion paper (Part 1, Wang et al., 2015). The effect of aerosols on solar and thermal radiation within the GRAPES/CUACE model is realized by implementing τ_a, ω_a and g_a into the CLIRAD radiation scheme. The radiative heating/cooling rates in the atmosphere, including aerosol absorption and scattering of solar and infrared radiation, were calculated and fed back to the thermal and dynamic processes at every radiation step in the

GRAPES/CUACE model. The online active interaction of "meteorology–aerosol–radiation" is completely achieved in the model, and the radiative feedback on the local PBL and the haze due to aerosols is studied using the model.

3 Experiment design

The control (CTL) experiment is the base simulation without calculating aerosol radiative feedback and impacts online as described in Part 1 (Wang et al., 2015). In this paper, the simulation experiment (online active interacting meteorology–aerosol–radiation) is referred to as the RAD experiment. The only difference between the RAD and CTL experiments is that in the RAD experiment, the aerosol radiation heating/cooling effect is calculated online and fed back to the model thermodynamic and dynamic processes.

In the following section, the simulation results of surface radiative fluxes from the RAD experiment are compared with those of the CTL simulation as a way to assess the aerosol impact on the local Earth-atmosphere radiation balance. The differences between the RAD and CTL experiments concerning the PBL meteorological fields, including PBL temperature, height, turbulence diffusion, meteorological pattern and pollutant particle loading, will be discussed as part of the study of aerosol radiative effects and feedback on local PBL thermal and dynamic processes. Finally, the aerosol impact on the haze episode itself is discussed.

The haze episode that occurred during 7–11 July 2008 was selected for this study. All model configuration options and model parameters adopted were the same as those used in the CTL experiment in Part 1 (Wang et al., 2015). The initial fields and lateral boundary data on the meteorology and tracers, together with the model domain, horizontal and vertical resolutions and both step and forecasting, also matched those used in the CTL experiment.

4 The impacts on the regional radiation budget

The solar radiation flux reaching the Earth's surface may change obviously due to aerosols absorbing and scattering of solar radiation during the haze episode. A large number of particles suspended in the atmosphere, as well as infrared radiation and the outgoing long-wave radiation at the top of atmosphere (TOA), may also be changed. This leads to the reforming of regional Earth-Atmosphere radiation budget. The key factor impacting radiation flux is the aerosol AOD. It can be seen in Fig. 1 that the averaged simulated AOD during 7–11 July shows an expected coherence with MODIS Deep Blue AOD at 550 in horizontal distribution, affected area, peak values and their geographical locations over 3JNS, and downwind area even though MODIS omits parts of the data in 3JNS. The land domain (111–119° E, 33–40° N named as LAND in Fig. 1) with the highest AOD values is regarded as the most representative of 3JNS; the aerosol impacts on meteorological fields are presented in the following sections. The three points labelled A (38.6° N, 119.5° E), B (35.0° N, 120.7° E) and C (38.4° N, 122.0° E) in Fig. 1 are selected to represent China's offshore region. SEA1 (32.0 to 36.8° N, 121.5 to 126.0° E) denotes the sea area from the eastern coast of China to the western edge of the Korean peninsula, while SEA2 (30.0 to 42.0° N, 130.0 to 139.5° E) represents the sea area to the east of the Korean peninsula.

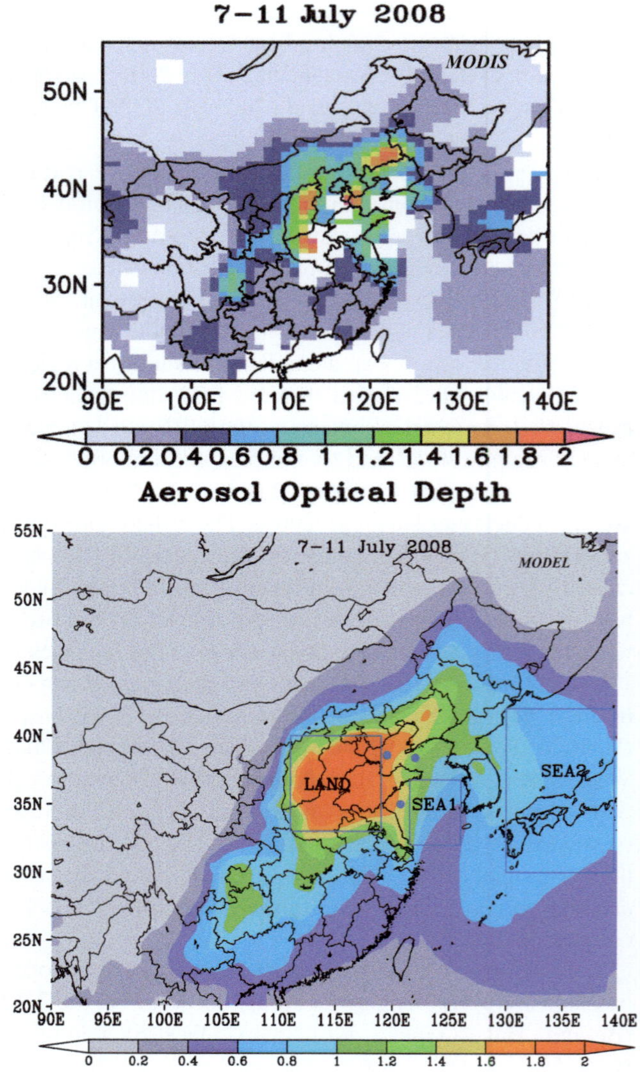

Figure 1. The averaged MODIS (top) and modelled AOD (bottom) during the period of 7–11 July 2008: LAND represents the polluted area in 3JNS; points A, B and C represent offshore China; domains SEA1 and SEA2 refer to China's Huang Sea and the Sea of Japan.

The percentage change in surface SW flux due to aerosol DRF at the surface (SFC) and change in LW at TOA are defined as

$$\Delta F_{\mathrm{SFC}} = (\mathrm{Flux}(\downarrow_{\mathrm{Solar,SFC}})_{\mathrm{RAD}} - \mathrm{Flux}(\downarrow_{\mathrm{Solar,SFC}})_{\mathrm{CTL}}) / \mathrm{Flux}(\downarrow_{\mathrm{Solar,SFC}})_{\mathrm{CTL}} \times 100\%, \quad (4)$$

$$\Delta F_{\mathrm{TOA}} = (\mathrm{Flux}(\uparrow_{\mathrm{IR,TOA}})_{\mathrm{RAD}} - \mathrm{Flux}(\uparrow_{\mathrm{IR,TOA}})_{\mathrm{CTL}}) / \mathrm{Flux}(\uparrow_{\mathrm{IR,TOA}})_{\mathrm{CTL}} \times 100\%, \quad (5)$$

where $\mathrm{Flux}(\downarrow_{\mathrm{Solar,SFC}})_{\mathrm{RAD}}$ and $\mathrm{Flux}(\downarrow_{\mathrm{Solar,SFC}})_{\mathrm{CTL}}$ represent the downward solar radiation flux (w/m^2) at the surface of the RAD and CTL experiment respectively. $\mathrm{Flux}(\uparrow_{\mathrm{IR,TOA}})_{\mathrm{RAD}}$ and $\mathrm{Flux}(\uparrow_{\mathrm{IR,TOA}})_{\mathrm{CTL}}$ are the infrared radiation fluxes emitted from Earth at TOA in the RAD and CTL experiments respectively. Figure 2a displays the averaged ΔF_{SFC} at 06:00 UTC from 7 to 11 July. It can be seen that aerosol DRF decreased more than 15 % of the solar radiation fluxes reaching the ground over most of 3JNS; the decrease reached up to 20–25 % in the most polluted area with the highest AOD values. This result indicates the important impact of aerosol DRF on ground and near-ground radiation budgets. Figure 2b shows the mean ΔF_{TOA} during 7–11 July, indicating that aerosol DRF reduced only 1–3 % of infrared emission at the TOA during this haze episode, which is far lower than the surface downward solar radiation flux change. This result suggests that aerosol DRF has more important impacts on the ground and near-Earth surface radiation budgets, i.e. on the PBL energy budget, than on TOA.

5 The radiative feedback on PBL meteorology due to aerosols

The remarkable reforming of the surface and PBL radiation energy budget by aerosols will certainly lead to changes in PBL thermodynamics, dynamics and physical processes, which result in changes in PBL meteorological fields and further the haze development. The impacts on air temperature, turbulence distribution, PBLH, wind speed, air pressure and $PM_{2.5}$ due to aerosols will be discussed in the following sections.

5.1 The impacts on temperature

The direct and initial change due to aerosol DRF is the temperature. The surface temperature change reached -1 to -3 K at 06:00 UTC on 7–11 July (Fig. 3a) in 3JNS, corresponding to the highest AOD values and substantial negative values of surface SW flux changes as shown in Fig. 1. A vertical cross section of temperature was drawn along latitude 38° N (black line in Fig. 3a) that shows the vertical temperature change due to aerosol DRF (Fig. 3b). Also shown is the reduction by aerosol DRF of surface and PBL temperature over the land surface. A PBL temperature decrease of 1

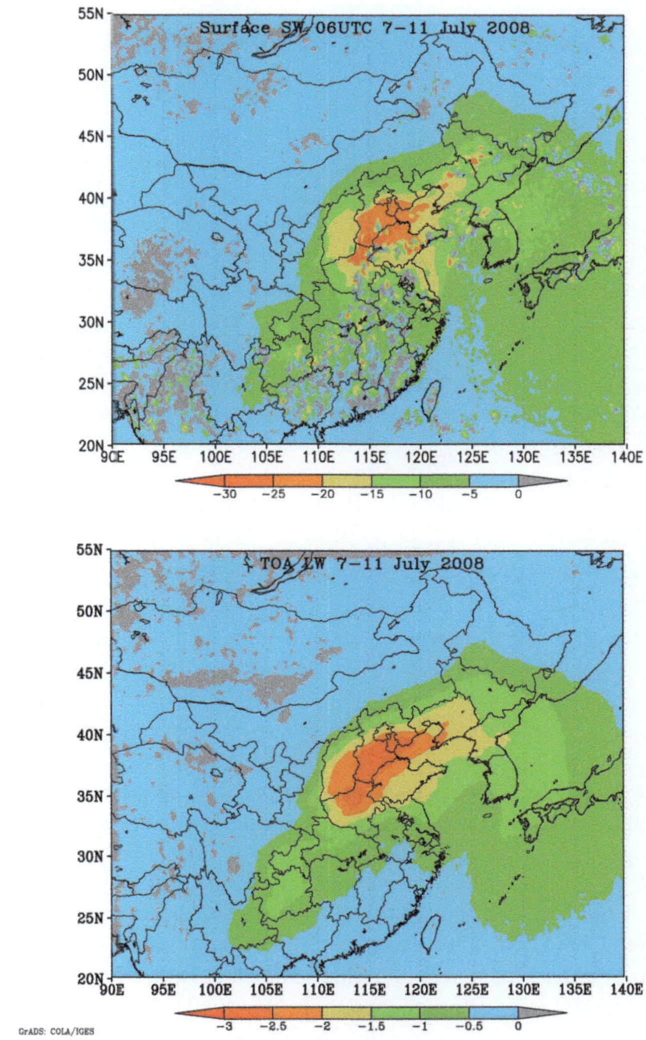

Figure 2. The percentage of change in the surface SW flux at 06:00 UTC **(a)** and in TOA outgoing LW flux **(b)** due to aerosol DRF during the 7–11 July period.

to 2 K occurred over the Chinese mainland (110–118° E) and 0.5 to 1 K over the Korean peninsula (125–128° E), while the aerosol impacts on the surface and PBL temperature changes were small or increased weakly over the oceanic area. Over this cooling atmospheric layer a weak warming layer exists with a vertical height ranging from 975 to 600 hPa along latitude 38° N. The vertical sections of regional average temperature change due to aerosols over LAND region (Fig. 3c): points A, B, C, SEA1 and SEA2 (Fig. 3d) display the vertical temperature changes over 3JNS with the highest pollution i.e. offshore China, the China Sea and the Sea of Japan. It is clear from Fig. 3c that temperature diminished from the surface to about 850 hPa over 3JNS and then increased above that level. This suggests the presence of aerosol cooling effects on the PBL atmosphere and warming effects on the atmosphere above it, which may lead to more stable stratification of the atmosphere over this region. Points A, B and C lie offshore

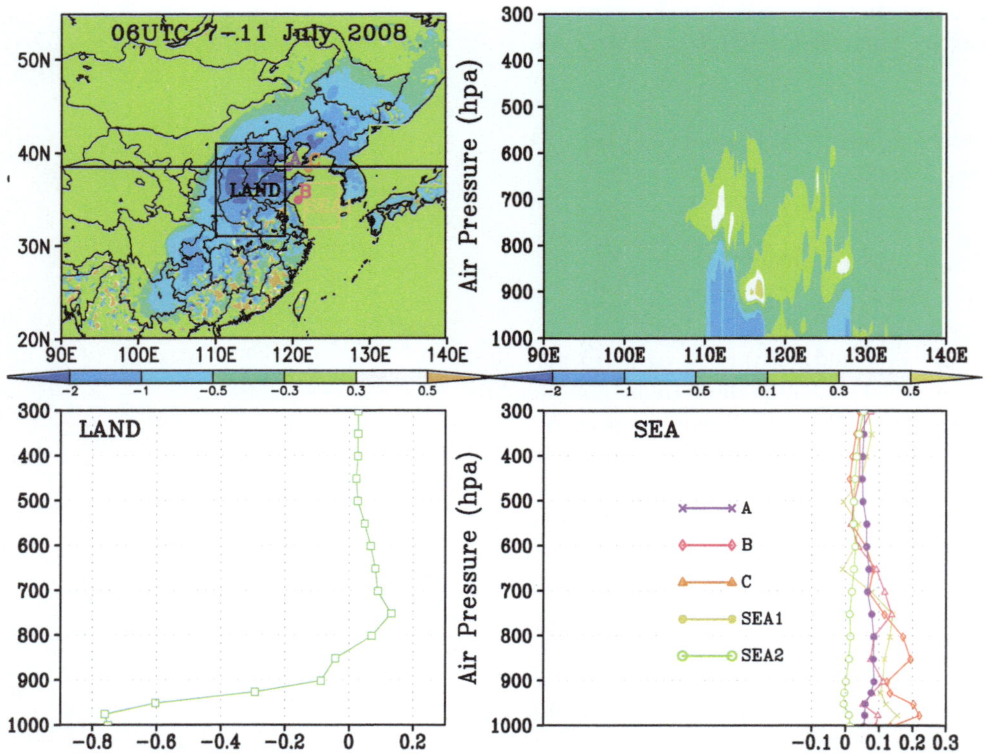

Figure 3. Mean temperature changes (K) at 06:00 UTC of 7–11 July due to aerosol DRF: **(a)** surface temperature; **(b)** vertical section at 38° N of **(a)**; **(c)** vertical section of domain LAND region; **(d)** vertical section of points A, B, C, SEA1 and SEA2.

of the Chinese coast, and SEA1 represents the region near the China Sea. The vertical profiles of temperature changing induced by the aerosols' radiative feedback effect over those areas are quite different than those over the LAND region due to the different surface albedo and the height and depth of aerosols layer. It can be seen from Fig. 3d that aerosols heat the atmosphere from the surface to a height of 600 hPa over these regions. This is especially so in the PBL atmosphere because the higher aerosol layer and the smaller AOD value may cause more unstable atmospheric stratification over the sea areas. Aerosol DRF has little impact on the surface and PBL temperatures in the SEA2 region, and only very weak warming can be found above a height of 750 hPa, owing to the lower AOD values in this region. These results and the discussion in Fig. 3 indicate that aerosol DRF led to more stable atmospheric stratification over 3JNS and to more unstable atmospheric stratification over offshore China and the China Sea regions during the haze episode of 7–11 July. This has an important influence on local PBL meteorology and the regional atmosphere circulation.

5.2 The impacts on PBL turbulence diffusion

Changes in regional atmospheric stratification positively result in varying turbulence diffusion. The turbulence diffusion coefficient (FKTM) used in Part 1 (Wang et al., 2015) of this study is a valid physical parameter that indicates the strength of turbulence diffusion. Figure 4 displays FKTM changes due to aerosol DRF. Figure 4a describes the regional distribution of mean impacts on turbulence diffusion in the haze from 7 to 11 July; it can be seen that low turbulence diffusion exists over the entire 3JNS region with mean FKTM values of 14–45 $m^2\ g^{-1}$ in the haze conditions during 7–11 July 2008. Aerosol DRF led to a mean 5 $m^2\ g^{-1}$ reduction of FKTM over most of the eastern Chinese mainland and a lessening of 10–15 $m^2\ g^{-1}$ in 3JNS, showing remarkable depression on the local atmospheric turbulence diffusion process from aerosol DRF. Figure 4b displays the daily changes in the regional averaged difference: FKTM difference between the RAD and CTL experiments over LAND and SEA1 in July 2008. It is clear from Fig. 4b that the averaged FKTM of the LAND region was reduced by aerosol DRF more or less during the whole of July 2008. As with the haze event during 7–11 July 2008, the FKTM declined by about 7–9 $g^2\ m^{-1}$ and 8–10 $g^2\ m^{-1}$ during another haze episode during 25–28 July 2008, which was also initiated by aerosol DRF. FKTM changes resulting from aerosol DRF also occurred over the SEA1 region but these were small to negligible in scale. These results suggest that the suppression of diffusion turbulence by aerosol DRF is both certain and significant over the middle and eastern Chinese mainland with its high pollutants while, in contrast, impact over the sea region is small and can be negligible during haze episodes.

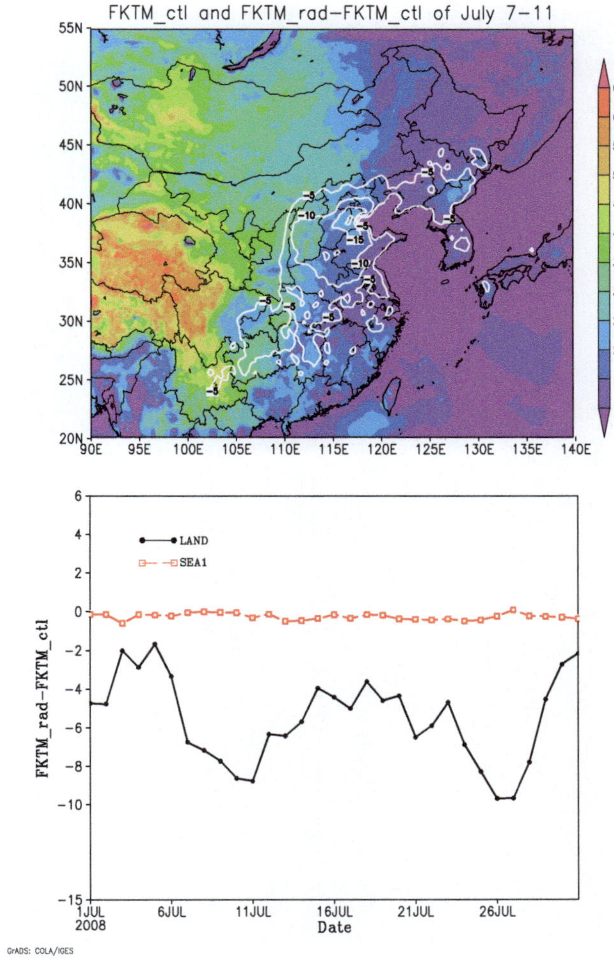

Figure 4. FKTM change ($m^2\,s^{-1}$) due to aerosol DRF: **(a)** mean FKTM by the CTL experiment (shaded) and FKTM difference between the RAD and CTL experiments (contour) during 7–11 July; **(b)** daily changes of LAND- and SEA1-averaged FKTM difference between the RAD and CTL experiments at the surface from 1 to 31 July.

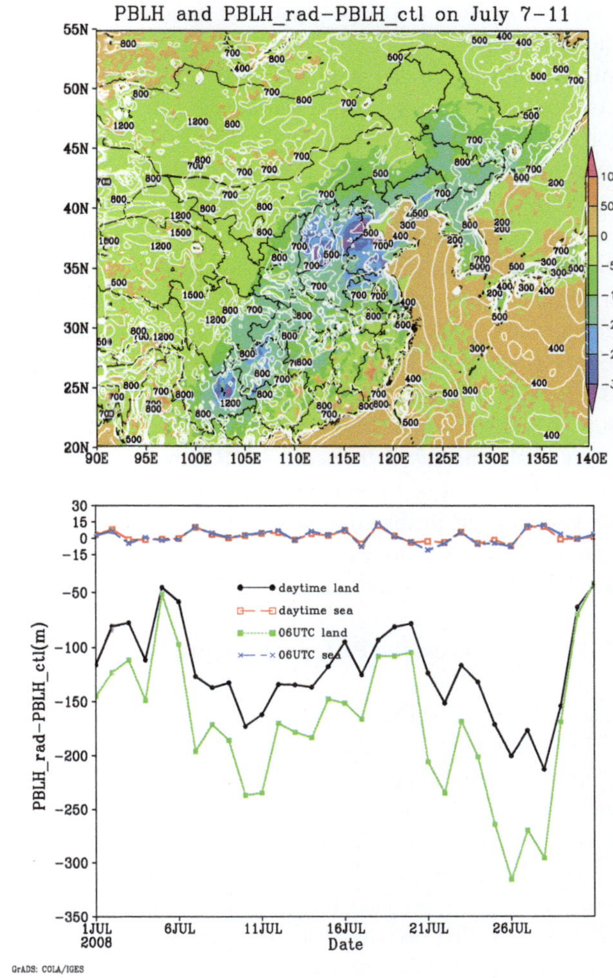

Figure 5. PBLH changes (m) due to aerosol DRF: **(a)** daytime mean PBLH of the CTL experiment (contour) and its difference between the RAD and CTL experiments (shading) during 7–11 July; **(b)** LAND- and SEA1-averaged PBLH difference between the RAD and CTL experiments from 1 to 31 July 2008.

5.3 The impacts on PBLH

PBLH is another key parameter to describe the PBL features closely related to haze and air pollution. Its impact on $PM_{2.5}$ and haze was discussed in Part 1 (Wang et al., 2015). Aerosol impacts on PBLH due to DRF during the haze episode during 7–11 July are discussed in this section. Figure 5 shows PBLH changes due to aerosol DRF. Figure 5a shows that the mean daytime PBLH was as low as 400–700 m over the eastern Chinese mainland during the haze episode during 7–11 July. PBLH declined by about 50–300 m generally in response to aerosol DRF over this region; the PBLH difference between the RAD and CTL experiments reaches up to 200–300 m in 3JNS. Figure 5b shows that daytime PBLH, especially PBLH at local noon-time (06:00 UTC), may have been diminished by aerosol DRF evidently and steadily in July 2008, although its reduction varies with time. The PBLH reduction may have reached about 250 m during 10–11 July and 250–300 m during another haze episode from 25 to 28 July. Figure 5b also shows that aerosol DRF inflicts very weak impacts on PBLH over the sea with slight increases or decreases of PBLH at different times.

5.4 The impacts on PBL wind

The influence of surface and PBL wind fields on haze pollution is as important as, or even more important than, that of PBLH and diffusion turbulence as discussed in Part 1 (Wang et al., 2015), but the impact on PBL winds from aerosol DRF is not so strong as its impact on PBLH and diffusion turbulence. PBL wind changes due to aerosol DRF is minor and may be neglected when haze pollution is weak. The focus is on the period from 9 to 11 July with the highest $PM_{2.5}$

and severest pollution to investigate the wind field changes due to aerosol DRF. Figure 6a shows the difference of PBL-averaged wind speed between the RAD and CTL experiments (shading) and wind vector (contour) of the CTL experiment. It can be seen from Fig. 6a that the whole PBL wind speed was increased by aerosol DRF over most of the middle and eastern Chinese mainland region, while it declined over the offshore and sea areas. Wind speed was increased from 0.4 to 0.8 $m\,s^{-1}$ by aerosol DRF in certain parts of 3JNS with high particle concentration. Figure 6b also indicates temporal changes in the LAND-averaged wind speed difference between the RAD and CTL experiments at the surface and PBL (950–850) hPa from 00:00 UTC on 9 July to 00:00 UTC on 12 July. Also shown is that both surface and PBL wind speed was obviously increased by aerosol DRF over this period; however, the extent of the increase in PBL wind speed was much greater than in the case of the surface wind, indicating that aerosols may impose much greater impacts on PBL winds than on surface winds.

5.5 The impacts on the PBL air pressure pattern

Figure 7a displays the PBL-averaged air pressure pattern during 7 to 11 July from the CTL experiment. It can be seen that subtropical high pressure controlled both eastern China and offshore regions of China. Eastern China was located on the west edge of the subtropical high with a weak southerly air flow controlling the area. This air pressure pattern is conducive to retention of haze (discussed in Part 1, Wang et al., 2015). The PBL-averaged air pressure changes due to aerosol DRF were calculated from the air pressure differences between the RAD and CTL experiments. It can be seen in Fig. 7b that the whole PBL air pressure was decreased by aerosol DRF over eastern China and its downwind region, especially over the offshore region, which resulted in the obvious weakening of the subtropical high over China's offshore and sea regions. The lessening and withdrawal eastward of the subtropical high sustained the eastward-moving cold air from the northwest and also delivered a downward flow of cold air together with some momentum from the upper atmosphere to the PBL. This seems to have helped the breaking down of the stable air pressure pattern that was controlling the retention of the haze.

5.6 The impacts on surface $PM_{2.5}$

The reforming of the local PBL meteorology structure by aerosol DRF in turn impacts the PBL and surface $PM_{2.5}$ spatial distribution, temporal changes and, perhaps, the duration time of the haze. The radiative feedback on $PM_{2.5}$ by aerosols consists of the synthesized results from the PBL meteorological parameters, involving temperature, turbulence diffusion, PBLH, wind, air pressure and others.

The averaged $PM_{2.5}$ loading within the PBL (contour, $kg\,m^{-2}$) of 7–11 July in the CTL experiment has been calculated and shown in Fig. 8 together with the surface $PM_{2.5}$ percentage changes attributable to aerosol DRF (shaded). It can be seen that the aerosol DRF generally increases the surface $PM_{2.5}$ over eastern China, the percentage change being > 10 % over most of 3JNS. The geographical location of the increasingly high percentage of $PM_{2.5}$ basically correlates with the location of the high PBL $PM_{2.5}$ loading. The $PM_{2.5}$ increasing percentage by aerosol DRF can reach up to more than 20 % over the region with the highest PBL $PM_{2.5}$ loading in 3JNS. The result indicates that the higher the PBL $PM_{2.5}$ loading, the more $PM_{2.5}$ might be concentrated at the surface due to aerosol DRF and in terms of the averaged condition of the haze episode. Surface $PM_{2.5}$ is enhanced by about 10–20 % due to aerosol DRF or even more over mid-eastern China.

The averaged temporal variations of surface $PM_{2.5}$ of 3JNS of the CTL and RAD experiments from 7 to 13 July are

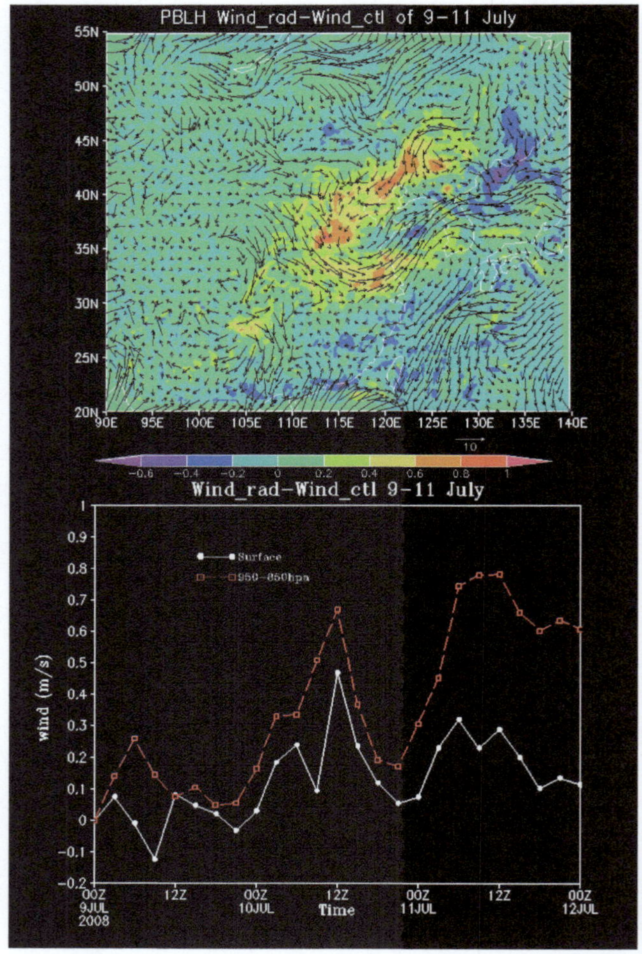

Figure 6. Wind field changes ($m\,s^{-1}$) due to aerosol DRF: **(a)** the mean PBL wind vector of the CTL experiment (contour) and PBL-averaged wind speed difference between the RAD and CTL experiments (shading) from 9 to 11 July. **(b)** Temporal changes of LAND-averaged wind speed difference between the RAD and CTL experiments at the surface and 950–850 hPa height from 9 to 11 July.

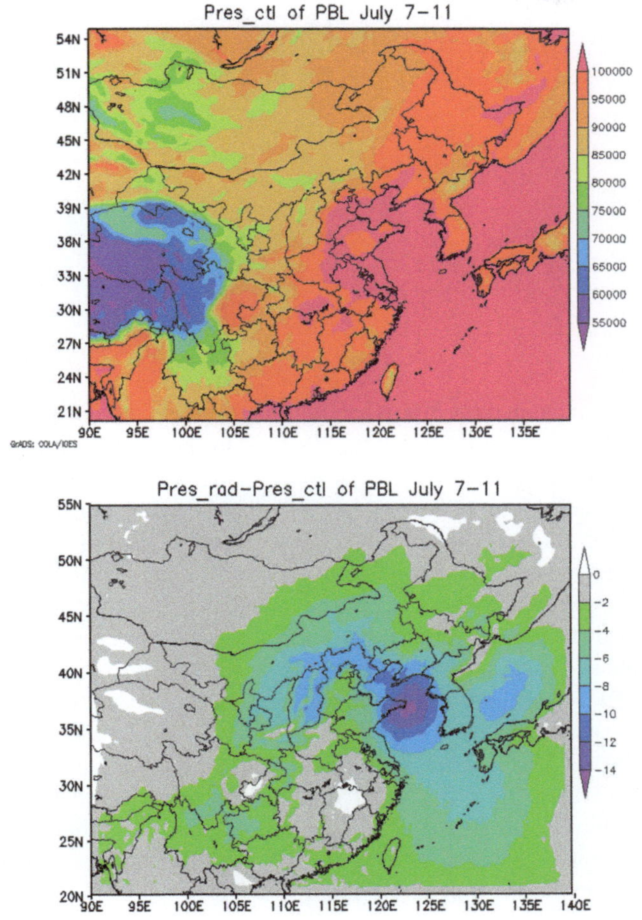

Figure 7. The PBL-averaged air pressure (hPa) from the CTL experiment (top) and its difference between the RAD and CTL experiments (bottom) of 7–11 July.

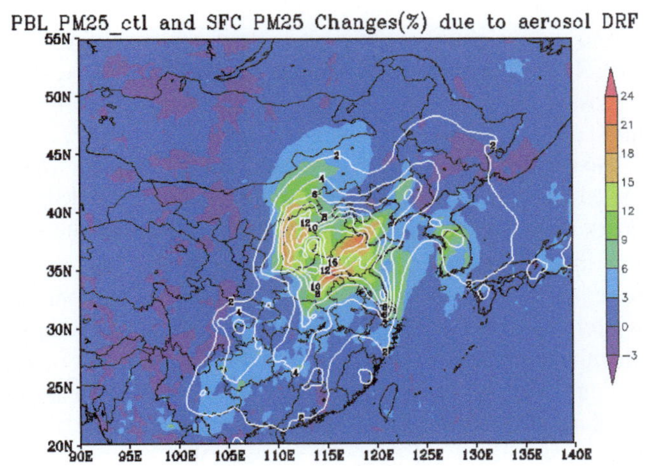

Figure 8. July during the CTL experiment and the surface $PM_{2.5}$ change percentage due to aerosol DRF for 7–11 July (shaded).

also displayed and compared in order to evaluate the impacts of aerosol DRF (Fig. 9). When the surface $PM_{2.5}$ concentration is regarded as the indicator of haze pollution, the obvious difference of PM2.5 values between the CTL and RAD experiments can be seen during the period from about 05:00 GTM on 7 July to about 18:00 GTM on 11 July. The LAND mean surface $PM_{2.5}$ also remains higher than $140\,\mu g\,m^{-3}$ during this period. The difference of LAND mean surface $PM_{2.5}$ between the CTL and RAD experiments is small before or after that period and, at the same time, the $PM_{2.5}$ values from both experiments are lower than$140\,\mu g\,m^{-3}$. This indicates that aerosol DRF may have very little impact on the haze-sustaining period or keeping time of the haze episode because when $PM_{2.5}$ declines below a certain level, the aerosol DRF may not be efficient enough to change the PBL meteorological circulation and then reform the $PM_{2.5}$ spatial and temporal distribution.

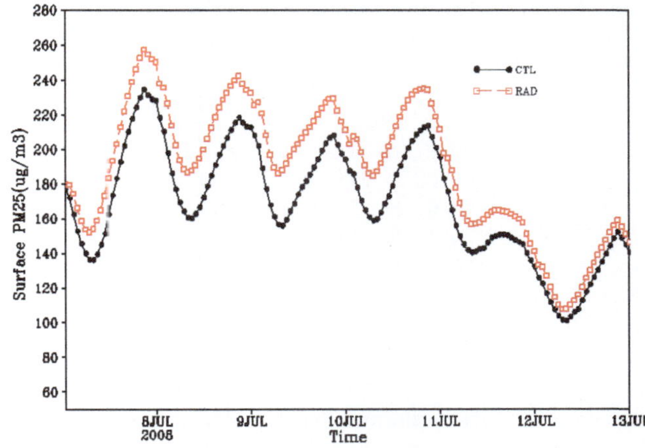

Figure 9. Temporal changes of LAND-averaged surface $PM_{2.5}$ by the CTL and RAD experiments.

The responses of PBL meteorology quantities to aerosol DRF relates, on the one hand, to the perturbation strength from aerosols and, on the other hand, to the thermodynamics and dynamic characteristics of these meteorological entities. In order to evaluate and order the sensitivity of these parameters to aerosol DRF, a weighting coefficient g_i is defined as follows:

$$g_{\mathrm{i\_LAND}} = \frac{\mathrm{var(i)_{rad\_LAND}} - \mathrm{var(i)_{ctl\_LAND}}}{\mathrm{var(i)_{ctl\_LAND}}}, \quad (6)$$

where var(i) stands for different meteorological variables involving radiation fluxes, wind speed, PBLH, FKTM and $PM_{2.5}$; the subscript ctl and rad identify the CTL and RAD experiments; and the subscript LAND means that all the variables are the averaged mean values of the LAND region and stand for the mean condition of 3JNS. With regard to air temperature and air pressure, the zero values have no physical meaning, g_i is not calculated here, and only the changes due to aerosol DRF are listed. Table 1 lists the daily g_i and the averaged g_i of the haze episode during 7–11 July. It can be seen, therefore, that the response of the meteorological parameters to aerosol DRF from high to low is FKTM, PBLH,

Table 1. Weighing coefficient of the response of meteorological parameters to aerosol DRF.

Time	$g_{flux\_sw\_sfc}$	$g_{flux\_lw\_toa}$	DT_{06} (K)	g_{difu}	g_{wind_PBL}	g_{PBLH}	DP_{06} (hPa)	$g_{PM_{2.5}}$
07:00–07:24 UTC	−0.14	−0.01	−0.93	−0.40	0.01	−0.30	−16	0.10
08:00–08:24 UTC	−0.18	−0.02	−1.02	−0.48	0.03	−0.29	−14	0.14
09:00–09:24 UTC	−0.18	−0.02	−1.20	−0.57	0.15	−0.31	−12	0.16
10:00–10:24 UTC	−0.20	−0.03	−1.13	−0.62	0.16	−0.39	−14	0.15
11:00–11:24 UTC	−0.18	−0.02	−0.6	−0.54	0.11	−0.36	−14	0.11
Averaged	−0.18	−0.02	−0.98	−0.52	0.09	−0.33	−15	0.13

ΔF_{SFC_Solar}, PBL wind and ΔF_{TOA}. The process-averaged g_{fktm} for 7–11 July is −0.54 daily, ranging from −0.40 to −0.62, and that of g_{PBLH} is −0.33, ranging from −0.29 to −0.39, showing that the most important impacting mechanism from aerosol DRF is the suppression of PBL turbulence diffusion, which may lead to increasing the surface $PM_{2.5}$ and to positive radiative feedback to haze pollution. g_{wind} is 0.09 with daily values ranging from 0.01 to 0.16. The PBL air pressure at 06:00 UTC fell to a mean of 15 hPa for the period 7–11 July and ranged from 0.12 to 0.16, which weakened the subtropical high. The changes in both wind and air pressure may result in negative feedback to haze development. Comparing g_{wind} with g_{fktm} and g_{PBLH} indicates that aerosol DRF may impose more important impacts on PBL height and turbulence diffusion than its impacts on PBL wind and air pressure. $g_{PM_{2.5}}$ ranged from 0.10 to 0.16 with a mean of 0.13 for the 7–11 July period and resulted from the synthesized influence of the two opposing sides, as mentioned above, showing the final positive feedback of surface $PM_{2.5}$ and haze pollution from aerosol DRF. $g_{flux\_sw\_sfc}$ is the weighing coefficient of change in downward solar radiation flux due to aerosols, ranging from 0.14 to 0.20 with a mean value of 0.18. The weighing coefficient of changing TOA long-wave radiation ($g_{flux\_lw\_TOA}$) is the smallest with a value of 0.02, showing that total impacts on regional TOA from aerosol DRF are minor and may be neglected during haze episodes.

6 Discussion and conclusion

Focusing on a haze episode from 7 to 11 July 2008, two model experiments (the control experiment without calculation of aerosol–radiation effects and the RAD experiment with online calculation of aerosol–radiation interaction) are designed to evaluate aerosol direct radiative effects and feedbacks on the regional PBL atmospheric circulation related to haze formation in general and the specific haze episode in July 2008. The study involves impacts on surface SW and TOA outgoing radiation flux, temperature, PBL turbulence diffusion, wind, PBLH, air pressure pattern and $PM_{2.5}$. A detailed discussion is summarized as follows:

Solar radiation flux reaching the ground is decreased by about 15 % in 3JNS and by 20–25 % in the region with the highest AOD. Only 1–3 % of long-wave outgoing flux is decreased at the TOA. Aerosol DRF has a greater impact on the ground and near-surface radiation budget than in the upper atmosphere. Aerosol cools the lower or entire PBL while warming the upper PBL or the atmosphere above it, which leads to stable stratification of the atmosphere over the middle and eastern Chinese region. In contrast, aerosol heats the PBL atmosphere weakly, causing unstable atmospheric stratification over the Chinese offshore area. On the one hand, aerosol DRF suppresses diffusion turbulence and decreases PBLH significantly over 3JNS, which enhances particle concentration on the PBL and the surface, intensifying the haze formation. On the other hand, aerosol DRF increases PBL wind speed and weakens subtropical high pressure, which contributes to the collapsing of haze pollution over this region. The impacts from the two opposite effects ultimately result in an averaged increase of 10–20 % in surface $PM_{2.5}$ over 3JNS by aerosol DRF but no change in the persistence time of the haze pollution. The ranking order of the impacts on meteorological parameters due to aerosol DRF according to the weighting coefficient is the turbulence diffusion, PBLH, short wave radiation flux at the surface, $PM_{2.5}$, PBL wind and the TOA long-wave outgoing flux when air temperature and air pressure are not considered.

Given that most of these discussions are based on a single case of haze that occurred during the period of 7–11 July 2008, there is clearly a need for research into more summertime haze episodes in order to support the conclusions. As haze pollution episodes occur very frequently in autumn and winter in eastern China, the PBL meteorological condition, chemical composition of aerosols and optical characteristics are quite different from those in summer, as is the radiative feedback. Finally, it should be noted that the response of different meteorological fields to aerosol DRF and their contributions to regional circulation changes also relate to their dynamic thermodynamic features.

Acknowledgements. This work is supported by the National Basic Research Program (973) (2011CB403404), the National Natural Scientific Foundation of China (nos. 41275007 and 41130104) and the CAMS key projects (no. 2013Z007).

Edited by: X.-Y. Zhang

References

Chapman, E. G., Gustafson Jr., W. I., Easter, R. C., Barnard, J. C., Ghan, S. J., Pekour, M. S., and Fast, J. D.: Coupling aerosol-cloud-radiative processes in the WRF-Chem model: Investigating the radiative impact of elevated point sources, Atmos. Chem. Phys., 9, 945–964, doi:10.5194/acp-9-945-2009, 2009.

Che, H., Zhang, X. Y., Li, Y., Zhou, Z., and Qu, J. J.: Horizontal visibility trends in China 1981–2005, Geophys. Res. Lett., 34, L24706, doi:10.1029/2007GL031450, 2007.

Che, H., Yang, Z. F., Zhang, X. Y., Zhu, C., Ma, Q. L., Zhou, H. G., and Wang, P.: Study on the Aerosol Optical Properties and their Relationship with Aerosol Chemical Compositions over three Regional Background stations in China, Atmos. Environ., 43, 1093–1099, 2009.

Che, H., Xia, X., Zhu, J., Li, Z., Dubovik, O., Holben, B., Goloub, P., Chen, H., Estelles, V., Cuevas-Agulló, E., Blarel, L., Wang, H., Zhao, H., Zhang, X., Wang, Y., Sun, J., Tao, R., Zhang, X., and Shi, G.: Column aerosol optical properties and aerosol radiative forcing during a serious haze-fog month over North China Plain in 2013 based on ground-based sunphotometer measurements, Atmos. Chem. Phys., 14, 2125–2138, doi:10.5194/acp-14-2125-2014, 2014.

Cheng, Y., Canuto, V. M., and Howard, A. M.: An improved model for the turbulent PBL, J. Atmos. Sci., 59, 1550–1565, 2002.

Chou, M. D., Suarez, M. J., Ho, C. H., Yan, M. M. H., and Lee, K. T.: Parameterizations for Cloud Overlapping and Shortwave Single-Scattering Properties for Use in General Circulation and Cloud Ensemble Models, J. Clim., 11, 202–214, 1998.

Chou, M. D., Suarez, M. J., Liang, X. Z., and Michael, M.-H. Y.: A Thermal Infrared Radiation Parameterization for Atmospheric Studies, Technical Report Series on Global Modeling and Data Assimilation, NASA/TM-2001-104606, 19, America, Goddard Space Flight Center, Greenbelt, Maryland, 55, 2001.

Fast, J. D., Gustafson Jr., W. I., Easter, R. C., Zaveri, R. A., Barnard, J. C., Chapman, E. G., Grell, G. A., and Peckham, S. E.: Evolution of Ozone, Particulates and Aerosol Direct Radiative Forcing in the Vicinity of Houston Using a Fully Coupled Meteorology-Chemistry-Aerosol Model, J. Geophys. Res., 111, D21305, doi:10.1029/2005JD006721, 2006.

Grell, G. A., Peckham, S. E., Schmitz, R., McKenn, S. A., Frost, G., Skamarock, W. C., and Eder, B.: Fully Coupled "Online" Chemistry within the WRF Model, Atmos. Environ., 39, 6957–6975, 2005.

Hansen, J., Sato, M., and Ruedy, R.: Radiative Forcing and Climate Response, J. Geophys. Res., 102, 6831–6864, 1997.

Heinold, B., Tegen, I., Schepanski, I. K., and Hellmuth, O.: Dust radiative feedback on Saharan boundary layer dynamics and dust mobilization, Geophys. Res. Lett., 35, L20817, doi:10.1029/2008GL035319, 2008.

Huang, J., Minnis, P., Lin, B., Wang, T., Yi, Y., Hu, Y., Sun-Mack, S., and Ayers, K.: Possible influences of Asian dust aerosols on cloud properties and radiative forcing observed from MODIS and CERES, Geophys. Res. Lett., 33, L06824, doi:10.1029/2005GL024724, 2006a.

Huang, J., Lin, B., Minnis, P., Wang, T., Wang, X., Hu, Y., Yi, Y., and Ayers, J.: Satellite-based assessment of possible dust aerosols semi-direct effect on cloud water path over East Asia, Geophys. Res. Lett., 33, L19802, doi:10.1029/2006GL026561, 2006b.

Huang, J., Minnis, P., Yi, Y., Tang, Q., Wang, X., Hu, Y., Liu, Z., Ayers, K., Trepte, C., and Winker, D.: Summer dust aerosols detected from CALIPSO over the Tibetan Plateau, Geophys. Res. Lett., 34, L18805, doi:10.1029/2007GL029938, 2007.

Huang, J., Minnis, P., Chen, B., Huang, Z., Liu, Z., Zhao, Q., Yi, Y., and Ayers, J.: Long-range transport and vertical structure of Asian dust from CALIPSO and surface measurements during PACDEX, J. Geophys. Res., 113, D23212, doi:10.1029/2008JD010620, 2008.

Huang, J., Fu, Q., Su, J., Tang, Q., Minnis, P., Hu, Y., Yi, Y., and Zhao, Q.: Taklimakan dust aerosol radiative heating derived from CALIPSO observations using the Fu-Liou radiation model with CERES constraints, Atmos. Chem. Phys., 9, 4011–4021, doi:10.5194/acp-9-4011-2009, 2009.

Li, Z. Q., Xia, X. A., Cribb, M., Mi, W., Holben, B., Wang, P. C., Chen, H. B., Tsay, S. C., Eck, T. F., Zhao, F. S., Dutton, E. G., and Dickerson, R. E.: Aerosol optical properties and their radiative effects in northern China, J. Geophys. Res., 112, D22S01, doi:10.1029/2006JD007382, 2007.

Liao, H., Chen, W. T., and Seinfeld, J. H.: Role of climate change in global predictions of future tropospheric ozone and aerosols, J. Geophys. Res., 111, D12304, doi:10.1029/2005JD006852, 2006.

Perez, C., Nickovic, S., Pejanovic, G., Maldasano, J. M., and Ozsoy, E.: Interactive dust-radiation modeling: A step to improve weather forecast, J. Geophys. Res., 111, D16206, doi:10.1029/2005JD006717, 2006.

Pleim, J.: A combined local and non-local closure model for the atmospheric boundary layer. Part II: Application and evaluation in a mesoscale meteorological model. J. Appl. Meteor. Clim., 46, 1396–1409, 2007.

Ramanathan, V., Crutzen, P. J., Kiehl, J. T., and Rosenfeld, D.: Aerosols, Climate and the Hydrological Cycle, Science, 294, 2119–2124, 2001.

Santanello Jr., J. A., Friedl, M. A., and Kustas, W. P.: An empirical investigation of convective planetary boundary layer evolution and its relationship with the land surface, J. Appl. Meteor., 44, 917–932, 2005.

Shindell, D. and Faluvegi, G.: Climate response to regional radiative forcing during the twentieth century, Nat. Geosci., 2, 294–300, doi:10.1038/ngeo473, 2009.

Vogelezang, D. H. P. and Holtslag, A. A. M.: Evaluation and model impacts of alternative boundary-layer height formulations, Bound.-Lay. Meteorl., 81, 245–269, doi:10.1007/BF02430331, 1996.

Wang, J. and Christopher, A.: Mesoscale modeling of Central American smoke transport to the United States: 2. Smoke radiative impact on regional surface energy budget and boundary layer evolution, J. Geophys. Res., 111, D14S92, doi:10.1029/2005JD006720, 2006.

Wang, H., Gong, S. L., Zhang, H. L., Chen, Y., Shen, X. S., Chen, D. H., Xue, J. S., Shen, Y. F., Wu, X. J., and Jin, Z. Y.: A new-generation sand and dust storm forecasting system GRAPES_CUACE/Dust: Model development, verification and numerical simulation, Chin. Sci. Bull., 55, 635–649, doi:10.1007/s11434-009-0481-z, 2010a.

Wang, H., Zhang, X. Y., Gong, S., Chen, Y., Shi, G., and Li, W.: Radiative feedback of dust aerosols on the East Asian dust storms, J. Geophys. Res., 115, D23214, doi:10.1029/2009JD013430, 2010b.

Wang, H., Shi, G. Y., Zhu, J., Chen, B., Che, H., and Zhao, T. L.: Case study of longwave contribution to dust radiative effects over East Asia, Chin. Sci. Bull., 30, 3673–3681, doi:10.1007/s11434-013-5752-z, 2013.

Wang, H., Tan, S. C., Wang, Y., Jiang, C., Shi, G. Y., Zhang, M., and Che, H. Z.: A multisource observation study of the severe prolonged regional haze episode over eastern China in January 2013, Atmos. Environ., 89, 807–815, 2014.

Wang, H., Xue, M., Zhang, X. Y., Liu, H. L., Zhou, C. H., Tan, S. C., Che, H. Z., Chen, B., and Li, T.: Mesoscale modeling study of the interactions between aerosols and PBL meteorology during a haze episode in China Jing-Jin-Ji and its near surrounding region – Part 1: Aerosol distributions and meteorological features, Atmos. Chem. Phys., 15, 3257–3275, doi:10.5194/acp-15-3257-2015, 2015

Yu, H., Kaufman, Y. J., Chin, M., Feingold, G., Remer, L. A., Anderson, T. L., Balkanski, Y., Bellouin, N., Boucher, O., Christopher, S., DeCola, P., Kahn, R., Koch, D., Loeb, N., Reddy, M. S., Schulz, M., Takemura, T., and Zhou, M.: A review of measurement-based assessments of the aerosol direct radiative effect and forcing, Atmos. Chem. Phys., 6, 613–666, doi:10.5194/acp-6-613-2006, 2006.

Zhang, X. Y., Sun, J. Y., Wang, Y. Q., Li, W. J., Zhang, Q., Wang, W. G., Quan, J. N., Cao, G. L., Wang, J. Z., Yang, Y. Q., and Zhang, Y. M.: Factors contributing to haze and fog in China, Chin. Sci. Bull., 58, 1178–1187, doi:10.1360/972013-150, 2013 (in Chinese).

14

Dicarboxylic acids, ketocarboxylic acids, α-dicarbonyls, fatty acids and benzoic acid in $PM_{2.5}$ aerosol collected during CAREBeijing-2007: an effect of traffic restriction on air quality

K. F. Ho[1], R.-J. Huang[2,3], K. Kawamura[4], E. Tachibana[4], S. C. Lee[5], S. S. H. Ho[1,6], T. Zhu[7], and L. Tian[8]

[1]The Jockey Club School of Public Health and Primary Care, The Chinese University of Hong Kong, Shatin, Hong Kong SAR, China
[2]Key Laboratory of Aerosol Chemistry and Physics, Institute of Earth Environment, Chinese Academy of Sciences, Xi'an, China
[3]Laboratory of Atmospheric Chemistry, Paul Scherrer Institute, 5232 Villigen, Switzerland
[4]Institute of Low Temperature Science, Hokkaido University, Sapporo, Japan
[5]Research Center of Urban Environmental Technology and Management, Department of Civil and Structural Engineering, The Hong Kong Polytechnic University, Hun Hom, Kowloon, Hong Kong SAR, China
[6]Division of Atmospheric Sciences, Desert Research Institute, 2215 Raggio Pathway, Reno, NV 89512, USA
[7]State Key Laboratory of Environmental Simulation and Pollution Control, College of Environmental Sciences and Engineering, Centre for Environment and Health, Peking University, Beijing 100871, China
[8]School of Public Health, The University of Hong Kong, Hong Kong SAR, China

Correspondence to: K. F. Ho (kfho@cuhk.edu.hk) and R.-J. Huang (rujin.huang@psi.ch)

Abstract. Thirty water-soluble organic species, including dicarboxylic acids, ketocarboxylic acids, α-dicarbonyls, fatty acids and benzoic acid were determined as well as organic carbon (OC), elemental carbon (EC) and water-soluble organic carbon (WSOC) in $PM_{2.5}$ samples collected during the Campaign of Air Quality Research in Beijing 2007 (CAREBeijing-2007) in the urban and suburban areas of Beijing. The objective of this study is to identify the influence of traffic emissions and regional transport to the atmosphere in Beijing during summer. $PM_{2.5}$ samples collected with or without traffic restriction in Beijing are selected to evaluate the effectiveness of local traffic restriction measures on air pollution reduction. The average concentrations of the total quantified bifunctional organic compounds (TQBOCs), total fatty acids and benzoic acid during the entire sampling period were 1184 ± 241, 597 ± 159 and 1496 ± 511 ng m^{-3} in Peking University (PKU), and 1050 ± 303, 475 ± 114 and 1278 ± 372 ng m^{-3} in Yufa, Beijing. Oxalic acid (C_2) was found as the most abundant dicarboxylic acid at PKU and Yufa followed by phthalic acid (Ph). A strong even carbon number predominance with the highest level at stearic acid ($C_{18:0}$), followed by palmitic acid ($C_{16:0}$) was found for fatty acids. According to the back trajectories modeling results, the air masses were found to originate mainly from the northeast, passing over the southeast or south of Beijing (heavily populated, urbanized and industrialized areas), during heavier pollution events, whereas they are mainly from the north or northwest sector (mountain areas without serious anthropogenic pollution sources) during less pollution events. The data with wind only from the same sector (minimizing the difference from regional contribution) but with and without traffic restriction in Beijing were analyzed to evaluate the effectiveness of local traffic restriction measures on the reduction of local air pollution in Beijing. The results suggested that the traffic restriction measures can reduce the air pollutants, but the decrease of pollutants is generally smaller in Yufa compared to that in PKU. Moreover, an enhancement of EC value indicates more elevated primary emissions in Yufa during restriction periods than in non-restriction periods. This study demonstrates that even when primary exhaust was controlled by traffic restriction, the contribution of secondary organic species formed from photochemical processes was critical with long-range atmospheric transport of pollutants.

1 Introduction

Organic aerosol (OA) typically constitutes 20–90 % of submicron aerosol (Huang et al., 2014; Jimenez et al., 2009) and is influencing Earth's climate directly by absorbing and scattering radiation and indirectly by acting as cloud condensation. OA also adversely affects air quality and human health. However, uncertainties exist in the effect of OA on health and climate, due in large part to the complexity of OA composition (Hallquist et al., 2009; Hoffmann et al., 2011; Pöschl, 2005). OA can have primary emissions or secondary production through photochemical reactions of gas-phase precursors. Due to polar functional groups formation (e.g., carbonyl, carboxyl and hydroxyl), a major fraction of the secondary organic aerosol (SOA) is thought to be water soluble which, together with some water-soluble primary organic aerosol (POA), accounts for about 40–80 % of the OA (Jaffrezo et al., 2005; Saxena and Hildemann, 1996).

Despite the dominant presence of water-soluble organic carbon (WSOC) in the atmosphere, there exist large uncertainties associated with sources, the chemical composition, removal mechanisms and atmospheric formation processing of aerosol WSOC. This is particularly evident in polluted megacities where multiple sources of local and regional origins may significantly change the chemical and physical properties of aerosol and therefore influence air quality, climate and human health. Dicarboxylic acids (diacids) are the most abundant organic compounds in OA, which can be derived from primary emissions and/or secondary formation from different precursor species via photochemical reactions (Glasius et al., 2000; Kawamura et al., 1996; Kundu et al., 2010; Legrand et al., 2007). Fossil fuel combustion and biomass burning (Falkovich et al., 2005; Ho et al., 2006; Huang et al., 2014; Kundu et al., 2010) are the major primary sources, whereas photochemical oxidation of volatile organic compounds (VOCs) from biogenic and anthropogenic emissions (Kawamura et al., 1996; Mkoma and Kawamura, 2013) is the major secondary sources.

Beijing is one of the largest metropolitan cities in Asia and has become a heavily polluted area due to the fast urbanization and industrialization over the past 2 decades. In 2009, more than 17.5 million residents and 4.0 million vehicles were reported in Beijing (BMBS, 2010). Besides local emissions, the air flowing into Beijing from polluted neighboring regions can have a significant impact on the air quality in Beijing (Hatakeyama et al., 2005; Luo et al., 2000; Mauzerall et al., 2000). Especially, the gas-to-particle partitioning of semi-volatile organic compounds (SVOCs) and their subsequent aging via photochemical processing during transport has been recognized to be a major air pollution source (Ding et al., 2008; Guttikunda et al., 2005). Atmospheric aerosols have been investigated extensively in China (An et al., 2007; Cao et al., 2003; Huang et al., 2014; Xu et al., 2008). However, relevant studies on organic acids are still very scarce. With such limited information available on organic acids despite the rapid urbanization and development (especially the increase in traffic density), it is essential to seek a better understanding of organic acids in Beijing. For the promised green Olympic games in 2008, many pollution control measures, such as controlling traffic, halting industrial/construction activities, and sweeping roads, were taken to improve the air quality. The traffic restriction measures, which only allowed vehicles to be on road in alternative business days according to their even and odd plate numbers, were proposed to reduce air pollution.

To investigate the effects of the traffic restriction on the air quality of Beijing and to accumulate experience and scientific evidence for the preparation of the 2008 Olympic games, we conducted aerosol ($PM_{2.5}$) monitoring at two sites in Beijing during 3–31 August 2007. In this study, $PM_{2.5}$ samples collected were analyzed by a gas chromatography flame ionization detector (GC-FID) and a gas chromatography mass spectrometry (GC-MS) to determine the composition of low molecular weight (MW) diacids (C_2–C_{12}), ketocarboxylic acids (ωC_2–ωC_9, pyruvic acid), α-dicarbonyls (C_2–C_3), benzoic acid and fatty acids (C_{12}–C_{25}). Moreover, organic carbon (OC), elemental carbon (EC), and WSOC were also analyzed. Through the intensive sampling campaign, the roles of regional transport, local emissions and secondary formations of particulate matter in the atmosphere of Beijing were investigated.

2 Experiment

2.1 $PM_{2.5}$ sampling

Two sampling locations, Peking University (PKU) (39.98° N, 116.35° E) and Yufa, Beijing (39.51° N, 116.31° E), were selected in this study. The detailed descriptions of the sampling locations were reported elsewhere (Ho et al., 2010). The air samplers were placed on the top floor of the buildings (PKU: a six-story building; Yufa: a four-story building). The meteorological data such as wind speed, wind direction, relative humidity and temperature were collected during the sampling period. The north and northwest of PKU are enclosed by mountains, whereas the south and southeast of Yufa are surrounded by heavily industrialized and urbanized areas such as Hebei province and Tianjin city.

Pre-heated (800 °C, 3 h) quartz-fiber filters (47 mm QM-A Whatman quartz filters) were used to collect 24 h integrated $PM_{2.5}$ samples by Airmetrics mini-volume $PM_{2.5}$ samplers at a flow rate of 5 L min^{-1}. A DryCal® flow meter (BIOS International, Butler, NJ, USA) was used to calibrate the sampling flows before and after the sampling. Sampling was carried out simultaneously from 09:00 onwards in a 24 hours interval at the two sampling locations from 3 to 31 August 2007. The samples were properly kept in a freezer (−20 °C) to prevent evaporation of semi-volatile components and microbial degradation of organics.

2.2 Chemical analysis

OC and EC were analyzed (on a 0.526 cm^2 punch) by thermal analysis with optical detection following the IMPROVE protocol on a Desert Research Institute (DRI) Model 2001 Thermal/Optical Carbon Analyzer (Atmoslytic Inc., Calabasas, CA, USA) (Cao et al., 2003; Chow et al., 2005). The method detection limit (MDL) of OC and EC analysis is 0.8 and 0.4 μg C cm^{-2}, respectively. To determine the WSOC, a total area of 2.63 cm^2 of the sample filter was cut from each filter and 5 mL of Milli-Q water (18 MΩ) was added into a 15 mL vial where the sample was placed. An ultra-sonic water bath was used to extract the particles on the filter for 1 h. Syringe filters (0.2 μm PTFE membrane) were used to remove the insoluble particles from the extracts. Filtered extract was then transferred into clean vials and analyzed total organic carbon (TOC) by using a Shimadzu TOC-V CPH Total Carbon Analyzer (Columbia, MD, USA). The MDL is 0.01 μg C m^{-3}, with a precision of ±5 %. The data reported in this study were all corrected by the blanks.

The analytical procedures for water-soluble organic species were well reported elsewhere (Kawamura and Yasui, 2005). Briefly, the sample was extracted with organic-free water (10 mL × 3) to isolate bifunctional organic compounds as well as fatty acids and benzoic acid. After the extracts were concentrated using a vacuum rotary evaporator, 14 % BF_3/*n*-butanol was added at 100 °C to convert the aldehyde groups to dibutoxy acetals and carboxyl groups to butyl esters. Homologous series of fatty acids were analyzed as butyl esters (Mochida et al., 2007). No serious contamination (< 5 % of real samples) was observed in our analysis. The data reported in this study were corrected by the field blanks. The derivatized samples were determined by a Agilent 6890GC/FID (Palo Alto, CA, USA) equipped with a split/splitless injector, HP-5 fused silica capillary column (25 m × 0.2 mm i.d. × 0.5 μm film thickness) and an FID detector. The retention time of authentic standard is the parameter of peak identification. ThermoQuest Trace MS (Austin, TX, USA) with a similar GC condition was used for mass spectral confirmation of the compounds. The reproducibility of the methods was < ±15 %; recoveries of the bifunctional organic compounds fatty acids and benzoic acid were > 70 % (Kawamura and Yasui, 2005; Mochida et al., 2007). Field blanks concentrations were < 15 % of real samples, except for phthalic acid (up to 30 %). The results shown in this study were all corrected by the field blanks.

3 Results and discussion

3.1 Overview of molecular compositions of bifunctional organic compounds in PKU and Yufa

Average OC, EC and WSOC concentrations in PKU and Yufa are illustrated in Table 1 and their levels during the entire sampling period were 14.9 ± 2.47, 6.21 ± 1.90 and 5.59 ± 1 49 μg C m^{-3} in PKU, and 11.1 ± 3.68, 5.6 ± 1.83 and 4.55 ± 1.79 μg C m^{-3} in Yufa. The WSOC accounted for 37 ± 7 and 40 ± 7 % of OC in PKU and Yufa, respectively. It was consistent with the WSOC / OC ratios (20–40 %) at other metropolitan cities (Ho et al., 2007; Yang et al., 2005), suggesting that WSOC is one of the main components in OA in China. Yufa is located in southern Beijing, which is close to the border of Beijing Municipality and Hebei province. Regional pollution from heavy industrialized and urbanized areas, like Hebei province and Tianjin city, has a great impact on the air quality of Yufa area.

The concentrations of bifunctional organic compounds measured in PKU and Yufa are presented in Table 1. The concentrations of total quantified bifunctional organic compounds (TQBOC) varied from 730 to 1455 ng m^{-3} (average concentration: 1184 ± 241 ng m^{-3}) in PKU, and from 554 to 1621 ng m^{-3} (average concentration: 1050 ± 303 ng m^{-3}) in Yufa. The results are higher than measurements (average 813 ng m^{-3} in PKU; average 771 ng m^{-3} in Yufa) reported in 2006 in same sampling locations (Ho et al., 2010), reflecting that there were continuous increases of primary emissions and more aging of aerosols in Beijing. However, the concentrations are close to other megacities studied recently (Ho et al., 2007).

Oxalic acid (C_2) was the most abundant diacid (435 ± 124 and 418 ± 130 ng m^{-3} at PKU and Yufa, respectively) determined in this study, followed by phthalic acid (Ph) (209 ± 28.8 and 176.3 ± 91.5 ng m^{-3}), and succinic acid (C_4) (89.9 ± 27.7 and 80.9 ± 26.9 ng m^{-3}). These three species accounted for 65 % of TQBOC in PKU and Yufa. Oxalic acid was also recognized as the predominant diacid in previous studies in China (Ho et al., 2010, 2011). C_2 can be either released from combustion processes (e.g., fossil fuel and biomass burning) (Kawamura and Kaplan, 1987; Narukawa et al., 1999) or produced through the secondary oxidation of VOCs (Carlton et al., 2006; Warneck, 2005).

The average Ph concentrations measured in this study are higher than those reported by other studies (Ho et al., 2007; Wang and Kawamura, 2005). Three phthalic acids (phthalic acid, *o*-isomer; terephthalic acid, *p*-isomer; and isophthalic acid, *m*-isomer) were determined and these isomer species distribution was dominated by *o*-isomer, followed by *p*-isomer and *m*-isomer, which are consistent with studies measured in Mt. Tai, China, and Pearl River Delta region (Fu et al., 2008; Ho et al., 2011). The abundant Ph can be released from incomplete combustion processes or formed by secondary oxidation of aromatic compounds (e.g., naphthalenes,

NAP) (Kawamura and Kaplan, 1987; Kawamura and Yasui, 2005). In some previous studies, high levels of NAP were found in Beijing urban areas (Liu et al., 2007; Tao et al., 2007) and it can be thus one of the potential precursors to Ph formation (Ho et al., 2007).

Besides diacids (C_2–C_4), azelaic acid (C_9) was the most abundant species among the saturated diacids in both sampling locations ($71.4 \pm 8.91\,ng\,m^{-3}$ in PKU, $49.2 \pm 8.99\,ng\,m^{-3}$ in Yufa). C_9 is recognized as a photochemical reaction product of biogenic unsaturated fatty acids, such as oleic ($C_{18:1}$) and linoleic ($C_{18:2}$) acids (the first number is the carbon number and the latter refers to the amount of the double bond) (Kawamura and Gagosian, 1987) and is generally abundant in the high molecular weight homologues. The unsaturated fatty acids are commonly determined in marine micro-organism or higher plant leaves. However, these acids could be released by meat charbroiling also (Rogge et al., 1991). During long-range transport, photochemical oxidation of $C_{18:1}$ to C_9 via oxidants (e.g., ozone and/or OH radicals) may occur in the air (Stephanou and Stratigakis, 1993). The $C_9 / C_{18:1}$ ratio determined was lower in Yufa (average value: 2.1) than that in PKU (average value: 5.1) suggesting that significant secondary production of C_9 occurred in urban areas of Beijing.

Diacids can be formed when ketocarboxylic acids, which are regarded as intermediate product of mono-carboxylic acids oxidation, react with other pollutants in the air (He et al., 2013; Kawamura et al., 1996). The concentrations of total measured ketocarboxylic acids varied from 87.4 to $169\,ng\,m^{-3}$, (average value: $122 \pm 28.8\,ng\,m^{-3}$) in PKU and from 52.0 to $131\,ng\,m^{-3}$ (average value: $97.0 \pm 22.9\,ng\,m^{-3}$) in Yufa (Table 1). The concentrations in both sampling sites are higher than those measured in a rural site in Gosan, South Korea ($53\,ng\,m^{-3}$), and megacities in China (summer: $37\,ng\,m^{-3}$) (Ho et al., 2007, 2010; Kawamura et al., 2004). These results reveal that the organic aerosols in PKU and Yufa were likely more photochemically aged than that in other urban sites caused by photochemical reaction during transportation (He et al., 2013). Glyoxylic acid (ωC_2) was found as the most abundant ketocarboxylic acid, followed by pyruvic acid (Pyr). Their concentration levels are similar to previous measurement in Tokyo, Japan (Kawamura and Yasui, 2005).

Concentrations of total measured α-dicarbonyls varied from 35.5 to $99.5\,ng\,m^{-3}$ (average value: $51.8 \pm 17.9\,ng\,m^{-3}$) in PKU and from 29.0 to $61.4\,ng\,m^{-3}$ (average value: $44.2 \pm 10.3\,ng\,m^{-3}$) in Yufa. The two simplest α-dicarbonyl compounds (glyoxal and methylglyoxal) have recently attracted much attention as potential SOA precursors. These compounds are formed by photochemical oxidation of both biogenic (e.g., isoprene and terpenes) and anthropogenic VOCs (e.g., toluene, xylene) (Fick et al., 2004; Volkamer et al., 2001). They have been identified as significant precursors in the heterogeneous processes for SOA formation (Kroll et al., 2005). High concentrations of glyoxal and methylglyoxal observed indicate the greater SOA formation potential in this region. α-Dicarbonyls levels measured in PKU and Yufa were higher than previous results in other cities of China (average value: $12\,ng\,m^{-3}$) (Ho et al., 2007). It indicates that biogenic sources such as oxidation of isoprene are more important than other urban cities in China.

3.2 Overview of molecular compositions of fatty acids and benzoic acid in PKU and Yufa

Table 1 presents the average concentrations of straight chain saturated fatty acids ($C_{12:0}$–$C_{25:0}$), unsaturated fatty acid and benzoic acid. Total measured fatty acid concentrations varied from 459 to $1003\,ng\,m^{-3}$ (average value: $597 \pm 159\,ng\,m^{-3}$) in PKU and from 375 to $684\,ng\,m^{-3}$ (average value: $475 \pm 114\,ng\,m^{-3}$) in Yufa. The distributions of fatty acids were dominated by even carbon numbers with a maximum at stearic acid ($C_{18:0}$) followed by palmitic acid ($C_{16:0}$). This finding is consistent with previous measurements reported in megacities of China (Fu et al., 2008; Ho et al., 2010). Both natural biogenic and anthropogenic emissions represent the major sources of fatty acids, whereas homologues $< C_{20}$ are partially released from microbial sources (Simoneit and Mazurek, 1982). Additionally, low MW fatty acids ($< C_{18}$) can be emitted by tire wear debris and traffic exhaust. Biomass burning also produces high fractions of fatty acids which are the major components of plant tissues and surface waxes. $C_{16:0}$ and $C_{18:0}$ were also the major organic compounds emitted from cooking meat (Schauer et al., 1999, 2002; Zhao et al., 2007a, b). Higher concentrations of fatty acids observed at PKU can be explained by the mixed contributions of regional and local emissions in the urban area. Interestingly the contributions of total quantified fatty acids to OC are similar in both sites (3.1 % in PKU and 3.2 % in Yufa, respectively).

The even-over-odd carbon number preference in fatty acid ($C_{12:0}$ to $C_{25:0}$) is measured by the carbon preference index (CPI):

$$CPI_{\text{fatty acid}} = \frac{\Sigma \text{ even carbon number fatty acids}}{\Sigma \text{ odd carbon number fatty acids}}.$$

CPI is a measure to differentiate anthropogenic and biogenic sources; the values are 43.3 in PKU and 45.9 in Yufa. High CPI values observed in this study indicate that biological sources such as vascular plants have significant influence in this region (Simoneit, 1984).

In this study, $C_{18:1}$ was detected in all samples which can be directly emitted from higher plants and soils. In urban areas, biomass burning and cooking are likely to be the main anthropogenic sources for this acid (Rogge et al., 1993). Its concentrations varied from 2.9 to $33.0\,ng\,m^{-3}$ (average value: $24.3 \pm 8.9\,ng\,m^{-3}$) and from 13.0 to $47.9\,ng\,m^{-3}$ (average value: $24.6 \pm 9.2\,ng\,m^{-3}$) in PKU and Yufa, respectively. Oleic acid is a good tracer for unsaturated organic aerosol and a representative compound for the reac-

Table 1. Concentrations of dicarboxylic acids, ketocarboxylic acids and α-dicarbonyls, fatty acids and benzoic acid in $PM_{2.5}$ samples during CAREBeijing 2007.

Compounds ($ng\ m^{-3}$)	PKU ($n = 10$) Range	Average	SD	Yufa ($n = 10$) Range	Average	SD
Dicarboxylic acids						
Oxalic, C2	212–586	435	124	226–632	418	130
Malonic, C3	30.0–73.5	54.9	14.0	17.1–68.6	43.5	15.0
Succinic, C4	52.8–147	89.9	27.7	44.8–129	80.9	26.9
Glutaric, C5	13.7–59.2	36.0	14.4	16.0–168	41.3	46.2
Adipic, C6	15.1–35.1	26.7	6.03	10.8–73.1	24.5	18.1
Pimeric, C7	MDLs–6.44	2.79	2.56	MDLs–3.38	0.34	1.07
Suberic, C8	MDLs	MDLs	–	MDLs	MDLs	–
Azelaic, C9	58.8–85.8	71.4	8.91	37.5–64.6	49.2	8.99
Sebacic, C10	MDLs–3.91	0.69	1.47	MDLs-2.78	0.28	0.88
Undecanedioic, C11	MDLs	MDLs	–	MDLs	MDLs	–
Dodecanedioic, C12	MDLs	MDLs	–	MDLs	MDLs	–
Methylmalonic, iC4	MDLs	MDLs	–	MDLs	MDLs	–
Methylsuccinic, iC5	MDLs-10.6	3.82	2.75	MDLs–5.21	3.05	2.22
2-Methylglutaric, iC6	MDLs–10.5	6.09	2.79	MDLs–8.92	4.80	2.40
Maleic, M	9.18–20.5	15.6	3.39	9.69–17.2	13.5	2.03
Fumaric, F	MDLs	MDLs	–	MDLs	MDLs	–
Methylmaleic, mM	7.52–12.2	9.60	1.40	5.75–8.83	7.21	0.92
Phthalic, Ph	171–250	209	28.9	80.6–415	176	91.5
Isophthalic, iPh	MDLs–15.1	7.27	5.76	MDLs–12.8	4.65	5.22
Terephthalic, tPh	12.2–51.8	30.4	10.1	6.83–53.2	30.9	13.8
Malic, hC4	MDLs–3.21	0.32	1.02	MDLs	MDLs	–
Ketomalonic, kC3	2.79–11.8	6.56	2.79	MDLs–10.4	6.67	3.07
4-Ketopimelic, kC7	MDLs–13.0	3.71	5.06	MDLs–11.0	3.21	4.33
Total diacids	599–1287	1010	220	473–1429	909	278
Ketocarboxylic acids						
Pyruvic	17.9–70.2	30.3	15.9	12.7–27.3	20.1	4.18
Glyoxylic, ωC2	49.2–105	72.9	16.1	32.8–81.7	59.7	14.5
3-Oxopropanoic, ωC3	MDLs–3.29	1.53	1.23	MDLs–1.31	0.59	0.62
4-Oxobutanoic, ωC4	5.91–25.0	15.8	5.55	6.59–23.3	15.0	4.83
9-Oxononaoic, ωC9	MDLs–7.28	1.70	2.50	MDLs–5.03	1.61	2.19
Total ketoacids	87.4–169	122	28.8	52.0–131	97.0	22.9
α-Dicarbonyls						
Glyoxal, Gly	1.40–21.4	13.1	5.72	2.93–17.3	11.5	4.04
Methylglyoxal, MeGly	23.3–81.3	38.7	16.2	21.3–51.3	32.7	8.76
Total dicarbonyls	35.5–99.5	51.8	17.9	29.0–61.4	44.2	10.3
Sum of bifunctional species	730–1455	1184	241	554–1621	1050	303
Fatty acids						
Tridecanoic acid, $C_{13:0}$	5.08–16.4	10.8	3.92	MDLs–11.8	7.56	3.31
Tetradecanoic acid, $C_{14:0}$	54.5–97.9	68.7	13.6	36.5–74.0	50.0	13.6
Hexadecanoic acid, $C_{16:0}$	199–393	249	57.0	149–310	201	48.9
Heptadecanoic acid, $C_{17:0}$	MDLs–13.3	4.32	4.53	MDLs–12.8	3.90	5.06
Octadecanoic acid, $C_{18:0}$	134–462	219	94.0	122–237	170	36.6
Octadecenoic acid, $C_{18:1}$	2.91–33.0	24.3	8.93	13.0–47.9	24.6	9.23
Eicosanoic acid, $C_{20:0}$	MDLs–7.84	4.01	3.51	MDLs–7.71	3.28	3.50
Docosanoic acid, $C_{22:0}$	5.69–13.6	9.24	2.43	MDLs–15.4	7.01	4.95
Tetracosanoic acid, $C_{24:0}$	MDLs–10.5	6.51	3.59	MDLs–12.1	7.11	3.09
Sum of fatty acids	459–1003	597	159	375–684	475	114
Benzoic acid	933–2754	1496	511	724–1930	1278	372
$PM_{2.5}$ ($\mu g\ m^{-3}$)	64.9–191	103	35.1	38.6–160	98.3	34.1
OC ($\mu g\ m^{-3}$)	12.1–19.3	14.9	2.47	4.63–17.0	11.1	3.68
EC ($\mu g\ m^{-3}$)	4.48–11.0	6.21	1.90	2.96-8.62	5.57	1.83
WSOC ($\mu g\ m^{-3}$)	2.69–8.20	5.59	1.49	1.72–7.16	4.55	1.79

MDL: method detection limit

tivity model (Rudich et al., 2007). The diagnostic ratio of $C_{18:1}$ / $C_{18:0}$ was used to determine the level of aerosol aging in this study. Low values indicate that the air masses are more aged. The ratios in PKU and Yufa were 0.12 and 0.14, respectively, which suggests that unsaturated fatty acids are depleted by the enhanced photochemical degradation in PKU (Wang et al., 2006). Moreover, the diagnostic ratio of $C_{18:0}$ / $C_{16:0}$ was applied as an indicator for source evaluation. Low ratios observed (<0.25) in $PM_{2.5}$ likely originated from wood smoke, waxy leaf surface abrasions and foliar vegetation combustion; ratios that ranged between 0.25 and 0.5 were indicated for vehicle exhausts; while ratios that ranged between 0.5 and 1 were obtained from hamburger charbroiling and paved/unpaved road dust (Oliveira et al., 2007; Rogge et al., 2006). The $C_{18:0}$ / $C_{16:0}$ ratios observed in this study had a range between 0.64 and 1.17 (average value: 0.85 in both locations) in PKU and Yufa, indicating that the contribution of cooking emissions and paved/unpaved road dust cannot be ruled out.

Almost all $PM_{2.5}$ samples collected contained benzoic acid which has been identified as a direct pollutant from the traffic emissions (Kawamura et al., 1985) and a indirect pollutant produced from photo-degradation of aromatic compounds (e.g., toluene) released from traffic exhausts (Suh et al., 2003). The average benzoic acid concentrations were 1496 ± 511 ng m^{-3} in PKU and 1278 ± 372 ng m^{-3} in Yufa. Moreover, benzoic acid is a semi-volatile organic species and is mainly found in gas phase (Fraser et al., 2003), it can be formed in particulate phase via gas-to-particle partitioning. During an ozone episode in August 2006, a high concentration of toluene was determined in Beijing (11.4μg m^{-3}) (Duan et al., 2008), which suggests that oxidation of toluene is one of the significant sources of benzoic acid in the air.

3.3 Significance of pollution events

Figure 1a and b show the temporal variation of mass concentrations of EC, OC and WSOC in PKU and Yufa, respectively, from 3 to 31 August 2007. Heavier air pollution events were observed during 3, 5, 9, 15 and 31 August, as reflected by the elevated $PM_{2.5}$ concentrations (i.e., range 96–191 μg m^{-3}, average 124μg m^{-3} in PKU and range 100–127μg m^{-3}, average 110μg m^{-3} in Yufa). The concentrations of OC, EC and WSOC significantly increased during these pollution events, but generally decreased for the less polluted air mass events on 7, 13, 21 and 27 August, consistent with lower $PM_{2.5}$ concentrations (i.e., a range of 65–77μg m^{-3}, average 71 μg m^{-3} in PKU and a range of 39–179μg m^{-3}, average 62μg m^{-3} in Yufa). Similar temporal variations in TQBOC and fatty acids were observed in both PKU and Yufa (Fig. 1c and d). However, the temporal variation of benzoic acid is different from the other compounds measured, indicating a different source or atmospheric processing for benzoic acid.

Air mass back trajectory analysis shows that the heavy pollution events were related to trajectories from the northeast, passing over the southeast or south of Beijing, whereas trajectories from the north or northwest sector were related to less pollution air (see Fig. 2). Areas south and southeast of Beijing are located close to heavily industrialized areas (e.g., Tianjin city, Shandong and Hebei province), whereas areas north and northwest of Beijing are enclosed by the massive mountain ranges with no impact from anthropogenic pollution sources (Ho et al., 2010). As seen in Fig. 3a and b, the concentration levels of EC, OC, WSOC, diacids and ketocarboxylic acids in PKU and Yufa are higher for heavier pollution episodes compared to the less polluted air events, suggesting that high emission of carbonaceous aerosols and their precursor gases from neighboring provinces and the subsequent transport to Beijing is one of the major sources responsible for the elevated particulate pollutants in Beijing.

The OC to EC ratio (OC / EC) was used to estimate the transformation and emission properties of carbonaceous aerosol. The average OC / EC ratios at less polluted air (PKU: 2.63; Yufa: 2.19) events were slightly higher than those found at the pollution episodes (PKU: 2.52; Yufa: 2.05) at both sites. The slightly lower OC / EC ratio during pollution episodes is likely associated with high combustion emissions, especially from traffic exhaust. The slightly higher OC / EC ratios observed during less polluted air events suggest that secondary formation of OA was critical during less polluted air events. Bendle et al. (2007) reported that the unsaturated to oversaturated C_{18} fatty acids ($C_{18:n}$ / $C_{18:0}$) ratio could be used as a good indicator to estimate the freshness of organic matter (OM) in marine samples. In this study, high ratios were recorded in samples associated with pollution episodes, whereas low ratios were observed in less polluted air events with air masses originating from the north and northwest of Beijing. Low ratios observed in less polluted air events represent an aged air mass, indicating longer residence time for particle transformation and transportation (Alves et al., 2007).

Moreover, malonic acid (C_3) can be a byproduct of photochemical breakdown of succinic acid (C_4) in the air. The C_3 / C_4 ratio, which was used as a tracer of the enhanced photochemical aging of OA (Kawamura and Ikushima, 1993), observed during less polluted air events was higher than pollution episodes in both sites (0.66 vs. 0.58 in PKU and 0.57 vs. 0.52 in Yufa). Higher C_3 / C_4 ratios in less polluted air events suggest that secondary formation of diacids are more significant in less polluted air events, which further indicates secondary photochemical formation of particulate diacids is also critical during less polluted air events.

It should be noted, however, that the concentrations of α-dicarbonyls and benzoic acid in both PKU and Yufa are higher during less polluted air episodes compared to pollution episodes. This indicates that local production or secondary formation could be an important source for these compounds. It is known that α-dicarbonyls are intermediate

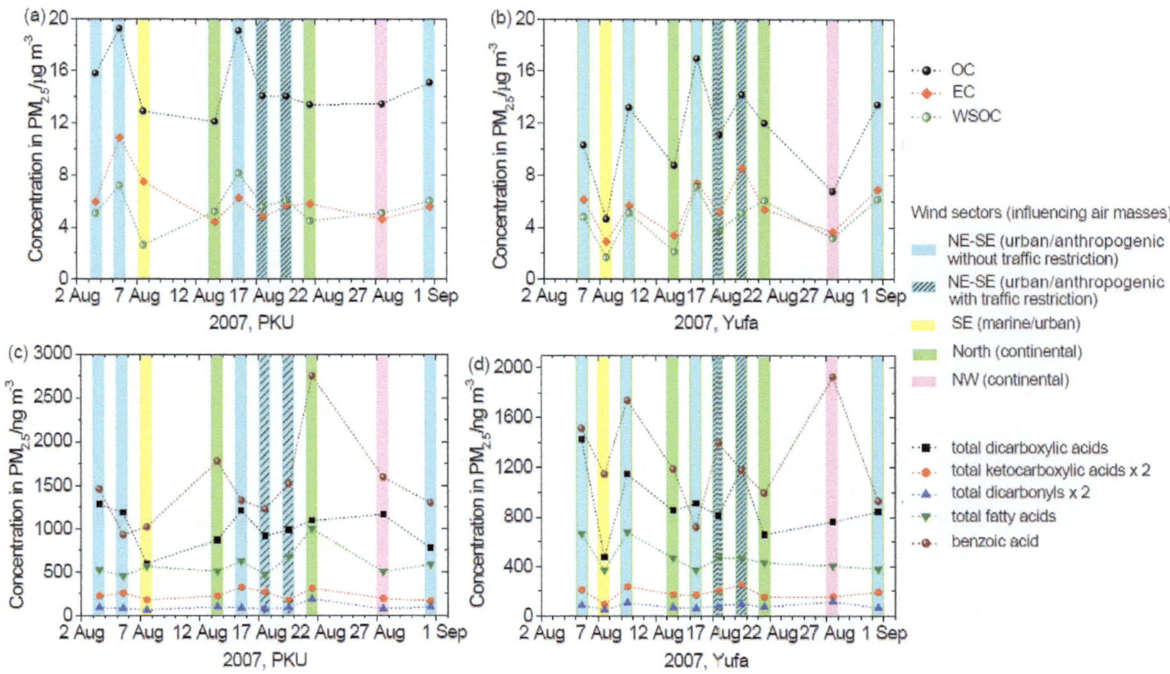

Figure 1. Temporal variations of EC, OC, WSOC and various organic compound classes at PKU and Yufa during CAREBeijing-2007 (note the difference in start time at PKU and Yufa, i.e., the first 3 days).

reaction products (via photochemical oxidation) of a wide range of biogenic and anthropogenic VOCs (Galloway et al., 2009). More distant sources lead to longer transport time and therefore increased chemical oxidation of glyoxal and methylglyoxal to their corresponding acids and other reaction products. This potentially reduces the local contribution of α-dicarbonyls in Beijing. Positive correlation was observed between α-dicarbonyls and benzoic acid ($R^2 = 0.82$ in PKU and $R^2 = 0.65$ in Yufa) at both sites (Fig. 4a and b), which further suggests that a major fraction of α-dicarbonyls and benzoic acid are most likely produced in the local atmosphere of Beijing through photochemical processing.

3.4 Influence of local traffic on air quality between restriction and non-restriction periods

One goal of this sampling campaign is to study the traffic controls influence on the air quality in Beijing given the use of a large number of vehicles and the resulting high emission of particulate matter and precursor gases. As described above, the level of particulate pollutants in Beijing is significantly influenced by regional transport depending on the wind sector. Therefore, in the following discussion, only events with wind from the same sector (minimizing the difference from regional contribution) but with and without traffic restriction in Beijing are selected to evaluate the effectiveness of local traffic restriction measures on air pollution reduction. Measurements taken on 17 and 19 August represent the restriction events and those taken on 3, 5, 9, 15 and 31 August represent the non-restriction events.

The concentration ratios of the restriction to the non-restriction periods (R / N) are shown in Fig. 5. A value of close to unity shows that the restriction does not have any impact in the pollution controls. In PKU, the R / N ratios of EC, OC, WSOC, total diacids, total ketocarboxylic acids and total α-dicarbonyls are much lower than 1, suggesting that these pollutants or their precursors are closely related to the traffic emissions and that the traffic restriction measures can reduce primary pollutants (e.g., EC) and the precursors of secondary pollutants (e.g., diacids and α-dicarbonyls). A previous study (Zhang et al., 2011) also indicated the reduction of anthropogenic elements in Beijing during the traffic restriction period of August, 2007. The average OC / EC ratios observed during traffic restriction periods (PKU: 2.69) were slightly higher than that found at non-restriction periods (PKU: 2.52). The slightly lower OC / EC ratio during non-restriction period was mainly due to the higher EC emissions from traffic exhaust, while EC emissions were reduced during traffic restriction periods. However, the R / N ratios of benzoic acid and total fatty acids are higher than 1. A possible explanation for this elevated R / N ratio is that these organics are mainly derived from regional emissions. An alternative is that they are mainly produced from sources other than vehicle emissions. For example, cooking emissions that were not controlled under the traffic restriction period are a significant source of fatty acids in the air. More household

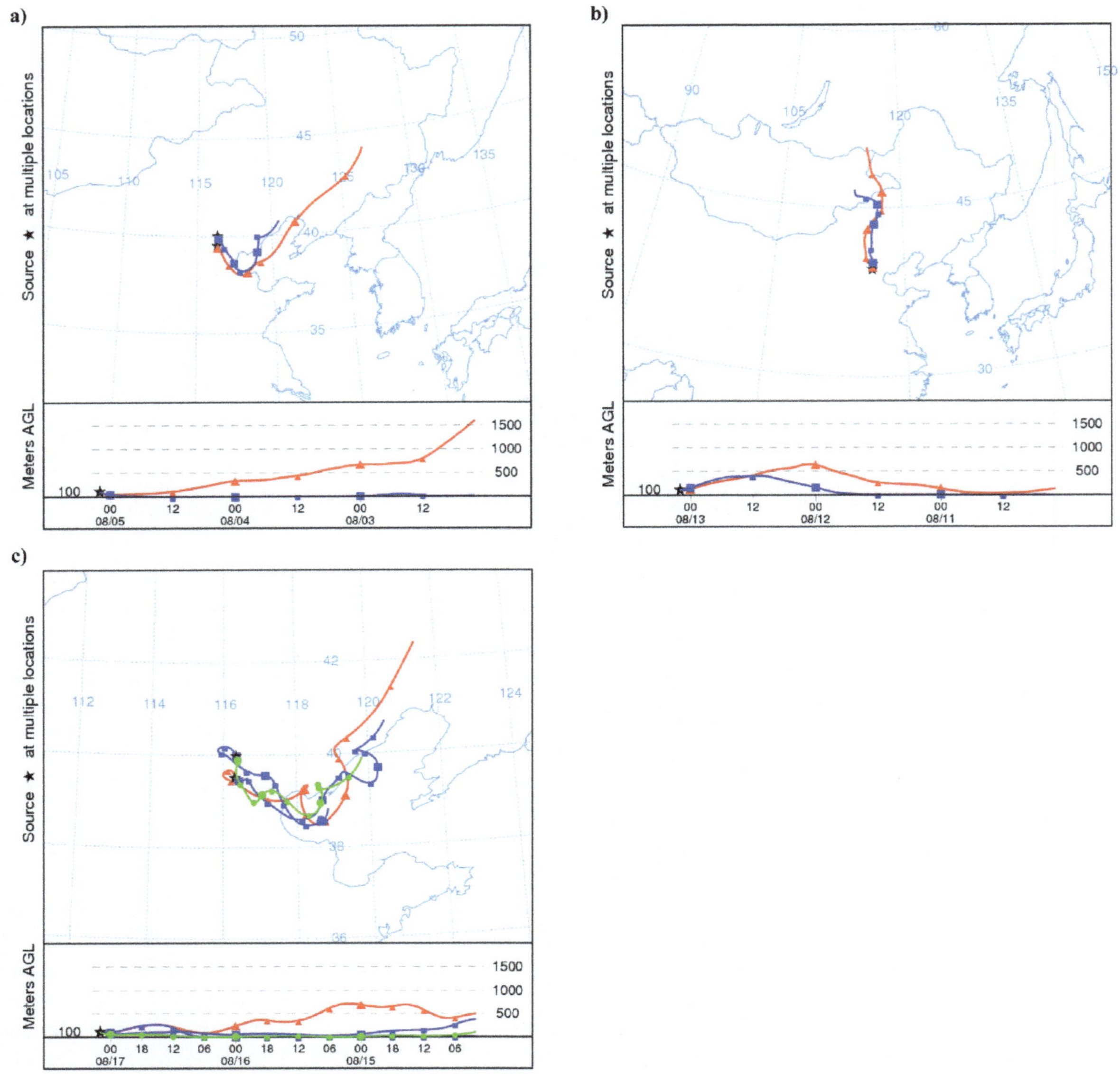

Figure 2. Three-day air mass back trajectories on **(a)** 05 August, **(b)** 13 August and **(c)** 17 and 19 August.

cooking activities can be found when the residents tended to stay home during the restriction period.

The profile of the R / N ratio in Yufa is different from that in PKU. The concentrations of OC, WSOC, total diacids and total fatty acids were lower during the restriction period than those during the non-restriction period, suggesting that the traffic restriction measures indeed reduced particulate pollutants. However, the decrease is generally smaller in Yufa compared to that in PKU, indicating that the contribution of local traffic emission to air pollution in Yufa is smaller. The R / N value > 1 occurred with EC, total ketocarboxylic acids, total α-dicarbonyls and benzoic acid. An enhanced EC value indicates more elevated primary emissions in Yufa during the restriction period than the non-restriction period. The potential contribution could be local rural emissions (e.g., biomass burning and coal burning) and/or regional transport from polluted neighboring provinces that are closer to Yufa. The average OC / EC ratios at traffic non-restriction period (Yufa: 2.05) events were slightly higher than those found at restriction period (Yufa: 1.89) events. The lower OC / EC ratios during restriction period further suggest the elevated emissions of EC from sources other than traffic at Yufa.

3.5 Ratios of selected species

The C_3 / C_4 ratios measured in this study varied from 0.28 to 0.84 (average value: 0.59), which are close to those measured in northern China (0.61) (Ho et al., 2007), but higher than that observed from traffic exhausts (0.3–0.5) (Kawamura and Kaplan, 1987). However, the ratios determined in this study are much lower than the marine particles measured

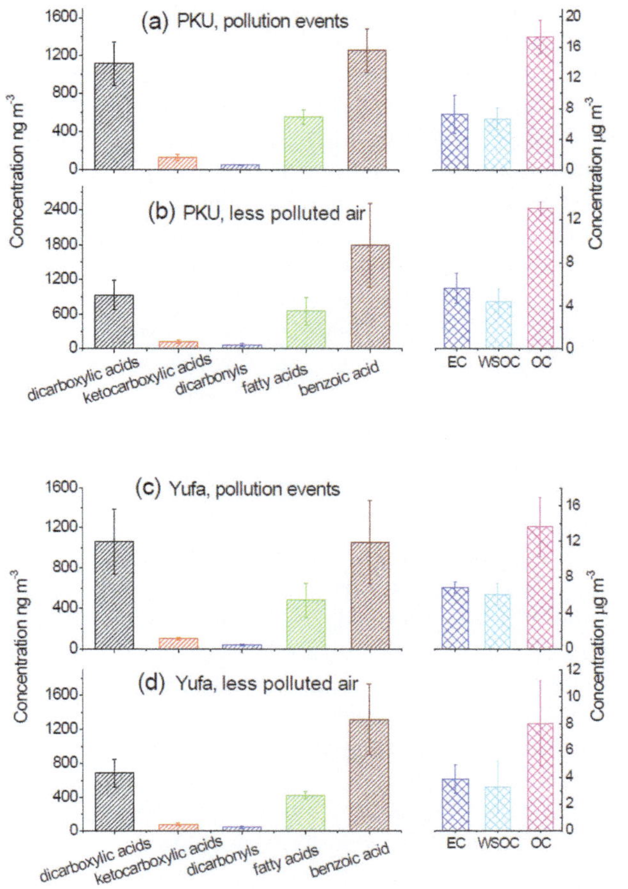

Figure 3. Pollution events vs. less polluted air at PKU (**a** and **b**) and Yufa (**c** and **d**), showing the variation of particulate pollutants.

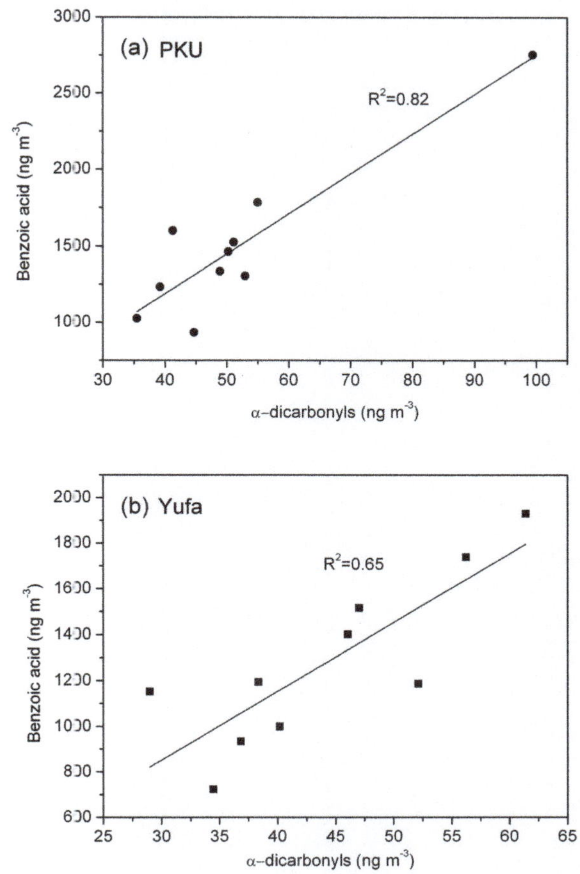

Figure 4. Positive correlation between α-dicarbonyls and benzoic acid observed at PKU (**a**) and Yufa (**b**).

from the Pacific Ocean, where photochemical processing is commonly more intensive (Kawamura and Sakaguchi, 1999). C_3 / C_4 ratios observed in PKU (0.62) were higher than in Yufa (0.56); additionally, the ratios observed during traffic restriction periods were higher than those in non-restriction periods at both sites (0.65 vs. 0.58 in PKU and 0.61 vs. 0.52 in Yufa). This result suggests that C_3 is vigorously produced in traffic restriction periods by photochemical reaction of C_4 (Kawamura and Ikushima, 1993). Even though variations of the ratio were small, these are sufficiently representative of any minor rotations and vibrations of emission sources. The results also suggested that secondary formation of diacids by photochemical oxidation was critical during traffic restriction periods even though primary exhaust was controlled.

Adipic acid (C_6) is considered a reaction product of the photochemical oxidation of cyclohexene, whereas C_9 is mainly emitted from unsaturated fatty acids (Hatakeyama et al., 1987; Kawamura and Gagosian, 1987). Therefore, a C_6 / C_9 ratio has been applied to evaluate the abundance of biogenic and anthropogenic sources to OA (Kawamura and Yasui, 2005). C_6 / C_9 ratios show higher values in non-restriction periods (PKU: 0.40; Yufa: 0.61) than in restriction periods (PKU: 0.36; Yufa: 0.38) in this study. Higher C_6 / C_9 ratios observed in non-restriction periods show that anthropogenic organic compounds, especially from vehicles, are the major source of OA during that period of time.

EC is a major component of vehicle exhaust, whereas C_2 is a major secondary organic species in the air. Therefore, a C_2 / EC ratio can be used to assess the aging of the air mass. The average C_2 / EC ratio (which has a range of 0.044 to 0.113) was 0.075 and 0.078 at PKU and Yufa, respectively, which is much higher than previously reported traffic exhaust ratio (0.0022), but similar to that measured in the air over Shenzhen (0.063 in summer) (Huang and Yu, 2007). The C_2 / EC ratio generally showed higher values in restriction periods (PKU: 0.081; Yufa: 0.077) than in non-restriction periods (PKU: 0.067; Yufa: 0.074). The results are consistent with the notion that the traffic restriction measures can reduce primary pollutants (e.g., EC).

Moreover, a C_2 / total diacids ratio can be applied as an indicator to assess the aging of OA (Kawamura and Sakaguchi, 1999). In this study, the abundances of C_2 in total diacids varied from less than 30 to 54 %. Interestingly, the ratios of C_2 / total diacids generally showed higher values in restriction periods than in non-restriction periods. The result indi-

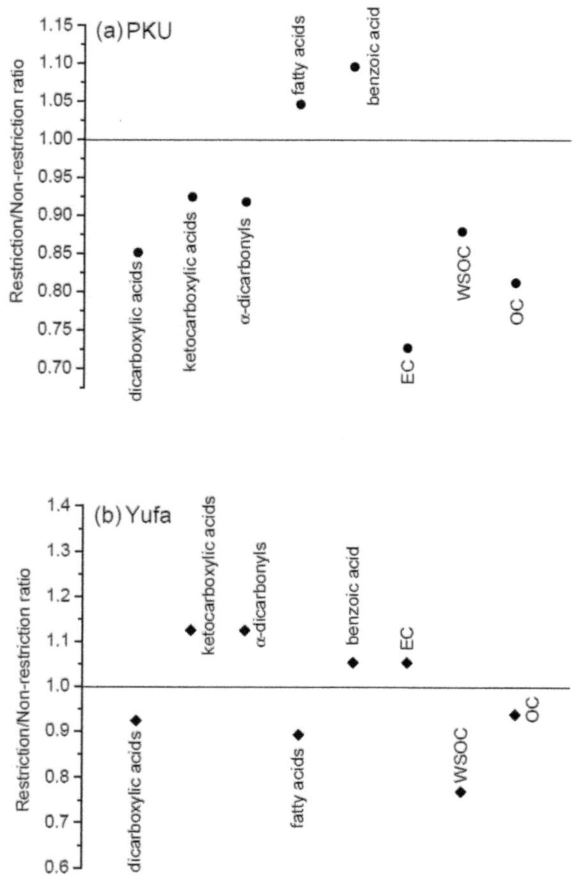

Figure 5. The R / N ratio of particulate compounds observed at PKU (**a**) and Yufa (**b**).

cates that oxalic acid is preferentially formed in restriction period by the oxidation of its precursors (other than anthropogenic VOCs, biogenic VOCs and their oxidation products may serve as important precursors in restriction periods) in the atmosphere. Further, ωC_9 is generated by biogenic unsaturated fatty acids oxidation, revealing higher concentrations in restriction periods (PKU: 3.47 ng m^{-3}; Yufa: 2.49 ng m^{-3}) than in non-restriction periods (PKU: 1.82 ng m^{-3}; Yufa: 2.12 ng m^{-3}) (Yokouchi and Ambe, 1986). This result indicates that biogenic emissions are important source for the formation of ωC_9 in restriction period, which can further breakdown to produce lower molecular weight diacids including C_4, C_3 and C_2. The results further indicate that secondary formation of diacids by atmospheric oxidation was also critical during traffic restriction periods even though primary exhaust was controlled.

4 Summary and conclusions

During the CAREBeijing-2007 in summer, molecular compositions of bifunctional organic compounds, fatty acids and benzoic acid were studied in Beijing. Oxalic acid (C_2) was detected as the most abundant diacid followed by phthalic (Ph) acid. Low MW bifunctional organic compounds were found as the major water-soluble organic fraction, accounting for more than 8.9 and 10.3 % of WSOC in PKU and Yufa, respectively. Additionally, total fatty acids and benzoic acid contributed, respectively, 3.1 and 7.2 % of OC in PKU and 3.2 and 9.3 % of OC in Yufa. Bifunctional organic compounds can be released from primary emissions (e.g., traffic exhaust and biomass burning) or formed by atmospheric oxidation of VOCs in the Beijing atmosphere. Both natural biogenic (e.g., microbial) and anthropogenic (e.g., traffic exhaust, cooking) sources provide the major inputs of fatty acids, whereas benzoic acid was mainly formed by the photo-degradation of aromatic compounds such as toluene from traffic emission.

The concentrations of OC, EC and WSOC significantly increased during the heavy pollution events, but generally decreased during the less pollution events. Results of back trajectory analyses indicated that the air masses originated mainly from the northeast, passing over heavily populated, urbanized and industrialized areas during the heavy pollution events, whereas they were mainly from mountainous, clean air areas during less pollution events.

In PKU, the restriction to non-restriction period (R / N) ratios of OC, EC, WSOC, total diacids, total ketocarboxylic acids and total α-dicarbonyls were much lower than 1, suggesting that the traffic restriction measures can reduce primary pollutants (e.g., EC) and the precursors of secondary pollutants (e.g., diacids and α-dicarbonyls). The R / N ratios of OC, WSOC, total diacids and total fatty acids in Yufa were lower than 1; however, the values are generally larger than those in PKU. Moreover, the R / N value >1 occurred for EC, total ketocarboxylic acids, total α-dicarbonyls and benzoic acid, indicating that there are higher contribution of local emissions (e.g., coal and biomass burning) and/or regional transport from polluted neighboring provinces than local traffic emission in Yufa.

The C_3 / C_4, C_2 / EC and C_2 / total diacids ratios observed during traffic restriction periods were higher than those of non-restriction periods at both sites. This result suggests that C_2 and C_3 are secondarily produced more in traffic restriction periods by the photochemical oxidation of their precursors, indicating that even when primary exhaust was controlled, secondary photochemical formation of particulate diacids was not controlled during traffic restriction periods. This study demonstrates that atmospheric oxidizing capability (photochemical aging) is enhanced by the reduction of atmospheric loading of aerosol particles during the traffic restriction periods possibly due to the increased solar radiation reaching the ground surface.

Acknowledgements. This study is partially supported by Research Grants Council of the Hong Kong Special Administrative Region (project no. 412612), the Strategic Priority Research Program of the Chinese Academy of Science (XDA05100401), and also by a grant-in-aid no. 19204055 from the Japan Society for the Promotion of Science.

Edited by: K. Schaefer

References

Alves, C., Oliveira, T., Pio, C., Silvestre, A. J. D., Fialho, P., Barata, F., and Legrand, M.: Characterisation of carbonaceous aerosols from the Azorean Island of Terceira, Atmos. Environ., 41, 1359–1373, doi:10.1016/j.atmosenv.2006.10.022, 2007.

An, X., Zhu, T., Wang, Z., Li, C., and Wang, Y.: A modeling analysis of a heavy air pollution episode occurred in Beijing, Atmos. Chem. Phys., 7, 3103–3114, doi:10.5194/acp-7-3103-2007, 2007.

Bendle, J., Kawamura, K., Yamazaki, K., and Niwai, T.: Latitudinal distribution of terrestrial lipid biomarkers and n-alkane compound-specific stable carbon isotope ratios in the atmosphere over the western Pacific and Southern Ocean, Geochim. Cosmochim. Ac., 71, 5934–5955, doi:10.1016/j.gca.2007.09.029, 2007.

Beijing statistical yearbook: http://www.bjstats.gov.cn/ (last access: February 2014), 2010.

Cao, J. J., Lee, S. C., Ho, K. F., Zhang, X. Y., Zou, S. C., Fung, K., Chow, J. C., and Watson, J. G.: Characteristics of carbonaceous aerosol in Pearl River Delta Region, China during 2001 winter period, Atmos. Environ., 37, 1451–1460, doi:10.1016/s1352-2310(02)01002-6, 2003.

Carlton, A. G., Turpin, B. J., Lim, H. J., Altieri, K. E., and Seitzinger, S.: Link between isoprene and secondary organic aerosol (SOA): Pyruvic acid oxidation yields low volatility organic acids in clouds, Geophys. Res. Lett., 33, L06822, doi:10.1029/2005gl025374, 2006.

Chow, J. C., Watson, J. G., Chen, L.-W. A., Paredes-Miranda, G., Chang, M.-C. O., Trimble, D., Fung, K. K., Zhang, H., and Zhen Yu, J.: Refining temperature measures in thermal/optical carbon analysis, Atmos. Chem. Phys., 5, 2961–2972, doi:10.5194/acp-5-2961-2005, 2005.

Ding, A. J., Wang, T., Thouret, V., Cammas, J.-P., and Nédélec, P.: Tropospheric ozone climatology over Beijing: analysis of aircraft data from the MOZAIC program, Atmos. Chem. Phys., 8, 1–13, doi:10.5194/acp-8-1-2008, 2008.

Duan, J. C., Tan, J. H., Yang, L., Wu, S., and Hao, J. M.: Concentration, sources and ozone formation potential of volatile organic compounds (VOCs) during ozone episode in Beijing, Atmos. Res., 88, 25–35, doi:10.1016/j.atmosres.2007.09.004, 2008.

Falkovich, A. H., Graber, E. R., Schkolnik, G., Rudich, Y., Maenhaut, W., and Artaxo, P.: Low molecular weight organic acids in aerosol particles from Rondônia, Brazil, during the biomass-burning, transition and wet periods, Atmos. Chem. Phys., 5, 781–797, doi:10.5194/acp-5-781-2005, 2005.

Fick, J., Nilsson, C., and Andersson, B.: Formation of oxidation products in a ventilation system, Atmos. Environ., 38, 5895–5899, doi:10.1016/j.atmosenv.2004.08.020, 2004.

Fraser, M. P., Cass, G. R., and Simoneit, B. R. T.: Air quality model evaluation data for organics. 6. C-3-C-24 organic acids, Environ. Sci. Technol., 37, 446–453, doi:10.1021/Es0209262, 2003.

Fu, P. Q., Kawamura, K., Okuzawa, K., Aggarwal, S. G., Wang, G. H., Kanaya, Y., and Wang, Z. F.: Organic molecular compositions and temporal variations of summertime mountain aerosols over Mt. Tai, North China Plain, J. Geophys. Res.-Atmos., 113, D19107, doi:10.1029/2008jd009900, 2008.

Galloway, M. M., Chhabra, P. S., Chan, A. W. H., Surratt, J. D., Flagan, R. C., Seinfeld, J. H., and Keutsch, F. N.: Glyoxal uptake on ammonium sulphate seed aerosol: reaction products and reversibility of uptake under dark and irradiated conditions, Atmos. Chem. Phys., 9, 3331–3345, doi:10.5194/acp-9-3331-2009, 2009.

Glasius, M., Lahaniati, M., Calogirou, A., Di Bella, D., Jensen, N. R., Hjorth, J., Kotzias, D., and Larsen, B. R.: Carboxylic acids in secondary aerosols from oxidation of cyclic monoterpenes by ozone, Environ. Sci. Technol., 34, 1001–1010, doi:10.1021/Es990445r, 2000.

Guttikunda, S. K., Tang, Y. H., Carmichael, G. R., Kurata, G., Pan, L., Streets, D. G., Woo, J. H., Thongboonchoo, N., and Fried, A.: Impacts of Asian megacity emissions on regional air quality during spring 2001, J. Geophys. Res-Atmos., 110, D20301, doi:10.1029/2004jd004921, 2005.

Hallquist, M., Wenger, J. C., Baltensperger, U., Rudich, Y., Simpson, D., Claeys, M., Dommen, J., Donahue, N. M., George, C., Goldstein, A. H., Hamilton, J. F., Herrmann, H., Hoffmann, T., Iinuma, Y., Jang, M., Jenkin, M. E., Jimenez, J. L., Kiendler-Scharr, A., Maenhaut, W., McFiggans, G., Mentel, Th. F., Monod, A., Prévôt, A. S. H., Seinfeld, J. H., Surratt, J. D., Szmigielski, R., and Wildt, J.: The formation, properties and impact of secondary organic aerosol: current and emerging issues, Atmos. Chem. Phys., 9, 5155–5236, doi:10.5194/acp-9-5155-2009, 2009.

Hatakeyama, S., Ohno, M., Weng, J., Takagi, H., and Akimoto, H.: Mechanism for the formation of gaseous and particulate products from ozone-cycloalkene reactions in air, Environ. Sci. Technol., 21, 52–57, doi:10.1021/es00155a005, 1987.

Hatakeyama, S., Takami, A., Wang, W., and Tang, D. G.: Aerial observation of air pollutants and aerosols over Bo Hai, China, Atmos. Environ., 39, 5893–5898, doi:10.1016/j.atmosenv.2005.06.025, 2005.

He, N., Kawamura, K., Okuzawa, K., Kanaya, Y., and Wang, Z. F.: Diurnal variations of total carbon, dicarboxylic acids, ketoacids and α-dicarbonyls in aerosols in the northern vicinity of Beijing, Atmos. Chem. Phys. Discuss., 13, 16699–16731, doi:10.5194/acpd-13-16699-2013, 2013.

Ho, K. F., Lee, S. C., Cao, J. J., Kawamura, K., Watanabe, T., Cheng, Y., and Chow, J. C.: Dicarboxylic acids, ketocarboxylic acids and dicarbonyls in the urban roadside area of Hong Kong, Atmos. Environ., 40, 3030–3040, 2006.

Ho, K. F., Cao, J. J., Lee, S. C., Kawamura, K., Zhang, R. J., Chow, J. C., and Watson, J. G.: Dicarboxylic acids, ketocarboxylic acids, and dicarbonyls in the urban atmosphere of China, J. Geophys. Res-Atmos., 112, D22S27, doi:10.1029/2006jd008011, 2007.

Ho, K. F., Lee, S. C., Ho, S. S. H., Kawamura, K., Tachibana, E., Cheng, Y., and Zhu, T.: Dicarboxylic acids, ketocarboxylic acids, α-dicarbonyls, fatty acids, and benzoic acid in urban aerosols collected during the 2006 Campaign of Air Quality Research in Beijing (CAREBeijing-2006), J. Geophys. Res., 115, D19312, doi:10.1029/2009jd013304, 2010.

Ho, K. F., Ho, S. S. H., Lee, S. C., Kawamura, K., Zou, S. C., Cao, J. J., and Xu, H. M.: Summer and winter variations of dicarboxylic acids, fatty acids and benzoic acid in $PM_2.5$ in Pearl Delta River Region, China, Atmos. Chem. Phys., 11, 2197–2208, doi:10.5194/acp-11-2197-2011, 2011.

Hoffmann, T., Huang, R.-J., and Kalberer, M.: Atmospheric analytical chemistry, Anal. Chem., 83, 4649–4664, 2011.

Huang, R.-J., Zhang, Y. L., Bozzetti, C., Ho, K. F., Cao, J. J., Han, Y. M., Dällenbach, K. R., Slowik, J. G., Platt, S. M., Canonaco, F., Zotter, P., Wolf, R., Pieber, S. M., Bruns, E. A., Crippa, M., Ciarelli, G., Piazzalunga, A., Schwikowski, M., Abbaszade, G., Schnelle-Kreis, J., Zimmermann, R., An, Z. S., Szidat, S., Baltensperger, U., El Haddad, I., Prévôt, A. S. H.: High secondary aerosol contribution to particulate pollution during haze events in China, Nature, 514, 218–222, 2014.

Huang, X. F. and Yu, J. Z.: Is vehicle exhaust a significant primary source of oxalic acid in ambient aerosols?, Geophys. Res. Lett., 34, L02808, doi:10.1029/2006gl028457, 2007.

Jaffrezo, J.-L., Aymoz, G., Delaval, C., and Cozic, J.: Seasonal variations of the water soluble organic carbon mass fraction of aerosol in two valleys of the French Alps, Atmos. Chem. Phys., 5, 2809–2821, doi:10.5194/acp-5-2809-2005, 2005.

Jimenez, J. L., Canagaratna, M. R., Donahue, N. M., Prevot, A. S. H., Zhang, Q., Kroll, J. H., DeCarlo, P. F., Allan, J. D., Coe, H., Ng, N. L., Aiken, A. C., Docherty, K. S., Ulbrich, I. M., Grieshop, A. P., Robinson, A. L., Duplissy, J., Smith, J. D., Wilson, K. R., Lanz, V. A., Hueglin, C., Sun, Y. L., Tian, J., Laaksonen, A., Raatikainen, T., Rautiainen, J., Vaattovaara, P., Ehn, M., Kulmala, M., Tomlinson, J. M., Collins, D. R., Cubison, M. J., Dunlea, E. J., Huffman, J. A., Onasch, T. B., Alfarra, M. R., Williams, P. I., Bower, K., Kondo, Y., Schneider, J., Drewnick, F., Borrmann, S., Weimer, S., Demerjian, K., Salcedo, D., Cottrell, L., Griffin, R., Takami, A., Miyoshi, T., Hatakeyama, S., Shimono, A., Sun, J. Y., Zhang, Y. M., Dzepina, K., Kimmel, J. R., Sueper, D., Jayne, J. T., Herndon, S. C., Trimborn, A. M., Williams, L. R., Wood, E. C., Middlebrook, A. M., Kolb, C. E., Baltensperger, U., and Worsnop, D. R.: Evolution of Organic Aerosols in the Atmosphere, Science, 326, 1525–1529, doi:10.1126/science.1180353, 2009.

Kawamura, K. and Gagosian, R. B.: Implications of [omega]-oxocarboxylic acids in the remote marine atmosphere for photo-oxidation of unsaturated fatty acids, Nature, 325, 330–332, 1987.

Kawamura, K. and Ikushima, K.: Seasonal-changes in the distribution of dicarboxylic-acids in the urban atmosphere, Environ. Sci. Technol., 27, 2227–2235, doi:10.1021/Es00047a033, 1993.

Kawamura, K. and Kaplan, I. R.: Motor exhaust emissions as a primary source for dicarboxylic-acids in Los-Angeles ambient air, Environ. Sci. Technol., 21, 105–110, doi:10.1021/Es00155a014, 1987.

Kawamura, K. and Sakaguchi, F.: Molecular distributions of water soluble dicarboxylic acids in marine aerosols over the Pacific Ocean including tropics, J. Geophys. Res-Atmos., 104, 3501–3509, doi:10.1029/1998jd100041, 1999.

Kawamura, K. and Yasui, O.: Diurnal changes in the distribution of dicarboxylic acids, ketocarboxylic acids and dicarbonyls in the urban Tokyo atmosphere, Atmos. Environ., 39, 1945–1960, doi:10.1016/j.atmosenv.2004.12.014, 2005.

Kawamura, K., Ng, L. L., and Kaplan, I. R.: Determination of organic-acids (C1-C10) in the atmosphere, motor exhausts, and engine oils, Environ. Sci. Technol., 19, 1082–1086, doi:10.1021/Es00141a010, 1985.

Kawamura, K., Kasukabe, H., and Barrie, L. A.: Source and reaction pathways of dicarboxylic acids, ketoacids and dicarbonyls in arctic aerosols: One year of observations, Atmos. Environ., 30, 1709–1722, doi:10.1016/1352-2310(95)00395-9, 1996.

Kawamura, K., Kobayashi, M., Tsubonuma, N., Mochida, M., Watanabe, T., and Lee, M.: Organic and inorganic compositions of marine aerosols from East Asia: Seasonal variations of water-soluble dicarboxylic acids, major ions, total carbon and nitrogen, and stable C and N isotopic composition, Geo Soc S P, 9, 243–265, doi:10.1016/S1873-9881(04)80019-1, 2004.

Kroll, J. H., Ng, N. L., Murphy, S. M., Varutbangkul, V., Flagan, R. C., and Seinfeld, J. H.: Chamber studies of secondary organic aerosol growth by reactive uptake of simple carbonyl compounds, J. Geophys. Res.-Atmos., 110, D23207, doi:10.1029/2005jd006004, 2005.

Kundu, S., Kawamura, K., Andreae, T. W., Hoffer, A., and Andreae, M. O.: Molecular distributions of dicarboxylic acids, ketocarboxylic acids and α-dicarbonyls in biomass burning aerosols: implications for photochemical production and degradation in smoke layers, Atmos. Chem. Phys., 10, 2209–2225, doi:10.5194/acp-10-2209-2010, 2010.

Legrand, M., Preunkert, S., Oliveira, T., Pio, C. A., Hammer, S., Gelencser, A., Kasper-Giebl, A., and Laj, P.: Origin of C-2-C-5 dicarboxylic acids in the European atmosphere inferred from year-round aerosol study conducted at a west-east transect, J. Geophys. Res- Atmos., 112, D23S07, doi:10.1029/2006jd008019, 2007.

Liu, Y. N., Tao, S., Yang, Y. F., Dou, H., Yang, Y., and Coveney, R. M.: Inhalation exposure of traffic police officers to polycyclic aromatic hydrocarbons (PAHs) during the winter in Beijing, China, Sci. Total Environ., 383, 98–105, doi:10.1016/j.scitotenv.2007.05.008, 2007.

Luo, C., John, J. C. S., Zhou, X. J., Lam, K. S., Wang, T., and Chameides, W. L.: A nonurban ozone air pollution episode over eastern China: Observations and model simulations, J. Geophys. Res-Atmos., 105, 1889–1908, doi:10.1029/1999jd900970, 2000.

Mauzerall, D. L., Narita, D., Akimoto, H., Horowitz, L., Walters, S., Hauglustaine, D. A., and Brasseur, G.: Seasonal characteristics of tropospheric ozone production and mixing ratios over East Asia: A global three-dimensional chemical transport model analysis, J. Geophys. Res-Atmos., 105, 17895–17910, doi:10.1029/2000jd900087, 2000.

Mkoma, S. L. and Kawamura, K.: Molecular composition of dicarboxylic acids, ketocarboxylic acids, α-dicarbonyls and fatty acids in atmospheric aerosols from Tanzania, East Africa during wet and dry seasons, Atmos. Chem. Phys., 13, 2235–2251, doi:10.5194/acp-13-2235-2013, 2013.

Mochida, M., Umemoto, N., Kawamura, K., Lim, H. J., and Turpin, B. J.: Bimodal size distributions of various organic acids and fatty acids in the marine atmosphere: Influence of anthropogenic aerosols, Asian dusts, and sea spray off the coast of East Asia, J. Geophys. Res.-Atmos., 112, D15209, doi:10.1029/2006jd007773, 2007.

Narukawa, M., Kawamura, K., Takeuchi, N., and Nakajima, T.: Distribution of dicarboxylic acids and carbon isotopic compositions in aerosols from 1997 Indonesian forest fires, Geophys. Res. Lett., 26, 3101–3104, doi:10.1029/1999gl010810, 1999.

Oliveira, C., Pio, C., Alves, C., Evtyugina, M., Santos, P., Goncalves, V., Nunes, T., Silvestre, A. J. D., Palmgren, F., Wahlin, P., and Harrad, S.: Seasonal distribution of polar organic compounds in the urban atmosphere of two large cities from the North and South of Europe, Atmos. Environ., 41, 5555–5570, doi:10.1016/j.atmosenv.2007.03.001, 2007.

Pöschl, U.: Atmospheric aerosols: Composition, transformation, climate and health effects, Angew Chem Int Edit, 44, 7520–7540, doi:10.1002/anie.200501122, 2005.

Rogge, W. F., Hildemann, L. M., Mazurek, M. A., Cass, G. R., and Simonelt, B. R. T.: Sources of fine organic aerosol .1. Charbroilers and meat cooking operations, Environ. Sci. Technol., 25, 1112–1125, doi:10.1021/Es00018a015, 1991.

Rogge, W. F., Hildemann, L. M., Mazurek, M. A., Cass, G. R., and Simoneit, B. R. T.: Sources of Fine Organic Aerosol. 2. Noncatalyst and Catalyst-Equipped Automobiles and Heavy-Duty Diesel Trucks, Environ. Sci. Technol., 27, 636–651, doi:10.1021/Es00041a007, 1993.

Rogge, W. F., Medeiros, P. M., and Simoneit, B. R. T.: Organic marker compounds for surface soil and fugitive dust from open lot dairies and cattle feedlots, Atmos. Environ., 40, 27–49, doi:10.1016/j.atmosenv.2005.07.076, 2006.

Rudich, Y., Donahue, N. M., and Mentel, T. F.: Aging of organic aerosol: Bridging the gap between laboratory and field studies, Annu. Rev. Phys. Chem., 58, 321–352, doi:10.1146/annurev.physchem.58.032806.104432, 2007.

Saxena, P. and Hildemann, L. M.: Water-soluble organics in atmospheric particles: A critical review of the literature and application of thermodynamics to identify candidate compounds, J. Atmos. Chem., 24, 57–109, doi:10.1007/Bf00053823, 1996.

Schauer, J. J., Kleeman, M. J., Cass, G. R., and Simoneit, B. R. T.: Measurement of emissions from air pollution sources. 1. C1 through C29 organic compounds from meat charbroiling, Environ. Sci. Technol., 33, 1566–1577, doi:10.1021/Es980076j, 1999.

Schauer, J. J., Kleeman, M. J., Cass, G. R., and Simoneit, B. R. T.: Measurement of emissions from air pollution sources. 4. C1-C27 organic compounds from cooking with seed oils, Environ. Sci. Technol., 36, 567–575, doi:10.1021/Es002053m, 2002.

Simoneit, B. R. T.: Organic-matter of the troposphere. 3. Characterization and sources of petroleum and pyrogenic residues in aerosols over the western United-States, Atmos. Environ., 18, 51–67, doi:10.1016/0004-6981(84)90228-2, 1984.

Simoneit, B. R. T. and Mazurek, M. A.: Organic-matter of the troposphere. 2. Natural background of biogenic lipid matter in aerosols over the rural western United-States, Atmos. Environ., 16, 2139–2159, doi:10.1016/0004-6981(82)90284-0, 1982.

Stephanou, E. G. and Stratigakis, N.: Oxocarboxylic and alpha,omega-dicarboxylic acids – Photooxidation products of biogenic unsaturated fatty-acids present in urban aerosols, Environ. Sci. Technol., 27, 1403–1407, doi:10.1021/Es00044a016, 1993.

Suh, I., Zhang, R., Molina, L. T., and Molina, M. J.: Oxidation mechanism of aromatic peroxy and bicyclic radicals from OH-toluene reactions, J. Am. Chem. Soc., 125, 12655–12665, doi:10.1021/ja0350280, 2003.

Tao, S., Wang, Y., Wu, S. M., Liu, S. Z., Dou, H., Liu, Y. N., Lang, C., Hu, F., and Xing, B. S.: Vertical distribution of polycyclic aromatic hydrocarbons in atmospheric boundary layer of Beijing in winter, Atmos. Environ., 41, 9594–9602, doi:10.1016/j.atmosenv.2007.08.026, 2007.

Volkamer, R., Platt, U., and Wirtz, K.: Primary and secondary glyoxal formation from aromatics: Experimental evidence for the bicycloalkyl-radical pathway from benzene, toluene, and p-xylene, J. Phys. Chem. A, 105, 7865–7874, doi:10.1021/Jp010152w, 2001.

Wang, G. H. and Kawamura, K.: Molecular characteristics of urban organic aerosols from Nanjing: A case study of a mega-city in China, Environ. Sci. Technol., 39, 7430–7438, doi:10.1021/Es051055+, 2005.

Wang, G. H., Kawamura, K., Lee, S., Ho, K., and Cao, J.: Molecular, seasonal, and spatial distributions of organic aerosols from fourteen Chinese cities, Environ. Sci. Technol., 40, 4619–4625, 2006.

Warneck, P.: Multi-phase chemistry of C2 and C3 organic compounds in the marine atmosphere, J. Atmos. Chem., 51, 119–159, doi:10.1007/s10874-005-5984-7, 2005.

Xu, J., Zhang, Y. H., Fu, J. S., Zheng, S. Q., and Wang, W.: Process analysis of typical summertime ozone episodes over the Beijing area, Sci. Total Environ., 399, 147–157, doi:10.1016/j.scitotenv.2008.02.013, 2008.

Yang, H., Yu, J. Z., Ho, S. S. H., Xu, J. H., Wu, W. S., Wan, C. H., Wang, X. D., Wang, X. R., and Wang, L. S.: The chemical composition of inorganic and carbonaceous materials in $PM_{2.5}$ in Nanjing, China, Atmos. Environ., 39, 3735–3749, doi:10.1016/j.atmosenv.2005.03.010, 2005.

Yokouchi, Y. and Ambe, Y.: Characterization of polar organics in airborne particulate matter, Atmospheric Environment (1967), 20, 1727–1734, doi:10.1016/0004-6981(86)90121-6, 1986.

Zhang, R. J., Shen, Z. X., Zhang, L. M., Zhang, M. G., Wang, X., and Zhang, K.: Elemental Composition of Atmospheric Particles during Periods with and without Traffic Restriction in Beijing: The Effectiveness of Traffic Restriction Measure, Sola, 7, 61–64, doi:10.2151/sola.2011-016, 2011.

Zhao, Y. L., Hu, M., Slanina, S., and Zhang, Y. H.: Chemical compositions of fine particulate organic matter emitted from Chinese cooking, Environ. Sci. Technol., 41, 99–105, doi:10.1021/Es0614518, 2007a.

Zhao, Y. L., Hu, M., Slanina, S., and Zhang, Y. H.: The molecular distribution of fine particulate organic matter emitted from Western-style fast food cooking, Atmos. Environ., 41, 8163–8171, doi:10.1016/j.atmosenv.2007.06.029, 2007b.

15

Analysis of the ozone profile specifications in the WRF-ARW model and their impact on the simulation of direct solar radiation

A. Montornès[1,2], B. Codina[1], and J. W. Zack[3]

[1]Department of Astronomy and Meteorology, University of Barcelona, Barcelona, Spain
[2]Information Services, AWS Truepower, Barcelona, Spain
[3]MESO Inc., Troy, NY, USA

Correspondence to: A. Montornès (amontornes@am.ub.es)

Abstract. Although ozone is an atmospheric gas with high spatial and temporal variability, mesoscale numerical weather prediction (NWP) models simplify the specification of ozone concentrations used in their shortwave schemes by using a few ozone profiles. In this paper, a two-part study is presented: (i) an evaluation of the quality of the ozone profiles provided for use with the shortwave schemes in the Advanced Research version of the Weather Research and Forecasting (WRF-ARW) model and (ii) an assessment of the impact of deficiencies in those profiles on the performance of model simulations of direct solar radiation. The first part compares simplified data sets used to specify the total ozone column in six schemes (i.e., Goddard, New Goddard, RRTMG, CAM, GFDL and Fu–Liou–Gu) with the Multi-Sensor Reanalysis data set during the period 1979–2008 examining the latitudinal, longitudinal and seasonal limitations in the ozone profile specifications of each parameterization. The results indicate that the maximum deviations are over the poles and show prominent longitudinal patterns in the departures due to the lack of representation of the patterns associated with the Brewer–Dobson circulation and the quasi-stationary features forced by the land–sea distribution, respectively. In the second part, the bias in the simulated direct solar radiation due to these deviations from the simplified spatial and temporal representation of the ozone distribution is analyzed for the New Goddard and CAM schemes using the Beer–Lambert–Bouguer law and for the GFDL using empirical equations. For radiative applications those simplifications introduce spatial and temporal biases with near-zero departures over the tropics throughout the year and increasing poleward with a maximum in the high middle latitudes during the winter of each hemisphere.

1 Introduction

The impact of the ozone variations in mesoscale numerical weather prediction (NWP) models has historically not been treated as a significant issue (Dudhia, 2014). This is primarily related to two factors. First, these models are not designed for stratosphere simulations because the typical timescales of mesoscale processes differ from the timescales of the interaction between the stratosphere and the troposphere. Second, the errors introduced into the solar irradiation by not considering ozone variations are typically much smaller than other sources of error such as those associated to the cloud distribution.

There is a growing interest in new applications of the mesoscale NWP models such as solar energy modeling (e.g., Ruiz-Arias et al., 2013) that require an accurate treatment of the radiative transfer equation (RTE) throughout the entire atmosphere as well as for the study of the stratosphere (e.g., Kim and Wang, 2011) that needs an accurate computation of the solar heating rate.

Together with water vapor, ozone is the most important absorber of the solar radiation in the Earth's atmosphere in cloudless and clear (i.e., without aerosols) sky conditions.

This gas is located in two atmospheric regions with a different impact on the radiative transfer (WMO, 2011). Most ozone ($\sim 90\,\%$) is located in the stratosphere. A layer with the highest ozone concentration that is typically found between 10 and 50 km above the surface is often referred to as the ozone layer. The remaining ozone (~10%) is found in the troposphere. The highest values in this layer are located near the surface and they are mainly related to human activities.

The absorption by ozone in the solar spectral region occurs in three spectral bands: Hartley, Huggins and Chappuis (Inn and Tanaka, 1953; Anderson and Mauersberger, 1992). The Hartley bands are the strongest covering the ultraviolet (UV) from 200 to 300 nm. This absorption of solar flux is located primarily in the upper stratosphere and in the mesosphere. The other two bands are weaker. The Huggins bands operate in a UV region from 300 to 360 nm. Energy absorption in this spectral range occurs in the lower stratosphere and in the troposphere. Finally, the Chappuis bands cover the photosynthetic active region (PAR) and the near-IR from 400 to 850 nm. The absorption by the Chappuis bands is mainly located in the troposphere.

The absorption of the solar flux by ozone produces a heating rate ranging from 10 to 30 K day^{-1} in the stratosphere. This absorbed energy is an important physical process in maintaining the stratospheric thermal structure (Ramanathan and Dickinson, 1979).

Stratospheric ozone is continuously created and destroyed by photochemical processes associated with solar UV radiation. Due to the annual solar variation as well as the Earth's sphericity, significant latitudinal and seasonal variations on the ozone distribution are observed. Since the tropics receive more insolation than the poles, those processes result in an ozone source in the tropics and a net poleward transport due to the large-scale air circulation in the stratosphere referred to as the Brewer–Dobson circulation (Brewer, 1949; Dobson, 1956). Consequently, the ozone layer in the tropics is thinner than at middle and higher latitudes where ozone is accumulated, increasing the thickness and, thus, the total ozone amount.

Seasonally, the total ozone in the tropics shows smaller variations than in the polar regions. The total ozone is maximum at high latitudes after the polar night because the ozone transport due to the Brewer–Dobson circulation is maximum during late fall and winter. In contrast, this circulation is weaker during summer and early fall, more in the Southern Hemisphere than in the Northern Hemisphere. In the polar summer, when daylight is continuous, the total ozone decreases gradually reaching the lowest value in early fall. This process is known as ozone depletion. Nevertheless, in Antarctica, an important minimum is observed in spring (September–October) as a result of chemical ozone destruction by other substances (i.e., the ozone hole).

Mesoscale NWP models do not consider prognostic or diagnostic equations for the ozone gas and its photochemical processes. In order to reduce the computational resources, the shortwave schemes in mesoscale NWP models simplify the ozone information. These simplifications include zonal averages and latitudinal, vertical and seasonal discretization that vary between shortwave parameterizations.

In the past few years, there has been considerable interest in improving the ozone representation within the WRF-ARW model. The version 3.5 (available since 2013) included a new option to share the ozone data sets between two schemes (see Sect. 2.1).

This paper presents an analysis of the strategies that are employed to specify the ozone profiles used as input into the shortwave radiation schemes in the WRF-ARW model. The analysis is split into two parts: (i) a study of the simplifications assumed in the ozone profiles and (ii) an analysis of the uncertainties associated with the computation of the direct solar radiation. In both, the idea is to show a global perspective (spatial and seasonal) of the impact of simplified specifications of the ozone distribution on the radiative transfer computation in the current solar parameterizations.

In the first part, spatial and temporal deviations over the total ozone column are discussed. Each ozone profile provided with the WRF-ARW package is vertically integrated and compared with monthly averaged values from the Multi-Sensor Reanalysis (MSR) data set (van der A et al., 2010) during the climate period 1979–2008. Surface conditions for the vertical integration are based on the ERA-Interim[1] reanalysis (Dee et al., 2011) for the same climate period.

In the second part, the effect of this error on the direct solar radiation at the surface is computed considering an atmosphere composed only of ozone. The analysis focuses on three shortwave schemes: the New Goddard (Chou and Suarez, 1999; Chou et al., 2001), the CAM (Collins et al., 2004) and the GFDL (Fels and Schwarzkopf, 1981). The first and second schemes use an ozone mass absorption coefficient that is independent of the temperature and pressure and consequently, the Beer–Lambert–Bouguer law may be computed as a function of the total ozone column calculated in the first part of the study. The GFDL scheme uses empirical equations as a function of the ozone amount that allows a similar treatment.

2 Methodology

2.1 Ozone absorption in the WRF-ARW model

Version 3.6.1 of the WRF-ARW model, available since 2014, includes seven shortwave schemes: Dudhia (available since 2000), Goddard (2000), New Goddard (2011), GFDL (2004), RRTMG (2009), CAM (2006) and FLG (2011).

The Dudhia scheme (Dudhia, 1989) is the simplest shortwave parameterization in the model without any consideration of the ozone absorption. For this reason, this parameterization is not considered in the following analyses.

The Goddard and the New Goddard schemes (Chou and Suarez, 1994, 1999; Chou et al., 2001) are similar because the second is an update of the first. The ozone treatment is

[1]ECMWF ERA-Interim data used in this study have been obtained from the ECMWF data server.

common for both schemes and is based on Chou and Suarez (1999). Both schemes will subsequently be denoted as G-NG. In these schemes the solar spectrum is divided into eleven spectral bands (seven in the ultraviolet, UV, one in the visible or photosynthetic active region, PAR, and three in the near-infrared, near-IR). In the UV+PAR spectral regions, G-NG neglect the pressure and temperature (i.e., height) dependence of the ozone absorption assuming a constant absorption coefficient in each spectral interval. These coefficients are obtained by dividing each band into 127 narrow sub-bands with a width of $\sim 0.003\,\mu$m and using the ozone absorption coefficient given in WMO (1986). The absorption in the near-IR is added by enhancing the absorption in the PAR region, reducing the computational time. The New Goddard scheme introduces a small correction for the ozone absorption coefficient in the PAR region, from 0.0539 (cm-atm) stp^{-1} to 0.0572 (cm-atm) stp^{-1}. The effect of this correction can be neglected for the purposes of this paper, which considers both schemes as one. All results are based on New Goddard values since it is the newest version.

The CAM scheme (Collins et al., 2004) splits the spectrum into nineteen bands (seven for the ozone, one in the visible or PAR, seven for the water vapor, three for the carbon dioxide and one for the near-IR). The ozone absorption is computed over the seven ozone bands and over the PAR region as well. As in the previous scheme, the CAM parameterization assumes a constant ozone absorption coefficient for each band. The procedure to compute these coefficients is described in Briegleb (1992).

The GFDL scheme (Fels and Schwarzkopf, 1981) divides the spectrum in two spectral bands: one in the UV+PAR and the other in the near-IR (composed by different subdivisions). The ozone absorption occurs in the UV+PAR region following the parametric formulas described in Lacis and Hansen (1974). It is noteworthy that it is the only scheme that considers the light scattering due to the ozone, explicitly. However, due to the small contribution it is not considered for the purposes of this paper.

The RRTMG (Iacono et al., 2008) parameterization divides the shortwave spectrum into fourteen bands covering the UV, PAR and near-IR regions. Each spectral band is divided in a set of sub-intervals (i.e., quadrature points) used to integrate the k-distributions for the correlated k-distribution (CKD) method detailed in Liou (1980) and Fu and Liou (1992). This scheme takes into account the ozone absorption coefficient dependence on pressure and temperature for each spectral band and interval. These values are stored in an external file called RRTMG_SW_DATA.

The Fu–Liou–Gu scheme (Fu and Liou, 1992; Gu et al., 2011) splits the solar spectrum into six spectral bands. As in the RRTMG parameterization, the interaction with the absorber gases is based on the CKD method.

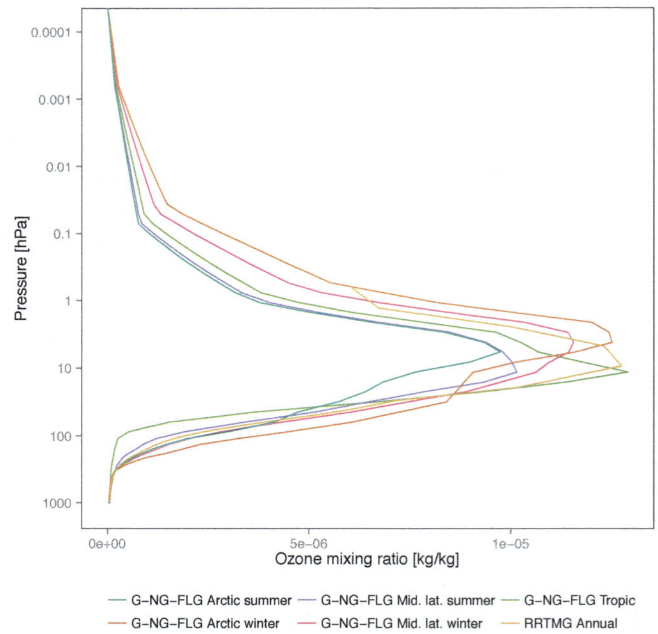

Figure 1. Ozone profile data sets available in the G-NG-FLG and RRTMG parameterizations.

Each aforementioned shortwave scheme has available different data sets reproducing ozone mixing ratio conditions in the atmosphere as a function of the pressure (i.e., vertical profile), the latitude and the season. The complexity of these data sets varies from one scheme to the other as it is summarized in Table 1 and illustrated in Figs. 1, 2 and 3. When a shortwave scheme is called by the model, the profiles are selected and interpolated into the sigma levels defined in the simulation domain.

G-NG and FLG schemes include five ozone profiles based on the same data sets (Fig. 1). These profiles simulate the ozone data for tropical, mid-latitude (summer/winter) and Arctic (summer/winter) atmospheres. The selection of a profile for use in the computations is based on the latitude of the center of the domain as well as on the day of the year. Tropical regions are assumed between 30° S and 30° N without seasonal variation. In this profile, the ozone mixing ratio is maximum at 11.417 hPa with a peak of $1.29 \times 10^{-5}\,\text{kg}\,\text{kg}^{-1}$. Mid-latitudes and Arctic regions are defined between 30–60 and 60–90°, respectively, considering winter and summer variations. In the Northern Hemisphere, winter is assumed between the 285th and the 80th days of the year and summer between the 81st and the 284th day of the year. In the Southern Hemisphere the dates associated with the winter and summer periods are reversed. The ozone layer in mid-latitudes and Arctic regions has a lower ozone concentration than in the tropics and is found at a lower altitude in summer than in winter. The highest level for all these profiles is set at 0.0006244 hPa while surface conditions depend on

Table 1. Description of the ozone profiles in the shortwave schemes in the WRF-ARW model. The Dudhia scheme is not analyzed in this study for reasons mentioned in the text.

Scheme	Profiles	Latitudes	Time	Levels	Location in code
Dudhia	None	–	–	–	–
Goddard	5	3	Summer/winter	75	Subroutine gsfcswrad in module_ra_gsfcsw.F
New Goddard	5	3	Summer/winter	75	Subroutine goddardrad in module_ra_goddard.F
GFDL	148	37	4 seasons	81	Subroutine o3clim in module_ra_gfdl.F
RRTMG	1	1	Annual	31	Subroutine o3data in module_ra_rrtmg_lw.F
CAM	768	64	12 months	59	Auxiliary file ozone.formatted
FLG	5	3	Summer/winter	75	Subroutine o3prof in module_ra_flg.F

the location and the season. In Sect. 3.1, they are referred as G-NG-FLG.

GFDL parameterization considers latitudinal and time variations based on the Eta/NMM model. Data sets are stored in a subroutine named O3CLIM. This routine generates a seasonal and spatial distribution considering four seasons (winter, spring, summer and fall) and 5° latitudinal distribution (Fig. 2). Ozone profiles are allocated at 81 vertical levels from 1013.25 to 0.0094 hPa. These data sets are interpolated to each grid-point of the user-defined domain. The interpolation is performed in a subroutine named as OZON2D considering the latitude and the day of the year. These data sets show a north–south symmetry between seasons with small differences that can not be appreciated in Fig. 2. Ozone in tropical regions is practically constant in all seasons while the higher latitudes experience substantial temporal variability.

The RRTMG scheme includes two ozone profiles as a function of the season (winter or summer). Nevertheless, this granularity is useless due to the fact that the final used profile is computed as a composition of both, without considering the day of the year. Therefore, only one profile is assumed for any latitude and season (Fig. 1). Details about this simplification can be found in subroutine O3DATA of module_ra_rrtmg_lw.F. The highest level is located at 0.647 hPa. In the region between ~50 hPa and ~10 hPa, this profile is similar to the G-NG-FLG tropical. Below ~50 hPa, this profile is practically the same as G-NG-FLG at mid-latitudes while above ~10 hPa it is similar to G-NG-FLG at mid-latitude and Arctic in winter. Since version 3.5, this scheme can utilize the ozone profiles available in the CAM scheme with the option o3input in the namelist.input file.

Finally, the CAM scheme includes several ozone profiles loaded from a binary auxiliary file called ozone_formatted with the ozone data and another two named ozone_lat.formatted and ozone_plev.formatted including latitude and pressure values, respectively. This data set covers 64 latitudes with a resolution ~2.28° and 59 pressure levels from 1003 to 0.28 hPa for each month of the year (Fig. 3). These data sets show the highest variations in time and latitude without north–south symmetries as in GFDL's profiles. Ozone values are latitudinally interpolated for each node of the domain.

Note that with the exception of CAM, all the ozone data sets show a bad representation of the ozone depletion in Antarctica throughout winter. This simplification will lead to large errors in this region as we will discuss in Sect. 3.

2.2 Part one: study of the simplifications assumed in the ozone profiles

In this part, ozone profiles of each shortwave parameterization are vertically integrated. Next, they are distributed over a regular 1° per 1° global domain for each month of a typical year using a bilineal interpolation. Then, these values are compared with the baseline typical year.

The reason to analyze the integrated profiles is the data availability because the real ozone profiles are limited in space and time. In general, these data sets are provided by ozone sounding stations located in a few sites around the world. Other data sets as the Binary Data Base of Profiles (Bodeker and Hassler, 2012) provide latitude and time variation profiles but neglect the longitudinal dependence and their values are the result of a regression model that fits real data. In contrast, satellite data provide a global covering but generally their algorithms compute the integrated amount.

First, let us assume one shortwave scheme. Given a vertical profile for the ozone mixing ratio called $q_{O_3}(z)$, the total ozone column TO_3, from the ground to the top of the atmosphere (TOA), is defined as

$$TO_3 = \int_0^\infty \rho q_{O_3} dz, \quad (1)$$

where ρ is the dry air density and z is the height with respect to the ground.

Under the assumption of a well-stratified atmosphere, the pressure and the geometric height are related by the hydrostatic equation given by

$$dp = -\rho g dz, \quad (2)$$

where g is the gravity acceleration, assumed as a constant value.

The hydrostatic equilibrium given by Eq. (2) leads Eq. (1) to

$$TO_3 = \frac{1}{g}\int_0^{p_s} q_{O_3}(p)\mathrm{d}p. \quad (3)$$

where pressure at TOA is zero by definition and the surface pressure is denoted by p_s.

Note that the integration covers the entire atmosphere including the upper levels (i.e., above 86 km) where the assumption of hydrostatic equilibrium progressively becomes less valid because the diffusion and vertical transport of the individual gas species become more important. In this context, the approach used in Eq. (3) is not valid leading to the need of a dynamically oriented model including the diffuse separation as shown in NOAA (1976). Notwithstanding, the dry air density and the ozone mixing ratio in those layers have an order of magnitude of 10^{-6} kg m^{-3} and 10^{-6} kg kg^{-1}, respectively, and are monotonically decreasing. Hence, non-hydrostatic effects may be neglected for the purposes of the current analysis.

Because the available ozone profiles in the shortwave schemes are not analytical functions, Eq. (3) in practice must be solved using a numerical integration scheme such as Simpson's method. Further, for an ozone profile composed by N vertical levels, Eq. (3) may be discretized such as

$$TO_3 = \frac{1}{6g}\sum_{k=1}^{N-1}(p_k - p_{k+1})(q_{O_3,k} + 0.5(q_{O_3,k} + q_{O_3,k+1}) + q_{O_3,k+1}), \quad (4)$$

where $q_{O_3,k}$ and p_k are the ozone mixing ratio and the pressure at a level k.

This vertical integration requires two boundary conditions: the ozone mixing ratio at the TOA and the surface pressure. The first one is assumed as zero (i.e., without ozone between the last available level and the TOA). The surface pressure requires a complex treatment since it varies by location and season. This boundary condition is computed using the ERA-Interim reanalysis covering the climate period from 1979 until 2008 (i.e., 30 years). This period is not arbitrary since it is consistent with the baseline data described below. Based on this period, monthly surface pressure averages are computed and used as surface conditions for the vertical integration of the ozone profiles.

From this procedure, the total ozone column for any location of the world and season can be computed. To quantify the geographical distribution of the errors, a global 1° per 1° grid is built using the latitudinal thresholds fixed in each shortwave scheme as described in Sect. 2.1. In order to examine the seasonal variability, values are computed throughout the twelve months of a year. Ozone profiles in some shortwave schemes like the New Goddard or the FLG are defined as a function of the day of the year instead of the month. In this situation, months are identified by the 15th day of the month. This means that January is the 15th day of the year, February is the 46th day of the year, etc.

These gridded results are compared with real data. The ozone data used as a baseline were derived from the Multi-Sensor Reanalysis, MSR (van der A et al., 2010), during the period 1979–2008 and are monthly averaged (this data set is provided with a monthly resolution).

The MSR was created from all available ozone column data measured by fourteen polar orbiting satellites in the near-ultraviolet Huggins band from November 1978 to December 2008, including TOMS (on the satellites Nimbus-7 and Earth Probe), SBUV (Nimbus-7, NOAA-9, NOAA-11 and NOAA-16), GOME (ERS-2), SCIAMACHY (Envisat), OMI (EOS-Aura), and GOME-2 (Metop-A). The data set processing includes two steps. In the first one, a bias correction scheme is applied over all satellite observations based on independent ground-based total ozone data from the World Ozone and Ultraviolet Data Center. In the second step, a data assimilation process is applied using a sub-optimal implementation of the Kalman filter method and based on a chemical transport model driven by ECMWF meteorological fields. This data set shows a bias departure less than 1 % with a root mean square standard deviation of around 2 % as compared to the corrected satellite observations used.

Therefore, for each node i (west–east direction) and j (south–north direction) and, month m, we have two data sets: one for each model under consideration, $TO_{3,\,\mathrm{sch},\,ij}(m)$, and the other describing the baseline data, $TO_{3,\,\mathrm{MSR},\,ij}(m)$. Both data sets may be compared node by node for the entire typical year. We define the relative error of the parameterization $\epsilon_{ij}(m)$ as

$$\epsilon_{\mathrm{sch},ij}(m) = TO_{3,\,\mathrm{sch},\,ij}(m) - TO_{3,\,\mathrm{MSR},\,ij}(m). \quad (5)$$

This metric will be used to analyze the spatial and temporal patterns of the deficiencies in the simplified profiles of ozone used by the shortwave schemes.

2.3 Part two: an analysis of the uncertainties added to the computation of the direct solar radiation

In the second part of the study, the previously computed total ozone columns are used to examine the ozone absorption over the direct solar radiation and to determine the introduced bias based on climate patterns.

Given one spectral band, the downward flux $F_\lambda^\downarrow$ can be divided in two components: the direct $F_{\lambda,\,\mathrm{dir}}^\downarrow$ and the diffuse $F_{\lambda,\,\mathrm{dif}}^\downarrow$ as

$$F_\lambda^\downarrow = F_{\lambda,\,\mathrm{dir}}^\downarrow + F_{\lambda,\,\mathrm{dif}}^\downarrow. \quad (6)$$

Each component requires a different mathematical treatment.

Considering a direct light beam from the Sun, traveling throughout a non-scattering isotropic plane-parallel atmosphere, the monochromatic downward solar flux density, covering the spectral interval $\Delta\lambda$, may be written as

$$F^{\downarrow}_{\lambda,\,\mathrm{dir}}(\tau_\lambda) = \mu_0 F_0(\lambda) e^{-\tau_\lambda/\mu_0}, \quad (7)$$

where τ_λ is denoted as the optical thickness for the spectral band λ and μ_0 is the cosine of the solar zenith angle. The derivation of Eq. (7) is extensively discussed in the literature such as in Chandrasekhar (1960) or in Liou (1980). This expression is commonly denoted as the Beer–Lambert–Bouguer law.

In contrast, the diffuse component requires solving the RTE in terms of a set of radiative variables as the optical thickness, the single scattering albedo and the asymmetry factor. Nevertheless, the common approximation in the solar parameterizations is to assume the ozone contribution in the Rayleigh scattering term. This term is computed as a function of the air mass without any consideration of the gas species. Therefore, the diffuse flux does not depend on the ozone data and is not analyzed in this paper.

In a non-scattering medium, when a solar beam travels throughout a layer, one part of the energy is absorbed A_λ by the medium and the other part is transmitted T_λ to the next layer (i.e., energy conservation). In other words, if we consider normalized values

$$1 = A_\lambda + T_\lambda. \quad (8)$$

As described in Liou (1980), due to the structure of the absorption lines, it is required to define the monochromatic absorptance covering the interval $\Delta\lambda$ as

$$A_{\bar{\lambda}}(\tau/\mu_0) = \int_{\Delta\lambda} (1 - e^{-\tau_\lambda/\mu_0}) \frac{\mathrm{d}\lambda}{\Delta\lambda}. \quad (9)$$

Then, assuming that the solar flux variation is small in $\Delta\lambda$, Eqs. (7) and (9) lead to

$$F^{\downarrow}_{\bar{\lambda},\,\mathrm{dir}}(\tau/\mu_0) \cong \mu_0 F_0(\lambda)(1 - A_{\bar{\lambda}}(\tau/\mu_0)). \quad (10)$$

Integrating Eq. (10) over the entire solar spectrum, the total flux $F^{\downarrow}_{\mathrm{dir}}(\tau/\mu_0)$ may be expressed as

$$F^{\downarrow}_{\mathrm{dir}}(\tau/\mu_0) = \int_0^\infty \mu_0 F_0(\lambda)(1 - A_{\bar{\lambda}}(\tau/\mu_0))\mathrm{d}\lambda. \quad (11)$$

Trivially, the radiation received at the TOA may be written as

$$\mu_0 F_0 = \int_0^\infty \mu_0 F_0(\lambda)\mathrm{d}\lambda. \quad (12)$$

Thus, Eq. (11) may be expressed as

$$F^{\downarrow}_{\mathrm{dir}}(\tau/\mu_0) = \mu_0 F_0 \int_0^\infty W(\lambda)(1 - A_{\bar{\lambda}}(\tau/\mu_0))\mathrm{d}\lambda, \quad (13)$$

where $W(\lambda)$ is the ratio of the extraterrestrial energy in a band $\mathrm{d}\lambda$ as

$$W(\lambda) = \frac{F_0(\lambda)}{F_0}. \quad (14)$$

Defining the total absorption $A(\tau/\mu_0)$ as

$$A(\tau/\mu_0) = \mu_0 F_0 \int_0^\infty W(\lambda) A_{\bar{\lambda}}(\tau/\mu_0)\mathrm{d}\lambda, \quad (15)$$

Eq. (13) may be written as

$$F^{\downarrow}_{\mathrm{dir}}(\tau/\mu_0) = \mu_0 F_0 - A(\tau/\mu_0). \quad (16)$$

Let us now consider the particular case in which the dependence of the optical thickness on wavelength in the interval $\Delta\lambda$ can be neglected. In that case, Eq. (9) may be written as

$$A_{\bar{\lambda}}(\tau/\mu_0) = 1 - e^{-\tau_\lambda/\mu_0}. \quad (17)$$

Leading Eq. (11) to

$$F^{\downarrow}_{\mathrm{dir}}(\tau/\mu_0) = \mu_0 F_0 \int_0^\infty W(\lambda) e^{-\tau_\lambda/\mu_0}\mathrm{d}\lambda. \quad (18)$$

Therefore, from Eqs. (16) and (18), the total absorption can be isolated and computed as

$$A(\tau/\mu_0) = \mu_0 F_0 - \mu_0 F_0 \int_0^\infty W(\lambda) e^{-\tau_\lambda/\mu_0}\mathrm{d}\lambda. \quad (19)$$

The optical thickness is a function of the gases and particles that compose the atmosphere and of the absorption coefficient cross section of each one. A widely used approximation is to assume that a set of absorbers are independent of one another. Therefore, the radiative schemes compute the total optical thickness τ_λ as the sum of different contributions such as ozone τ_{λ,O_3}, water vapor $\tau_{\lambda,\mathrm{wv}}$, clouds $\tau_{\lambda,\mathrm{cld}}$, aerosols $\tau_{\lambda,\mathrm{aer}}$ and others:

$$\tau_\lambda = \tau_{\lambda,O_3} + \tau_{\lambda,\mathrm{wv}} + \tau_{\lambda,\mathrm{cld}} + \tau_{\lambda,\mathrm{aer}} + \ldots \quad (20)$$

Thus, considering Eq. (20), the contribution of each absorber to the absorption can be analyzed independently in Eq. (19).

In the particular case of the ozone, the optical thickness defined from the TOA to a level z may be expressed as

$$\tau_{\lambda,O_3}(z) = \int_z^\infty k_\lambda \rho q_{O_3} \mathrm{d}z, \quad (21)$$

where k_λ denotes the mass absorption cross section and ρ is the dry air density. Note that this integral requires the vertical information of the ozone mixing ratio and the dry air density.

Moreover, k_λ is dependent on temperature and pressure as can be demonstrated by virtue of the kinetic theory of gases. Hence, as τ_{λ,O_3} is a function of the height z and this is a function of the temperature and pressure, the integral can not be computed without detailed information about k_λ.

Regarding the spectral computation given by Eq. (19), the most accurate method is the line-by-line (LBL) calculation. However, this method is not computationally feasible because it would require many thousands of computations at each grid-point. Instead of this, some approximations are assumed in terms of the gas and its spectral behavior.

Ozone absorption coefficient cross section shows a smooth variation with the wavelength (Inn and Tanaka, 1953). Hence, an effective k_λ is defined for each spectral band. This coefficient is previously computed using the LBL at a reference pressure and temperature and then, scaled to the pressure and temperature of each value in order to consider the dependency on these magnitudes.

Goddard, New Goddard and CAM follow this approach without scaling k_λ as detailed in Chou and Suarez (1999) and Briegleb (1992).

Thus, the absorption coefficient becomes temperature and pressure independent and Eq. (21) may be expressed by

$$\tau_{\lambda,O_3}(z) = k_\lambda \int_z^\infty \rho q_{O_3} \mathrm{d}z. \tag{22}$$

Extending the integral over the entire atmosphere and assuming the hydrostatic equilibrium given by Eq. (2), Eq. (22) may be written as

$$\tau_{\lambda,O_3}(p_s) = \frac{k_\lambda}{g} \int_0^{p_s} q_{O_3} \mathrm{d}p. \tag{23}$$

In virtue of Eq. (3), the optical thickness may be expressed as

$$\tau_{\lambda,O_3}(p_s) = \frac{k_\lambda}{g} TO3(p_s). \tag{24}$$

Substituting Eq. (24) into Eq. (19), the total absorption may be written as

$$A(\tau_{\lambda,O_3}/\mu_0) = \mu_0 F_0 \left(1 - \int_0^\infty W(\lambda) e^{-\frac{k_\lambda}{g\mu_0} TO_3} \mathrm{d}\lambda\right). \tag{25}$$

The necessary information to compute the $A(\tau/\mu_0)$ in Eq. (25) are the TO_3, $W(\lambda)$, k_λ, and μ_0. Information about the TO_3 can be obtained from Sect. 2.2. The $W(\lambda)$, k_λ are data available in the source code of each shortwave scheme (i.e., New Goddard and CAM). Finally, the cosine of the solar zenith angle μ_0 may be computed as a function of the latitude, the longitude, the hour, and the day of the year.

From the expression (25), we can conclude that, given a fixed wavelength, there are two variables that may change the ozone absorption over the globe. The first is the cosine of the solar zenith angle which determines the length of a solar beam's path through the atmosphere. The beam traverses a longer path when the Sun is near the horizon than when it is higher in the sky. The second is the total ozone column which modulates the opacity of the atmosphere. Higher opacity results in more absorption and less transmission of radiative energy.

Regarding RRTMG and FLG, both consider the absorption coefficient dependence on pressure and temperature as we explained in Sect. 2. Therefore, the approximation assumed in Eq. (22) is not valid. For this reason these schemes are not considered in the Part two.

The aforementioned procedure can not be applied to GFDL because this scheme does not calculate Beer's law explicitly. Instead of that, this scheme uses empirical relationships proposed by Lacis and Hansen (1974). Following these expressions, the ozone absorption is computed as

$$A_{O_3}^{uv}(x) = \frac{1.082x}{(1+138.6x)^{0.805}} + \frac{0.0658x}{1+(103.6x)^3} \tag{26}$$

in the UV spectral region, and

$$A_{O_3}^{vis}(x) = \frac{0.02118x}{1+0.042x+0.000323x^2} \tag{27}$$

in the PAR region.

In Eqs. (26) and (27), x is defined as the ozone amount u_{O_3} traversed by the solar beam in a defined layer as

$$x = u_{O_3} M, \tag{28}$$

where M is the magnification factor proposed by Rodgers (1979) as

$$M = \frac{135}{(1224\mu_0^2+1)^{1/2}}. \tag{29}$$

Therefore, considering the UV and PAR bands as a single one, the total absorption is

$$A(x) = A_{O_3}^{uv}(x) + A_{O_3}^{par}(x). \tag{30}$$

Note that if we integrate the layer values (i.e., u_{O_3}) over the entire atmosphere, the result is TO_3 as in schemes New Goddard and CAM. Thus, although this parameterization does not compute Beer's law, an analogous procedure can be applied.

Under these considerations, given a shortwave scheme, Eq. (25)[2] is applied over each node of the grid for all months. To calculate the bias, the absorption is computed using TO_3 from the model and MSR data sets. For a given month m, let us assume $A_{\mathrm{sch}}(i,j)(m)$ and $A_{\mathrm{MSR}}(i,j)(m)$ the absorption

[2]Eq. (30) in GFDL.

result for the ozone data set of the scheme and the MSR, respectively, for a node at i (west–east direction) and j (south–north direction). The bias of the parameterization $\mathrm{BIAS}_{ij}(m)$ may be defined as

$$\mathrm{BIAS}_{ij}(m) = A_{\mathrm{sch},\,ij}(m) - A_{\mathrm{MSR},\,ij}(m). \tag{31}$$

To avoid day/night problems throughout the zonal direction, all longitudes assume midday in local time (i.e., the minimum slant path). In the meridional direction, those latitudes showing a solar zenith angle greater than 80° are considered as night (i.e., polar night).

Considering Eq. (8), an overestimation (underestimation) in the ozone absorption implies that the modeled atmosphere is too opaque (transparent) and consequently, the direct flux is underestimated (overestimated) in the same magnitude with an opposite sign.

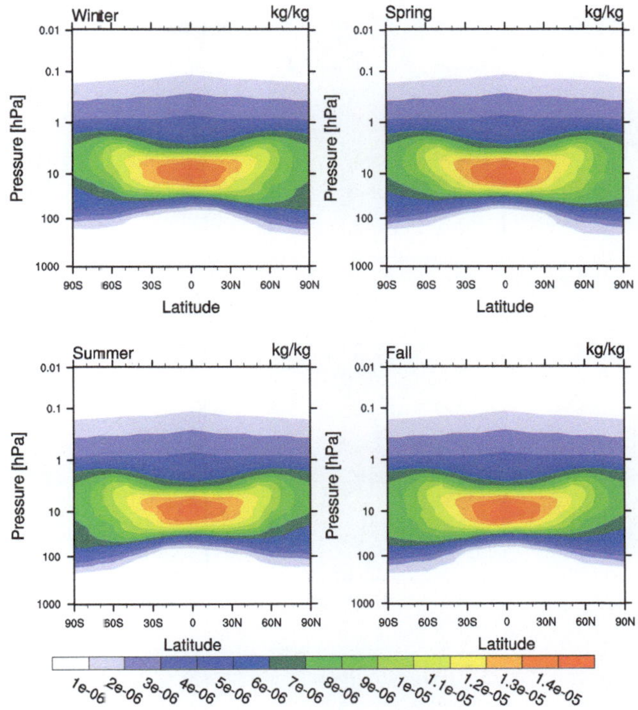

Figure 2. Ozone profile data sets available in GFDL.

3 Results

3.1 Part one: study of the simplifications assumed in the ozone profiles

The results of the evaluation of the ozone profiles provided with the WRF-ARW model using the procedures description in Sect. 2.2 are presented in this section. We first focus on detailing results for each data set (i.e., RRTMG, G-NG-FLG, CAM and GFDL) over the globe. After this, there is a general discussion of the spatial and temporal deviation patterns.

In the RRTMG scheme, shown in Fig. 4, the lowest deviations of the total ozone column are observed along the mid-latitudes of each hemisphere during the respective (i.e., Northern or Southern Hemisphere) winter and spring, lower in the Northern than in the Southern. The global minimum is reached in May, in Siberia (−30 DU) and across Europe (between −20 and +20 DU). With the exception of the ozone hole, the largest departures in the total ozone column are observed along the tropics (+100 to +150 DU) as a function of the month. An asymmetry with respect to the Equator line is observed, reaching the highest overestimation in the winter hemisphere due to the low ozone production. The global maximum is observed over Antarctica in September (+170 to +190 DU) and October (+180 to +200 DU).

The G-NG-FGL results, Fig. 5, show a strong difference between the tropics, mid-latitudes and high latitudes. The tropics show the best accuracy over the year with values between −20 and +20 DU with a tendency to underestimate the total ozone column values. Seasonally, positive and negative departures are observed over the summer and winter hemispheres, respectively. Mid-latitudes and high latitudes show a high seasonal variability due to the ozone profiles being limited to winter or summer (Table 1). The mid-latitude winter profile shows positive deviations over both hemispheres, larger in the Southern Hemisphere from March to May (+40 to +120 DU) and lower in the Northern Hemisphere from December to February (−60 to −0 DU over the eastern side of Asia, −20 to 0 DU over the northern side of the United States and 0 to +100 DU over the rest). In January a near-zero belt around 60° N is observed, which expands southward during February and March. In contrast, the mid-latitude summer profile drifts from slightly negative departures during the spring to slightly positive deviations in summer of each hemisphere. The Arctic winter profile shows positive deviations in both hemispheres. The greatest values in the Northern Hemisphere are observed in October (around +200 DU in Greenland and the Scandinavian Peninsula) while the largest deviations in the Southern Hemisphere are reached in September (+260 to +300 DU over Antarctica). The Arctic summer profile shows different patterns in each hemisphere. In the Northern Hemisphere, negative deviations (−80 to −20 DU) are observed during the spring drifting into positive departures in summer (+20 to +40 DU in July, +40 to +80 DU in August and +60 to +80 DU in September) with near-zero values in June. The Southern Hemisphere shows positive departures during all the months, reaching the minimum value in December in Antarctica (+20 to +60 DU).

In the CAM case, Fig. 6, the latitudinal and seasonal variations are well represented with typical departures between the −40 and the +40 DU with a clear overestimation. The largest departures are a result of the lack of consideration of the longitudinal ozone variations in the data sets available to the shortwave radiation schemes. Departures between +60 and +80 DU are reached over January and February in a de-

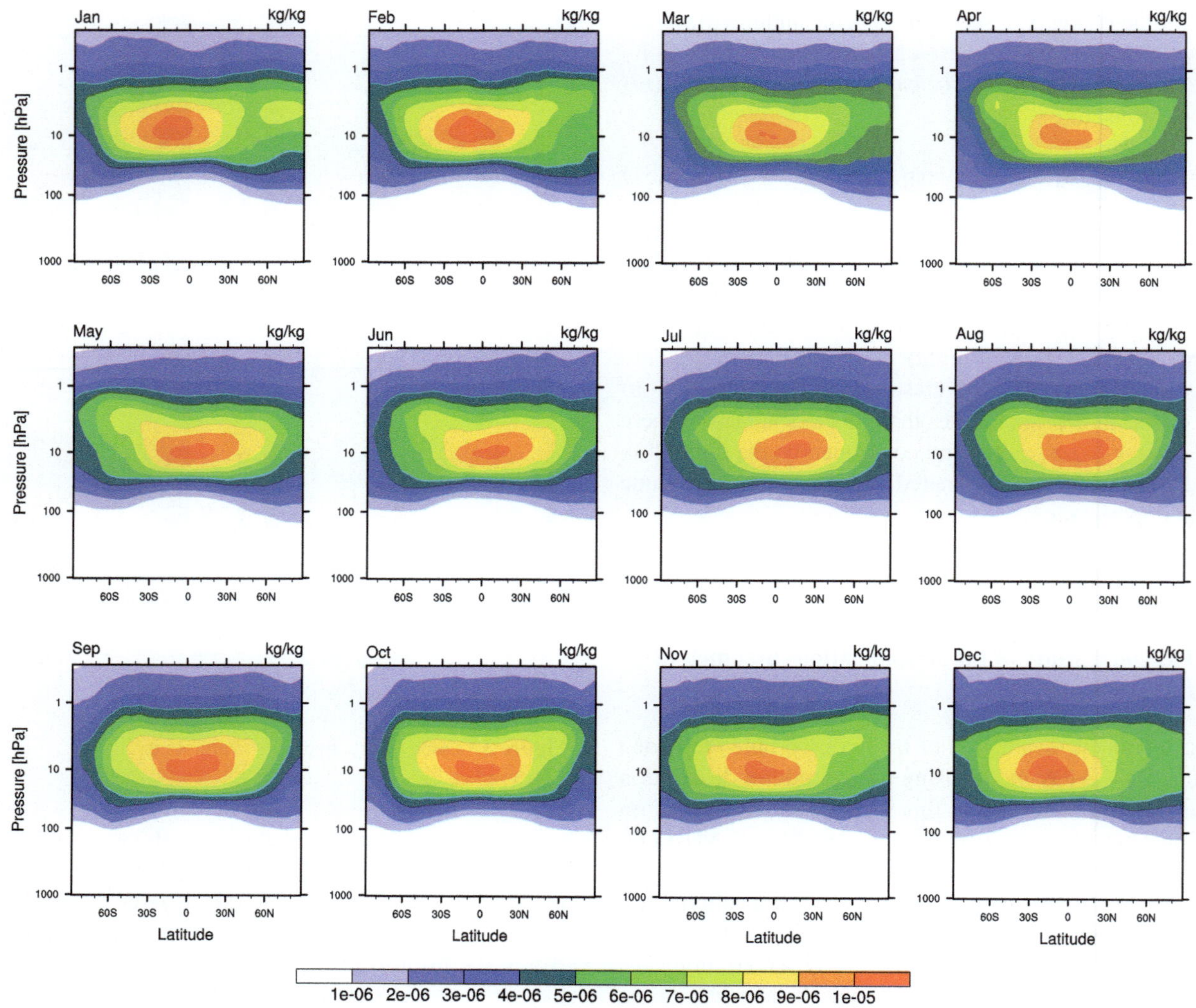

Figure 3. Ozone profile data sets available in CAM.

fined region between Greenland and the Scandinavian Peninsula. Moreover, the east–west variations due to the ozone hole produce a high overestimated region (between +60 and +80 DU with peaks above +90 DU) around 0° E and a high underestimation region (−60 to −20 DU) in the opposite side of Antarctica (i.e., 180 ° E). In November and December, an overestimated region (+0 to +20 DU) is observed over the Mediterranean Basin and over the Sahara.

Finally, in GFDL scheme (Fig. 7), the total ozone column is slightly underestimated in the tropics and strongly overestimated poleward. The lowest negative biases are observed from December to February (−20 to 0 DU) with a maximum at the Equator in a region between Africa and South America (−60 to −20 DU). During the Northern Hemisphere spring and summer, the negative values increase in magnitude as well as in areas reaching higher latitudes (−50° S to 50° N) with peaks reaching −80 DU at the end of summer. From October to December, this underestimated area is progressively weakened. In mid-latitudes, the pattern differs from the Northern Hemisphere to the Southern. In the first, a dipole between the Atlantic and eastern Asia is observed from December to March. A positive area is observed in the Atlantic region with values from +60 to +100 DU. In contrast, eastern Asia shows negative values from −80 to −40 DU. This pattern is weaker during the Northern Hemisphere spring and fall, disappearing in summer. In the Southern Hemisphere, positive errors are observed without a longitudinal dependency. The lowest biases are produced from January to March (+20 to +100 DU), drifting to higher values from April to December (+80 to +120 DU) during the Southern Hemisphere spring and are maximum in summer and fall reaching peaks greater than +140 DU. The polar regions show a strong seasonal pattern with a positive bias. In both hemispheres, the lowest values are reached in the respective summer while the highest values are produced during the polar night. In the Arctic, the bias is lower than in the Antarctica. From December to February, the Arctic regions reach errors between +100 and +180 DU. In contrast, from

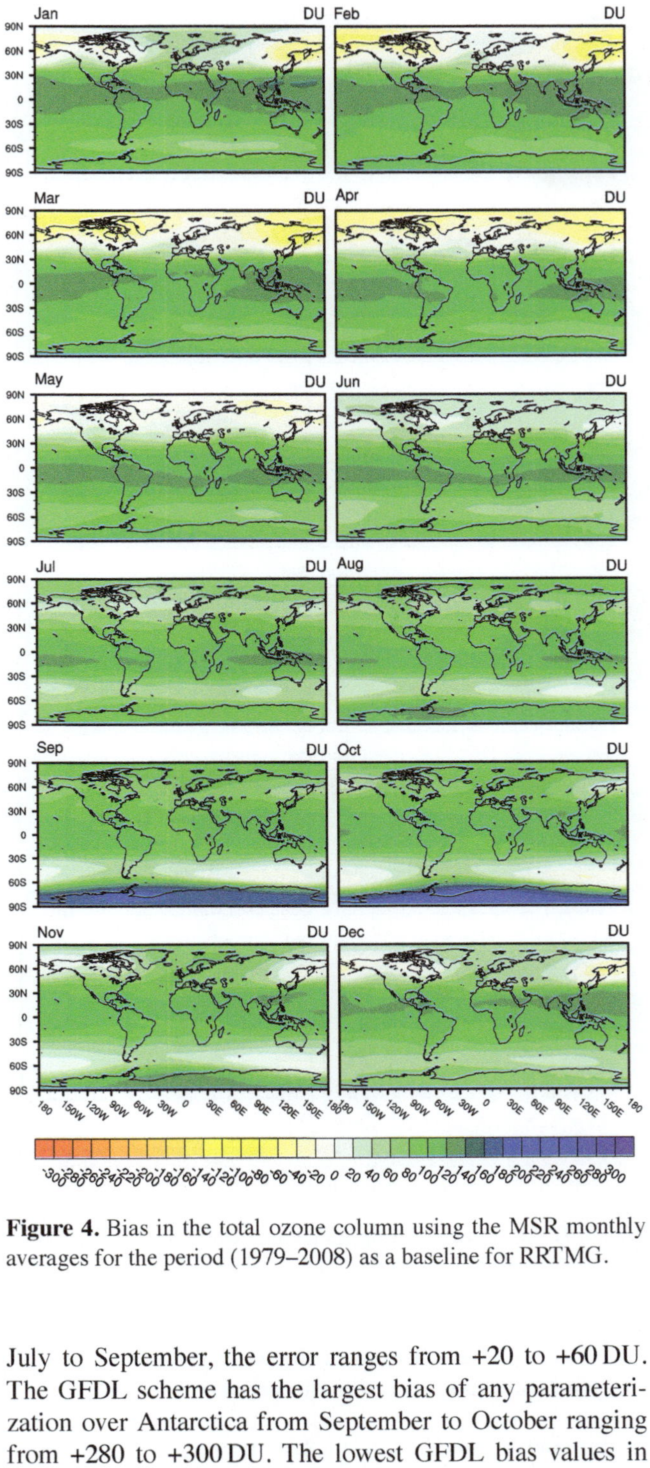

Figure 4. Bias in the total ozone column using the MSR monthly averages for the period (1979–2008) as a baseline for RRTMG.

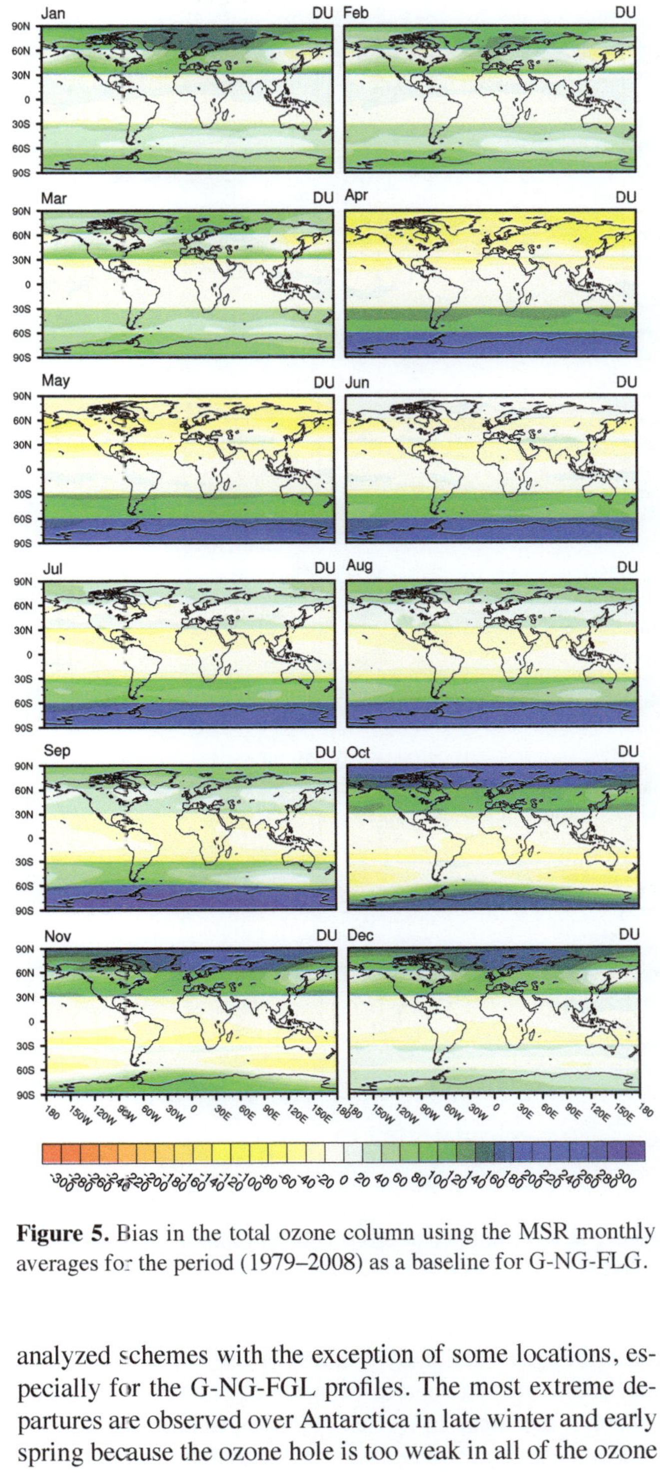

Figure 5. Bias in the total ozone column using the MSR monthly averages for the period (1979–2008) as a baseline for G-NG-FLG.

July to September, the error ranges from +20 to +60 DU. The GFDL scheme has the largest bias of any parameterization over Antarctica from September to October ranging from +280 to +300 DU. The lowest GFDL bias values in Antarctica are experienced during the Southern Hemisphere summer and are larger than those in the Arctic region during its summer.

Latitudinally and seasonally, the distribution of the departures shows a logical coherence with the quality of the ozone profiles available in each shortwave scheme. Thus, the ozone data set in the CAM scheme shows the lowest deviations while the largest global errors are observed in the RRTMG. Generally, the total ozone column is overestimated by all the analyzed schemes with the exception of some locations, especially for the G-NG-FGL profiles. The most extreme departures are observed over Antarctica in late winter and early spring because the ozone hole is too weak in all of the ozone data sets.

Longitudinally, similar distribution patterns can be observed for all the shortwave schemes because all of them assume meridional averages in the ozone mixing ratio. Two zones may be discussed. Firstly, during the Northern Hemisphere fall and winter, an underestimated region is observed between the northeastern side of Asia and the northwestern side of Canada as well as an overestimated region between Greenland and the Scandinavian Peninsula. This pattern re-

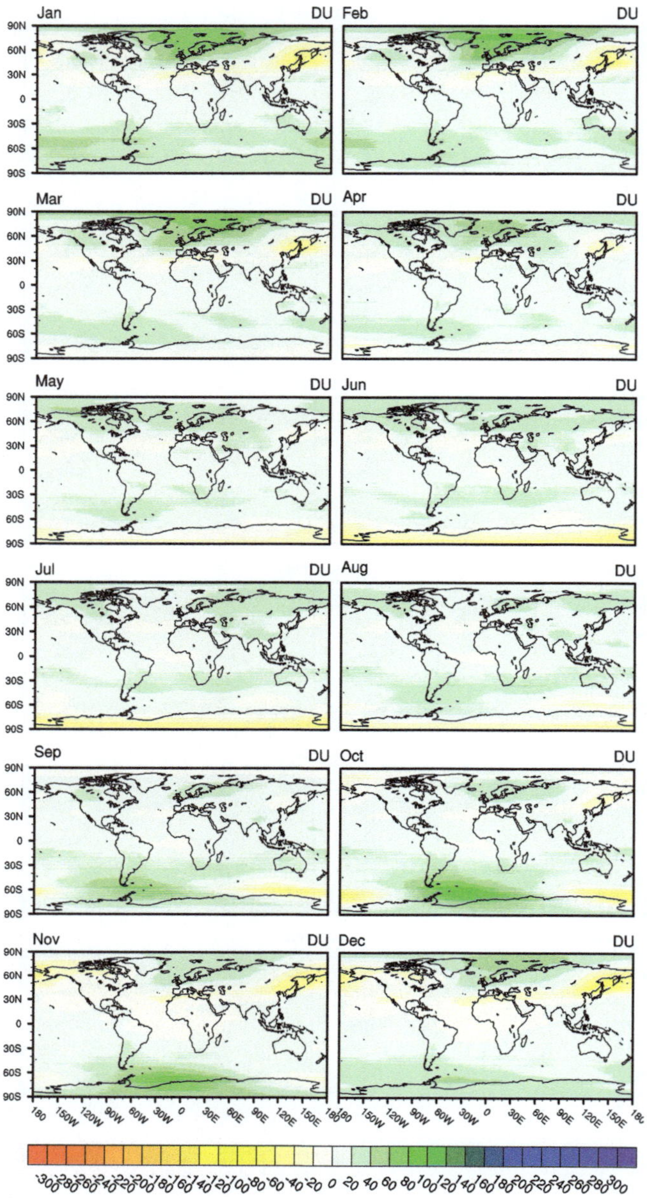

Figure 6. Bias in the total ozone column using the MSR monthly averages for the period (1979–2008) as a baseline for CAM.

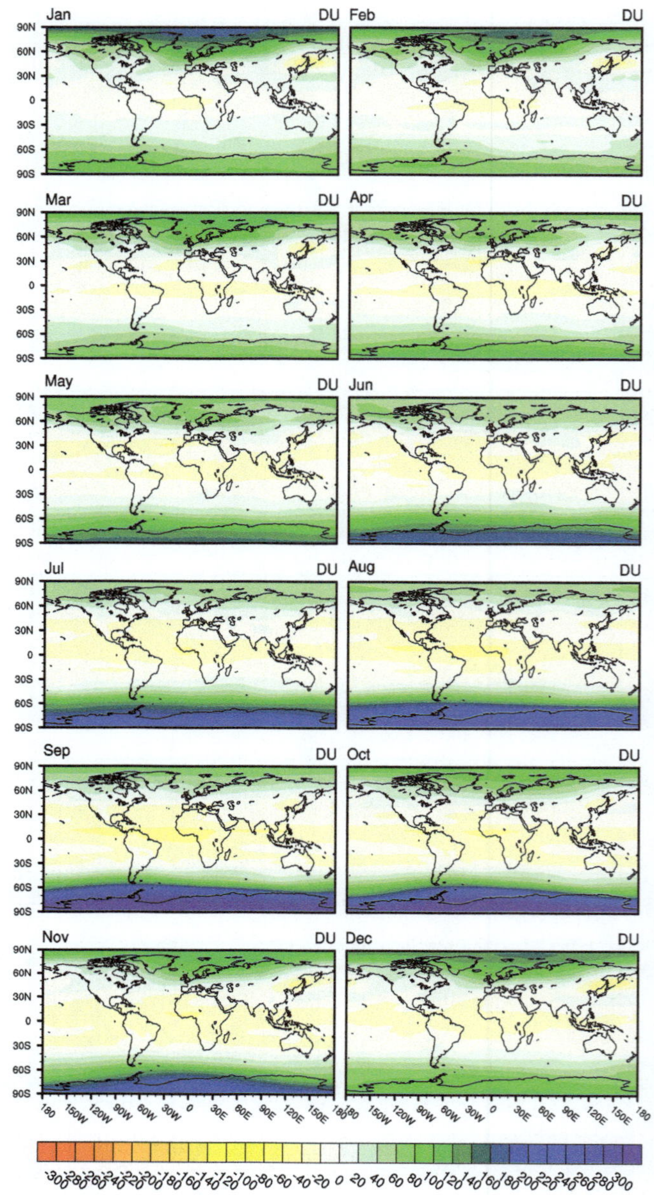

Figure 7. Bias in the total ozone column using the MSR monthly averages for the period (1979–2008) as a baseline for GFDL.

flects the quasi-stationary features of the upper-air circulation due to the sea–land distribution in the Northern Hemisphere as discussed in Dütsch (1974) or in Fusco and Salby (1999). Secondly, strong longitudinal gradients in the distribution of the errors are observed over Antarctica due to the ozone hole in September and October. In the other locations, the east–west distribution of the errors may be neglected.

3.2 Part two: an analysis of the uncertainties added to the computation of the direct solar radiation

As previously noted in Sect. 2.3, the errors in the specification of the ozone profiles are propagated in the shortwave radiation results. In this section, systematic biases introduced in the modeling of the direct solar radiation are discussed, focusing on three schemes: New Goddard, CAM and GFDL. First, a detailed description of the uncertainties over the globe is shown. Then, a general view of those limitations and their implications is discussed.

In the New Goddard scheme (Fig. 8), the bias in the total absorption ranges from -3 to $+3\,\mathrm{Wm^{-2}}$ reaching peaks near $+6\,\mathrm{Wm^{-2}}$ values close to the poles. The absorption is slightly underestimated in the tropics for the entire year. Midlatitude regions show overestimated values in winter, more homogeneous and higher in the Southern Hemisphere than in the Northern, ranging from $+0$ to $+2\,\mathrm{Wm^{-2}}$. Near-zero values are observed over North America and Asia and extended over Europe in March. An exception of this winter pattern occurs over the eastern side of Asia where a bias greater than $-1\,\mathrm{Wm^{-2}}$ is observed. During the spring, negative and near-zero departures in the bias are observed over both hemispheres, higher in the Northern (from $-2\,\mathrm{Wm^{-2}}$ in April to near-zero values in June) than in the Southern (from slightly negative in October to near-zero in December). In summer, the departures drift from near-zero biases during the early season to positive values at the end of the summer. As in the winter season, the bias is higher in the Southern Hemisphere. The largest negative bias is observed over both hemispheres during the fall, with a peak in the early autumn.

The CAM scheme (Fig. 9) shows the lowest biases in the total absorption of the solar beam due to the ozone. The bias is clearly positive during the whole year with the exception of the northern side of the Pacific Ocean from October to March reaching deviations of $-1\,\mathrm{Wm^{-2}}$ in December over the Asian coast and, in some regions over Antarctica from August to October, reaching departures around $-1\,\mathrm{Wm^{-2}}$. Two positive maxima are observed, one over the Arctic from January to March with peaks around $+2\,\mathrm{Wm^{-2}}$ and, the other over Antarctica from September to November reaching a bias of $+0.5\,\mathrm{Wm^{-2}}$. Although this scheme has the highest latitudinal resolution, the largest deviations are observed throughout high-latitudes because the ozone profiles simplify the meridional variations produced by the seasonal ozone depletion that appears from winter until near spring (Southern and Northern Hemisphere).

In GFDL (Fig. 10), absorption biases show a different pattern between the tropics and the other latitudes. In the tropics, the bias is slightly negative throughout the year. The values range from $-3\,\mathrm{Wm^{-2}}$ to $0\,\mathrm{Wm^{-2}}$. In contrast, the absorption is strongly overestimated in mid-latitudes and polar regions with typical departures between $0\,\mathrm{Wm^{-2}}$ and $+5\,\mathrm{Wm^{-2}}$. The highest departures are observed over Antarctica from September to November with values between $+6\,\mathrm{Wm^{-2}}$ and $+12\,\mathrm{Wm^{-2}}$ and in the Arctic from February to April with departures between $+3\,\mathrm{Wm^{-2}}$ and $+6\,\mathrm{Wm^{-2}}$ being maxima in the Scandinavian Peninsula. Mid-latitude regions show a similar distribution than CAM. Whereas the Southern Hemisphere shows a low dependence on longitude, the Northern shows a pattern with a dependence on space (i.e., meridional) and time. From October to April, two different regions, positive and negative, are observed between the Atlantic and Pacific, respectively. During the other months, the meridional differences are lower and the bias tends to be slightly negative in all latitudes.

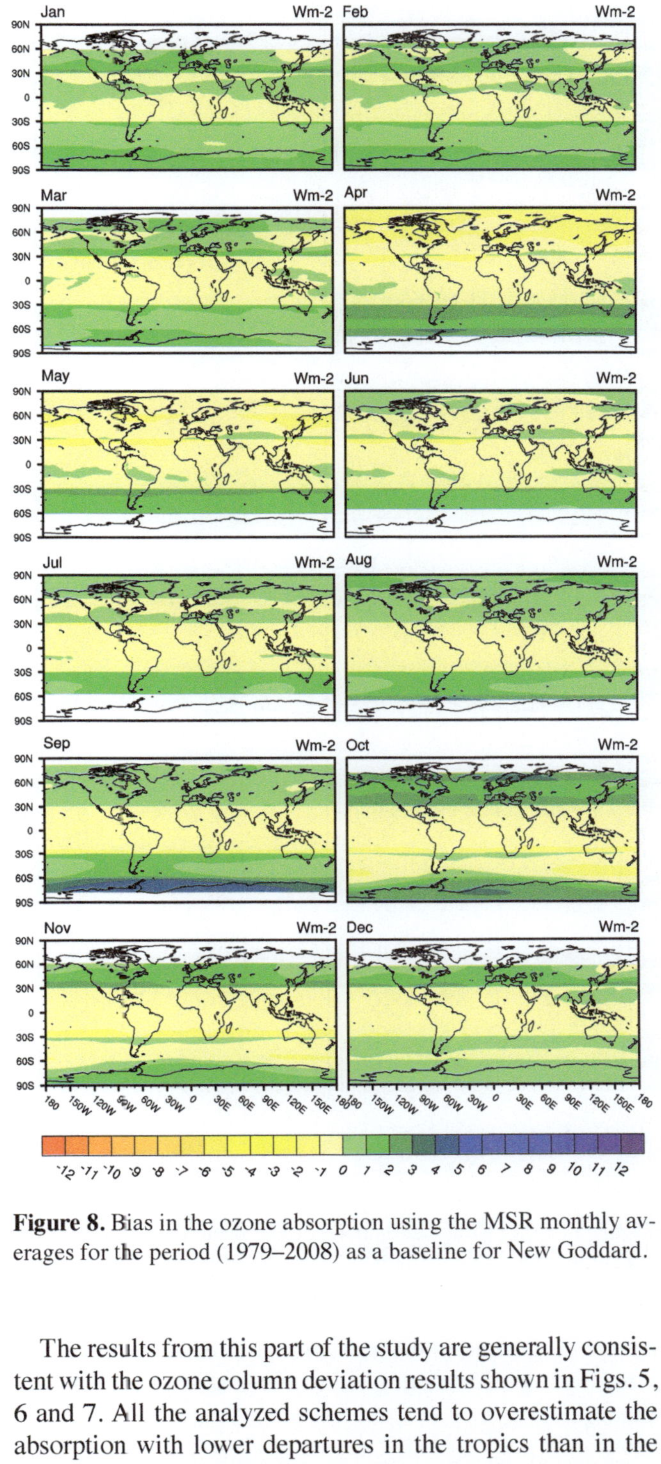

Figure 8. Bias in the ozone absorption using the MSR monthly averages for the period (1979–2008) as a baseline for New Goddard.

The results from this part of the study are generally consistent with the ozone column deviation results shown in Figs. 5, 6 and 7. All the analyzed schemes tend to overestimate the absorption with lower departures in the tropics than in the middle or high latitudes and a maximum over Antarctica during the early Southern Hemisphere spring. As opposed to the results in Sect. 3.1, the impact of these errors on the simulation of the shortwave irradiance at the surface is linked to the solar elevation angle. The highest ozone biases in the poles are masked by their coincidence with the polar night. However, the low solar elevation angles at high latitudes results in a higher sensitivity to the ozone data sets in these latitudes. These factors combine to produce the largest meridional gra-

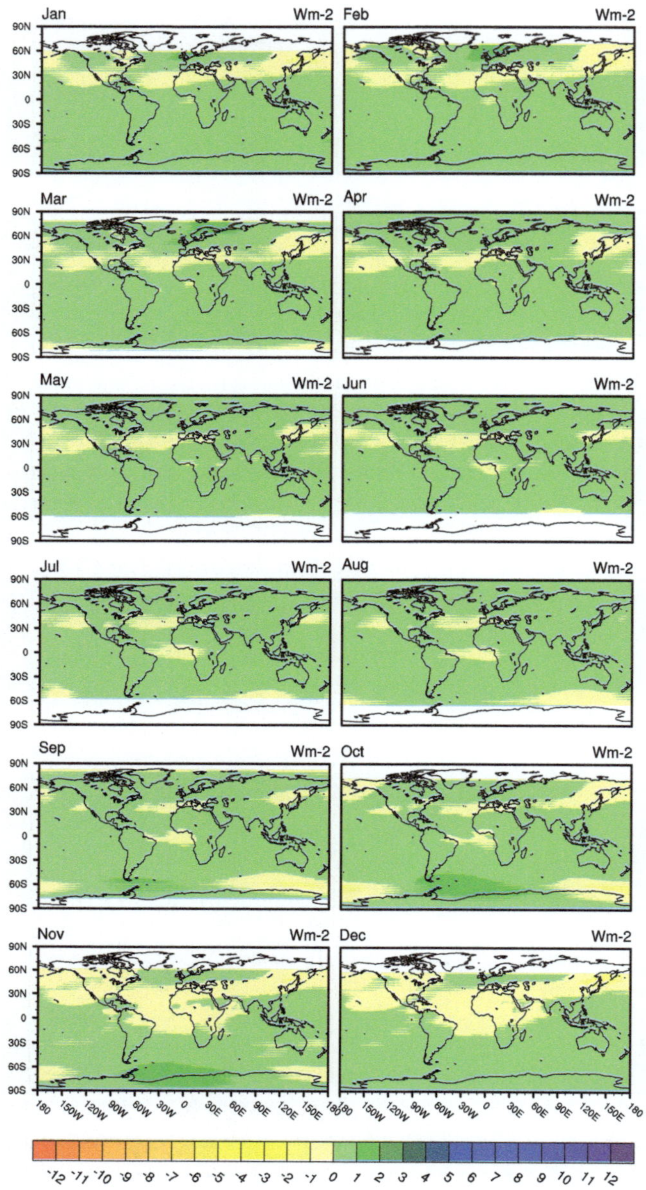

Figure 9. Bias in the ozone absorption using the MSR monthly averages for the period (1979–2008) as a baseline for CAM.

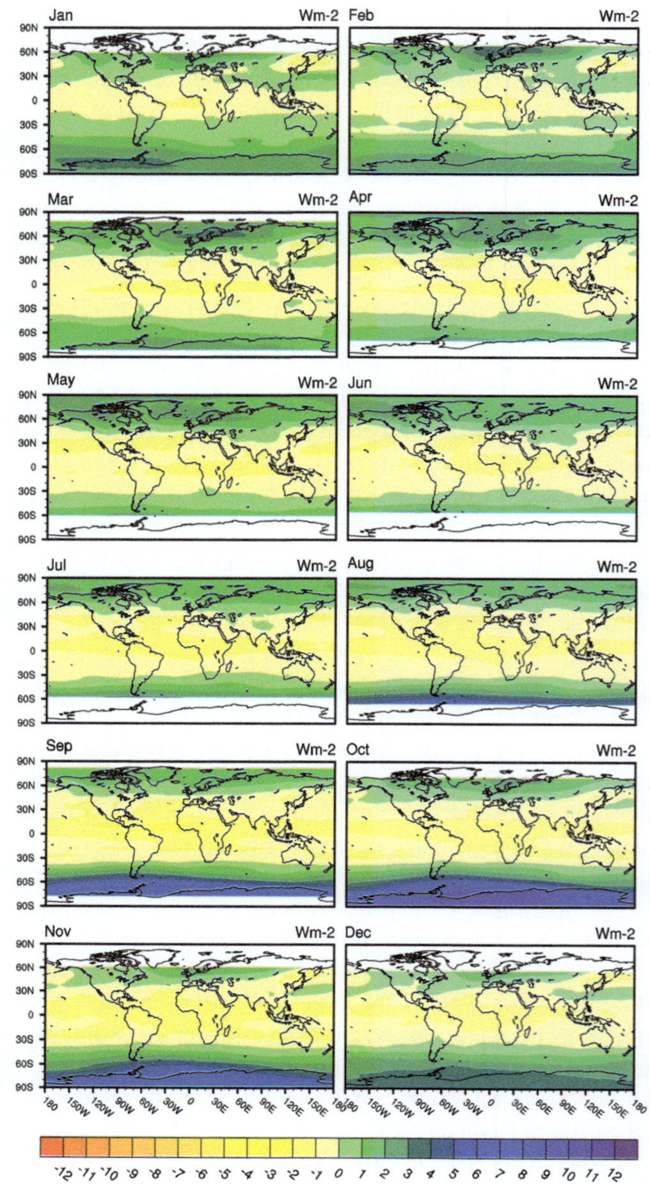

Figure 10. Bias in the ozone absorption using the MSR monthly averages for the period (1979–2008) as a baseline for GFDL.

dients in the errors in the modeling of direct solar radiation in the high latitudes during the winter season of each hemisphere.

4 Conclusions

Two sets of conclusions can be derived from the results of the analysis presented in this paper. The first set is related to the quality of the ozone concentration data sets available to the WRF-ARW mesoscale model and the second set is associated with the impact of deficiencies in representing the spatial and temporal variations of the ozone profiles on the performance of the shortwave radiation schemes available to WRF-ARW model users.

The key point is that the analysis indicates that the ozone profiles available to the WRF-ARW package are a poor representation of the ozone distribution over the planet during a typical year. These data sets assume zonal averages in the ozone mixing ratio and describe the anomalies in latitude and in time with a low resolution.

In general, the largest deviations are observed over the polar latitudes during the winter of each hemisphere due to the ozone depletion, greater in Antarctica than in the Arctic.

All the WRF-ARW ozone data sets that were analyzed in this study exhibited similar longitudinal error patterns. The error patterns were more prominent in the Northern Hemisphere due to the quasi-stationary features associated with the land–sea distribution that are not captured in the ozone profiles. As a consequence, a systematic underestimation of the total ozone column is observed in a region between the east of Asia (i.e., eastern Russia) and the west of North America (i.e., Alaska and western Canada) during the Northern Hemisphere winter and near spring. In contrast, a systematic overestimation occurs in a region defined between Greenland and the Scandinavian Peninsula during the Southern Hemisphere fall and near winter.

The RRTMG, with a single ozone profile for all the latitudes and seasons, is the shortwave scheme with the poorest ozone resolution and the largest departures relative to the climatology. Only the mid-latitudes in the Northern Hemisphere show small deviations for being calibrated in these latitudes.

The ozone profiles used by the Goddard, New Goddard and the Fu–Liou–Gu consider five ozone profiles: tropical, mid-latitude (winter/summer) and Arctic (winter/summer) for both hemispheres. This discretization shows better results in the Northern Hemisphere than in the Southern Hemisphere. The tropical profile shows a systematic underestimation of the ozone amount over any longitude, greater in the summer hemisphere, near-zero in the winter hemisphere and practically homogeneous during the equinoxes. This underestimation pattern is directly linked to the obliquity of the ecliptic and the available insolation producing more ozone in summer than in winter. Positive departures are observed over the mid-latitudes in winter and summer, better for the second one in both hemispheres. Negative deviations are observed during spring while the worst results of the year are obtained during fall. A similar pattern is observed in the polar regions with greater differences between the northern and the southern as discussed at the beginning of this section.

Finally, the CAM shortwave parameterization shows the lowest departures in the total ozone column. This scheme, composed of 64 ozone profiles with a monthly temporal resolution, captures a large part of the ozone variations over the globe. The largest deviations are observed throughout the longitudes because of the zonal averages in the profile data sets. The highest zonal gradients in the errors are observed over the poles during the winter season of each hemisphere.

The second set of conclusions addresses the impact of the deficiencies in the specification of the ozone distribution on the simulation of the shortwave radiation. A key point is that the impact of errors in the representation of the spatial and temporal distribution of ozone on the model's simulation of shortwave radiation is determined by multiple factors and it is not a simple function of the errors in the ozone profiles. For example, the largest errors in the ozone profiles were determined to be in the polar regions during winter. However, the impact of these errors on the simulation of shortwave radiation are masked by the coincidence of these errors with the polar night. On the other hand, the low solar elevation angles at high latitudes result in a higher sensitivity of the shortwave radiation schemes to the ozone profiles in these latitudes. These factors combine to produce the largest meridional gradients in the errors in the simulations of shortwave radiation in the high latitudes during the winter season of each hemisphere.

The lowest biases in the absorption of the solar direct beam occur over the tropics (Figs. 8, 9, 10) with near-zero departures. In contrast, the largest biases are observed poleward during the winter of each hemisphere. Longitudinally, the underestimated ozone region over the northern Pacific produces important biases in the absorption.

The CAM parameterization shows lower biases (-1 to $1\,\mathrm{Wm^{-2}}$) than the New Goddard scheme (-3 to $3\,\mathrm{Wm^{-2}}$) and GFDL (-2 to $2\,\mathrm{Wm^{-2}}$) with the same spatial and temporal distribution found in the total ozone errors as expected.

The greatest overestimation for all schemes is over Antarctica during the September–November period. The GFDL is the worst parameterization in this situation (8 to $12\,\mathrm{Wm^{-2}}$) followed by New Goddard (1 to $5\,\mathrm{Wm^{-2}}$) and then CAM (0 to $1\,\mathrm{Wm^{-2}}$).

In conclusion, the ozone profiles provided with the WRF-ARW package have significant limitations because of their simplified representation of spatial and temporal variability of ozone concentrations. These limitations introduce systematic biases in the modeling of shortwave radiation at the surface that can be easily reduced with more sophisticated data sets. The results of this study suggest that there would be value in having a reference ozone climatology that is dependent on the month and longitude and in a format that could be used by all of the radiation schemes in the WRF-ARW model. However, there is a substantial obstacle that must be overcome to achieve this objective. This is the fact that most of the global climatology data sets (e.g., MSR) are vertically integrated and the other sources of data (e.g., Binary Data Base of Profiles (Bodeker and Hassler, 2012)) provide latitude and vertical profiles without considering the longitudinal dependence, such as in the data sets used in the CAM parameterization. Therefore, the performed improvement made since version 3.5 (Sect. 2.1), which enables the RRTMG scheme to utilize the ozone profiles available in CAM, should be extended to the other solar parameterizations.

In view of the conclusions presented in this paper, a future study of the daily variation in the deviations could be valuable for solar short-term forecasting, since introduced biases could be corrected by using different statistical post-processing approaches (e.g., model output statistics, MOS). Furthermore, the ozone profiles could be validated using real ozone soundings in order to determine the impact of deficiencies in the vertical profiles on the calculation of the vertical profile of the solar heating rate, which is important for the modeling of stratospheric processes.

Acknowledgements. The research leading to these results received funding from the Departament d'Economia i Coneixement de la Generalitat de Catalunya in the frame of the Talent Empresa programme (grant: 2010-TEM-49).

Edited by: J.-Y. Christine Chiu

References

Anderson, S. M. and Mauersberger, K.: Laser measurements of ozone absorption cross sections in the Chappuis band, J. Geophys. Res. Lett., 19, 933–936, doi:10.1029/92GL00780, 1992.

Bodeker, G. and Hassler, B.: Bodeker Scientific Global Vertically Resolved Ozone Database, (Internet), NCAS British Atmospheric Data Centre, 2012, 2014-12-04, available at: http://catalogue.ceda.ac.uk/uuid/9a2438602c2b534f54d81e54b9d98d75, 2012.

Brewer, A.: Evidence for a world circulation provided by the measurements of helium and water vapour distribution in the stratosphere, Q. J. Roy. Meteor. Soc., 75, 351–363, 1949.

Briegleb, B. P.: Delta–Eddington approximation for solar radiation in the NCAR Community Climate Model, J. Geophys. Res., 97, 7603–7612, 1992.

Chandrasekhar, S.: Radiative Transfer, Dover Publications, New York, 1960.

Chou, M.-D. and Suarez, M. J.: An efficient thermal infrared radiation parameterization for use in general circulation models, NASA Tech. Memo, NASA/GSFC, 104606, 3, 1994.

Chou, M.-D. and Suarez, M. J.: A Solar Radiation Parameterization for Atmospheric Studies, NASA Tech. Memo, NASA/GSFC, 104606, 40, 1999.

Chou, M.-D., Suarez, M. J., Liang, X.-Z., and Yan, M. M.-H.: A thermal infrared radiation parameterization for atmospheric studies, NASA Tech. Memo, 104606, 56, available at: http://ntrs.nasa.gov/search.jsp?R=20010072848 (last access: 1 August 2014), 2001.

Collins, W. D., Rasch, P. J., Boville, B. A., Hack, J. J., McCaa, J. R., Williamson, D. L., Kiehl, J. T., Briegleb, B., Bitz, C., Lin, S., Zhang, M., and Dai, Y.: Description of the NCAR Community Atmosphere Model (CAM 3.0), NCAR Tech. Note NCAR/TN-464+ STR, Boulder, Colorado, 2004.

Dee, D. P., Uppala, S. M., Simmons, A. J., Berrisford, P., Poli, P., Kobayashi, S., Andrae, U., Balmaseda, M. A., Balsamo, G.,Bauer, P., Bechtold, P., Beljaars, A. C. M., Berg, L., Bidlot, J., Bormann, N., Delsol, C., Dragani, R., Fuentes, M., Geer, A. J., Haimberger, L., Healy, S. B., Hersbach, H., Hólm, E. V., Isaksen, L., Kållberg, P., Köhler, M., Matricardi, M.,McNally, A. P., Monge-Sanz, B. M., Morcrette, J.-J., Park, B.-K., Peubey, C., Rosnay, P., Tavolato, C., Thépaut, J.-N., and Vitart, F.: The ERA-Interim reanalysis: configuration and performance of the data assimilation system, Q. J. Roy. Meteor. Soc., 137, 553–597, doi:10.1002/qj.828, 2011.

Dobson, G. M.: Origin and distribution of the polyatomic molecules in the atmosphere, Proc. R. Soc. Lon. Ser.-A, 236, 187–193, 1956.

Dudhia, J.: Numerical study of convection observed during the winter monsoon experiment using a mesoscale two-dimensional model, J. Atmos. Sci., 46, 3077–3107, doi:10.1175/1520-0469(1989)046<3077:NSOCOD>2.0.CO;2, 1989.

Dudhia, J.: A history of mesoscale model development, Asia-Pac. J. Atmos. Sci., 50, 121–131, doi:10.1007/s13143-014-0031-8, 2014.

Dütsch, H.: The ozone distribution in the atmosphere, Can. J. Chem., 52, 1491–1504, 1974.

Fels, S. and Schwarzkopf, M.: An efficient, accurate algorithm for calculating CO2 15 μm band cooling rates, J. Geophys. Res., 86, C2, 1205–1232, 1981.

Fu, Q. and Liou, K., N.: On the correlated k-distribution method for radiative transfer in nonhomogeneous atmospheres, 49, 22, 2139–2156, doi:10.1175/1520-0469(1992)049<2139:OTCDMF>2.0.CO;2, 1992.

Fusco, A. and Salby, M.: Interannual variations of total ozone and their relationship to variations of planetary wave activity, J. Climate, 12, 1619–1629, 1999.

Gu, Y., Liou, K., Ou, S., and Fovell, R.: Cirrus cloud simulations using WRF with improved radiation parameterization and increased vertical resolution, J. Geophys. Res., 116, D06119, doi:10.1029/2010JD014574, 2011.

Iacono, M. J., Delamere, J. S., Mlawer, E. J., Shephard, M. W, Clough, S. A. and Collins, W. D.: Radiative forcing by long lived greenhouse gases: Calculations with the AER radiative transfer models, J. Geophys. Res., 113, D13103, doi:10.1029/2008JD009944, 2008.

Inn, E. C. and Tanaka, Y.: Absorption coefficient of ozone in the ultraviolet and visible regions, J. Opt. Soc. Am., 43, 10, 870–872, 1953.

Kim, H. J. and Wang, B.: Sensitivity of the WRF model simulation of the East Asian summer monsoon in 1993 to shortwave radiation schemes and ozone absorption, Asia-Pac. J. Atmos. Sci., 47, 2, 167–180, 2011.

Lacis, A. A. and Hansen, J.: A parameterization for the absorption of solar radiation in the Earth's atmosphere, J. Atmos. Sci., 31, 118–133, 1974.

Liou, K. N.: An Introduction to Atmospheric Radiation, Vol. 84, International Geophysics Series, Academic Press, New York, 1980.

NOAA: US Standard Atmosphere, 1976, Tech. rep., NOAA-S/T, U.S. Government Printing Office, Washington, D.C., 1976.

Ramanathan, V. and Dickinson, R. E.: The Role of Stratospheric Ozone in the Zonal and Seasonal Radiative Energy Balance of the Earth-Troposphere System, J. Atmos. Sci., 36, 1084–1104, 1979.

Rodgers, C. and Dickinson, R. E.: The radiative heat budget of the troposphere and lower stratosphere, Planetary Circulation Project, Rep, A2, 1967.

Ruiz-Arias, J. A., Dudhia, J., Santos-Alamillos, F. J. and Pozo-Vázquez, D.: Surface clear-sky shortwave radiative closure intercomparisons in the Weather Research and Forecasting model, J. Geophys. Res. Atmos., 118, 9901–9913, doi:10.1002/jgrd.50778, 2013.

van der A, R. J., Allaart, M. A. F., and Eskes, H. J.: Multi sensor reanalysis of total ozone, Atmos. Chem. Phys., 10, 11277–11294, doi:10.5194/acp-10-11277-2010, 2010.

WMO: Atmospheric Ozone, Tech. Rep. 16, Global Ozone Research and Monitoring Project, Geneva, 1986.

WMO: Scientific Assessment of Ozone Depletion: 2010, Tech. Rep. 16, Global Ozone Research and Monitoring Project, Geneva, 2011.

Permissions

All chapters in this book were first published in ACP, by Copernicus Publications; hereby published with permission under the Creative Commons Attribution License or equivalent. Every chapter published in this book has been scrutinized by our experts. Their significance has been extensively debated. The topics covered herein carry significant findings which will fuel the growth of the discipline. They may even be implemented as practical applications or may be referred to as a beginning point for another development.

The contributors of this book come from diverse backgrounds, making this book a truly international effort. This book will bring forth new frontiers with its revolutionizing research information and detailed analysis of the nascent developments around the world.

We would like to thank all the contributing authors for lending their expertise to make the book truly unique. They have played a crucial role in the development of this book. Without their invaluable contributions this book wouldn't have been possible. They have made vital efforts to compile up to date information on the varied aspects of this subject to make this book a valuable addition to the collection of many professionals and students.

This book was conceptualized with the vision of imparting up-to-date information and advanced data in this field. To ensure the same, a matchless editorial board was set up. Every individual on the board went through rigorous rounds of assessment to prove their worth. After which they invested a large part of their time researching and compiling the most relevant data for our readers.

The editorial board has been involved in producing this book since its inception. They have spent rigorous hours researching and exploring the diverse topics which have resulted in the successful publishing of this book. They have passed on their knowledge of decades through this book. To expedite this challenging task, the publisher supported the team at every step. A small team of assistant editors was also appointed to further simplify the editing procedure and attain best results for the readers.

Apart from the editorial board, the designing team has also invested a significant amount of their time in understanding the subject and creating the most relevant covers. They scrutinized every image to scout for the most suitable representation of the subject and create an appropriate cover for the book.

The publishing team has been an ardent support to the editorial, designing and production team. Their endless efforts to recruit the best for this project, has resulted in the accomplishment of this book. They are a veteran in the field of academics and their pool of knowledge is as vast as their experience in printing. Their expertise and guidance has proved useful at every step. Their uncompromising quality standards have made this book an exceptional effort. Their encouragement from time to time has been an inspiration for everyone.

The publisher and the editorial board hope that this book will prove to be a valuable piece of knowledge for researchers, students, practitioners and scholars across the globe.

List of Contributors

C. J. Hennigan
Department of Chemical, Biochemical and Environmental Engineering, University of Maryland, Baltimore County, Baltimore, MD, USA

J. Izumi
Department of Chemical, Biochemical and Environmental Engineering, University of Maryland, Baltimore County, Baltimore, MD, USA

A. P. Sullivan
Department of Atmospheric Science, Colorado State University, Fort Collins, CO, USA

R. J. Weber
School of Earth and Atmospheric Sciences, Georgia Institute of Technology, Atlanta, GA, USA

A. Nenes
School of Earth and Atmospheric Sciences, Georgia Institute of Technology, Atlanta, GA, USA
School of Chemical and Biomolecular Engineering, Georgia Institute of Technology, Atlanta, GA, USA Foundation for Research and Technology – Hellas, Patras, Greece

E. Hammann
University of Hohenheim, Institute of Physics and Meteorology, Garbenstrasse 30, 70599 Stuttgart, Germany

A. Behrendt
University of Hohenheim, Institute of Physics and Meteorology, Garbenstrasse 30, 70599 Stuttgart, Germany

F. Le Mounier
Laboratoire de Météorologie Dynamique, C.N.R.S, Paris, France

V. Wulfmeyer
University of Hohenheim, Institute of Physics and Meteorology, Garbenstrasse 30, 70599 Stuttgart, Germany

C. Vigouroux
Department of Atmospheric Composition, Belgian Institute for Space Aeronomy (BIRA-IASB), Brussels, Belgium

T. Blumenstock
Karlsruhe Institute of Technology (KIT), Institute for Meteorology and Climate Research (IMK-ASF), Karlsruhe, Germany

M. Coffey
Atmospheric Chemistry Division, National Center for Atmospheric Research (NCAR), Boulder, Colorado, USA

Q. Errera
Department of Atmospheric Composition, Belgian Institute for Space Aeronomy (BIRA-IASB), Brussels, Belgium

O. García
Izaña Atmospheric Research Centre (IARC), Agencia Estatal de Meteorología (AEMET), Santa Cruz de Tenerife, Spain

N. B. Jones
Centre for Atmospheric Chemistry, University of Wollongong, Wollongong, Australia

J. W. Hannigan
Atmospheric Chemistry Division, National Center for Atmospheric Research (NCAR), Boulder, Colorado, USA

F. Hase
Karlsruhe Institute of Technology (KIT), Institute for Meteorology and Climate Research (IMK-ASF), Karlsruhe, Germany

B. Liley
Department of Atmosphere, National Institute of Water and Atmospheric Research Ltd (NIWA), Lauder, New Zealand

E. Mahieu
Institute of Astrophysics and Geophysics, University of Liège (ULg), Liège, Belgium

J. Mellqvist
Department of Earth and Space Science, Chalmers University of Technology, Göteborg, Sweden

J. Notholt
Institute of Environmental Physics, University of Bremen, Bremen, Germany

M. Palm
Institute of Environmental Physics, University of Bremen, Bremen, Germany

G. Persson
Department of Earth and Space Science, Chalmers University of Technology, Göteborg, Sweden

M. Schneider
Izaña Atmospheric Research Centre (IARC), Agencia Estatal de Meteorología (AEMET), Santa Cruz de Tenerife, Spain

C. Servais
Institute of Astrophysics and Geophysics, University of Liège (ULg), Liège, Belgium

D. Smale
Department of Atmosphere, National Institute of Water and Atmospheric Research Ltd (NIWA), Lauder, New Zealand

L. Thölix
Climate Research, Finnish Meteorological Institute (FMI), Helsinki, Finland

M. De Mazière
Department of Atmospheric Composition, Belgian Institute for Space Aeronomy (BIRA-IASB), Brussels, Belgium

F. Pacifico
College of Engineering, Mathematics and Physical Sciences, University of Exeter, Exeter, UK

G. A. Folberth
Met Office Hadley Centre, Exeter, UK

S. Sitch
Geography, College of Life and Environmental Sciences, University of Exeter, Exeter, UK

J. M. Haywood
College of Engineering, Mathematics and Physical Sciences, University of Exeter, Exeter, UK
Met Office Hadley Centre, Exeter, UK

L. V. Rizzo
Department of Earth and Exact Sciences, Institute of Environmental, Chemical and Pharmaceutics Sciences, Federal

F. F. Malavelle
College of Engineering, Mathematics and Physical Sciences, University of Exeter, Exeter, UK

P. Artaxo
Department of Applied Physics, Institute of Physics, University of Sao Paulo, Sao Paulo, Brazil

Y. Liu
State Joint Key Laboratory of Environmental Simulation and Pollution Control, College of Environmental Sciences and Engineering, Peking University, Beijing, China
Chinese Research Academy of Environmental Sciences, Beijing, China

B. Yuan
State Joint Key Laboratory of Environmental Simulation and Pollution Control, College of Environmental Sciences and Engineering, Peking University, Beijing, China

X. Li
State Joint Key Laboratory of Environmental Simulation and Pollution Control, College of Environmental Sciences and Engineering, Peking University, Beijing, China

M. Shao
State Joint Key Laboratory of Environmental Simulation and Pollution Control, College of Environmental Sciences and Engineering, Peking University, Beijing, China

S. Lu
State Joint Key Laboratory of Environmental Simulation and Pollution Control, College of Environmental Sciences and Engineering, Peking University, Beijing, China

Y. Li
State Joint Key Laboratory of Environmental Simulation and Pollution Control, College of Environmental Sciences and Engineering, Peking University, Beijing, China

C.-C. Chang
Research Center for Environmental Changes, Academia Sinica, Taipei, Taiwan

Z. Wang
State Joint Key Laboratory of Environmental Simulation and Pollution Control, College of Environmental Sciences and Engineering, Peking University, Beijing, China

W. Hu
State Joint Key Laboratory of Environmental Simulation and Pollution Control, College of Environmental Sciences and Engineering, Peking University, Beijing, China

X. Huang
School of Environment and Energy, Peking University Shenzhen Graduate School, Shenzhen, China
now at: Earth System Research Laboratory, National Oceanic and Atmospheric Administration (NOAA), Boulder, Colorado, USA

L. He
School of Environment and Energy, Peking University Shenzhen Graduate School, Shenzhen, China
now at: Earth System Research Laboratory, National Oceanic and Atmospheric Administration (NOAA), Boulder, Colorado, USA

L. Zeng
State Joint Key Laboratory of Environmental Simulation and Pollution Control, College of Environmental Sciences and Engineering, Peking University, Beijing, China

M. Hu
State Joint Key Laboratory of Environmental Simulation and Pollution Control, College of Environmental Sciences and Engineering, Peking University, Beijing, China

T. Zhu
State Joint Key Laboratory of Environmental Simulation and Pollution Control, College of Environmental Sciences and Engineering, Peking University, Beijing, China

J.-E. Petit
Institut National de l'Environnement Industriel et des Risques, Verneuil-en-Halatte, France
Laboratoire des Sciences du Climat et de l'Environnement (CNRS-CEA-UVSQ), CEA Orme des Merisiers, Gif-sur-Yvette, France

O. Favez
Institut National de l'Environnement Industriel et des Risques, Verneuil-en-Halatte, France

J. Sciare
Laboratoire des Sciences du Climat et de l'Environnement (CNRS-CEA-UVSQ), CEA Orme des Merisiers, Gif-sur-Yvette, France

V. Crenn
Laboratoire des Sciences du Climat et de l'Environnement (CNRS-CEA-UVSQ), CEA Orme des Merisiers, Gif-sur-Yvette, France

R. Sarda-Estève
Laboratoire des Sciences du Climat et de l'Environnement (CNRS-CEA-UVSQ), CEA Orme des Merisiers, Gif-sur-Yvette, France

N. Bonnaire
Laboratoire des Sciences du Climat et de l'Environnement (CNRS-CEA-UVSQ), CEA Orme des Merisiers, Gif-sur-Yvette, France

G. Moˇcnik
Aerosol d.o.o., Ljubljana, Slovenia

J.-C. Dupont
Laboratoire de Météorologie Dynamique, Institut Pierre Simon Laplace, Ecole Polytechnique, Palaiseau, France
Air Lorraine, 20 rue Pierre Simon Laplace 57000 Metz, France

M. Haeffelin
Laboratoire de Météorologie Dynamique, Institut Pierre Simon Laplace, Ecole Polytechnique, Palaiseau, France
Air Lorraine, 20 rue Pierre Simon Laplace 57000 Metz, France

E. Leoz-Garziandia
Institut National de l'Environnement Industriel et des Risques, Verneuil-en-Halatte, France

L. Brégonzio-Rozier
Laboratoire Interuniversitaire des Systèmes Atmosphériques (LISA), UMR7583, CNRS, Université Paris-Est-Créteil
(UPEC) et Université Paris Diderot (UPD), Institut Pierre Simon Laplace (IPSL), Créteil, France

F. Siekmann
Aix-Marseille Université, CNRS, LCE FRE 3416, 13331, Marseille, France

C. Giorio
Department of Chemistry, University of Cambridge, Cambridge CB2 1EW, UK
Dipartimento di Scienze Chimiche, Università degli Studi di Padova, Padova, 35131, Italy

E. Pangui
Laboratoire Interuniversitaire des Systèmes Atmosphériques (LISA), UMR7583, CNRS, Université Paris-Est-Créteil (UPEC) et Université Paris Diderot (UPD), Institut Pierre Simon Laplace (IPSL), Créteil, France

S. B. Morales
Laboratoire Interuniversitaire des Systèmes Atmosphériques (LISA), UMR7583, CNRS, Université Paris-Est-Créteil
(UPEC) et Université Paris Diderot (UPD), Institut Pierre Simon Laplace (IPSL), Créteil, France

B. Temime-Roussel
Aix-Marseille Université, CNRS, LCE FRE 3416, 13331, Marseille, France

A. Gratien
Laboratoire Interuniversitaire des Systèmes Atmosphériques (LISA), UMR7583, CNRS, Université Paris-Est-Créteil
(UPEC) et Université Paris Diderot (UPD), Institut Pierre Simon Laplace (IPSL), Créteil, France

V. Michoud
Laboratoire Interuniversitaire des Systèmes Atmosphériques (LISA), UMR7583, CNRS, Université Paris-Est-Créteil
(UPEC) et Université Paris Diderot (UPD), Institut Pierre Simon Laplace (IPSL), Créteil, France

S. Ravier
Aix-Marseille Université, CNRS, LCE FRE 3416, 13331, Marseille, France

M. Cazaunau
Laboratoire Interuniversitaire des Systèmes Atmosphériques (LISA), UMR7583, CNRS, Université Paris-Est-Créteil
(UPEC) et Université Paris Diderot (UPD), Institut Pierre Simon Laplace (IPSL), Créteil, France

A. Tapparo
Dipartimento di Scienze Chimiche, Università degli Studi di Padova, Padova, 35131, Italy

A. Monod
Aix-Marseille Université, CNRS, LCE FRE 3416, 13331, Marseille, France

J.-F. Doussin
Laboratoire Interuniversitaire des Systèmes Atmosphériques (LISA), UMR7583, CNRS, Université Paris-Est-Créteil
(UPEC) et Université Paris Diderot (UPD), Institut Pierre Simon Laplace (IPSL), Créteil, France

A. Ripoll
Institute of Environmental Assessment and Water Research (IDAEA-CSIC), Jordi Girona 18–26, 08034, Barcelona, Spain
Department of Astronomy and Meteorology, Faculty of Physics, University of Barcelona, Martí i Franquès 1, 08028, Barcelona, Spain

M. C. Minguillón
Institute of Environmental Assessment and Water Research (IDAEA-CSIC), Jordi Girona 18–26, 08034, Barcelona, Spain

J. Pey
3Aix-Marseille Université, CNRS, LCE FRE 3416, Marseille, 13331, France

J. L. Jimenez
Department of Chemistry and Biochemistry, and Cooperative Institute for Research in the Environmental Sciences (CIRES), University of Colorado at Boulder, CO 80309, USA

D. A. Day
Department of Chemistry and Biochemistry, and Cooperative Institute for Research in the Environmental Sciences (CIRES), University of Colorado at Boulder, CO 80309, USA

Y. Sosedova
Paul Scherrer Institute, Laboratory of Atmospheric Chemistry, 5232 Villigen PSI, Switzerland

F. Canonaco
Paul Scherrer Institute, Laboratory of Atmospheric Chemistry, 5232 Villigen PSI, Switzerland

A. S. H. Prévôt
Paul Scherrer Institute, Laboratory of Atmospheric Chemistry, 5232 Villigen PSI, Switzerland

X. Querol
Institute of Environmental Assessment and Water Research (IDAEA-CSIC), Jordi Girona 18–26, 08034, Barcelona, Spain

A. Alastuey
Institute of Environmental Assessment and Water Research (IDAEA-CSIC), Jordi Girona 18–26, 08034, Barcelona, Spain

M. D. Jolleys
Centre for Atmospheric Science, School of Earth, Atmospheric and Environmental Science, University of Manchester, Manchester, UK

H. Coe
Centre for Atmospheric Science, School of Earth, Atmospheric and Environmental Science, University of Manchester, Manchester, UK

G. McFiggans
Centre for Atmospheric Science, School of Earth, Atmospheric and Environmental Science, University of Manchester, Manchester, UK

J. W. Taylor
Centre for Atmospheric Science, School of Earth, Atmospheric and Environmental Science, University of Manchester, Manchester, UK

S. J. O'Shea
Centre for Atmospheric Science, School of Earth, Atmospheric and Environmental Science, University of Manchester, Manchester, UK

M. Le Breton
Centre for Atmospheric Science, School of Earth, Atmospheric and Environmental Science, University of Manchester, Manchester, UK

S. J.-B. Bauguitte
Facility for Airborne Atmospheric Measurements, Bedford, UK

S. Moller
National Centre for Atmospheric Science (NCAS), Department of Chemistry, University of York, York, UK

P. Di Carlo
Centre of Excellence CETEMPS, University of L'Aquila, L'Aquila, Italy
Department of Physical and Chemical Sciences, University of L'Aquila, L'Aquila, Italy

E. Aruffo
Centre of Excellence CETEMPS, University of L'Aquila, L'Aquila, Italy
Department of Physical and Chemical Sciences, University of L'Aquila, L'Aquila, Italy

P. I. Palmer
School of Geosciences, University of Edinburgh, Edinburgh, UK

J. D. Lee
National Centre for Atmospheric Science (NCAS), Department of Chemistry, University of York, York, UK

C. J. Percival
Centre for Atmospheric Science, School of Earth, Atmospheric and Environmental Science, University of Manchester, Manchester, UK

M. W. Gallagher
Centre for Atmospheric Science, School of Earth, Atmospheric and Environmental Science, University of Manchester, Manchester, UK

F. Tummon
ETH Zurich, Zurich, Switzerland

B. Hassler
Cooperative Institute for Research in Environmental Sciences, University of Colorado, Boulder, Colorado, USA
Chemical Sciences Division, NOAA Earth System Research Laboratory, Boulder, Colorado, USA

N. R. P. Harris
University of Cambridge, Chemistry Department, Cambridge, UK

J. Staehelin
ETH Zurich, Zurich, Switzerland

W. Steinbrecht
Deutscher Wetterdienst, Hohenpeissenberg, Germany

J. Anderson
Department of Atmospheric and Planetary Science, Hampton University, Hampton, Virginia, USA

G. E. Bodeker
Bodeker Scientific, Alexandra, New Zealand

A. Bourassa
Institute of Space and Atmospheric Studies, University of Saskatchewan, Saskatoon, Saskatchewan, Canada

S. M. Davis
Cooperative Institute for Research in Environmental Sciences, University of Colorado, Boulder, Colorado, USA
Chemical Sciences Division, NOAA Earth System Research Laboratory, Boulder, Colorado, USA

D. Degenstein
Institute of Space and Atmospheric Studies, University of Saskatchewan, Saskatoon, Saskatchewan, Canada

S. M. Frith
Science Systems and Applications, Inc., Lanham, Maryland, USA

L. Froidevaux
Jet Propulsion Laboratory, California Institute of Technology, Pasadena, California, USA

E. Kyrölä
Finnish Meteorological Institute, Helsinki, Finland

M. Laine
Finnish Meteorological Institute, Helsinki, Finland

C. Long
NOAA/NWS/NCEP/Climate Prediction Center, College Park, Maryland, USA

A. A. Penckwitt
Bodeker Scientific, Alexandra, New Zealand

C. E. Sioris
Institute of Space and Atmospheric Studies, University of Saskatchewan, Saskatoon, Saskatchewan, Canada

K. H. Rosenlof
Chemical Sciences Division, NOAA Earth System Research Laboratory, Boulder, Colorado, USA

C. Roth
Institute of Space and Atmospheric Studies, University of Saskatchewan, Saskatoon, Saskatchewan, Canada

H.-J. Wang
Georgia Institute of Technology, Atlanta, Georgia, USA

J. Wild
NOAA/NWS/NCEP/Climate Prediction Center, College Park, Maryland, USA
Innovim, Greenbelt, Maryland, USA

L. Millán
Joint Institute for Regional Earth System Science and Engineering, University of California, Los Angeles, California, USA
Jet Propulsion Laboratory, California Institute of Technology, Pasadena, California, USA

S. Wang
Jet Propulsion Laboratory, California Institute of Technology, Pasadena, California, USA

N. Livesey
Jet Propulsion Laboratory, California Institute of Technology, Pasadena, California, USA

D. Kinnison
National Center for Atmospheric Research, Boulder, Colorado, USA

H. Sagawa
National Institute of Information and Communications Technology, Koganei, Tokyo, Japan

Y. Kasai
National Institute of Information and Communications Technology, Koganei, Tokyo, Japan

H. Wang
Institute of Atmospheric Composition, Key Laboratory of Atmospheric Chemistry of the China Meteorological Administration, Chinese Academy of Meteorological Sciences (CAMS), Beijing 100081, China
Collaborative Innovation Center on Forecast and Evaluation of Meteorological Disasters, Nanjing University of Information Science & Technology, Nanjing 210044, China

G. Y. Shi
State Key Laboratory of Numerical Modeling for Atmospheric Sciences and Geophysical Fluid Dynamics, Institute of Atmospheric Physics, Chinese Academy of Sciences, Beijing 100029, China

X. Y. Zhang
Institute of Atmospheric Composition, Key Laboratory of Atmospheric Chemistry of the China Meteorological Administration, Chinese Academy of Meteorological Sciences (CAMS), Beijing 100081, China

S. L. Gong
Institute of Atmospheric Composition, Key Laboratory of Atmospheric Chemistry of the China Meteorological Administration, Chinese Academy of Meteorological Sciences (CAMS), Beijing 100081, China

S. C. Tan
State Key Laboratory of Numerical Modeling for Atmospheric Sciences and Geophysical Fluid Dynamics, Institute of Atmospheric Physics, Chinese Academy of Sciences, Beijing 100029, China

B. Chen
State Key Laboratory of Numerical Modeling for Atmospheric Sciences and Geophysical Fluid Dynamics, Institute of Atmospheric Physics, Chinese Academy of Sciences, Beijing 100029, China

H. Z. Che
Institute of Atmospheric Composition, Key Laboratory of Atmospheric Chemistry of the China Meteorological Administration, Chinese Academy of Meteorological Sciences (CAMS), Beijing 100081, China

T. Li
School of Atmospheric Physics, Nanjing University of Information Science & Technology, Nanjing 210044, China

H. Wang
Institute of Atmospheric Composition (IAC), Key Laboratory of Atmospheric Chemistry (LAC) of China Meteorological Administration (CMA), Chinese Academy of Meteorological Sciences (CAMS), Beijing, 100081, China
Collaborative Innovation Center on Forecast and Evaluation of Meteorological Disasters, Nanjing University of Information Science & Technology, Nanjing, 210044, China

M. Xue
Institute of Atmospheric Composition (IAC), Key Laboratory of Atmospheric Chemistry (LAC) of China Meteorological Administration (CMA), Chinese Academy of Meteorological Sciences (CAMS), Beijing, 100081, China

X. Y. Zhang
Institute of Atmospheric Composition (IAC), Key Laboratory of Atmospheric Chemistry (LAC) of China Meteorological Administration (CMA), Chinese Academy of Meteorological Sciences (CAMS), Beijing, 100081, China

H. L. Liu
Institute of Atmospheric Composition (IAC), Key Laboratory of Atmospheric Chemistry (LAC) of China Meteorological Administration (CMA), Chinese Academy of Meteorological Sciences (CAMS), Beijing, 100081, China

C. H. Zhou
Institute of Atmospheric Composition (IAC), Key Laboratory of Atmospheric Chemistry (LAC) of China Meteorological Administration (CMA), Chinese Academy of Meteorological Sciences (CAMS), Beijing, 100081, China

S. C. Tan
State Key Laboratory of Numerical Modeling for Atmospheric Sciences and Geophysical Fluid Dynamics (LASG), Institute of Atmospheric Physics, Chinese Academy of Sciences, Beijing, 100029, China

H. Z. Che
Institute of Atmospheric Composition (IAC), Key Laboratory of Atmospheric Chemistry (LAC) of China Meteorological Administration (CMA), Chinese Academy of Meteorological Sciences (CAMS), Beijing, 100081, China

B. Chen
State Key Laboratory of Numerical Modeling for Atmospheric Sciences and Geophysical Fluid Dynamics (LASG), Institute of Atmospheric Physics, Chinese Academy of Sciences, Beijing, 100029, China

T. Li
School of Atmospheric Physics, Nanjing University of Information Science & Technology, Nanjing, 210044, China

K. F. Ho
The Jockey Club School of Public Health and Primary Care, The Chinese University of Hong Kong, Shatin, Hong Kong SAR, China

R.-J. Huang
Key Laboratory of Aerosol Chemistry and Physics, Institute of Earth Environment, Chinese Academy of Sciences, Xi'an, China
Laboratory of Atmospheric Chemistry, Paul Scherrer Institute, 5232 Villigen, Switzerland

K. Kawamura
Institute of Low Temperature Science, Hokkaido University, Sapporo, Japan

E. Tachibana
Institute of Low Temperature Science, Hokkaido University, Sapporo, Japan

S. C. Lee
Research Center of Urban Environmental Technology and Management, Department of Civil and Structural Engineering, The Hong Kong Polytechnic University, Hun Hom, Kowloon, Hong Kong SAR, China

S. S. H. Ho
he Jockey Club School of Public Health and Primary Care, The Chinese University of Hong Kong, Shatin, Hong Kong SAR, China
Division of Atmospheric Sciences, Desert Research Institute, 2215 Raggio Pathway, Reno, NV 89512, USA

T. Zhu
State Key Laboratory of Environmental Simulation and Pollution Control, College of Environmental Sciences and Engineering, Centre for Environment and Health, Peking University, Beijing 100871, China

L. Tian
School of Public Health, The University of Hong Kong, Hong Kong SAR, China

A. Montornès
Department of Astronomy and Meteorology, University of Barcelona, Barcelona, Spain
Information Services, AWS Truepower, Barcelona, Spain

B. Codina
Department of Astronomy and Meteorology, University of Barcelona, Barcelona, Spain

J. W. Zack
MESO Inc., Troy, NY, USA